VOLUME ONE HUNDRED AND TEN

ADVANCES IN PARASITOLOGY

National Institute of Parasitic Diseases, China: 70 Years and Beyond

VOLUME ONE HUNDRED AND TEN

ADVANCES IN PARASITOLOGY

National Institute of Parasitic Diseases, China: 70 Years and Beyond

Edited by

DIRK ENGELS

Uniting to Combat NTDs Secretariat, Geneva, Switzerland

YU WANG

Chinese Center for Disease Control and Prevention, People's Republic of China

XIAO-NONG ZHOU

National Institute of Parasitic Diseases, Chinese Center for Disease Control and Prevention; Chinese Center for Tropical Diseases Research; WHO Collaborating Centre for Tropical Diseases; National Center for International Research on Tropical Diseases, Ministry of Science and Technology; Key Laboratory of Parasite and Vector Biology, Ministry of Health; School of Global Health, Chinese Center for Tropical Diseases Research, Shanghai Jiao Tong University School of Medicine, Shanghai, People's Republic of China

ELSEVIER

ACADEMIC PRESS

An imprint of Elsevier

Academic Press is an imprint of Elsevier
125 London Wall, London EC2Y 5AS, United Kingdom
The Boulevard, Langford Lane, Kidlington, Oxford OX5 1GB, United Kingdom
50 Hampshire Street, 5th Floor, Cambridge, MA 02139, United States
525 B Street, Suite 1650, San Diego, CA 92101, United States

First edition 2020

Notices

Knowledge and best practice in this field are constantly changing. As new research and experience broaden our understanding, changes in research methods, professional practices, or medical treatment may become necessary.

Practitioners and researchers must always rely on their own experience and knowledge in evaluating and using any information, methods, compounds, or experiments described herein. In using such information or methods they should be mindful of their own safety and the safety of others, including parties for whom they have a professional responsibility.

To the fullest extent of the law, neither the Publisher nor the authors, contributors, or editors, assume any liability for any injury and/or damage to persons or property as a matter of products liability, negligence or otherwise, or from any use or operation of any methods, products, instructions, or ideas contained in the material herein.

ISBN: 978-0-12-820752-9
ISSN: 0065-308X

For information on all Academic Press publications
visit our website at https://www.elsevier.com/books-and-journals

Publisher: Zoe Kruze
Acquisitions Editor: Ashlie M. Jackman
Editorial Project Manager: Ana Claudia Garcia
Production Project Manager: Abdulla Sait
Cover Designer: Victoria Pearson

Typeset by SPi Global, India

Contents

Contributors *xi*
Preface *xxiii*

1. Contributions and achievements on schistosomiasis control and elimination in China by NIPD-CTDR **1**

Chun-Li Cao, Li-Juan Zhang, Wang-Ping Deng, Yin-Long Li, Chao Lv, Si-Min Dai, Ting Feng, Zhi-Qiang Qin, Li-Ping Duan, Hao-Bing Zhang, Wei Hu, Zheng Feng, Jing Xu, Shan Lv, Jia-Gang Guo, Shi-Zhu Li, Jian-Ping Cao, and Xiao-Nong Zhou

1. Introduction 2
2. The endemic status of schistosomiasis in P.R. China 3
3. Transmission patterns and evolution of the control strategy on schistosomiasis 6
4. Diagnostics and vaccine development 20
5. Research and development of drugs and molluscicides 31
6. Monitoring and evaluation 40
7. Technical guidelines and expertise support to the control programme 46
8. Conclusion and Perspectives 50

References 51

2. The contributions and achievements on malaria control and forthcoming elimination in China over the past 70 years by NIPD-CTDR **63**

Xinyu Feng, Zhi-Gui Xia, Jun Feng, Li Zhang, He Yan, Linhua Tang, Xiao-Nong Zhou, and Shuisen Zhou

1. Introduction 64
2. Epidemic situation of malaria in China 66
3. Monitoring and information management 74
4. Technical support to the malaria control and elimination programmes 76
5. Experimental research in the biology of the malaria pathogen 79
6. Biological investigations on malaria vectors 84
7. Research and development of control tools, including diagnostics, drugs, and insecticides 88
8. Future perspectives 98
9. Conclusion 101

References 102

3. Contribution to the echinococcosis control programme in China by NIPD-CTDR **107**

Meihua Fu, Shuai Han, Chuizhao Xue, Xu Wang, Baixue Liu, Ying Wang, Liying Wang, Sihui Wei, Xiaoyu Cui, Ting Zhang, Hao-Bing Zhang, Bin Zheng, Tian Tian, Shijie Yang, Chun-Hua Gao, Zhisheng Dang, Bin Xu, Qing Yu, and Weiping Wu

1. Introduction 108
2. Active participation in the national echinococcosis control programme 110
3. Promoting the pilot project on echinococcosis control 113
4. Role in the surveillance of echinococcosis 121
5. Contribution to scientific research and personnel training 123
6. Conclusions 139

References 141

4. Contributions to the lymphatic filariasis elimination programme and post-elimination surveillance in China by NIPD-CTDR **145**

De-Jian Sun, Yuan Fang, Yun Huang, and Yi Zhang

1. Introduction 146
2. The prevalence patterns of lymphatic filariasis in China 147
3. Epidemiological and experimental investigations in China 151
4. Research and development of control tools, including diagnostics and MDA drugs 155
5. Monitoring and evaluation supported by information management both in elimination and post-elimination stages 172
6. Technical supports to the elimination and post-elimination programmes 175
7. Conclusion 176

References 178

5. Contributions of the National Institute of Parasitic Diseases to the control of visceral leishmaniasis in China **185**

Zheng-Bin Zhou, Jun-Yun Wang, Chun-Hua Gao, Shuai Han, Yuan-Yuan Li, Yi Zhang, and Xiao-Nong Zhou

1. Introduction 186
2. Epidemiological patterns of visceral leishmaniasis 187
3. Vectors and their distribution 190
4. Visceral leishmaniasis in China: Then and now 196
5. Conclusion and way forward 210

References 212

6. Engagement of the National Institute of Parasitic Diseases in control of soil-transmitted helminthiasis in China **217**

Hui-Hui Zhu, Chang-Hai Zhou, Mi-Zhen Zhang, Ji-Lei Huang, Ting-Jun Zhu, Men-Bao Qian, Ying-Dan Chen, Shi-Zhu Li, and Xiao-Nong Zhou

1. Introduction 218
2. Epidemiological surveys and national surveillance system 219
3. Control pilots and new strategy development 229
4. Applied research on diagnostics and drugs 233
5. Future perspectives 236
6. Conclusions 238

References 241

7. From awareness to action: NIPD's engagement in the control of food-borne clonorchiasis **245**

Men-Bao Qian, Chang-Hai Zhou, Hui-Hui Zhu, Ting-Jun Zhu, Ji-Lei Huang, Ying-Dan Chen, and Xiao-Nong Zhou

1. Introduction 246
2. Implementation of national epidemiological surveys 247
3. Establishment of a national surveillance system 251
4. Evaluation of disease burden 252
5. Capturing social determinants 254
6. Applied research on diagnostics and drugs 256
7. Exploration of an optimal control strategy and interventional measures 260
8. Conclusions and perspectives 262

References 263

8. Control of eosinopilic meningitis caused by *Angiostrongylus cantonensis* in China **269**

Shan Lv, Yun-Hai Guo, Fu-Rong Wei, Yi Zhang, Ning Xiao, and Xiao-Nong Zhou

1. Introduction 270
2. Research on life cycle 272
3. Detection and diagnosis 275
4. Epidemiology and surveillance of eosinopilic meningitis 277
5. Population genetics 280
6. Transmission modelling 282
7. Conclusions 283

References 285

9. Control of taeniasis and cysticercosis in China **289**

Men-Bao Qian, Ning Xiao, Shi-Zhu Li, Bernadette Abela-Ridder, Hélène Carabin, Anna Sophie Fahrion, Dirk Engels, and Xiao-Nong Zhou

1. Introduction 290
2. Historical records 291
3. Epidemiological map 292
4. Socio-ecological determinants 296
5. Clinical manifestations 298
6. Diagnosis 299
7. Treatment 303
8. Control 305
9. Gap analysis 307
10. Conclusions and future perspectives 308

References 309

10. Epidemiological big data and analytical tools applied in the control programmes on parasitic diseases in China: NIPD's sustained contributions in 70 years **319**

Shang Xia, Jin-Xin Zheng, Xin-Yi Wang, Jing-Bo Xue, Jian-Hong Hu, Xue-Qiang Zhang, Xiao-Nong Zhou, and Shi-Zhu Li

1. Introduction 320
2. Empirical analysis 322
3. Theoretical and mathematical analysis 333
4. Computational analysis 335
5. Remaining challenges and future opportunities 339
6. Conclusion 344

References 344

11. Construction and application of surveillance and response systems for parasitic diseases in China, led by NIPD-CTDR **349**

Yu-Wan Hao, Qiang Wang, Chun-Li Cao, Tian Tian, Ze-Lin Zhu, Jing Xu, Shuisen Zhou, Weiping Wu, Ying-Dan Chen, Yi Zhang, Jia-Xu Chen, Shi-Zhu Li, Ning Xiao, and Xiao-Nong Zhou

1. Background 350
2. History of the parasitic disease surveillance system in China 352
3. Surveillance work organized by NIPD and its role 359
4. Emergency response mechanisms for parasitic diseases 363
5. Role of the emergency response system in managing parasitic diseases 364
6. Conclusion 367

References 368

12. Establishment and application of the National Parasitic Resource Center (NPRC) in China **373**

Shao-Hong Chen, Hai-Mo Shen, Yan Lu, Lin Ai, Jia-Xu Chen, Xue-Nian Xu, Peng Song, Yu-Chun Cai, and Xiao-Nong Zhou

1. Background 374
2. History for the NPRC development 376
3. Main function of the NPRC 379
4. Achievements of the NPRC 386
5. Looking forward 395
6. Conclusion 396

References 397

13. Contribution of NIPD-CTDR to the parasitic diseases control and elimination in China: Memory of the 70th anniversary for NIPD-CTDR **401**

Ning Xiao, Shi-Zhu Li, Men-Bao Qian, Zhi-Gui Xia, Qing Yu, Qin Liu, Shan Lv, and Xiao-Nong Zhou

1. Introduction 402
2. Achievements of parasite disease control programmes in China 404
3. Challenges for parasitic disease control 409
4. Roadmap for parasitic diseases control 413
5. Role and how to influence the global situation of the new roadmap from China 420
6. Conclusion and perspectives 422

References 423

Contributors

Bernadette Abela-Ridder
Department of Control of Neglected Tropical Diseases, World Health Organization, Geneva, Switzerland

Lin Ai
National Institute of Parasitic Diseases, Chinese Center for Disease Control and Prevention; WHO Collaborating Centre for Tropical Diseases; National Center for International Research on Tropical Diseases, Ministry of Science and Technology; Key Laboratory of Parasite and Vector Biology, Ministry of Health, Shanghai, People's Republic of China

Yu-Chun Cai
National Institute of Parasitic Diseases, Chinese Center for Disease Control and Prevention; WHO Collaborating Centre for Tropical Diseases; National Center for International Research on Tropical Diseases, Ministry of Science and Technology; Key Laboratory of Parasite and Vector Biology, Ministry of Health, Shanghai, People's Republic of China

Chun-Li Cao
National Institute of Parasitic Diseases, Chinese Center for Disease Control and Prevention; Chinese Center for Tropical Diseases Research; WHO Collaborating Centre for Tropical Diseases; National Center for International Research on Tropical Diseases, Ministry of Science and Technology; Key Laboratory of Parasite and Vector Biology, Ministry of Health; School of Global Health, Chinese Center for Tropical Diseases Research, Shanghai Jiao Tong University School of Medicine, Shanghai, People's Republic of China

Jian-Ping Cao
National Institute of Parasitic Diseases, Chinese Center for Disease Control and Prevention; Chinese Center for Tropical Diseases Research; WHO Collaborating Centre for Tropical Diseases; National Center for International Research on Tropical Diseases, Ministry of Science and Technology; Key Laboratory of Parasite and Vector Biology, Ministry of Health; School of Global Health, Chinese Center for Tropical Diseases Research, Shanghai Jiao Tong University School of Medicine, Shanghai, People's Republic of China

Hélène Carabin
Department of Biostatistics and Epidemiology, College of Public Health, University of Oklahoma Health Sciences Center, Oklahoma City, OK, United States

Jia-Xu Chen
National Institute of Parasitic Diseases, Chinese Center for Disease Control and Prevention; Chinese Center for Tropical Diseases Research; WHO Collaborating Centre for Tropical Diseases, Shanghai, China; National Center for International Research on Tropical Diseases, Ministry of Science and Technology; Key Laboratory of Parasite and Vector Biology, Ministry of Health, Shanghai, People's Republic of China

Shao-Hong Chen
National Institute of Parasitic Diseases, Chinese Center for Disease Control and Prevention; WHO Collaborating Centre for Tropical Diseases; National Center for International Research on Tropical Diseases, Ministry of Science and Technology; Key Laboratory of Parasite and Vector Biology, Ministry of Health, Shanghai, People's Republic of China

Ying-Dan Chen
National Institute of Parasitic Diseases, Chinese Center for Disease Control and Prevention; Chinese Center for Tropical Diseases Research; Key Laboratory of Parasite and Vector Biology, Ministry of Health; National Center for International Research on Tropical Diseases, Ministry of Science and Technology; WHO Collaborating Centre for Tropical Diseases, Shanghai, People's Republic of China

Xiaoyu Cui
National Institute of Parasitic Diseases, Chinese Center for Disease Control and Prevention; Chinese Center for Tropical Diseases Research; WHO Collaborating Centre for Tropical Diseases; National Center for International Research on Tropical Diseases, Ministry of Science and Technology; Key Laboratory of Parasite and Vector Biology, Ministry of Health, Shanghai, People's Republic of China

Si-Min Dai
National Institute of Parasitic Diseases, Chinese Center for Disease Control and Prevention; Chinese Center for Tropical Diseases Research; WHO Collaborating Centre for Tropical Diseases; National Center for International Research on Tropical Diseases, Ministry of Science and Technology; Key Laboratory of Parasite and Vector Biology, Ministry of Health; School of Global Health, Chinese Center for Tropical Diseases Research, Shanghai Jiao Tong University School of Medicine, Shanghai, People's Republic of China

Zhisheng Dang
National Institute of Parasitic Diseases, Chinese Center for Disease Control and Prevention; Chinese Center for Tropical Diseases Research; WHO Collaborating Centre for Tropical Diseases; National Center for International Research on Tropical Diseases, Ministry of Science and Technology; Key Laboratory of Parasite and Vector Biology, Ministry of Health, Shanghai, People's Republic of China

Wang-Ping Deng
National Institute of Parasitic Diseases, Chinese Center for Disease Control and Prevention; Chinese Center for Tropical Diseases Research; WHO Collaborating Centre for Tropical Diseases; National Center for International Research on Tropical Diseases, Ministry of Science and Technology; Key Laboratory of Parasite and Vector Biology, Ministry of Health; School of Global Health, Chinese Center for Tropical Diseases Research, Shanghai Jiao Tong University School of Medicine, Shanghai, People's Republic of China

Li-Ping Duan
National Institute of Parasitic Diseases, Chinese Center for Disease Control and Prevention; Chinese Center for Tropical Diseases Research; WHO Collaborating Centre for Tropical Diseases; National Center for International Research on Tropical Diseases, Ministry of Science and Technology; Key Laboratory of Parasite and Vector Biology, Ministry of Health; School of Global Health, Chinese Center for Tropical Diseases Research, Shanghai Jiao Tong University School of Medicine, Shanghai, People's Republic of China

Dirk Engels
Uniting to Combat NTDs Secretariat, Geneva, Switzerland

Anna Sophie Fahrion
Department of Control of Neglected Tropical Diseases, World Health Organization, Geneva, Switzerland

Yuan Fang
National Institute of Parasitic Diseases, Chinese Center for Disease Control and Prevention; Chinese Center for Tropical Diseases Research; WHO Collaborating Centre for Tropical Diseases; National Center for International Research on Tropical Diseases, Ministry of Science and Technology; Key Laboratory of Parasite and Vector Biology, Ministry of Health, Shanghai, People's Republic of China

Jun Feng
National Institute of Parasitic Diseases, Chinese Center for Disease Control and Prevention; Chinese Center for Tropical Diseases Research; WHO Collaborating Centre for Tropical Diseases; National Center for International Research on Tropical Diseases, Ministry of Science and Technology; Key Laboratory of Parasite and Vector Biology, Ministry of Health, Shanghai, People's Republic of China

Ting Feng
National Institute of Parasitic Diseases, Chinese Center for Disease Control and Prevention; Chinese Center for Tropical Diseases Research; WHO Collaborating Centre for Tropical Diseases; National Center for International Research on Tropical Diseases, Ministry of Science and Technology; Key Laboratory of Parasite and Vector Biology, Ministry of Health; School of Global Health, Chinese Center for Tropical Diseases Research, Shanghai Jiao Tong University School of Medicine, Shanghai, People's Republic of China

Xinyu Feng
National Institute of Parasitic Diseases, Chinese Center for Disease Control and Prevention; Chinese Center for Tropical Diseases Research; WHO Collaborating Centre for Tropical Diseases; National Center for International Research on Tropical Diseases, Ministry of Science and Technology; Key Laboratory of Parasite and Vector Biology, Ministry of Health, Shanghai, People's Republic of China

Zheng Feng
National Institute of Parasitic Diseases, Chinese Center for Disease Control and Prevention; Chinese Center for Tropical Diseases Research; WHO Collaborating Centre for Tropical Diseases; National Center for International Research on Tropical Diseases, Ministry of Science and Technology; Key Laboratory of Parasite and Vector Biology, Ministry of Health; School of Global Health, Chinese Center for Tropical Diseases Research, Shanghai Jiao Tong University School of Medicine, Shanghai, People's Republic of China

Meihua Fu
National Institute of Parasitic Diseases, Chinese Center for Disease Control and Prevention; Chinese Center for Tropical Diseases Research; WHO Collaborating Centre for Tropical Diseases; National Center for International Research on Tropical Diseases, Ministry of Science and Technology; Key Laboratory of Parasite and Vector Biology, Ministry of Health, Shanghai, People's Republic of China

Chun-Hua Gao
National Institute of Parasitic Diseases, Chinese Center for Disease Control and Prevention; Chinese Center for Tropical Diseases Research; WHO Collaborating Centre for Tropical Diseases; National Center for International Research on Tropical Diseases, Ministry of Science and Technology; Key Laboratory of Parasite and Vector Biology, Ministry of Health; School of Global Health, Chinese Center for Tropical Diseases Research, Shanghai Jiao Tong University School of Medicine, Shanghai, People's Republic of China

Jia-Gang Guo
National Institute of Parasitic Diseases, Chinese Center for Disease Control and Prevention; Chinese Center for Tropical Diseases Research; WHO Collaborating Centre for Tropical Diseases; National Center for International Research on Tropical Diseases, Ministry of Science and Technology; Key Laboratory of Parasite and Vector Biology, Ministry of Health, Shanghai, People's Republic of China

Yun-Hai Guo
National Institute of Parasitic Diseases, Chinese Center for Disease Control and Prevention; Chinese Center for Tropical Diseases Research; WHO Collaborating Centre for Tropical Diseases; National Center for International Research on Tropical Diseases, Ministry of Science and Technology; Key Laboratory of Parasite and Vector Biology, Ministry of Health, Shanghai, People's Republic of China

Shuai Han
National Institute of Parasitic Diseases, Chinese Center for Disease Control and Prevention; Chinese Center for Tropical Diseases Research; WHO Collaborating Centre for Tropical Diseases; National Center for International Research on Tropical Diseases, Ministry of Science and Technology; Key Laboratory of Parasite and Vector Biology, Ministry of Health; School of Global Health, Chinese Center for Tropical Diseases Research, Shanghai Jiao Tong University School of Medicine, Shanghai, People's Republic of China

Yu-Wan Hao
National Institute of Parasitic Diseases, Chinese Center for Disease Control and Prevention; National Center for International Research on Tropical Diseases, Ministry of Science and Technology; Key Laboratory of Parasite and Vector Biology, Ministry of Health; WHO Collaborating Centre for Tropical Diseases, Shanghai, People's Republic of China

Jian-Hong Hu
National Institute of Parasitic Diseases, Chinese Center for Disease Control and Prevention; Chinese Center for Tropical Diseases Research; Key Laboratory of Parasite and Vector Biology, Ministry of Health; National Health Commission of China; WHO Collaborating Centre for Tropical Diseases, Shanghai, People's Republic of China

Wei Hu
National Institute of Parasitic Diseases, Chinese Center for Disease Control and Prevention; Chinese Center for Tropical Diseases Research; WHO Collaborating Centre for Tropical Diseases; National Center for International Research on Tropical Diseases, Ministry of Science and Technology; Key Laboratory of Parasite and Vector Biology, Ministry of Health, Shanghai, People's Republic of China

Ji-Lei Huang
National Institute of Parasitic Diseases, Chinese Center for Disease Control and Prevention; Chinese Center for Tropical Diseases Research; WHO Collaborating Centre for Tropical Diseases; National Center for International Research on Tropical Diseases, Ministry of Science and Technology; Key Laboratory of Parasite and Vector Biology, Ministry of Health, Shanghai, People's Republic of China

Yun Huang
National Institute of Parasitic Diseases, Chinese Center for Disease Control and Prevention; Chinese Center for Tropical Diseases Research; WHO Collaborating Centre for Tropical Diseases; National Center for International Research on Tropical Diseases, Ministry of Science and Technology; Key Laboratory of Parasite and Vector Biology, Ministry of Health, Shanghai, People's Republic of China

Shi-Zhu Li
National Institute of Parasitic Diseases, Chinese Center for Disease Control and Prevention; Chinese Center for Tropical Diseases Research; Key Laboratory of Parasite and Vector Biology, Ministry of Health; National Center for International Research on Tropical Diseases, Ministry of Science and Technology; WHO Collaborating Centre for Tropical Diseases; School of Global Health, Chinese Center for Tropical Diseases Research, Shanghai Jiao Tong University School of Medicine, Shanghai, People's Republic of China

Yin-Long Li
National Institute of Parasitic Diseases, Chinese Center for Disease Control and Prevention; Chinese Center for Tropical Diseases Research; WHO Collaborating Centre for Tropical Diseases; National Center for International Research on Tropical Diseases, Ministry of Science and Technology; Key Laboratory of Parasite and Vector Biology, Ministry of Health; School of Global Health, Chinese Center for Tropical Diseases Research, Shanghai Jiao Tong University School of Medicine, Shanghai, People's Republic of China

Yuan-Yuan Li
National Institute of Parasitic Diseases, Chinese Center for Disease Control and Prevention; Chinese Center for Tropical Diseases Research; WHO Collaborating Centre for Tropical Diseases; National Center for International Research on Tropical Diseases, Ministry of Science and Technology; Key Laboratory of Parasite and Vector Biology, Ministry of Health; School of Global Health, Chinese Center for Tropical Diseases Research, Shanghai Jiao Tong University School of Medicine, Shanghai, People's Republic of China

Baixue Liu
National Institute of Parasitic Diseases, Chinese Center for Disease Control and Prevention; Chinese Center for Tropical Diseases Research; WHO Collaborating Centre for Tropical Diseases; National Center for International Research on Tropical Diseases, Ministry of Science and Technology; Key Laboratory of Parasite and Vector Biology, Ministry of Health, Shanghai, People's Republic of China

Qin Liu
National Institute of Parasitic Diseases, Chinese Center for Disease Control and Prevention; Chinese Center for Tropical Diseases Research; Key Laboratory of Parasite and Vector Biology, Ministry of Health; National Center for International Research on Tropical Diseases, Ministry of Science and Technology; WHO Collaborating Centre for Tropical Diseases, Shanghai, People's Republic of China

Yan Lu
National Institute of Parasitic Diseases, Chinese Center for Disease Control and Prevention; WHO Collaborating Centre for Tropical Diseases; National Center for International Research on Tropical Diseases, Ministry of Science and Technology; Key Laboratory of Parasite and Vector Biology, Ministry of Health, Shanghai, People's Republic of China

Chao Lv
National Institute of Parasitic Diseases, Chinese Center for Disease Control and Prevention; Chinese Center for Tropical Diseases Research; WHO Collaborating Centre for Tropical Diseases; National Center for International Research on Tropical Diseases, Ministry of Science and Technology; Key Laboratory of Parasite and Vector Biology, Ministry of Health; School of Global Health, Chinese Center for Tropical Diseases Research, Shanghai Jiao Tong University School of Medicine, Shanghai, People's Republic of China

Shan Lv
National Institute of Parasitic Diseases, Chinese Center for Disease Control and Prevention; Chinese Center for Tropical Diseases Research; Key Laboratory of Parasite and Vector Biology, Ministry of Health; National Center for International Research on Tropical Diseases, Ministry of Science and Technology; WHO Collaborating Centre for Tropical Diseases; School of Global Health, Chinese Center for Tropical Diseases Research, Shanghai Jiao Tong University School of Medicine, Shanghai, People's Republic of China

Men-Bao Qian
National Institute of Parasitic Diseases, Chinese Center for Disease Control and Prevention; Chinese Center for Tropical Diseases Research; Key Laboratory of Parasite and Vector Biology, Ministry of Health; National Center for International Research on Tropical Diseases, Ministry of Science and Technology; WHO Collaborating Centre for Tropical Diseases, Shanghai, People's Republic of China

Zhi-Qiang Qin
National Institute of Parasitic Diseases, Chinese Center for Disease Control and Prevention; Chinese Center for Tropical Diseases Research; WHO Collaborating Centre for Tropical Diseases; National Center for International Research on Tropical Diseases, Ministry of Science and Technology; Key Laboratory of Parasite and Vector Biology, Ministry of Health; School of Global Health, Chinese Center for Tropical Diseases Research, Shanghai Jiao Tong University School of Medicine, Shanghai, People's Republic of China

Hai-Mo Shen
National Institute of Parasitic Diseases, Chinese Center for Disease Control and Prevention; WHO Collaborating Centre for Tropical Diseases; National Center for International Research on Tropical Diseases, Ministry of Science and Technology; Key Laboratory of Parasite and Vector Biology, Ministry of Health, Shanghai, People's Republic of China

Peng Song
National Institute of Parasitic Diseases, Chinese Center for Disease Control and Prevention; WHO Collaborating Centre for Tropical Diseases; National Center for International Research on Tropical Diseases, Ministry of Science and Technology; Key Laboratory of Parasite and Vector Biology, Ministry of Health, Shanghai, People's Republic of China

De-Jian Sun
National Institute of Parasitic Diseases, Chinese Center for Disease Control and Prevention; Chinese Center for Tropical Diseases Research; WHO Collaborating Centre for Tropical Diseases; National Center for International Research on Tropical Diseases, Ministry of Science and Technology; Key Laboratory of Parasite and Vector Biology, Ministry of Health, Shanghai, People's Republic of China

Linhua Tang
National Institute of Parasitic Diseases, Chinese Center for Disease Control and Prevention; Chinese Center for Tropical Diseases Research; WHO Collaborating Centre for Tropical Diseases; National Center for International Research on Tropical Diseases, Ministry of Science and Technology; Key Laboratory of Parasite and Vector Biology, Ministry of Health, Shanghai, People's Republic of China

Tian Tian
National Institute of Parasitic Diseases, Chinese Center for Disease Control and Prevention; Chinese Center for Tropical Diseases Research; WHO Collaborating Centre for Tropical Diseases; National Center for International Research on Tropical Diseases, Ministry of Science and Technology; Key Laboratory of Parasite and Vector Biology, Ministry of Health, Shanghai, People's Republic of China

Jun-Yun Wang
National Institute of Parasitic Diseases, Chinese Center for Disease Control and Prevention; Chinese Center for Tropical Diseases Research; WHO Collaborating Centre for Tropical Diseases; National Center for International Research on Tropical Diseases, Ministry of Science and Technology; Key Laboratory of Parasite and Vector Biology, Ministry of Health; School of Global Health, Chinese Center for Tropical Diseases Research, Shanghai Jiao Tong University School of Medicine, Shanghai, People's Republic of China

Liying Wang
National Institute of Parasitic Diseases, Chinese Center for Disease Control and Prevention; Chinese Center for Tropical Diseases Research; WHO Collaborating Centre for Tropical Diseases; National Center for International Research on Tropical Diseases, Ministry of Science and Technology; Key Laboratory of Parasite and Vector Biology, Ministry of Health, Shanghai, People's Republic of China

Qiang Wang
National Institute of Parasitic Diseases, Chinese Center for Disease Control and Prevention; National Center for International Research on Tropical Diseases, Ministry of Science and Technology; Key Laboratory of Parasite and Vector Biology, Ministry of Health; WHO Collaborating Centre for Tropical Diseases, Shanghai, People's Republic of China

Xin-Yi Wang
National Institute of Parasitic Diseases, Chinese Center for Disease Control and Prevention; Chinese Center for Tropical Diseases Research; Key Laboratory of Parasite and Vector Biology, Ministry of Health; National Health Commission of China; WHO Collaborating Centre for Tropical Diseases, Shanghai, People's Republic of China

Xu Wang
National Institute of Parasitic Diseases, Chinese Center for Disease Control and Prevention; Chinese Center for Tropical Diseases Research; WHO Collaborating Centre for Tropical Diseases; National Center for International Research on Tropical Diseases, Ministry of Science and Technology; Key Laboratory of Parasite and Vector Biology, Ministry of Health, Shanghai, People's Republic of China

Ying Wang
National Institute of Parasitic Diseases, Chinese Center for Disease Control and Prevention; Chinese Center for Tropical Diseases Research; WHO Collaborating Centre for Tropical Diseases; National Center for International Research on Tropical Diseases, Ministry of Science and Technology; Key Laboratory of Parasite and Vector Biology, Ministry of Health, Shanghai, People's Republic of China

Fu-Rong Wei
National Institute of Parasitic Diseases, Chinese Center for Disease Control and Prevention; Chinese Center for Tropical Diseases Research; WHO Collaborating Centre for Tropical Diseases; National Center for International Research on Tropical Diseases, Ministry of Science and Technology; Key Laboratory of Parasite and Vector Biology, Ministry of Health, Shanghai, People's Republic of China

Sihui Wei
National Institute of Parasitic Diseases, Chinese Center for Disease Control and Prevention; Chinese Center for Tropical Diseases Research; WHO Collaborating Centre for Tropical Diseases; National Center for International Research on Tropical Diseases, Ministry of Science and Technology; Key Laboratory of Parasite and Vector Biology, Ministry of Health, Shanghai, People's Republic of China

Weiping Wu
National Institute of Parasitic Diseases, Chinese Center for Disease Control and Prevention; Chinese Center for Tropical Diseases Research; WHO Collaborating Centre for Tropical Diseases; National Center for International Research on Tropical Diseases, Ministry of Science and Technology; Key Laboratory of Parasite and Vector Biology, Ministry of Health, Shanghai, People's Republic of China

Shang Xia
National Institute of Parasitic Diseases, Chinese Center for Disease Control and Prevention; Chinese Center for Tropical Diseases Research; Key Laboratory of Parasite and Vector Biology, Ministry of Health; National Health Commission of China; WHO Collaborating Centre for Tropical Diseases, Shanghai, People's Republic of China

Zhi-Gui Xia
National Institute of Parasitic Diseases, Chinese Center for Disease Control and Prevention; Chinese Center for Tropical Diseases Research; Key Laboratory of Parasite and Vector Biology, Ministry of Health; National Center for International Research on Tropical Diseases, Ministry of Science and Technology; WHO Collaborating Centre for Tropical Diseases, Shanghai, People's Republic of China

Ning Xiao
National Institute of Parasitic Diseases, Chinese Center for Disease Control and Prevention; Chinese Center for Tropical Diseases Research; Key Laboratory of Parasite and Vector Biology, Ministry of Health; National Center for International Research on Tropical Diseases, Ministry of Science and Technology; WHO Collaborating Centre for Tropical Diseases, Shanghai, People's Republic of China

Bin Xu
National Institute of Parasitic Diseases, Chinese Center for Disease Control and Prevention; Chinese Center for Tropical Diseases Research; WHO Collaborating Centre for Tropical Diseases; National Center for International Research on Tropical Diseases, Ministry of Science and Technology; Key Laboratory of Parasite and Vector Biology, Ministry of Health, Shanghai, People's Republic of China

Jing Xu
National Institute of Parasitic Diseases, Chinese Center for Disease Control and Prevention; Chinese Center for Tropical Diseases Research; WHO Collaborating Centre for Tropical Diseases; National Center for International Research on Tropical Diseases, Ministry of Science and Technology; Key Laboratory of Parasite and Vector Biology, Ministry of Health; School of Global Health, Chinese Center for Tropical Diseases Research, Shanghai Jiao Tong University School of Medicine, Shanghai, People's Republic of China

Xue-Nian Xu
National Institute of Parasitic Diseases, Chinese Center for Disease Control and Prevention; WHO Collaborating Centre for Tropical Diseases; National Center for International Research on Tropical Diseases, Ministry of Science and Technology; Key Laboratory of Parasite and Vector Biology, Ministry of Health, Shanghai, People's Republic of China

Chuizhao Xue
National Institute of Parasitic Diseases, Chinese Center for Disease Control and Prevention; Chinese Center for Tropical Diseases Research; WHO Collaborating Centre for Tropical Diseases; National Center for International Research on Tropical Diseases, Ministry of Science and Technology; Key Laboratory of Parasite and Vector Biology, Ministry of Health, Shanghai, People's Republic of China

Jing-Bo Xue
National Institute of Parasitic Diseases, Chinese Center for Disease Control and Prevention; Chinese Center for Tropical Diseases Research; Key Laboratory of Parasite and Vector Biology, Ministry of Health; National Health Commission of China; WHO Collaborating Centre for Tropical Diseases, Shanghai, People's Republic of China

He Yan
National Institute of Parasitic Diseases, Chinese Center for Disease Control and Prevention; Chinese Center for Tropical Diseases Research; WHO Collaborating Centre for Tropical Diseases; National Center for International Research on Tropical Diseases, Ministry of Science and Technology; Key Laboratory of Parasite and Vector Biology, Ministry of Health, Shanghai, People's Republic of China

Shijie Yang
National Institute of Parasitic Diseases, Chinese Center for Disease Control and Prevention; Chinese Center for Tropical Diseases Research; WHO Collaborating Centre for Tropical Diseases; National Center for International Research on Tropical Diseases, Ministry of Science and Technology; Key Laboratory of Parasite and Vector Biology, Ministry of Health, Shanghai, People's Republic of China

Qing Yu
National Institute of Parasitic Diseases, Chinese Center for Disease Control and Prevention; Chinese Center for Tropical Diseases Research; Key Laboratory of Parasite and Vector Biology, Ministry of Health; National Center for International Research on Tropical Diseases, Ministry of Science and Technology; WHO Collaborating Centre for Tropical Diseases, Shanghai, People's Republic of China

Hao-Bing Zhang
National Institute of Parasitic Diseases, Chinese Center for Disease Control and Prevention; Chinese Center for Tropical Diseases Research; WHO Collaborating Centre for Tropical Diseases; National Center for International Research on Tropical Diseases, Ministry of Science and Technology; Key Laboratory of Parasite and Vector Biology, Ministry of Health; School of Global Health, Chinese Center for Tropical Diseases Research, Shanghai Jiao Tong University School of Medicine, Shanghai, People's Republic of China

Li Zhang
National Institute of Parasitic Diseases, Chinese Center for Disease Control and Prevention; Chinese Center for Tropical Diseases Research; WHO Collaborating Centre for Tropical Diseases; National Center for International Research on Tropical Diseases, Ministry of Science and Technology; Key Laboratory of Parasite and Vector Biology, Ministry of Health, Shanghai, People's Republic of China

Li-Juan Zhang
National Institute of Parasitic Diseases, Chinese Center for Disease Control and Prevention; Chinese Center for Tropical Diseases Research; WHO Collaborating Centre for Tropical Diseases; National Center for International Research on Tropical Diseases, Ministry of Science and Technology; Key Laboratory of Parasite and Vector Biology, Ministry of Health; School of Global Health, Chinese Center for Tropical Diseases Research, Shanghai Jiao Tong University School of Medicine, Shanghai, People's Republic of China

Mi-Zhen Zhang
National Institute of Parasitic Diseases, Chinese Center for Disease Control and Prevention; Chinese Center for Tropical Diseases Research; WHO Collaborating Centre for Tropical Diseases; National Center for International Research on Tropical Diseases, Ministry of Science and Technology; Key Laboratory of Parasite and Vector Biology, Ministry of Health, Shanghai, People's Republic of China

Ting Zhang
National Institute of Parasitic Diseases, Chinese Center for Disease Control and Prevention; Chinese Center for Tropical Diseases Research; WHO Collaborating Centre for Tropical Diseases; National Center for International Research on Tropical Diseases, Ministry of Science and Technology; Key Laboratory of Parasite and Vector Biology, Ministry of Health, Shanghai, People's Republic of China

Xue-Qiang Zhang
National Institute of Parasitic Diseases, Chinese Center for Disease Control and Prevention; Chinese Center for Tropical Diseases Research; Key Laboratory of Parasite and Vector Biology, Ministry of Health; National Health Commission of China; WHO Collaborating Centre for Tropical Diseases, Shanghai, People's Republic of China

Yi Zhang
National Institute of Parasitic Diseases, Chinese Center for Disease Control and Prevention; Chinese Center for Tropical Diseases Research; WHO Collaborating Centre for Tropical Diseases; National Center for International Research on Tropical Diseases, Ministry of Science and Technology; Key Laboratory of Parasite and Vector Biology, Ministry of Health; School of Global Health, Chinese Center for Tropical Diseases Research, Shanghai Jiao Tong University School of Medicine, Shanghai, People's Republic of China

Bin Zheng
National Institute of Parasitic Diseases, Chinese Center for Disease Control and Prevention; Chinese Center for Tropical Diseases Research; WHO Collaborating Centre for Tropical Diseases; National Center for International Research on Tropical Diseases, Ministry of Science and Technology; Key Laboratory of Parasite and Vector Biology, Ministry of Health; School of Global Health, Chinese Center for Tropical Diseases Research, Shanghai Jiao Tong University School of Medicine, Shanghai, People's Republic of China

Jin-Xin Zheng
National Institute of Parasitic Diseases, Chinese Center for Disease Control and Prevention; Chinese Center for Tropical Diseases Research; Key Laboratory of Parasite and Vector Biology, Ministry of Health; National Health Commission of China; WHO Collaborating Centre for Tropical Diseases, Shanghai, People's Republic of China

Chang-Hai Zhou
National Institute of Parasitic Diseases, Chinese Center for Disease Control and Prevention; Chinese Center for Tropical Diseases Research; WHO Collaborating Centre for Tropical Diseases; National Center for International Research on Tropical Diseases, Ministry of Science and Technology; Key Laboratory of Parasite and Vector Biology, Ministry of Health, Shanghai, People's Republic of China

Shuisen Zhou
National Institute of Parasitic Diseases, Chinese Center for Disease Control and Prevention; Chinese Center for Tropical Diseases Research; WHO Collaborating Centre for Tropical Diseases; National Center for International Research on Tropical Diseases, Ministry of Science and Technology; Key Laboratory of Parasite and Vector Biology, Ministry of Health, Shanghai, People's Republic of China

Xiao-Nong Zhou
National Institute of Parasitic Diseases, Chinese Center for Disease Control and Prevention; Chinese Center for Tropical Diseases Research; Key Laboratory of Parasite and Vector Biology, Ministry of Health; National Center for International Research on Tropical Diseases, Ministry of Science and Technology; WHO Collaborating Centre for Tropical Diseases; National Health Commission of China; School of Global Health, Chinese Center for Tropical Diseases Research, Shanghai Jiao Tong University School of Medicine, Shanghai, People's Republic of China

Zheng-Bin Zhou
National Institute of Parasitic Diseases, Chinese Center for Disease Control and Prevention; Chinese Center for Tropical Diseases Research; WHO Collaborating Centre for Tropical Diseases; National Center for International Research on Tropical Diseases, Ministry of Science and Technology; Key Laboratory of Parasite and Vector Biology, Ministry of Health; School of Global Health, Chinese Center for Tropical Diseases Research, Shanghai Jiao Tong University School of Medicine, Shanghai, People's Republic of China

Hui-Hui Zhu
National Institute of Parasitic Diseases, Chinese Center for Disease Control and Prevention; Chinese Center for Tropical Diseases Research; WHO Collaborating Centre for Tropical Diseases; National Center for International Research on Tropical Diseases, Ministry of Science and Technology; Key Laboratory of Parasite and Vector Biology, Ministry of Health, Shanghai, People's Republic of China

Ting-Jun Zhu
National Institute of Parasitic Diseases, Chinese Center for Disease Control and Prevention; Chinese Center for Tropical Diseases Research; WHO Collaborating Centre for Tropical Diseases; National Center for International Research on Tropical Diseases, Ministry of Science and Technology; Key Laboratory of Parasite and Vector Biology, Ministry of Health, Shanghai, People's Republic of China

Ze-Lin Zhu
National Institute of Parasitic Diseases, Chinese Center for Disease Control and Prevention; National Center for International Research on Tropical Diseases, Ministry of Science and Technology; Key Laboratory of Parasite and Vector Biology, Ministry of Health; WHO Collaborating Centre for Tropical Diseases, Shanghai, People's Republic of China

Preface

Development strategy of NIPD-CTDR in the new era

Xiao-Nong Zhou[a,b], Xue-Ning Xu[a,b], Jian-Ping Cao[a,b], Ning Xiao[a,b], Shi-Zhu Li[a,b], Ru-Bo Wang[a,b]

[a]National Institute of Parasitic Diseases, Chinese Center for Diseases Control and Prevention; Chinese Center for Tropical Diseases Research, Shanghai, People's Republic of China

[b]WHO Collaborating Centre for Tropical Diseases; National Center for International Research on Tropical Diseases, Ministry of Science and Technology; Key Laboratory of Parasite and Vector Biology, Ministry of Health, Shanghai, People's Republic of China

1. Foreword

Parasitic and tropical diseases have seriously endangered the health of the Chinese people in history, and affected economic and social development (Wang and Zhou, 2020). After the founding of the People's Republic of China in 1949, great efforts on control of parasitic and tropical diseases were deployed by the government of China by launching national disease control programmes. In the initial stage of the national parasitic disease control programme in the early 1950s, it was estimated that there were more than 10 million schistosomiasis patients, 30 million malaria cases annually, 30 million filariasis patients, 530,000 kala-azar patients and more than 200 million people infected with hookworm in China (Chen et al., 2018; Hu et al., 2016; Wu et al., 2018). Since the government attached great importance to those diseases of poverty, it implemented programmes to control those diseases and mobilized participation both by the public at community level as well as by health professionals who provided technical support to the programmes. Consequently, China has made remarkable achievements in the control of parasitic diseases over the last 7 decades. For example, lymphatic filariasis has been eliminated in the whole country in 2006 and this elimination was certified by the World Health Organization (WHO) in 2007, which has set an example for other countries and promoted the global filariasis elimination programme. The goal of no indigenous malaria cases was achieved in China and the country is currently working to obtain validation for this achievement from WHO in 2020. The criteria of transmission control for schistosomiasis were achieved in China in 2016, and transmission interruption is expected to be achieved by 2025.

The nationwide infection rate of soil-transmitted helminthiasis has dropped to 4.8% in China according to the third national survey completed in 2013, and the endemic situation of echinococcosis has been brought under control after the launch of special actions on echinococcosis control since 2016.

The National Institute of Parasitic Diseases at Chinese Center for Disease Control and Prevention (NIPD), and Chinese Center for Tropical Diseases Research (CTDR), is the only national-level research institution for control of parasitic and tropical diseases in China (Chen et al., 2020). Under the guidance of the National Health Commission and the Chinese Center for Disease Control and Prevention, NIPD-CTDR carried out disease prevention and control, scientific research, emergency response, global health cooperation, education and training, health promotion and other activities relevant to parasitic and tropical diseases.

The earliest precursor of NIPD-CTDR, the East China Branch of the National Institute of Health, was founded in Nanjing in November 1950. It was named as the Institute of Parasitic Diseases (IPD) under the Chinese Academy of Medical Sciences (CAMS) in 1956, and moved to Shanghai after having merged with the Hainan Anti-malaria Station of CAMS in 1957. In 1983, the Chinese Academy of Preventive Medicine (CAPM) was created by separating institutions from the CAMS and the IPD became part of it. In 2002, it was renamed as the National Institute of Parasitic Diseases (NIPD) under the Chinese Center for Disease Control and Prevention (China CDC), due to an institutional reform of CAPM that changed its name to China CDC (Zhou et al., 2005). An additional name, the 'Chinese Center for Tropical Diseases Research (CTDR)' was awarded by the central government of China in 2017. Moreover, in recognition of its importance both in disease control and research in the country, several titles have been granted to NIPD-CTDR by various departments of the government as well as international agencies, such as the *World Health Organization Collaborating Centre for Malaria, Schistosomiasis and Filariasis* awarded by WHO in 1980, which was renamed as the *World Health Organization Collaborating Centre for Tropical Diseases* in 2015 (Guan et al., 2020), the *Key Laboratory of Parasite and Vector Biology* by the Ministry of Health in 1992, the *National Center for Medical Malacology* awarded by the Ministry of Health in 1994, the *National Center for International Research for Tropical Diseases* awarded by the Ministry of Science and Technology in 2014, the *Center for Postdoctoral Programmes* awarded by the Shanghai Municipality Government in 2015, and the *National Parasitic Resources Center* awarded by the Ministry of Science and Technology in 2019.

In order to celebrate the 70th anniversary of the founding of the NIPD-CTDR, this special issue features 13 articles that summarize the historical achievements of the NIPD-CTDR over the last 7 decades, and put forward the key orientations for future development, so as to enhance contributions to control/elimination and scientific research of tropical diseases around the world.

2. Major achievements

Since its founding 70 years ago, the NIPD-CTDR has made great achievements in scientific research on epidemiology, surveillance and early warning, aetiology, diagnosis, drug development and vector control in the fields of schistosomiasis, malaria, echinococcosis, lymphatic filariasis, kala-azar and other protozoal blood infections, soil-transmitted helminthiasis, fascioliasis, angiostrongyliasis, taeniasis and cysticercosis, and intestinal protozoa infections. Those research achievements have been transferred to and applied in national parasitic diseases control programmes in China and have encouraged those programmes to move from control to elimination. They have also supported more innovation in China's medical parasitology and have boosted global health cooperation in tropical diseases. The major achievements are reflected in the following three aspects.

2.1 Promoting parasitic diseases from control to elimination

The NIPD-CTDR has completed three nationwide surveys on the endemic status of major parasitic diseases in humans, four nationwide surveys on the endemic status of schistosomiasis, and three national surveys on the prevalence of echinococcosis, so as to not only clearly understand the prevalence of parasitic diseases in different periods in China, but also to transfer the results of the surveys into policy making and the formulation of national workplans for the control programmes. The NIPD-CTDR has also contributed to national disease control programmes in the following aspects: (i) establishment of the national parasitic disease surveillance network, the diagnostic reference laboratory network and the control research base, together with the diseases endemic provinces; (ii) development of the national demonstration zone for the control of soil-transmitted and food-borne diseases and carrying out the pilot project on comprehensive control of echinococcosis in Shiqu County, Sichuan Province; (iii) participation in the formulation of national control plans, technical solutions and standards for the prevention and control of major parasitic diseases, such as schistosomiasis, malaria and echinococcosis; (iv) putting forward integrated control strategies and technical measures for schistosomiasis and other parasitic diseases, that have contributed to moving forward from disease control to elimination; (v) development of new technologies and products for control and surveillance; (vi) organization of national training as well as related activities, e.g. national competition on specific technical skills of parasitic diseases; and (vii) efforts to solve the technical problems in programme implementation in the field. In addition, the NIPD-CTDR has established the surveillance, early warning and emergency response system for parasitic diseases, so as to effectively have information on emerging or re-emerging parasitic diseases, such as trypanosomiasis, fascioliasis, babesiosis and angiostrongyliasis (External Evaluation Committee, 2015).

2.2 Promoting technological innovation in national medical parasitology

The NIPD-CTDR has contributed to the development of human parasitology and medical entomology in China, and carried out operational research on epidemic characteristics, control strategies, surveillance and response, diagnostic techniques, vaccines and preventive drug development, health education and related fundamental research, so as to provide new approaches and technologies for the control and elimination of parasitic diseases in China. This has allowed NIPD-CTDR to play a leading role in the development of medical parasitology and tropical medicine in China (External Evaluation Committee, 2015). For example, NIPD-CTDR has won more than 80 science and technology awards, including 1 first-grade prize, 2 second-grade prizes and 2 third-grade prizes for the national scientific and technological awards, more than 30 patents, 2 new class-I drugs, and more than 20 diagnostics.

The project of the 'Research on Strategies and Technical Measures to Block the Transmission of Lymphatic Filariasis in China' has won the first prize for the national scientific and technological award, which has provided impetus for initiating the elimination of lymphatic filariasis around the world. The results of the 'Research on the Drugs for Treatment of Schistosomiasis', 'Countermeasures to Control and Block the Prevalence of Schistosomiasis in the Mountainous Regions', 'Application and Basic Research of Artemether in Prevention of Schistosomiasis japonica, Schistosomiasis mansoni and Schistosomiasis haematobia', 'Integrated

Control Strategies for Schistosomiasis with Emphasis on the Control of Infection Source', and 'Establishment and Application of the Surveillance and Early Warning System for Schistosomiasis in China' have provided the new approach and technologies for comprehensive control and elimination of schistosomiasis in China. The results of the 'First National Survey on the Distribution of Human Parasites', 'Comprehensive Prevention and Control Model of Parasitic Diseases in China', 'Epidemiological Characteristics and Surveillance Schemes in the Later Period of Malaria Control', and 'Control Strategies for Soil-transmitted and Food-borne Parasitic Diseases and Their Application Effectiveness' have promoted the national control programmes on parasitic diseases in China.

The NIPD-CTDR started projects on the National Scientific and Technological Resource Sharing Service Platform and on the National Center for Pathogenic Microbial Species Preservation in 2004 and 2018, respectively, supported by the Ministry of Science and Technology, in cooperation with 10 other institutions and universities. With more than 10,000 specimens of parasites both from humans and animals being preserved in NIPD-CTDR, the Ministry of Science and Technology positively confirmed this platform as a 'National Resource Pool for Parasites' in 2019, one of 30 national resource centres in China. The NIPD-CTDR also contributed to information exchanges by creating national and international journals. For instance, the *Chinese Journal of Parasitology and Parasitic Diseases* was launched in 1983. This journal publishes in Chinese, with English abstract. *Infectious Diseases of Poverty* is an international journal, published by the publisher of BioMed Centre at Springer Nature in cooperation with Chinese Medical Journal Publisher, that was launched in 2012. Its current SCI impact factor is maintained above 3.0. Both journals have had great impact on national parasitology and international tropical medicine communities (Xia et al., 2013).

2.3 Promoting the disciplinary development in tropical medicine and global health

With the open-up policy and social economic development in China, more international cooperation opportunities have been provided in the fields of parasitology and tropical medicine. In order to take advantage of these developments, a Global Health Center was established in the NIPD-CTDR, which is in charge of all international activities and global health projects in the following five domains: (i) Under the 'Belt and Road' Initiative and China-Africa Cooperation Forum, the NIPD-CTDR has taken the leadership in establishment of south–south co-operative networks, such as the Institution-based Network for China-Africa Cooperation on Elimination of Schistosomiasis (INCAS), the Asian Network on Schistosomiasis and other Helminthic Zoonoses (RNAS+), and the Asia-Pacific Network on Drugs and Diagnostics Innovation for Tropical Diseases (AP-NDI), etc. (Xu et al., 2016). (ii) The NIPD-CTDR carried out the China-Tanzania Demonstration Project on Malaria Control, that aims to tailor Chinese experiences to local settings in Tanzania in 4 communities covering 200,000 population. Another project, China-Australia-Papua New Guinea Trilateral Cooperation on Malaria Control Project, was carried out aiming to assist and strengthen the laboratory capacities on malaria control both at national and regional levels. (iii) The NIPD-CTDR has hosted 4 series of symposiums on surveillance-response systems leading to the elimination of tropical diseases, and organized more than a dozen of international training courses. (iv) The NIPD-CTDR recruited international postdoctoral fellows and young scientists from Asia and Africa and more than a dozen international postdoctoral fellows have completed their research projects in the NIPD-CTDR. (v) The NIPD-CTDR has signed more than a dozen of Memorandums of Understanding (MOUs) to cooperate with national and international institutions and universities in the field of global health. The School of Global Health was established in June 2019, jointly sponsored by the Chinese Center for Tropical Diseases Research and Shanghai Jiao Tong University School of Medicine. In summary, the NIPD-CTDR has carried out several projects in global health to cooperate with scientists from other developing countries, aiming to tailor Chinese knowledge and technologies to local settings with support from various sources, such as the International Development Research Centre of Canada, the Department for International Cooperation of UK, the National Institutes of Health of USA, the Bill & Melinda Gates Foundation, WHO, etc. (Chen et al., 2020; Wang et al., 2019).

3. Future orientations

With the rapid development of China's social economy and the trend of global economic integration, the Chinese population's mobility is much greater than before. As a result more imported, new and rare parasitic and tropical diseases have occurred year by year, posing also great challenges for the control and elimination of parasitic diseases in China, such as schistosomiasis and malaria. Therefore, Chinese institutions like NIPD-CTDR will engage more in global control programmes of tropical diseases in collaboration with international communities. Meanwhile, parasitic and tropical diseases in western China are still important public health problems. For example, a total of 38.59 million people were infected with major parasites, the overall infection rate of helminthiasis was 5.1% as shown in the third national survey on major human parasitic diseases completed in 2013 (Wang and Zhou, 2020). In some local areas, the infection rate of soil-transmitted helminth infection was still more than 20% in some regions of western China. There are still nearly 12 million people infected with *Clonorchis sinensis*, and echinococcosis is still a serious burden on the health of people in Tibetan pastoral areas. Therefore, the NIPD-CTDR has to take the dual task of control and elimination of parasitic and tropical diseases both at the national and international level.

In 2010, the NIPD-CTDR had implemented its first medium and long-term development strategy with the vision of 'leading in China, first-class in Asia and advanced in the world'. The strategic plan consists of two rounds of the 5-year plans, that has led to two great achievements during the past decades. One is the fact that NIPD-CTDR provided strong technical support for the national parasitic diseases control programmes in China which made the programmes shifting from control to elimination in China. The other one is the fact that NIPD-CTDR has launched global health programmes which promotes the institute's capacity on global health security and development, and strengthens the staff's knowledge, approaches and policy on global health. The NIPD-CTDR has now initiated the proposed development of a roadmap for the control of parasitic diseases in China from 2020 to 2030. An 'enhanced development strategy' for the NIPD-CTDR has been conceptualized and another medium and long-term development strategy will soon be completed and implemented in the institute from 2021 onwards. Its principal strategy will be to build the institute as an international advanced tropical disease research centre, to fully integrate scientific research, disease control and emergency response, education and training, information exchange and global health by 2030. It is expected that through another two rounds of 5-year development plans, the NIPD-CTDR will significantly contribute to promoting the construction of a 'Healthy China 2030', transform scientific and technological innovation of parasitic and tropical diseases, and deepen global health cooperation.

In order to achieve the goal of promoting health services for people in China and around the world in the new development era, the NIPD-CTDR has been recommended to maintain the high-quality and high-level development trend with the following three aspects:

(i) *Contribute more to the construction of* 'Healthy China 2030': According to the 'Healthy China 2030' planning outline: by 2030, 'all endemic counties in China will meet the standard of schistosomiasis elimination. Continue to consolidate the efforts of national malaria elimination … all endemic counties in China will basically control the prevalence of echinococcosis and other major parasitic diseases'. In order to achieve this, the NIPD-CTDR will provide technical support, organize the formulation of technical programmes and guidelines for the control of parasitic diseases, and carry out technical guidance and assessment activities for the control of parasitic diseases in China, organize the national surveillance activities and improve the surveillance programmes, and strengthen the operation and management of surveillance systems, information building and analysis of surveillance data. We also need to train ourselves to carry out emergency response to outbreaks in major epidemic diseases and public health emergencies, improve the national laboratory testing and diagnosis network for parasitic and tropical diseases, and provide diagnostic reference and consultation on diagnosis and treatment. These important activities will provide evidence-based references to formulation of national policies, workplans and standards for the control and elimination of parasitic and tropical diseases.

(ii) *Strengthening transformation of scientific and technological innovation in parasitic and tropical diseases*: Establish the biological technology, species resource, large population-based cohort platform, AI decision-making and other research and development platforms; carry out basic research on multi-disciplinary issues such as epidemiology, pathogen and vector biology, pathology, immunology, molecular biology, biochemistry, pharmacy and pharmacology, as well as applied research on disease prevention and control, immunologic diagnosis, detection and testing, and vaccines and drug development; put forward theoretical innovation especially in the pathogenic mechanism, pathogenic mutation and vector control and realize technical innovation especially in the diagnosis, vaccine, drug targets and early warning, so as to strive to build a 'national key laboratory'; build pilot and demonstration areas for the transformation of achievements, so as to promote the application and industrial transformation of technological achievements in the detection and surveillance of important parasitic and tropical diseases, drugs for parasites and vectors, and prevention and control.

(iii) *Strengthening global health cooperation*: Since China has rich experience and mature approaches and technologies in the control and elimination of tropical diseases, the WHO expects China to transfer its experience and technology to Africa and the developing world as a leading force. With the aim to build the world-class global health discipline, the NIPD-CTDR will carry out the following important activities: (a) in capacity strengthening, more technical training, health education and public information exchanges on the control and elimination of parasitic and tropical diseases based on the concept of One Health will be performed, so as to cultivate global leading health talents; (b) in engagement in global health, more cooperation with international institutions, particularly from developing countries, will be carried out with the aim to help other developing countries to control and eliminate schistosomiasis, malaria and other tropical diseases which finally will contribute to achieving the UN 2030 agenda of sustainable development goals (SDGs) (Wang and Zhou, 2020; Engels and Zhou, 2020), and to promote the 'Belt and Road' Initiative and China-Africa Cooperation Forum; (c) in transferring Chinese approaches and technologies, more activities on transferring Chinese approach or technologies to other developing countries in the local disease control programmes. These actions will not only support China's participation in global health governance, but also realize the basic principles of equitable development and healthy poverty alleviation advocated by the UN agenda for SDGs (Engels and Zhou, 2020), and advance the building of an international community with a shared future for mankind.

References

Chen, J., Xu, J., Bergquist, R., Li, S.Z., Zhou, X.N., 2018. "Farewell to the God of Plague": the importance of political commitment towards the elimination of schistosomiasis. Trop. Med. Infect. Dis. 3 (4), 108.

Chen, J., Ding, W., Li, Z., Zhou, D.D., Yang, P., et al., 2020. From parasitic disease control to global health: new orientation of the National Institute of Parasitic Diseases, China CDC. Acta Trop. 201, 105219.

Engels, D., Zhou, X.N., 2020. Neglected tropical diseases: an effective global response to local poverty-related disease priorities. Infect. Dis. Poverty 9, 10.

Guan, Y.Y., Xiao, N., Zhou, X.N., 2020. Role of WHO Collaborating Centre of Tropical Diseases in China. CCDC Weekly 2, 44–47.

Hu, T., Liu, Y.B., Zhang, S.S., Xia, Z.G., Zhou, S.S., et al., 2016. Shrinking the malaria map in China: measuring the progress of the National Malaria Elimination Programme. Infect. Dis. Poverty 5, 52.

Wang, Y.P., Zhou, X.N., 2020. The year 2020, a milestone in breaking the vicious cycle of poverty and illness in China. Infect. Dis. Poverty 9, 11.

Wang, D., Chaki, P., Mlacha, Y., Gavana, T., Michael, M.G., et al., 2019. Application of community-based and integrated strategy to reduce malaria disease burden in southern Tanzania: the study protocol of China-UK-Tanzania pilot project on malaria control. Infect. Dis. Poverty 8, 4.

WHO External Evaluation Committee, 2015. External Evaluation Report on the Twelfth Five-Year Strategic Development Plan (2011–2015) of the National Institute of Parasitic Diseases. China CDC, Shanghai.

Wu, W.P., Wang, H., Wang, X., Zhou, X.N., Wang, L.Y., Zheng, C.J., 2018. A nationwide survey on echincoccosis in China during 2012–2016. Chin. J. Parasitol. Parasit. Dis. 36, 1–14.

Xia, S., Allotey, P., Reidpath, D.D., Yang, P., Sheng, H.F., Zhou, X.N., 2013. Combating infectious diseases of poverty: a year on. Infect. Dis. Poverty 2, 27.

Xu, J., Yu, Q., Tchuente, L.A., Bergquist, R., Sacko, M., et al., 2016. Enhancing collaboration between China and African countries for schistosomiasis control. Lancet Infect. Dis. 16, 376–383.

Zhou, X.N., Chen, J.X., Chen, M.G., Bergquist, R., 2005. The National Institute of parasitic diseases, Chinese Center for Disease Control and Prevention: a new administrative structure for schistosomiasis control. Acta Trop. 96, 296–302.

CHAPTER ONE

Contributions and achievements on schistosomiasis control and elimination in China by NIPD-CTDR

Chun-Li Cao[a,b,c,†], Li-Juan Zhang[a,b,c,†], Wang-Ping Deng[a,b,c], Yin-Long Li[a,b,c], Chao Lv[a,b,c], Si-Min Dai[a,b,c], Ting Feng[a,b,c], Zhi-Qiang Qin[a,b,c], Li-Ping Duan[a,b,c], Hao-Bing Zhang[a,b,c], Wei Hu[a,b], Zheng Feng[a,b,c], Jing Xu[a,b,c], Shan Lv[a,b,c], Jia-Gang Guo[a,b], Shi-Zhu Li[a,b,c], Jian-Ping Cao[a,b,c], Xiao-Nong Zhou[a,b,c,*]

[a]National Institute of Parasitic Diseases, Chinese Center for Disease Control and Prevention; Chinese Center for Tropical Diseases Research, Shanghai, People's Republic of China

[b]WHO Collaborating Centre for Tropical Diseases; National Center for International Research on Tropical Diseases, Ministry of Science and Technology; Key Laboratory of Parasite and Vector Biology, Ministry of Health, Shanghai, People's Republic of China

[c]School of Global Health, Chinese Center for Tropical Diseases Research, Shanghai Jiao Tong University School of Medicine, Shanghai, People's Republic of China

*Corresponding author: e-mail address: zhouxn1@chinacdc.cn

Contents

1. Introduction 2
2. The endemic status of schistosomiasis in P.R. China 3
3. Transmission patterns and evolution of the control strategy on schistosomiasis 6
   3.1 Epidemiological patterns 6
   3.2 Strategy evolution 14
4. Diagnostics and vaccine development 20
   4.1 Development and application of diagnostics 20
   4.2 Vaccines development 28
5. Research and development of drugs and molluscicides 31
   5.1 Drug development for human treatment 31
   5.2 Molluscicides development 36
6. Monitoring and evaluation 40
   6.1 National sampling surveys 40
   6.2 Routine control programme 42
   6.3 Schistosomiasis surveillance system 43
   6.4 Early warning and modelling 44

† Contributed equally as the first author.

Advances in Parasitology, Volume 110
ISSN 0065-308X
https://doi.org/10.1016/bs.apar.2020.04.002

7. Technical guidelines and expertise support to the control programme 46
7.1 Providing technical support as major member agency of the Expert Committee on Schistosomiasis Control 47
7.2 Involving in the design and implementation of the national control strategy and work plan 47
7.3 Technical support to the response systems 48
7.4 Formulating national criteria related to schistosomiasis control 49
7.5 Attending field investigation to provide accurate information for policy makers 49
8. Conclusion and Perspectives 50
References 51

Abstract

Being a zoonotic parasitic disease, schistosomiasis was widely spread in 12 provinces of Southern China in the 1950s, severly harming human health and hindering economic development. The National Institute of Parasitic Diseases at the Chinese Center for Diseases Control and Prevention, and Chinese Center for Tropical Diseases Research (NIPD-CTDR), as the only professional institution focussing on parasitic diseases at the national level, has played an important role in schistosomiasis control in the country. In this article, we look back at the changes of schistosomiasis endemicity and the contribution of NIPD-CTDR to the national schistosomiasis control programme. We review NIPD-CTDR's activities, including field investigations, design of control strategies and measures, development of diagnostics and drugs, surveillance-response of endemic situation, and monitoring & evaluation of the programme. The NIPD-CTDR has mastered the transmission status of schistosomiasis, mapped the snail distribution, and explored strategies and measures suitable for different types of endemic areas in China. With a good understanding of the life cycle of *Schistosoma japonicum* and transmission patterns of the disease, advanced research carried out in the NIPD-CTDR based on genomics and modern technology has made it possible to explore highly efficient and soft therapeutic drugs and molluscicides, making it possible to develop new diagnostic tools and produce vaccine candidates. In the field, epidemiological studies, updated strategies and targeted intervention measures developed by scientists from the NIPD-CTDR have contributed significantly to the national schistosomiasis control programme. This all adds up to a strong foundation for eliminating schistosomiasis in China in the near future, and recommendations have been put forward how to reach this goal.

1. Introduction

Schistosomiasis japonica caused by *Schistosoma japonicum* is endemic in the People's Republic of China (P.R. China) and has had a long history. In 1972, *S. japonicum* eggs found in the liver and intestinal tissues of a female corpse unearthed from No. 1 Han Tomb in Mawangdui, Changsha, Hunan province, which proved that schistosomiasis has been prevalent in P.R. China for more than 2100 years (Hunan Medical College, 1980). Human beings and more than 40 kinds of mammals serve as definitive host, and the *Oncomelania* snail is the only intermediate host. Various infection sources and recurrent infection have made the prevention and control of schistosomiasis difficult. In the early 1950s, just before the national schistosomiasis control programme was started. It was estimated that over 10 million people and 1 million cattle were infected. Schistosomiasis was therefore one of the major infectious diseases in P.R. China, imposing a huge burden on human and veterinary health. After the founding of the P.R. China in 1949, the Chinese government has started to pay attention to controlling the disease. The Central Committee of Schistosomiasis Control was established in 1955 and a national campaign against schistosomiasis was launched (Lv et al., 2019).

During the past seven decades, researchers from the National Institute of Parasitic Disease, affiliated to the Chinese Center for Disease Control and Prevention (China CDC) and the Chinese Center for Tropical Diseases Research (NIPD-CTDR) have devoted themselves to the national schistosomiasis control programme and relevant research activities. These activities included the provision of technical support and the development and implemention of new control strategies in P.R. China. These activities include research and production of advanced tools and technologies for schistosomiasis control. We review here the contribution of the NIPD-CTDR to the national schistosomiasis control and elimination programme and put forward recommendations for eliminating schistosomiasis in the near future.

2. The endemic status of schistosomiasis in P.R. China

According to a nation-wide investigation in the mid-1950s, schistosomiasis was prevalent in 400 counties of 12 provinces, autonomous regions and municipalities in P.R. China, namely Jiangsu, Zhejiang, Anhui, Jiangxi, Fujian, Shanghai, Hunan, Hubei, Guangdong, Guangxi, Yunnan and Sichuan in the south of the Yangtze River Basin. The endemic area of schistosomiasis in P.R. China stretched as far north as Baoying County in Jiangsu Province (latitude 33°15′) ， and as far south as Yulin County in Guangxi Zhuang Autonomous Region (latitude 22°5′), as far east as Nanhui County in Shanghai Municipality (longitude 121°51′), and as far west as

Yunlong County in Yunnan Province (longitude 99°50′). The lowest altitude of the endemic area was 0 m (Shanghai) and the highest altitude was about 3000 m above the sea level (Yunnan Province). In the 1950s, about 11.6 million people and 1.2 million cattle were infected by *S. japonicum* in P.R. China, while 100 million people were at risk. The land area infested with *Oncomelania* spp. snails had a surface of 14.5 billion m^2 (Zhou et al., 2004). Before the founding of the People's Republic of China in 1949, the prevention and treatment on schistosomiasis was inadequate due to poverty, war, and lack of effective technical tools that had let to very high prevalence, incidence and mortality. According to surveys at that time, the infection rates with schistosomiasis in some villages was over 70%.

In order to carry out large-scale treatment and prevention, a national committee for schistosomiasis control was set up to organize the treatment of patients in known endemic areas, and organize schistosomiasis control groups in a short period of time (Wang et al., 1989). In 1950, the East-China Branch of the National Institute of Health—the earliest precursor of the NIPD-CTDR was established in Nanjing (Wang, 2011). This institute was later named as Institute of Parasitic Diseases (IPD) affiliated to the Chinese Academy of Medicine, before it moved from Nanjing to Shanghai in 1957, and became a national centre for technical support to the national schistosomiasis control programme and carry out research on various innovative tools and strategies.

In 1953, after reading the report on the endemic situation of schistosomiasis in Southern China, Chairman Mao Zedong, pointed out that "the harm of schistosomiasis is very serious, and we must focus on prevention and control of schistosomiasis" (Zhou et al., 2007c). In 1955, Chairman Mao Zedong issued a great call to "eliminate schistosomiasis" after visiting the epidemic areas and listening to professional opinions widely. As proposed by Chairman Mao Zedong's, the Central Committee of the Communist Party of P.R. China quickly established a central committee of schistosomiasis control (commonly known as nine-member leading group for the control of schistosomiasis) (Lv et al., 2019). Meanwhile, the local leading groups for schistosomiasis control composited of major leaders in charge of the party and government were set up. Professional organizations and teams for control of Schistosomiasis were subsequently established at all levels. This initiated the national schistosomiasis control programme in P.R. China.

The seven decades' efforts by the national schistosomiasis control programme have resulted in outstanding achievements. Among the 12 provinces initially endemic of schistosomiasis japonica, 5 provinces, namely Shanghai, Zhejiang, Fujian, Guangdong and Guangxi, continued to consolidate status of elimination of schistosomiasis. Sichuan Province achieved transmission interrupt and six other provinces namely Yunnan, Jiangsu, Hubei, Anhui, Jiangxi and Hunan maintained the status of transmission control by the end of 2018. Out of the all endemic counties, city, district with 58.44% (263/450), 27.56% (124/450) and 14% (63/450) achieving the criteria of elimination, transmission interruption and transmission control, respectively (Fig. 1). The endemic situation of schistosomiasis in P.R. China has been reduced to the lowest level in history. In 2018, only eight people and two cattle were still testing positive. Seventy years of schistosomiasis control have reached these remarkable achievements thanks to technically support by professionals and the leadership of the party and the government in cooperation with various departments. At present, the national schistosomiasis control programme in P.R. China is gradually entering the process of elimination.

The national control programme has moved the schistosomiasis situation in stages from being widely endemic, to endmic control, morbidity control, transmission control, transmission interruption to finally being on the way towards elimination. The NIPD-CTDR, as a national scientific research institution, has played an important role on the improvement of control technology and has promoted the transition from control to elimination of schistosomiasis in P.R. China.

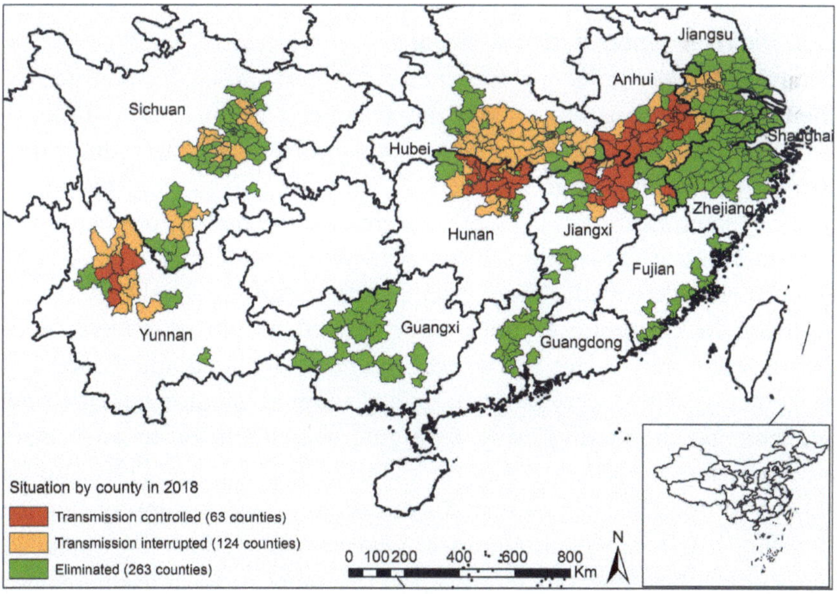

Fig. 1 The epidemiology situation of schistosomiasis by county in P.R. China in 2018.

3. Transmission patterns and evolution of the control strategy on schistosomiasis

3.1 Epidemiological patterns

Through field investigations and in-depth studies, the NIPD-CTDR classified the endemic areas of schistosomiasis in P.R. China into three different categories according to the ecological features of snails habitats, namely plains & water network regions, marshland & lake regions, and hilly & mountainous regions. The NIPD-CTDR conducted a wide variety of pilot studies in different regions of P.R. China, such as Guichi County in Anhui Province in the 1950s, 1960s and 1980s; Changshu County in Jiangsu Province in the 1960s; Poyang,Yugan, Xingzi, Pengze counties in Jiangxi Province in the 1970s and 1980s; and Weishan County in Yunnan Province in the 1980s and 1990s. Based on those research results, it put forward that interventions and control strategies be adapted to local settings. In the 1980s, with the achievement of schistosomiasis control, most of the endemic areas in the plains & water network regions have reached the so-called "eradication standard" formulated at that time, while the control programme in marshlands & lakes regions, and hilly & mountainous regions faced more difficulties . The NIPD-CTDR continued to conduct pilot studies on epidemiology and control strategies in these two areas at that time.

3.1.1 Plains & water network regions

Changshu County in Jiangsu Province is a typical area with water network where schistosomiasis was once highly prevalent. From 1961 to 1966, the NIPD-CTDR cooperated with the local institutes to carry out clinical research on treating large numbers of patients with acute, chronic and advanced schistosomiasis. In-depth research on the pathogenesis and treatment methods of advanced schistosomiasiswere conducted and scientific achievements were made. Since 1965, in view of the task to eliminate schistosomiasis in Changshu County, pilot control studies on schistosomiasis control were carried out in two communes, Mocheng and Zhitang. In the former, a wide variety of measures had adopted, including treating positive patients, eliminating snails, managing excrement, protecting citizens and promoting health education. In addition, the NIPD-CTDR also formulated proposals and suggested constructing water conservancy and improving environment to eliminate *Oncomelania hupensis*. To implement these control measures efficiently, a large number of local health workers were trained well and attended the project. The NIPD-CTDR invested substantial human and material resources in the general survey, general treatment and control as well as research for eliminating snails. A hundred technical and administrative staff from the NIPD-CTDR participated in this work. This practice involved an ordered system involving soil burial and chemical externination to eliminate the snails according to local conditions. After long and arduous efforts, the whole Changshu County reached the standard of schistosomiasis elimination in the 1980s. In Zhitang commune, on the other hand, the work mainly focussed on investigation and analysis of acute schistosomiasis infection and the implementation of personal self-protection. The results provided a reference for the prevention of acute schistosomiasis infection in areas where *Oncomelania hupensis* had not been eradicated.

3.1.2 Marshlands & lakes regions

Guichi County in Anhui Province is adjacent to the East Lake and the South Lake where snails are widely present in large areas (more than 50,000 acre). After a rigorous survey on topography of the snail habitats and analysing with hydrological data, several snail control methods have been proposed which have tailored to local settings, such as changing the water course from lake-type to river-type, building dikes on boundary between the Yangtze River and the lakes to stop river water flowing back, transforming the East Lake and the South Lakes to fertile farmlands, etc. After the implement of these measures, both lakes witnessed an earthshaking change, with large scales of agricultural production bases sprouting out. Such kind of changes not only help to control schistosomiasis through eliminating *Oncomelania* snails, but also increased agricultural output with more than 500 million tons of grain produced annually.

The NIPD-CTDR continued to investigate and analyse the relationship between snail distributions and infested lake water in other areas, such as Shima Lake in Huaining County, Anhui Province, Danyang Lake and the Jiangxin marshlands in Dangtu County, Anhui Province, as well as Bali Lake in Jiujiang City, Jiangxi Province. Combining with these results, the NIPD-CTDR put forward that efforts should focus on eliminating the snails near residential areas and boat mooring sites, as well as other measures to establish safe zones. This concept and measures have been adopted by many endemic areas in the following years.

In the 1980s, when praziquantel had been discovered as a highly safe medicine for schistosomiasis, the NIPD-CTDR established a research pilot at Tuniu Village in Xinzi County of Jiangxi Province along the banks of Poyang Lake. The coverage of human and cattle treatment with praziquantel was expanded in epidemic area to thoroughly eliminate the source of infection, and proposed intervention measures were subjected to long-term monitoring. One year later, the effectiveness of schistosomiasis control had significantly improved. However, since the infection sources and infected snails still existed, it was difficult to prevent the re-infection in human and animals, and the achievements were this difficult to sustain.

3.1.2.1 Combination of sluice construction and snail control in the Poyang Lake

From 1970 to 1972, many professionals engaged in research on schistosomiasis control in Poyang and Yugan counties of Jiangxi Province. Based on observations in the field, the NIPD-CTDR's scientists believed that control and elimination of snails is essential to control schistosomiasis in marshland and lakes region (He, 2005). Because *Oncomelania* is an amphibious snail the concept of "snail-drowning" has proved successful to eliminate snail in other endemic areas. To that end, the technical staff from the NIPD-CTDR put forward the proposal to the local government to build sluices at the intersection of the Poyang Lake that can store water and maintain water level to eliminate snails. These measures could also be integrated with other comprehensive development engineering projects and better utilization of other resources. With the support and cooperation with leading departments of the local government, the NIPD-CTDR's professionals visited relevant departments, communes and farms in more than 10 counties and cities around Poyang Lake, held more than 60 seminar sessions and listened to opinions. They collected data on industry, agriculture, water conservancy, hydrology, aquatic products, meteorology, land and water transportation, geology and schistosomiasis control weighing up the advantages and disadvantages of eliminating snails by building sluices to store water. The opinion of the NIPD-CTDR staff was that the way of building sluices inPoyang Lake would change the lifecycle patterns of *Oncomelania* snails that spent "winter on land and summer in water" on the grass beaches of the lake. This would solve the dilemma of having to chose between water conservation and snail elimination, after building sluice to control the water level, the lake resources could be developed. With the active support of the relevant departments in Jiangxi provincial government, the NIPD-CTDR's professionals completed the proposal on "building sluice connection in the Poyang Lake to save water and eliminate snails in combination with better utilization of resources and thus contribute to the comprehensive development of the area". After investigation for more than 2 years, the proposal was submitted to the Leading Group on Schistosomiasis Control in the State Council, Jiangxi Provincial Committee for Schistosomiasis Control.

3.1.2.2 Test for snail control by aircraft spraying

In 1970, given the difficulties in snail elimination due to the snail specific ecological patterns, such as the wide range of snail distribution, short season for snail living in the marshland above the water, the Jiangxi Provincial Government decided to carry out the filled trial on snail control by spraying pentachlorophenol sodium, a molluscicide, by airplane. The NIPD-CTDR professionals were invited to participate in this field trial, and has successively carried out airplane spraying of molluscicide to kill the snails of 40,000 acre in the marshland. The results showed that under appropriate conditions, i.e. high temperature, appropriate water level, long immersion time, and flat grass beaches, etc., good results could be obtained to control snail populations. The effectiveness of eliminating snails was better in the Spring than that the autumn (Guo et al., 2002). In the 1970s, in spite of the adverse effects on the environment, it was considered that large-scale airplane spraying to kill snails had the advantages of saving manpower and time, particularly in areas far away from settlement and where it is difficult to eliminate the snails by manpower.

3.1.3 Hilly & mountainous regions

More than 50% of the schistosomiasis endemic counties (cities) in P.R. China are located in the hilly & mountainous regions. According to the natural geographical environment and the epidemic characteristics of schistosomiasis, these areas can be divided into three subtypes: hilly areas, canyon and flatland. Guichi County in Anhui Province is one of the highly prevalent counties of schistosomiasis in P.R. China, where most of endemic areas are belong to the marshlands & hill. In the hilly area of Guichi County, professionals chose two pilot villages (Pailou and Douxi) to conduct a study combining praziquantel treatment with comprehensive snail elimination in Pailou Village or snail elimination in hot spots in Douxi village. Two years later, schistosomiasis transmission had initially been interrupted in both villages. However, in Douxi, several new cases of infection and positive snails appeared in the next year, spinning the situation out of control. Therefore, it was believed that the spread of schistosomiasis could be controlled by mass dug distribution (MDA) in combination with the overall elimination of snails in those hill-endemic areas that have become independent epidemic units.

As a key scientific and technological research project during the seventh five-year plan period (1986–1990), pilot studies were carried out in Weishan County, Yunnan Province and Xichang City, Sichuan Province. The ecology, biological, natural and social factors affecting schistosomiasis, prevalence in mountainous areas were investigated by using epidemiological methods and satellite images systematically and comprehensively (Cao et al., 2004; Zheng, 2000; Zheng et al., 2000). Based on the data collected, different intervention measures were implemented to provide evidence for improving the control strategy of schistosomiasis in mountainous areas. The Weishan County pilot represented the canyon area, which is the long-term and systematic observation area of the NIPD-CTDR. The Xichang City pilot was in flatlands, which was the key observation area of Sichuan Provincial Institute of Parasitic Diseases Prevention and Control. According to a systematic analysis with the probability accumulation method, six main factors were found to affect schistosomiasis prevalence in the mountainous areas, namely temperature, rainfall, the number of infection sources, the number of positive snail points, water source for agriculture production and the use of ditch water, which served as important reference for formulating intervention measures.

Three measures adopted in three different pilot areas of Weishan County, namely, synchronous chemotherapy for humans and animals with elimination of snails in infected spots/areas (group A), synchronous chemotherapy for human and animal (group B), and routine control (group C). Two years later, the positive rate in faecal tests of residents in A and B groups decreased by 80% and 55%, respectively, while there was no significant change in C group. This indicated that the two intervention programmes have certain effects on controlling local schistosomiasis prevalence even under continued transmission.

Based on research carried out during the seventh five-year plan, the eighth five-year plan focussed on controlling and interrupting the transmission of schistosomiasis in the mountainous areas. The study took 4 years. According to the epidemic factors, such as social, economic, behavioural and psychological factors, a series of prevention and control measures were formulated, emphasizing chemotherapy and health education, supplemented by key environmental control measures. Pilot studies were carried out in Zhonghe Village of Weishan County, Yunnan Province and Chuanxing Village of Xichang City, Sichuan Province, and satisfactory results were achieved (Qin et al., 2012). This study shed light on the characteristics of faecal pollution, snail distribution and ecological characteristics of positive snail in the mountainous area. According to the peak of schistosomiasis infection, the life history of schistosomiasis and the pharmacodynamics of praziquantel treatment in the mountainous area, scientists from the NIPD-CTDR proposed the optimal time of chemotherapy would to be 1 month after the peak of infection with the number of peak were equal to the number of treatment. According to the obvious family aggregation of local schistosomiasis, MDA carried out in families, which clarified the distribution rule of infected *O. hupensis* and helped to establish a simple and effective evaluation method for hazardous environmental improvement in hazardous areas. In addition, the method of comprehensive molluscicidal replaced by the method of niclosamide buried in soil, with less environmental impact. Systematic investigation of residents' attitudes towards schistosomiasis control and health education contributed to make health education more targeted, and triggered the establishment of a comprehensive evaluation systems for schistosomiasis prevention and control.

According to the results of epidemiological investigation, six different control strategies were carried out, mainly including human and animal chemotherapy, crucial environment control by eliminating snail, and health education. After 3–4 years of prevention and control, satisfactory results have been obtained. In Weishan County pilot, the infection rate of residents in one observation group decreased from 24.9% before the intervention to 0.2%, and that of livestock decreased from 4.6% to 0%. No positive snails found and there were no new infections in humans, animals or sentinel rat in water. All these improvements initially interrupted the transmission of schistosomiasis. In the other two groups, the infection rate of human and animal dropped to less than 3%, with no new infection detected in human and animal. The natural infection rate and density of snails declined to very low levels, which has controlled the spread of schistosomiasis. Similar results were obtained in the Xichang City pilot. Specific interruption of transmission was obtained in one strategy group, and transmission control in the other two groups. These results clearly showed that after 3–4 years of control, each strategy group in Weishan and Xichang pilots had successfully interrupted the transmission of schistosomiasis, and two strategy groups controlled the transmission of schistosomiasis. They therefore have guided the formulation of prevention and control plans or their improvement in mountainous areas.

3.1.4 The influence of Three Gorges Dam on schistosomiasis transmission

Many reports have suggested that the construction of water conservancy projects can change the ecological environment so as to intensify or spread the epidemic of schistosomiasis (Guo and Zheng, 1999; Hunter et al., 1982; Wang and Zheng, 2003).

The Three Gorges Water Conservancy Project is a large-scale comprehensive project to develop and harness the Yangtze River, whose functions mainly involving flood control, power generation, shipping, sand drainage and environmental protection. There has been no history of schistosomiasis in the area before or after the construction of The Three Gorges Dam (TGD). Yet, with the social and economic development and ecological environment changes, many researches have focussed on whether schistosomiasis infection sources and *Oncomelania* can imported into the reservoir area where was no schistosomiasis in history after the construction of TGD. In order to understand the risk factors of schistosomiasis transmission in the TGD area and put forward the corresponding control measures, the NIPD-CTDR researchers participated in the design and implementation of the research to understand the impact of the TGD on the epidemic of schistosomiasis (Zheng et al., 2003).

In the 1990s, based on hydrology, sediment and other materials provided by the Yangtze River Water Resources Commission and prediction results of water and sediment changes after the dam built, the NIPD-CTDR's scientists conducted a series of field surveys and analysed the possible impact of ecological environment changes on schistosomiasis transmission. Also, the scientists took the principles and methods of epidemiology, ecology and social medicine were taken into consideration and investigated the infection situation in human and animals, as well as the social and economic behaviour factors in the TGD reservoir area and the middle and lower reaches of the Yangtze River. The growth and development of *O. hupensis* were investigated through simulating the conditions of similar reservoirs. Systematic analysis was conducted to predict the relationship between TGD and schistosomiasis transmission. The results showed that the natural conditions in the TGD area were suitable for snails' breeding and schistosomiasis transmission. After the TGD was built, the water flow in the reservoir area slows down and silt deposition will form many beaches. In the area from 145 to 175 m high, the beaches are below the flood outsourcing line and at the intersection of each tributary river all have the necessary conditions for snail breeding (Zheng et al., 2002). In addition, irrigation ditches in settlements may were also suitable for snails breeding.

An analogy survey showed that the vegetation, soil and physical and chemical properties in the reservoir area were like those of meadow of lakes in the endemic area. The results showed that the riff-shell snails from Hubei and Sichuan could grow, develop and reproduce under simulated conditions. Epidemiological investigation also confirmed that the snail and the source of infection might be imported into the reservoir area. Therefore, the reservoir area has become a potential schistosomiasis endemic area, and the monitoring and intervention measures should be strengthened to prevent the spread of schistosomiasis from endemic areas to the reservoir area. Jianghan Plain, the head of the TGD was one of the most serious endemic area. After the construction of the TGD, the water level of the Yangtze River increased at the turn of spring and summer, which may lead to the increase of underground-water level in some areas of Jianghan Plain and possibly enlarge the snail area and change the planting schedule. To prevent the spread of snail, the production pattern should be changed, water modification and rotation should be promoted, to control the spread of snail and reduce the infection of humans and animals.

The influence of the TGD on Dongting Lake is mainly due to the decrease in sediment entering the lake, and the gradual development of grass and reed areas, conducive to the breeding of snail (Dai et al., 2019a). In autumn, the water recession happened earlier, which may increase the infection risk of human and animals due to an increased frequency with water contact. The construction of TGD has a relatively small impact on the area of Poyang Lake and below. The result from Jiangsu province's survey suggested that early spring flooding is not conducive to the growth and development of *O. hupensis*. At the same time, the scour and siltation of the Yangtze River section were changed. In some areas, erosion is not conducive to snails' breeding, which can reduce the distribution area. In other areas situation may increase snails habitats.

Over the past century, floods have occurred in the middle and lower sections of the Yangtze River. After the construction of the dam, the flood will be effectively controlled, which will not only protect people's lives and property, but also reduce the spread of *Oncomelania* snails and the occurrence of human and animal infection by schistosomes. The construction of TGD was a large-scale project. The ecological changes after its completion will be a gradual process. At present, the relationship between TGD and schistosomiasis can only be analysed based on prediction. Some results need to be further investigation and verification. According to the results obtained, TGD has both disadvantages and advantages to the spread of schistosomiasis. Therefore, it is urgent to explore effective and feasible countermeasures to prevent the spread of schistosomiasis to minimize the negative effects.

3.2 Strategy evolution

In the early days after the founding of P.R. China, there was a tragic scene described in the poem "because of the harm of schistosomiasis, many people died, leaving only empty villages, cold and desolate" (Wang, 2006). The Communist Party Central Committee and the State Council have always attached great importance to the prevention and control of schistosomiasis. In 1955, the Central People's Government of the Communist Party established a leading group for schistosomiasis control. In 1956, the central government issued a call for "Do eliminate schistosomiasis", setting off a vigorous mass campaign for the prevention and control of schistosomiasis. In 1984, Comrade Deng Xiaoping gave instructions for the "prevention and control of endemic diseases for the benefit of the people". In 1989, Comrade Jiang Zemin pointed out in his letter to the conference of schistosomiasis control in five provinces of lake area that "the control and elimination of schistosomiasis is the binding duty of all levels of the governments in the epidemic area". This upsurged schistosomiasis prevention and control work in the early 1990s (Zhou et al., 2010).

With the efforts of several generations, the strategy of schistosomiasis control in P.R. China has also undergone a transformation process from "snail eradication oriented" to "human and animal chemotherapy oriented" to "infection source control oriented" comprehensive control strategy. Finally, the prevention and control of schistosomiasis in P.R. China has made remarkable achievements (Zhang, 2009). During this process, many scientists of NIPD-CTDR contributed their intelligence and involvement in the formulation and integration of new strategy into national control programmes.

3.2.1 The first stage, from the mid-1950s to the early 1980s

A comprehensive prevention and control strategy based on the elimination of *Oncomelania*, the combination of human and animal diagnosis and treatment, and the rescue of critical patients.

At this stage, the focus of prevention and control schistosomiasis is to eliminate *Oncomelania* and treat patients actively according to local conditions. As *Oncomelania* is the only intermediate host of *S. japonicum,* the life cycle of schistosomes will be blocked and the transmission will stop once the snails are eliminated. At that time, there were no good methods to control pathogens and avoid contact with epidemic water, and no safe, cheap and effective drugs to treat schistosomiasis cases. To reduce the harm of schistosomiasis, mobilization of the masses and resources were good news to control the snails at that time. In general, it can be concluded that large scope and large-scale molluscicidal projects focussed on water conservancy and farmland infrastructure construction, as well as large-scale drug molluscicidal projects, large-scale investigation and treatment of patients and infected livestock, were supplemented by water and faeces management, individual protection, publicity and education, etc. At this stage, great achievements have been made in the prevention and control of schistosomiasis in P.R. China. The area of the epidemic area has been greatly reduced, and the disease condition has been significantly reduced. In places with a relatively developed economy and high population density, where snail breeding area was relatively limited, such as in Guangdong, Shanghai, Guangxi, Fujian and Zhejiang etc., snail control and chemotherapy were taken as the main measures, combined with other control measures, to eliminate schistosomiasis and effectively block the spread of schistosomiasis. However, snail eradication not only costs a lot of money, but also needs long-term and repeated prevention and control. In addition, molluscicicing will cause environmental pollution. It is very difficult to interrupt the transmission of schistosomiasis by taking the above measures before the environment has been completely transformed in seven epidemic areas and provinces of Hunan, Hubei, Jiangsu, Jiangxi, Anhui, Sichuan and Yunnan, where snails are widely distributed, the breeding environment is complex and the water level is also difficult to control (Zhang, 2009). By the end of 1984, more than 11 million patients had been treated and more than 11 billion m^2 of snails had been eradicated. Seventy six of the 370 endemic counties (cities) in P.R. China had reached the "standard of schistosomiasis elimination", 193 of them had reached the threshold of "basic elimination of schistosomiasis", and 4 provinces (cities) in the East, Shanghai, Fujian and Guangxi, mainly mountainous or water network type, had eliminated Schistosomiasis (Zhou et al., 2010).

3.2.2 The second stage, from the 1980s to the beginning of the 21st century

In the mid-1980s, with the advent of praziquantel, an effective, cheap and safe anti-schistosome drug, the World Health Organization (WHO) recommended the implementation of a strategy of "morbidity control". The prevention and control strategy focusses on chemotherapy on human beings and a change in their behaviours. It was believed that the transmission of schistosomiasis does not caused by oncomelanade snails, but by human behaviour of contacting with infected water containing cercariae. Therefore, only focussing on human and human behaviour, can we prevent water pollution and reduce the contact frequency between human and contaminated water, to achieve the purpose of preventing and reducing infection (Zhou et al., 2010). For this reason, WHO suggested that schistosomiasis control should be focussed on health education and the role of human behaviour, in addition to chemotherapy. The prevention and control of schistosomiasis can be divided into three stages and implemented gradually: (a) to reduce the number of severe cases, especially acute and advanced patients with obvious symptoms; (b) to reduce the infection rate and infection intensity; (c) to control and interrupt the transmission. In 1980–1985, the research on control strategy was mainly focus on the chemotherapy supplemented by eliminating *Oncomelania* in susceptible areas, to realize the goal of morbidity control. Thus, researchers from the NIPD-CTDR coordinated and conducted numerous pilot studies, based on four pillars: (i) to eliminate the source of infection, and at the same time to eliminate snails in the susceptible areas; (ii) to eliminate snails in a large area, and eliminate the source of infection simultaneously; (iii) to expand the coverage of chemotherapy to eliminate the source of infection; (iiii) to eradicate infectious sources in mountainous areas and conduct molluscicidal trials in key areas (Guo, 2006; Zhou et al., 2010).

The NIPD-CTDR researchers have studied the epidemic features and patterns of schistosomiasis in the mountainous area, and put forward that health education and environmental snail and cercariae control should be placed in a very important position while carrying out the expanded chemotherapy for human and livestock in this area (Zheng et al., 1996, 2001). Our country also began to adjust the strategy of schistosomiasis control to morbidity control strategy with "taking simultaneous chemotherapy of human and animal as the main intervention, eliminating snails in susceptible areas and conducting health education as the supplementary methods". This strategy was strongly supported by the World Bank Loan Project (WBLP) for schistosomiasis control in China between 1992 and 2001, with a total investment of 890 million RMB yuan (approximately US$152 million), of which 416 million RMB yuan (US$71 million) was a loan provided by the World Bank and 474 million RMB yuan (US$ 81 million) came from counterpart funds provided by the Chinese government at all levels (Xianyi et al., 2005). The overall goal of the WBLP project was to reduce and control the prevalence of schistosomiasis and to block the transmission in some areas. Through the implementation of the project, the infection in human and animals was effectively controlled (Zhou et al., 2010). The number of people infected with schistosomes decreased from 1.638 million in 1989 to 928 thousand in 1995. By the end of 2002, the infection rate of humans and livestock in P.R. China had decreased by 55% and 50%, respectively, the density of infected snails in the national epidemic area declined by 75% compared with that before the implementation of the project in 1992, and Zhejiang province had reached the "standard of schistosomiasis elimination" in 1995 (Zhou et al., 2010). In order to understand the cost-effectiveness of the World Bank Loan Project, a retrospective economic evaluation was done in 2001. Six representative counties, i.e., Huarong in Hunan province, Qianjiang in Hubei province, Yugan in Jiangxi province, Tongling in Anhui province, Xichang in Sichuan province and Dali in Yunnan province, were selected for the study. The total financial input in these counties from 1992 to 2000 was RMB Yuan 90.334 million with the World Bank loan accounting for 40.9%. Control efforts resulted in reduction of human prevalence rates in the six counties from 0.7%–9.0% in 1992 to 0.1%–2.7% in 2000. The net benefit-cost ratio was 6.20, which means that this project gained US$ 6.20 for every dollar spent. The correlation coefficients of the net benefit-cost ratio to the human and bovine infection rates at the beginning of the project were 0.55 and 0.66, respectively. It is conceivable that further progress in schistosomiasis control is an important feature for sustained growth of the local economy, particularly in areas where control of the disease has been most challenging (Zhou et al., 2005c). However, although chemotherapy based disease control measures can quickly control the epidemic situation of schistosomiasis, it is difficult to control reinfection and consolidate the control effect in the lake region (Zhou et al., 2004, 2012). By the end of 2003, schistosomiasis rebounded in 17 of 150 counties (cities, districts) which had reached transmission interruption. Among the 63 counties (cities, districts) in the stage of transmission control, 21 counties (cities, districts) showed a significant increase in snailsor infections in humans or livestock (Guo, 2006).

3.2.3 The third stage, from 2004 to present

At the beginning of the 21st century, the number of people infected with schistosomes was on the rise. For example, in some areas where had reached the standards of transmission control or transmission interruption had the acute infection and infected snails, indicating schistosomiasis transmission reestablished in those areas; new epidemic areas were occurred in some non-endemic areas; and the disease was spreaded from rural areas to cities. There are several reasons for this kind of resurgence as follows: (i) the environment changes or continues to deteriorate, and the floods occurred more frequent. In addition, affected by the national policy of returning farmland to Lake after serious flooding in 1989, the distribution of *Oncomelania* snail was expanding in wider scale (Lin et al., 2002). (ii) the existing molluscicides have a certain degree of pollutive impacts on the environment, so that more and more marshland could not apply the molluscicides any more due to its side effect on toxicity. (iii) with the development of a socialist market economy and animal husbandry in P.R. China, livestock trade had been frequent, and the role of the animal host in the whole process of schistosomiasis transmission had become more prominent, with increased opportunities for exposure of people and animals (Wu et al., 2007). (iv) by the beginning of the 21st century, the national schistosomiasis control programme had lasted for quite a long time, and the compliance and coverage of chemotherapy have decreased year by year (Guo, 2006).

In order to strengthen the national schistosomiasis control programme to protect the health of the people and promote the economic and social development in the endemic areas, the State Council established a leading group for schistosomiasis control again in 2004. A series of important documents, such as "The Notice of the State Council on Further Strengthening the Schistosomiasis Control Programme", "The National Medium- and Long-Term Strategic Plan for Schistosomiasis Control Programme (2004–2015)" and "The Planning Framework of the Key Project on Integrated Control on Schistosomiasis (2004–2008)", were issued by ministries in various sectors, including National Commission of Planning, Ministry of Finance, Ministry of Agriculture, Ministry of Water Conservation, Ministry of Forestry, Ministry of Health, etc. The national conferences on schistosomiasis control were held and facilitated the coordination and collaboration of multi-sectoral ministries on the national schistosomiasis control programme (Wu et al., 2007; Zhang, 2009). The National Med- and Long-Term Strategic Plan for Schistosomiasis Control Programme in P.R. China (2004–2015) put forward the following goals: (1) by the end of 2008, all the epidemic counties (cities and districts) reach the threshold of infection control, and no or only a very few outbreaks occur, (2) by the end of 2015, all endemic counties (cities and districts) reach the transmission control, and all counties (cities and districts) that have reached transmission control more than 10 years interrupted the transmission of schistosomiasis while other historical epidemic areas and areas with potential transmission risks (such as the Three Gorges Reservoir Area) consolidate and expand prevention and control achievements (Guo, 2006; Zhou et al., 2004).

An integrated control strategy focussing on the control of infectious sources had been formulated for the national programme in 2006 after pilots studies, and became the new control strategy in P.R. China. The strategy was based on the transmission chain of schistosomiasis, the eggs of adults are the most critical link, so how to prevent the eggs from polluting the environment is the most important step to reach the transmission blocking standard of schistosomiasis. The key control measures for the new strategy are as follows: (i) to "replace cattle with machines", that is, to vigorously develop modern agriculture, apply modern industrial civilization to agriculture, gradually eliminate the backward agricultural production mode dominated by cattle, and reduce the number of infectious sources of livestock. (ii) the government to strengthen the infrastructure construction of schistosomiasis endemic areas, for example, to build a new harmless sanitary toilet to directly kill the eggs in the faeces and prevent the faeces from polluting the environment. (iii) continue to strengthen advertising and education, popularize the knowledge of disease detection and treatment, snail detection and egg killing among people and livestock in epidemic areas, and raise the awareness of the people in disease prevention and treatment (Wang, 2006; Wang et al., 2009c).

In 2005, the leading group of the national schistosomiasis control programme of the State Council set up five counties to verify the effectiveness of the new strategy and the feasibility of its popularization and application. The five counties were Anxiang county of Hunan province, Hanchuan county of Hubei province, Puge county of Sichuan province, Jinxian county of Jiangxi province and Guichi District of Anhui province. According to the actual situation of each county and local control programme, with the purpose of controlling the source of infection, new strategies had been adopted along with the routine activities. In 3 years, all the pilots have achieved good results, and the effects have been continuously consolidated, which shows that the new strategies have significant effects in epidemic areas and can be popularized and applied nationwide (Chen et al., 2011). It was the fundamental way out for the prevention and control of schistosomiasis in the lakes and marsh areas in P.R. China, and improve living standards. Scientific and technological progress and long-term compliance are two important cornerstones for significant achievements of schistosomiasis control. Next, the work of schistosomiasis control in P.R. China should focus on scientific research, rely on the comprehensive control strategy of infectious source control, continue to strengthen the publicity and education of the masses in the epidemic area, so that all the epidemic areas of schistosomiasis can reach the standard of transmission interruption as soon as possible, and truly deliver Take the "God of plague" to realize the sustainable development of the economy in the epidemic area (Guo, 2006; Li et al., 2012; Utzinger et al., 2005; Wang et al., 2009b,c).

In summary, the NIPD-CTDR has actively carried out research on the epidemic environment and characteristics of schistosomiasis in P.R. China in order to determine scientific control strategies in the different type of endemic regions. Through 70 years of continuous exploration, the NIPD-CTDR has established control strategies corresponding at different control stages, and gradually defined control strategies and measures for different environmental settings. In addition, it was verified and popularized through the piloting studies, which provided the epidemiological theoretical reference for decision making in the national schistosomiasis control programme in P.R. China. Particularly in 2005–2015, several pilot studies involved young staff from the NIPD-CTDR which fostered the capacity building for young generation of the institute, and set milestones for the scaling up of the new strategy and promotion of the national schistosomiasis control programme to reach the transmission control criteria with prevalence less than 1% in the whole country.

4. Diagnostics and vaccine development

4.1 Development and application of diagnostics

Diagnosis provides the necessary information and scientific evidence for schistosomiasis control programme in P.R. China. Over the past 70 years, Chinese scholars, especially NIPD-CTDR researchers, have been working on the development, evaluation and quality control of the diagnostic techniques for schistosomiasis. During this period, the parasitological methods have been improved, and a variety of immunoassay techniques have been developed and are commercially available. Meanwhile, many innovative ideas have been derived. These tools have played an important role in the prevention and control of schistosomiasis, which has effectively promoted the process of schistosomiasis elimination in P.R. China (Zhang et al., 2016a). The main contribution of NIPD-CTDR in the field of schistosomiasis diagnosis is elaborated below.

4.1.1 Parasitological diagnosis

Detection of *Schistosoma* eggs in faeces or tissue biopsies is the direct way to confirm the infection of schistosome and some parasitological tests have been most widely used for the diagnosis of schistosomiasis in the endemic areas. In spite of reduced sensitivity especially for low infection intensity, Kato-Katz thick smear (Katz et al., 1972) and the miracidium-hatching test (Qiu and Xue, 1990) remain the gold standard for the diagnosis of schistosomiasis. Over the years, the NIPD-CTDR has made a number of improvements in the method of faecal examination, including nylon silk egg hatching test and improved Kato-katz thick smear combined with faecal hatching, which has contributed to the improvement of the detection rate of schistosomiasis.

4.1.1.1 Hatching test

The main method for the diagnosis of schistosomiasis by faecal examination was the combination of faecal sedimentation and miracidium-hatching test (sedimentation-hatching) in P.R. China before 1970. However, the infection intensity of patients was greatly reduced due to the strengthened control, and hence it was still difficult to detect infections with 3 times sedimentation-hatching. Therefore, NIPD-CTDR developed the nylon-bag-collection method. Compared to the conventional method of sedimentation-hatching, the method is able to reduce the loss of eggs, simplify the operation steps, improve the work efficiency, reduce the labour intensity, save the water consumption, and increase the detection rate. This diagnostic method is very popular among field workers and has been widely used in the endemic areas for investigation of schistosomiasis infection and assessment of the treatment efficacy. Currently, this method has been further improved and used as the way for evaluating the infection rate of schistosomiasis in human beings in areas where aiming for transmission interruption and elimination.

4.1.1.2 Improved Kato-Katz method combined with faecal hatching evaluation

In 2005, the NIPD-CTDR researchers evaluated the effect of modified Kato-Katz method on large-scale on-site screening of Schistosomiasis (Zhu et al., 2005). They found that modified Kato-Katz method alone has a high rate of misdiagnosis. The detection rate further significantly increased when in combination with hatching techniques. This research provided a useful reference for setting diagnosis criteria and application strategy of schistosomiasis. The diagnostic strategy of combing Kato-Katz thick smear method and miracidia hatching technique applied to sero-positive human was firstly integrated into the national surveillance project in 2011, and then widely implemented in the national schistosomiasis elimination programme since 2018.

4.1.2 Immunological diagnosis

Since the 1950s, NIPD-CTDR has contributed to the development and implementation of new immunological diagnostics for schistosomiasis. The antigen used is involved in various stages of the life cycle of schistosomes. The methods used include intradermal test (ID), Cercarien-Hullen-Reaktion (CHR), circumoval precipitin test (COPT), latex agglutination test (LAT), indirect hemagglutination assay (IHA), enzyme-linked immuno-electron transfer blot assay (EITB), enzyme-linked immunosorbent assay (ELISA), rapid diagnostic tests (RDT) and so on (Wu, 2002, 2005; Yan et al., 2006; Zhu, 2005).

4.1.2.1 The intradermal test

Gan (Kan, 1936) first introduced the intradermal test in the diagnosis of schistosomiasis in P.R. China. In the early 1950s, we used cold soak liquid of adult worm frozen dry powder as the antigen for ID, however, the yield of adult antigens obtained from infected rabbits is limited. In order to meet the needs of control work, NIPD-CTDR replaced adult antigens with egg antigens collected from infected rabbit liver, which achieved similar effects, and made mass production available (Li and Yan, 1956). In addition, NIPD-CTDR also develops antigens derived from adult metabolites of *S. japonicum*, that is, metabolic antigens, and presented high sensitivity and specificity (Mao et al., 1956).

From 1956 to 1957, according to the national investigation requirements, NIPD-CTDR produced 58 million copies of antigens for intradermal testing, which played an active role in determining the epidemic area, infection status and prevalence of schistosomiasis in P.R. China. In the 1980s, due to the simplicity and sensitivity of this method, but the antibody persisting a long time in the infected people after treatment, NIPD-CTDR only used this method for comprehensive investigation, and screening those population without history of treatment (Mao and Shao, 1982). Later, it was soon replaced by other tests due to its lower specificity.

4.1.2.2 Cercarien-Hullen-Reaktion

The cercarien-hullen-reaktion established in 1951 has high sensitivity to schistosomiasis patients and early diagnostic value. However, the need for fresh cercariae as an antigenic material limited its practical applications. NIPD-CTDR researchers explored the use of freeze-dried cercariae to replace fresh cercariae. Through animal experiments and field observations, freeze-dried cercariae achieved the same sensitivity as fresh cercariae but lower false positive rate. The freeze-dried cercariae can be stored at room temperature or in the refrigerator (4 °C ∼ 6 °C) for 4 to 5 weeks; it was more suitable for field application.

4.1.2.3 The circumoval precipitin test

Because of the high sensitivity and appropriate specificity of the circumoval precipitation test, it was regarded as one of the effective serological diagnostic methods for schistosomiasis. However, it is not easy to preserve fresh eggs. To this end, the scientific researchers have studied the method of replacing the fresh eggs with freeze-dried eggs, but because the eggshell is easily broken and affecting the reaction effect. Then, the NIPD-CTDR researchers treated the fresh eggs with formaldehyde and then frozen them with decompression. This method overcomes the phenomenon that the eggshell is easily broken. After field tests, the efficiency of dry egg antigen detection on schistosome infection was no less than that of fresh egg antigen (Qiu, 1975; Yan, 1976). The use of freeze-dried technology has made the COPT the most widely used immunodiagnostic method for schistosomiasis in the 1960s and 1980s (Chen, 1980; Li, 1976), but because of the high labour intensity of COPT operation, which limits its large-scale applications on site, this method currently was no longer used in endemic areas.

There was no report on immunological diagnosis of *S. mansoni* infection using *S. japonicum* antigens before the 1970s. From 1974 to 1975, NIPD-CTDR first used three immunological methods including ID, CHR and COPT for the diagnosis of foreigners returning from Guinea where is the endemic area of schistosomiasis mansoni. A faecal test and immunological examination were performed, and the positive rate of faecal test was 54.0%. The positive rate of CHR and COPT in the faecal test positive group was 97.0% and 94.0%, respectively, while the positive rate in the ID was only 42.9%, suggesting that the COPT and CHR using antigens from *S. japonicum* can also diagnose the infection of *S. masoni* (Yu et al., 1979).

4.1.2.4 Latex agglutination test

In the 1960s and 1970s, experimental studies on the detection of antibodies against *Schistosoma* by latex agglutination test was carried out at home and abroad, but the physical method of sensitizing latex has poor stability and low sensitivity. Since the 1980s, research teams from the NIPD-CTDR have collaborated with the Shanghai Medical Laboratory and the US CDC to apply the chemical cross-linking method to prepare schistosome antigen according to the principle and method of latex pregnancy test. This method characterized with good stability, simple and fast operation without special instruments, and ease-of-use for field application. Animal experiments, clinical case tests and field applications have shown that the method achieves similar efficacy with indirect hemagglutination assay, enzyme-linked immunosorbent assay and circumoval precipitation test, in terms of stability, specificity and sensitivity. This method has been applied within a certain range of field (Yan et al., 1985, 1986, 1987).

4.1.2.5 Enzyme-linked immunoelectron transfer blot assay

Enzyme-linked immunoelectrotransfer blot assay (EITB), also known as Western blot, is a new diagnostic technique developed in the late 1970s and early 1980s. Qiu et al. from NIPD-CTDR first established an enzyme-linked immunoelectron transfer blotting technique for serum antibody analysis in patients with schistosomiasis in 1988 (Qiu et al., 2000). Results showed that the sera from acute, chronic and advanced schistosomiasis patients recognize different adult antigen strips. The specific antibodies decreased or disappeared rapidly after treatment with praziquantel in acute infection. This method is helpful to distinguish between chronic and acute schistosomiasis, and has certain reference value for the evaluation of the efficacy of schistosomiasis after treatment (Yan et al., 1988).

4.1.2.6 Enzyme-linked immunosorbent assay

In the late 1970s, Enzyme linked immunosorbent assay (ELISA) was used for the detection of schistosomiasis in Japan. After that, the NIPD-CTDR scholars developed various improved techniques based on classical ELISA. Yan et al. from the NIPD-CTDR first applied the enzyme-linked immunosorbent assay for the detection of schistosomiasis mansoni with good sensitivity and specificity (Yan, 1979; Yu et al., 1979). Because the ELISA method recommended by the World Health Organization Tropical Diseases Programme requires 6h of reaction time, the reagents need to be refrigerated and stored, and the results need to be red by the instrument, so there are certain restrictions on the field application.

For this reason, Yan et al. from the NIPD-CTDR developed a rapid-ELISA method in the 1990s for the detection of antibodies against *Schistosoma* with a sensitivity of 94% and a specificity of 96%, which is identical to the performance recommended by the World Health Organization. Fast-ELISA only takes 30–40 min. The antigen plate and reagent can be stored at room temperature for more than 4 weeks, and can be used separately for the detection of sporadic samples. The results can be visually observed, which is very suitable for field application (Yan et al., 1996). However, because antibody-based detection is difficult to distinguish between active and past infections, this method cannot be used for efficacy evaluation.

The detection of schistosomes' circulating antigen can reflect the host active infection, the load of the worm and evaluate the effect of drug treatment. With the advent of hybridoma technology, NIPD-CTDR has conducted in-depth research on the detection of schistosoma antigens. In the late 1980s and early 1990s, Yan et al. from NIPD-CTDR collaborated with the researchers from the University of Leiden in the Netherlands. They discovered high-efficiency monoclonal antibody 3D8A against *Schistosoma* intestinal-associated cathodic antigens and monoclonal antibody SM21-3 against *Schistosoma* egg antigen through preparation and screening of more than 30 monoclonal antibodies (Yan and Deelder, 1988; Yan et al., 1990b). In this way, monoclonal antibody-based sandwich ELISA and Dot-ELISA were developed to detect the corresponding circulating antigen in the serum of patients with Schistosomiasis (Yan et al., 1990a,c, 1992a,b), and the sensitivity was 90.6% and 83.2% for acute and chronic schistosomiasis patients, respectively. The cross-reaction with other worm is very low. In 1990, Xue et al. improved the Dot-ELISA method, and the NC film with the antigen was fixed in a PVC plate for testing. The improved operation was simple and convenient, and it was more suitable for field application (Xue et al., 1990). In 2000, Qiu et al. (2000) reported the detection of circulating antigen in the schistosomiasis patients' sera based on polyclonal and monoclonal antibody sandwich ELISA, compared the detection effect of dot ELISA with polyclonal-monoclonal antibody sandwich ELISA. The results of field test showed that both methods performed high sensitivity and specificity, and had a certain cure efficacy evaluation value. Dot-ELISA is more sensitive, simple, low-cost, with less blood and do not need special equipment. Itcan be visually observed and is suitable for field applications.

4.1.2.7 Rapid diagnostic kits

Rapid diagnostic kits based on immunofiltration or immunochromatography principles with dyes or colloid gold as markers have been developed and applied in field because of its rapid, sensitive and simple advantages. The rapid diagnostic tests commonly used in P.R. China including the dot immunogold filtration assay (DIGFA) where the probe is anti-human IgG labelled with colloidal gold, the dipstick dye immunoassay (DDIA) where it consists of the antigen labelled with a blue colloidal dye and the latex immuno-chromatographic assay (DLIA). The results of the test are revealed by anti-human IgG tagged with red latex particles (Zhu, 2005).

Hu et al. from the NIPD-CTDR has constructed a cDNA library of different stages of schistosomes since 2000 through the research of *S. japonicum* functional genomics (Hu et al., 2003, 2004), and obtained effective recombinant antigen by cDNA library immunoscreening (Lu et al., 2012). Then, they developed a dynamic flow method based on gold standard immunochromatography strip for anti-*S. japonicum* IgG based on the recombinant antigen. The laboratory evaluation showed the assay performed high sensitivity and specificity, and the operation was simple and stable. Currently, this assay has got patent and now transferred to a company to apply certificates from NFDA, considering its potential value i for large-scale screening of schistosomiasis in field settings in future.

4.1.3 Assessment of diagnostic assays and quality control

4.1.3.1 Assessment of diagnostic assays

Assessment of the efficacy, reliability and operational characteristics of diagnostic assays in use are important to ensure the diagnostic accuracy. Being a national institute, NIPD –CTDR has conducted many studies to evaluate the immunodiagnosis assays to provide scientific information for national control programme.

In 1995, the National Center for Schistosomiasis Immunodiagnosis and Testing, funded by the World Bank loan-financed Schistosomiasis Control Project Joint Scientific Research Management Committee, was established. The centre is based on the National Institute of Parasitic Diseases of the Chinese Academy of Preventive Medicine (the former name of the NIPD-CTDR). One of the tasks is to monitor and evaluate the existing diagnostic reagents for schistosomiasis in P.R. China, in order to maintain the quality supervision of existing diagnostic reagents, and select several test systems recommended for on-site use. In 1995 and 1996, the Centre evaluated 12 circulating antigen kits and other nine antibody detection reagents collected by the various units (Xu et al., 2005). These evaluations demonstrated that the CAg kits evaluated were inappropriate for use by the schistosomiasis control programme at that time, especially in endemic areas characterized by low-level infections. On the contrary, most antibody-based detection kits had specificities over 90% and sensitivities higher than 80% in chronic infection. The results of these evaluations suggest that the detection of antibodies should consider the value of their efficacy assessment, and further researches are needed on component antigens, short-range antibodies and their quantitative detection (Feng et al., 1998; Guan and Shi, 1996).

At the beginning of the 21st century, Xu et al. from the NIPD-CTDR conducted a study on the detection of *S. japonicum* antibodies by colloidal gold immuno-infiltration (DIGFA) and on-site evaluation. The results of laboratory and on-site evaluation showed that the colloidal gold diafiltration method and F-ELISA method were sensitive. There was no significant difference between the sensitivity and the specificity ($P > 0.05$), and the two methods had high consistency (Xu et al., 2006). Subsequently, the NIPD-CTDR combined with the provincial schistosomiasis control system to evaluate the performance of the colloidal dye test strip method (DDIA) in the low-endemic area (Xu et al., 2011a). The results showed high sensitivity, but the specificity is only 53.08%, because these two methods are simple, fast and sensitive, they are suitable for the screening of schistosomiasis patients.

In 2005, the NIPD-CTDR organized a comprehensive evaluation of nine reagents from eight institutions used for national schistosomiasis surveillance. The sensitivities and specificities of some kits have reached the field application standards, and a kit with better comprehensive evaluation indicators was recommended as a screening tool in the national surveillance sites. (Xu et al., 2007).

In 2008, funded by TDR/WHO, the NIPD-CTDR carried out a two-step evaluation for existing immunodiagnostic reagents for schistosomiasis diagnostic. In laboratory evaluation stage, 10 kinds of immunodiagnostic reagent widely used in the field, such as ELISA, IHA, DDIA, DIGFA, etc., were assessed using archived serum samples. The results show that the sensitivity and specificity of most reagents reached more than 90%. The field evaluation for five excellent diagnostic reagents selected through laboratory evaluation proved that these antibody- based assays were acceptable as tools for community survey, targeting chemotherapy, monitoring and elimination verification (Xu et al., 2011b; Zhou et al., 2011).

Before 2008, the State Food and Drug Administration (SFDA) in P.R. China used many assays for schistosomiasis diagnosis without having passed certification. However, the evaluations carried out previously provided data for the manufacturer to apply for licence from NFDA. Until now, at least 9 kits have been commercial available in P.R. China market, which indicates that the quality of the assays is improving under the NIPD-CTDR's monitoring and management.

It should finally mentioned that in 2015, the NIPD-CTDR and Leiden University jointly conducted a small-scale field trial for the evaluation of UCP-LF to detect CCA of *S. japonicum* in urine samples in low endemic areas in P.R. China (van Dam et al., 2015). The assay exhibited a much higher sensitivity than that of the Kato-Katz technique and detected a significant number of egg-negative cases.

4.1.3.2 Quality control

In order to strengthening the capacity of control stations/agencies for schistosomiasis diagnostics, since 2009, the NIPD-CTDR organized a diagnostic network platform aiming at establishing a systematically perfect system with disease control institutions at all levels. This platform run through a set of reference or sentinel laboratories for external quality control activities, and also holds training courses to ensure the quality of schistosomiasis prevention and control work in the country (Qin et al., 2013). So far, a national diagnostic centre established in the NIPD-CTDR, it is responsible for assay assessments; external control and improvement of diagnostic quality control scheme; planning and organizing training activities, providing technical support for subordinate laboratories, exploration of the feasibility of molecular biology technology for schistosomiasis monitoring. The isothermal amplification technique developed by the NIPD-CTDR based on this platform has been used in the rapid assessment of the transmission risk of schistosomiasis in the country (Qin et al., 2018). The establishment and improvement of the schistosomiasis diagnosis network system is of great significance to improve the ability of monitoring and early warning of schistosomiasis in P.R. China and to promote the elimination process of schistosomiasis.

4.2 Vaccines development

Vaccines are an important means of controlling infectious diseases. In the 20th century, schistosomiasis vaccines were envisaged to prevent from infection with *S. japonicum*. In view of the repeated infections, the possibility of drug-resistant strains of long-term use of praziquantel, and the existence of a large number of mammalian host, schistosomiasis vaccine development could reduce infection rate, worm and eggs burden by inhibiting the invasion, development and reproduction of schistosoma is considered important approach in the control of schistosomiasis. It is expected that vaccine prevention as a long-term control measure to solve the problem of schistosomiasis. It is of great significance for controlling and blocking the transmission of schistosomiasis and finally eliminating schistosomiasis. The development of schistosomiasis vaccine has gone through the process of exploration from dead vaccine, weak live vaccine, subunit vaccine to genetic engineering vaccine. However, due to the complexity of schistosomes, there is still no effective schistosomiasis vaccine entering the clinic or apply for on-site prevention from infection. Since the 1990s, NIPD-CTDR has carried out a large number of vaccine candidate antigen screening, development, laboratory evaluation and field trials. The main contributions are as follows:

4.2.1 Vaccine candidate antigens

From the biological point of view and the actual situation of vaccine research, the development of *S. japonicum* vaccine is more difficult than *S. mansoni*. Since the 1990s, with the support of the National "863" Programme, the Prime Minister's Fund and international cooperation projects, NIPD-CTDR has made great progress in screening, gene cloning, expression and immune protection of schistosomiasis vaccine candidate antigens. NIPD-CTDR researchers firstly carried out the research in collaboration with domestic and abroad experts, on glutathione-S-transferase of the *S. japonicum,* detoxification enzyme-related antigen (Liu et al., 1993; Liu et al., 1997), the triose phosphate isomerase involved in the glycolysis of *S. japonicum*, the hypoxanthine, the natural antigens such as ribose transferase, thioredoxin, more than 90% muscle-related muscle structural proteins, paramyosin and tropomyosin, and the full sequence of the coding genes and recombinant antigens. In addition, other candidate vaccine recombinant antigens, such as myosin, fatty acid binding protein, Mr. 10,000 dynein light chain antigen, Mr. 38,000 muscle-associated protein antigen, glyceraldehyde triphosphate dehydrogenase, Serine protein inhibitors, heat shock proteins, cathepsin B, myosin-like protein, and T, B cell epitope protein antigens of *S. mansoni* Mr. 20,000 were studied (Cao et al., 2005; Chen et al., 2006; Hu et al., 2010; Li et al., 2005; Qi and Liu, 1996; Tong et al., 2007; Yan et al., 2005).

4.2.2 Multivalent vaccines

S. japonicum worm is a compound cell parasite with complex life history and multiple developmental stages. Antigens can be expressed as species-specific or phase-specific. Different antigen molecules have different antigenic determinants, and the types and mechanisms of host immune responses are incomplete same. The use of multivalent vaccines prepared with different antigens or different epitopes to synergistically induce different immune insecticidal mechanisms may kill schistosomes at multiple developmental stages, possibly overcoming the problem of low levels of immune protection induced by single antigen molecules. In particular, schistosomiasis vaccines can be designed for multiple stages of schistosomes life cycle. The NIPD-CTDR researchers obtained a batch of multivalent vaccine recombinants, and the protective level of multivalent vaccines was significantly improved. In addition, the Sj26GST gene was successfully expressed in combination with the *Escherichia coli* (ETEC) cilia antigen gene (encoding the pili-antigen K99 gene that causes acute diarrhoea in livestock), which can prevent not only diarrhoea in young animals caused by ETEC but also schistosomiasis and has a broad application prospect.

4.2.3 Immunization test on cattle

Infected livestock is an important source of infection, and it is importance of development of vaccine against schistosomiasis japonica in animals. During the "Tenth Five-Year Plan" period (2001–2005), the multivalent genetic engineering vaccine candidates (multivalent protein vaccine and multivalent DNA vaccine) against *S. japonicum* were developed in the laboratory of the NIPD-CTDR. Based on the preclinical experimental research on the safety, effectiveness and production process of the vaccine, the effectiveness of multivalent vaccine on the on-site immunization of buffalo to interrupt the transmission was observed (He et al., 2002a,b, 2003). Liu et al. vaccinated the recombinant 26 kDa GST (reSjc26GST) in mice and pigs, with a 22.3% reduction in worm numbers, and 50% reduction in faecal egg output and eggs deposited in host tissues. In addition to the anti-fecundity effect, reSjc26GST reduced by nearly 40% the egg-hatching capacity of *S. japonicum* eggs into viable miracidia (Liu et al., 1995). Therefore, the NIPD-CTDR's researchers applied the reSj26GST to 96 Bos buffalo as experimental group vs 90 buffalo without vaccine as control group in the field to observe its impacts on the endemicity of schistosomiasis. As a result, 20 months after immunization, the infection rate of *S. japonicum* in vaccinated buffalo was 60.4% lower than that before immunization (infection rate reduced from 13.54% to 5.36%), and 67.9% lower than that of the unimmunized bovine control group. After the buffalo was immunized with the reSj26GST antigen, it had a significant egg-reducing effect within the first 3 months of infection, and then the ovulation volume gradually decreased to negative, which was helpful to reduce the spread of schistosomiasis (He et al., 2002a,b).

In summary, during last 7 decades, experts on immunodiagnosis from NIPD-CTDR have carried out long-term research and development of schistosomiasis parasitological and immunological detection technologies suitable for on-site application in P.R. China, continuously improving detection specificity and sensitivity, and optimizing various detection technologies for pathogens and immunity. They have provided accurate diagnostic techniques for on-site diagnosis of schistosomiasis in P.R. China.

5. Research and development of drugs and molluscicides

5.1 Drug development for human treatment

Drug treatment has always been one of the main means of schistosomiasis control measures. Since 1980, the NIPD-CTDR has carried out extensive and in-depth research on clinical pharmacology studies on praziquantel, artemether and artesunate and screening of other anti-schistosomiasis drugs. A fruitful progress obtained initially contributed greatly to the successful implementation of schistosomiasis control in P.R. China.

5.1.1 Praziquantel

The advent of praziquantel in the 1970s has had a profound impact on the global prevention and control of schistosomiasis (Balen et al., 2007; Davis and Wegner, 1979; Sleigh et al., 1998; Tallo et al., 2008; Taylor, 2008). Praziquantel is currently the only effective drug for five *Schistosoma* spp. infections in humans. However, for more than 40 years, the global schistosomiasis prevention and treatment has long relied on the application of praziquantel. People have worried about whether repeated use of praziquantel would induce resistance against schistosomes (Viana et al., 2017; William et al., 2001). Due to the lack of new effective drug replacement drugs, it is necessary to understand the mechanism of action of praziquantel against schistosomiasis. In the past 40 years, scientists and technicians represented by Prof. Xiao Shuhua from the NIPD-CTDR have conducted systematic and in-depth research on praziquantel against Schistosomiasis.

5.1.1.1 Effect of praziquantel on the nervous system activity of schistosomes

Xiao Shuhua et al. explored the mechanism of action of praziquantel-exciting schistosomiasis by using neural media related to worm activity and its blockers. It found that praziquantel neither increased the endogenous 5-HT of the worm nor enhanced its exogenous 5-HT intake, either in vivo or in vitro. Thus, praziquantel-induced anti-schistosomiasis does not act through 5-HT, but possibly a 5-HT-like or a 5-HT receptor agonist (Xiao et al., 1984). Xiao et al. found in the initial experiment that the action model of praziquantel against schistosomiasis depends on the presence of Ca^{2+} that was restricted by Mg^{2+}(Xiao et al., 1980). Therefore, under the action of a high concentration of praziquantel, it is difficult to explain the accumulation of Ca^{2+} in the worm for a short time (Xiao et al., 1987). Praziquantel causes muscle contracture and cortical damage of schistosomiasis depending on the presence of Ca^{2+}, but there is no consensus on how praziquantel acts through Ca^{2+}.

5.1.1.2 Effect of praziquantel on the metabolism of schistosomes

For the effect of praziquantel on the metabolism of schistosomes after treated by praziquantel, the study showed that praziquantel not only inhibits the glucose uptake and glycogen synthesis of the worm, but also promotes the glycogenolysis of the worm (Xiao et al., 1985a). Further observation of histochemical methods found that glycogen, alkaline phosphatase and ribonucleic acid was significantly reduced or disappeared, acid phosphatase and basic protein were significantly increased, while deoxyribonucleic acid, tryptophan, histidine, tyrosine Acid, etc. didn't change significantly (Xiao et al., 1985b).

5.1.1.3 Effect of praziquantel on the ultrastructure of different stages of schistosomes

Praziquantel has a strong killing effect on adult worms and cercariae of *S. japonicum*. Two hours after treated by praziquantel at a concentration of 1 μg/mL, the glycocalyx in the outer surface of tegument decreased markedly or even disappeared and the matrix of tegument became sparse and indistinct with a decrease in membranous vesicles. Meanwhile, the muscle layer beneath the tegument exhibited swelling or lysis, and the mitochondria distributed in the muscle and parenchymal cells showed swelling and degeneration. After 4 h, an extensive lysis of cercariae tissues which resulted in the formation of numerous vacuoles with varying sizes could be seen. No apparent alteration in the ultrastructure of cercariae exposed to praziquantel in normal saline was observed. The ultrastructural changes of schistosomula within mouse skin in worm tegument, muscle and parenchymal tissues were similar to those mentioned above 0.5–24 h after intragastric praziquantel 600 mg/kg to mice infected with cercariae for 3 h (Xiao et al., 1988).

In another experiment, the results obtained in the mice infected with *S. japonicum* at different intensities and treated intragastric with single doses of praziquantel at 300 mg/kg at different durations were similar to the results in the rabbits. The correlation between efficacy and specific antibody level was confirmed by calculating the coefficient of correlation which was $R=0.45$–0.98 ($P<0.01$). The results also indicated that heavy infectious with *S. japonicum* were still more responsive to praziquantel treatment. In addition, the efficacy of praziquantel was closely related to the specific antibody response in the host (Yue et al., 1989). After praziquantel treatment, the sensitivity of the cercariae to the metabolic inhibitor NaF increased, making NaF easy to infiltrate into the worm body, if it will be treated with praziquantel in water. The cercariae were transferred to the drug-free HBSS for 2h, and the tegument of the most cercariae returned to normal (Xiao et al., 1991b). When the cercariae is cultured in the serum of the isotonic salt balance solution after praziquantel is removed, the worm body not only can survive but also could develop into schistosomula. Praziquantel has no obvious effect on the worms that developed into schistosomula for 3, 7, 14, and 21 days (Xiao and Catto, 1989). However, a larger dose of praziquantel has a direct killing effect on 21-day larvae (Xiao and Catto, 1989). Praziquantel has a strong killing effect on adult schistosomes (Xiao et al., 1991a). The changes in the surface of schistosomes after praziquantel treatment are not completely consistent, even if the same worm can be mixed with normal tegument and damaged tegument. Body damage caused by damaging the sucker between the mouth is happened later and lighter (Xiao et al., 1992; Xiao et al., 1995). The tegument of schistosomes is rapidly damaged by the action of praziquantel, and the surface antigens of the worm are exposed to the microenvironment of the host which making the worm susceptible attacked by the host immune system. The white blood cells of the host can quickly attach to the damaged surface of the worm, indicating that the specific antibody is involved in action together with praziquantel (Xiao et al., 1997; Zhai et al., 2002). Based on the all research results, the action process of praziquantel includes two aspects, including the direct action of the drug on the worm as well as the immune effect of the host on the worm. The above two aspects are indispensable, but only taking effect when the worm body has been damaged through the action of praziquantel, then the immune effects of the host can play a role in action, thus showing that the direct action of praziquantel on the worms is particularly important in its action process (Xiao et al., 2002b).

5.1.1.4 Comparison of the effects of different structures of praziquantel on schistosomes

Praziquantel is a racemic mixture of levo-praziquantel (L-praziquantel) and dextro- praziquantel (D-praziquantel). Xiao et al. determined the effects of racemic praziquantel, including L-praziquantel and D-praziquantel, on *S. japonicum* in different developmental stages, and found that in L-praziquantel is an active ingredient against *S. japonicum*. Scanning electron microscopic observation on the damage of *S. japonicum* tegument by L-praziquantel and D-praziquantel showed that the actions of the two optical enantiomers of praziquantel were different in quantity but not in quality. The effect of L-praziquantel on worm was better than that in D-praziquantel (Xiao and Shen, 1994).

5.1.2 Artemether

Artemether is a new type of anti-malarial sialoin derivative. Since its discovery in the last century, it has demonstrated artemether is effective against *S. japonicum* (Huang et al., 1999; Xiao et al., 2002a). Scientists from NIPD-CTDR have carried out systematic research on the mechanism and treatment trials in the endemic area.

5.1.2.1 Mechanism of action

Artemether has a wide-ranging effect on the glucose metabolism of *S. japonicum* (Huang et al., 1999; Xiao et al., 1987). Hexokinase, phosphoglucose isomerase, phosphofructokinase, glyceraldehyde phosphate dehydrogenase, phosphoglycerate light kinase, acetone involved in glycolysis various enzyme activities such as acid kinase and lactate dehydrogenase were inhibited in various degrees, resulting in a significant decrease in the amount of lactic acid produced. Researches also indicated that artemether has a glycogenolytic action that inhibits schistosome and inhibits its glycolysis (Xiao et al., 1987, 2002a). Subsequent mouse experiments suggest that phosphofructokinase (PFK) may be one of the target enzymes of artemether on the enzymatic pathway of schistosomes (Xiao et al., 2002a). In vitro, the schistosomes were cultured in medium containing artemether and hemin for 24h, or infected mice were treated with artemether 300mg/kg for 6–24h in vivo. Results showed that in vitro, 50μmol/L of artemether and hemin caused a significant decrease in the total antioxidant capacity of female worms. In vivo, artemether acts on *S. japonicum* for 6h, and the total antioxidant capacity of female worms decreased significant.

Artemether has no effect on the total antioxidant capacity of males in vivo and in vitro (Xiao et al., 2003).

Xiao et al. also found that artemether had significant effects on the ultrastructure of 7 d's and 35 d's adults of *S. japonicum*, which was characterized by fuzzy, loose, dissolved and vacuolar formation in the cortical matrix (Xiao et al., 1988). The outer membrane of cortical cytoplasmic processes was blurred, fused, increased electron density and rupture, as well as degeneration of sensory structures. Muscle bundles, syncytial cells, and intestinal epithelial cells under the cortex can be found to have extensive swelling, dissolution, and vacuolation. Female yolk cells also showed changes in nuclear vacuoles, reduced rough endoplasmic reticulum, and yolk globule lysis. At 14–28 days after treatment, the ultrastructural changes of some surviving worms have been restored, but some worms still show damage. Artemether treatment of schistosomes-infected mice, infected rabbits, dogs, and cattle showed that it has the effect of treating early infection of schistosomes (Xiao and Catto, 1989; Xiao et al., 1987, 1995, 1997, 2002a,b, 2003, 2011).

5.1.2.2 Field trials

In order to evaluate the effect of artemether-preventing schistosomes infection, Xiao et al. first conduct a field study in August–October 1994 in Minglang Village, a schistosomiasis endemic area in the south bank of Dongting Lake located in Hunan P rovince (Xianyi et al., 2005). Preliminary results showed that oral artemether preventive treatment could significantly reduce the infection rate and intensity, and might prevent the occurrence of acute schistosomiasis. Subsequently, the research team further conducted therapeutic trials in Guichi City, Anhui province (Liu et al., 1997), and in Hanshou county, Hunan Province in 1996 (Tian et al., 1997), and in the severely endemic areas of the Poyang Lake, Jiangxi province (Chen et al., 2006). It indicated that artemether had a good preventive effect on the population of severe schistosomiasis in lake areas, and thus provided a new measure for prevention and control of schistosomiasis in lake region. Furthermore, Xiao et al. further selected Dali County and other mountainous counties to observe the effect of oral artemether to prevent the infection of schistosomes (Xiao et al., 1996). Studies proved that oral artemether had a good preventive effect on schistosomiasis in mountainous endemic areas and could prevent acute schistosomiasis (Wang et al., 1997). Huang et al. (1999) observed that artemether also has a good effect on preventing schistosomiasis in people who were working on boats.

5.1.3 Other new anti-schistosomiasis drugs

Artemisinin-based artemether and artesunate are anti-malarial drugs but show effect against *S. mansoni* and *S. japonicum*. Xue et al. compared the effects of seven other antimalarial drugs against *S. japonicum* in vivo and in vitro, and found that quinine and hemin can significantly enhance the anti-schistosome effect in vitro, while pyronaridine in combination with hemin shows antagonism (Xue et al., 2013). Mefloquine has a strong direct insecticidal effect in vitro, and animal experiments have further observed that mefloquine has an inhibitory effect on the formation of granuloma caused by *S. japonicum* eggs (Tao et al., 2015; Xue et al., 2013, 2014). Xiao et al. observed that the synthetic snail adamantane odour oxide (OZ78) has extensive damage to the cortical and subcortical tissues of *S. japonicum* adult worms, including cortical cells, parenchymal tissues, intestinal epithelial cells and yolk cells (Xiao et al., 2018).

5.2 Molluscicides development

Oncomelania snail is the only intermediate host of *S. japonicum*. Eliminating snails is one of the important measures to control and block the transmission of schistosomiasis (Zhou, 2005). Therefore, snail control is an important component in the national schistosomiasis control programme in P.R. China.

5.2.1 Bromoacetamide

In 1980, the NIPD-CTDR first screened and found a new molluscicide, namely bromoacetamide (Zhu et al., 1984). Zhu et al. did experimental studies showed that bromoacetamide has an effective role in killing snails with low toxicity, easily dissolved in water and ease for application in the field (Yin et al., 1986; Zhu and Yin, 1986). The field trials have shown that at the concentration 0.5 ppm of bromoacetamide, the compound had good molluscicidal effect with the means of immersion at 1 ppm, spraying at 1 g/m^2, and immersion with shoving the canal banks at 1 g/m (Zhu and Yin, 1986). No toxicity effects to fish were observed when the concentration of bromoacetamide used in the field at 6 ppm that was 6 times higher than that used for molluscicidng. Spraying the chemical with molluscicidal dosage could damage to plants, but no toxicity to rice at 10 ppm (Zhu and Yin, 1986). In order to clarify the action mechanism of bromoacetamide, observation was conducted to explore the effects at different concentrations of bromoacetamide on different stages of *Oncomelania* snail. After bromoacetamide treatment, the fecundity of female snails was significantly reduced, and the mature snail ovary was obviously atrophied. When snail was immersioned with 3H-labelled bromoacetamide, it is observed that bromoacetamide could enter into the snail eggs, with its radioactive content became higher when increased drug concentration (Zhang and Gao, 1992). Experimental results provided an evidence to select appropriate period of season and appropriate concentration of the molluscicide to control snails on site (Zhu et al., 1998). Application of bromoacetamide at the double concentrations of the conventional mollusciciding dose of bromoacetamide sprayed to the fishponds, the clams that were suspended in the ponds were well survived (Yin et al., 1994). Therefore, the drug can be used to kill snails in fish-raising water body (Zhu et al., 1995). The acute oral toxicity of bromoacetamide to mammals is moderately toxic, acute dermal toxicity is low toxicity. The chemical is also no sensitization to the skin, no mutagenic, teratogenic and carcinogenic effects, but irritate to the skin and eye mucosa (Hu et al., 1996). Bromoacetamide is highly degradable in soil and water. Residues in the environment are easily degradable and short-residue pesticides. The use of bromoacetamide to kill snails will not cause residual harm to the environment. Bromoacetamide was included in the World Health Organization's pesticide evaluation plan in 1986 and listed as a scientific and technological research top project in the national R&D programme (1980–1990) (Bao et al., 2011; Zhu et al., 1998).

5.2.2 Niclosamide

Niclosamide is the only chemical molluscicide that WHO has retained and recommended since 1972. It is low in toxicity to mammals but poor in water solubility and highly toxic to fish (Li and Wang, 2017; Souza, 1995; Yang et al., 2010). Zhang et al. used niclosamide paste in water body where *Oncomelania* snails habitated, and found that it not only kill some snails but also reduced the infection rate of the snails. Zhu et al. (Zhu et al., 2001) from NIPD-CTDR tested the value of niclosamide combination with local insecticide (shachongding, [dimethylamine] trithacyclohexane hydrochloride) in 1997–1999. When combined use of niclosamide at 0.2 mg/L and shachongding at 0.1 mg/L, the LC90 for *Oncomelania* snail was 0.198 mg/L, which was slightly better than the molluscicidal effect of niclosamide alone, in addition, this combinated practice with two molluscicides can inhibit 92.7% of snail climbing, so as to improve the molluscicidal effect and reduce the cost of mollusciciding. It is a potential alternative to the molluscicides, such as sodium pentachlorophenol and niclosamide, but the mechanism of inhibition of snail climbing by shachongding remains further study. In recent years, niclosamide ethanolamine salt wettable powder has been gradually promoted and applied in P.R. China, which has improved the molluscicidal effect, but the toxicity of niclosamide ethanolamine salt to fish is difficult to overcome, which restricts its field application (Yang et al., 2010).

5.2.3 Development of novel chemical molluscicides

5.2.3.1 Pyrrolidine

Duan et al. synthesized a series of compounds and studied on the chemical mechanism of niclosamide (Duan et al., 2014). It was found that the molluscicidal activity of salicylic acid compound 31, namely pyrrolidine, with its chemical structure as 4-chloro-2-(morpholine-4-carbonyl)phenyl-4′-methoxybenzoate ($C_{19}H_{18}ClNO_5$, $Mr=375.79$), is better than other molluscicides. Bioassays experiments showed its $IC50=0.15\,g/m^2$, lower than that of niclosamide ($IC50=0.25\,g/m^2$), and its ($IC50=10\,g/m^2$) toxicity to zebrafish is only 1/40 times lower than that of niclosamide ($IC50=0.25\,g/m^2$), the compound was proved to be less toxic to mammals as the acute toxicity test of mice $LD50>5000\,mg/kg$ (Wang et al., 2015). This pyrrolidine, a 5-chlorosalicylic acid derivative incorporating morpholine moiety, is conveniently synthesized through two steps and crystallized. Duan et al. further developed pyrrolidine into a formulation of 25% piros tose sulphate wettable powder, and tested the molluscicidal effect in schistosomiasis endemic area in the lake and the mountain region, respectively. The field results showed that the molluscicidal effect of pyrrolidine by using immersion and spraying methods at the concentration of 1.0 mg/L and $2.0\,g/m^2$ respectively, is comparable to that of niclosamide ethanolamine salt wettable powder, and it is low-toxic to fish as no dead fish found in the field trail (Luo et al., 2018). Pyrrolidine has several advantages such as simplicity to prepare, high solubility and good dispersibility, and easily mixing with water evenly on site. Further multi-site expansion trials should be conducted to formulate field application technical specifications, and it is expected to develop into a new high efficient, low toxic, inexpensive molluscicides (Luo et al., 2019).

5.2.3.2 Meta-Li

Meta-Li is a low-toxicity and environmental friendly pesticide produced by Swiss Longsha (Hong Kong) Co., Ltd. Meta-Li is mainly composed of tetraacetaldehyde, which is characterized by rapid degradation and no residue in the environment after administration. Scientists from NIPD-CTDR conducted laboratory and field tests using domestic bioassay method, In on-site spraying tests of Meta-Li, it was showed that the mortality of snails was over 90% at a concentration of 2 g/m^2 7 days post spraying. The amine 1 g/m^2 is equivalent and has a significant inhibition of the snail climbing (Zhu et al., 2006). On this basis, the mechanism of mollusciciding was investigated by observing the changes of enzyme activity of snail tissue treated with Meta-Li (40% tetraacetaldehyde). The results show that inhibiting the energy supply of snails which eventually caused the death of snails may be the molluscicidal mechanisms of Meta-Li, so that the relatively long process of energy depletion is one of the reasons for the slower molluscicidal effect after application of Meta-Li (Zhu et al., 2007). Subsequent studies have obtained the safe concentration of Meta-Li on zebrafish, carp fry, *Eriocheir sinensis*, *Macrobrachium nipponense*, *Hyriopsis cumingii*, and river carp, with their LD50 are >10 mg/L in all of them. Based on the requirements for pesticide using in freshwater aquaculture species in P.R. China that needs all low-grade toxicity (LC50 > 10 mg/L), it was indicated that Meta-Li is a safe molluscicide for aquatic organisms when it is used mollusciciding. Due to its low-toxicity and friendly use in environment, it is warranted to conduct further studies on being a mulluscicidal candidate (Zhu et al., 2010).

5.2.4 Plant molluscicides

Chemical mollucicides limit their use due to environmental pollution and toxicity to non-target organisms. The use of phytochemical ingredients to kill snails and organic synthesis of new biomolecular molluscicides is currently encouraged with focus on plant molluscicides (Li et al., 2016). *Eucalyptus* tree is a native to Australia and also known as white diesel. Chen et al. tested the molluscicidal effect of the *Eucalyptus* leaves and found that the mortality rates of *Oncomelania hupensis* snails reached 80% with the volatile oil extracted from five species of *Eucalyptus* leaves immersed at the concentrations of 100 mg/L for 48 h, and the mortality rates were both 93.3% with the volatile oil from *Corymbia citriodora* and *Eucalyptus urophylla* leaves. The mortality rate was up to 95% with the chloroform extract, and the mortality rate reached 60% at the concentrations of 10 mg/L for 48 h (Chen et al., 2012). They also found the active components against snails are mainly found in extracts with petroleum ether and chloroform solutions, and the molluscicidal effect of chloroform extracts is better than other organic solvent extracts. The correlation of the main components in 12 extracts of *Eucalyptus camaldulensis* from different month of a year and their molluscicidal activities was performed by the grey relative correlation analysis model with an assistance of the software GTMS 3.0. Results showed that all the dichloromethane extracts of eucalyptus leaves showed excellent molluscicidal activities, the highest LC50 was 0.257 mg/L and 0.242 mg/L for the samples in June and July, and the lowest LC50 was 6.802 mg/L and 5.406 mg/L in December and January, respectively. The structures of 16 main chemical components were demonstrated by gas chromatography-electron impact-mass spectrometry (GC–MS) and NIST Mass Spectral Library, most of which were monoterpenes and sesquiterpenoids. The most of main ingredients in the dichloromethane extracts of *E. camaldulensis* leaves show good correlations with the molluscicidal activity' which suggests that the molluscicidal effects are synergistically made by the multiple components together (Yin et al., 2014). According to WHO's standards for screening plant molluscicides, the five eucalyptus leaves extracts collected from Guangdong province contained higher molluscicidal components. Wang et al. from the NIPD-CTDR also have found that tea seed was effective regent against *Oncomelania* snails through damaging the gonads and liver of snails (Song et al., 1992; Wang and Song, 1990).

In order to provide safe and effective control methods for control of the intermediate snail host of *Schistosoma japonicum*, the NIPD-CTDR actively carried out the research on the development of molluscicides, in order to provide safe molluscicides for interrupting the transmission of schistosomiasis in P.R. China. Because of the long-term existence of snails in the complex environment in the lakes and mountainous regions in P.R. China, mollusciciding has the characteristics of quick effect, low cost compared to other approaches, but its environmental pollution has been concerned. Over the years, several generations of the NIPD-CTDR researchers have done a lot of pioneering research on application of molluscicides development, which has made a significant contribution to the national schistosomiasis control programme in P.R. China.

6. Monitoring and evaluation

The NIPD-CTDR takes a leading role in monitoring and evaluation for the national schistosomiasis control programme. The institute conducted four times of the national sampling surveys and established a surveillance system to support the control programme (Fig. 2).

6.1 National sampling surveys

Four national sampling surveys on schistosomiasis were conducted by NIPD-CTDR in 1989, 1995, 2004 and 2016, respectively (Department of Diseases Control, 2006; Department of Endemic Diseases Control, 1993, 1998).

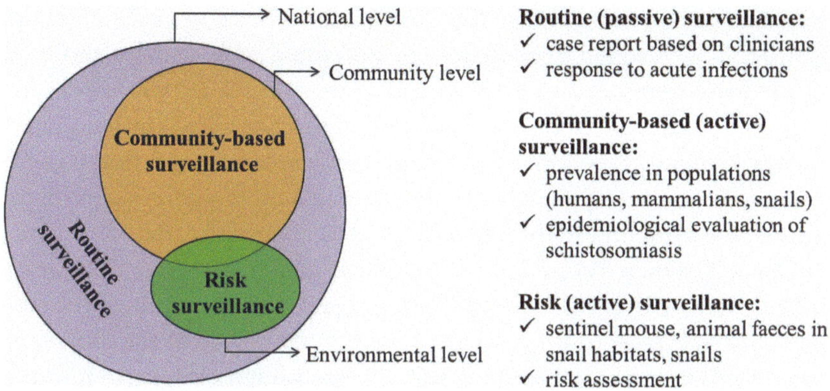

Fig. 2 The surveillance system of schistosomiasis in P.R. China.

Successful control of schistosomiasis before 1980 was owing to mass campaign in the era of the planned economic system. However, the reform in rural area at the end of 1970s changed the situation. Schistosomiasis resurged in some areas where the disease transmission had controlled. In order to evaluate the national status of schistosomiasis in time which provide the reference to the decision making in the national control programme, the first national survey was initiated by Ministry of Health in 1989. The estimated prevalence in humans was 10.2% and the inferred infections of human was 1,638,103 people. The survey provided very informative results for formulating the "Eighth Five-year Plan of the national schistosomiasis control programme.

The second national survey was carried out in 1995 in order to evaluate the effect of new control strategy since 1992 when the World Bank Load Project on schistosomiasis control started. Before the initiation of the project the control strategy focussed on shrinking the snail infested areas. The massive drug administration with praziquantel replaced the previous strategy with the advent of praziquantel. The overall prevalence was 4.9% and the estimated number of human cases was 865,084. The survey demonstrated that the new strategy was effective at that time.

The third national survey performed in 2004 under the circumstance that the prevalence in some area increased and acute infections re-occurred, after the World Bank Load Project on schistosomiasis control terminated. The Ministry of Health organized the third survey in order to understand the situation and prepare for the long-term action plan. The survey revealed an overall prevalence of 2.5% and the inferred human infections were 726,112. The decline pattern of the prevalence had slowed down from 1995 to 2004, which requiring the new strategy to be adjusted to the new situation, particularly in the lake region where resurgence happed seriously. This resulted in the effort to develop a new strategy and finally proposed the integrated strategy focussing on control sources of infection (Wang et al., 2009b). Both the results of the third national sampling survey on schistosomiasis prevalence and the new integrated strategy had provided the baseline data and approaches to the national schistosomiasis control programme, which achieved the expected goal to control the transmission with prevalence less than 1% by 2015. The fourth survey, however, focussed on the intermediate host snails due to following reasons: the disease prevalence has been at low level (<1%), potential risk closely related to the distribution of snail habitats remained, and the surveillance system on human infections has fully established since 2004.

The fourth survey had produced a snail distribution map for the first time, showing that appropriately 400 thousand of snail habitats in the whole country, including all historical and existing habitats. The results will be informative to assess the emergence and re-emergence of *O. hupensis* and thus to prepare the further control plan.

6.2 Routine control programme

The tasks for the NIPD-CTDR in routine control programme on schistosomiasis varied in different stages. In the mid-1950s when the national control programme commenced, the institute was in charge of collecting annual data pertaining infections and treatments of humans, cattle and snails as well as the elimination process at county level. After 1985, the institute also took the responsibility to draft the annual report on schistosomiasis prevalence in China every year, based on the national schistosomiasis database established. Such database is of importance to provide the basic information for control policy making of the national schistosomiasis control programme.

For a long time before 2011, the data on schistosomiasis prevalence was submitted to the NIPD-CTDR by provincial department of schistosomiasis control in paper or later in electronic sheets manually. The NIPD-CTDR constructed a Reporting System for Parasitic Disease Prevention and Control to improve the timeliness and accuracy of data, and realized the network report of the data (Zhang et al., 2016b). The information system was officially launched in 2011, which mainly included data on schistosomiasis control and schistosomiasis surveillance work. The reporting system covered the users from different levels, from provincial and municipality users to counties (or districts) and townships users. Level-by-level management mode was adopted to manage the data systems and users. The system is divided into four modules according to its functions: data collection, statistical analysis, data management and system management. Data on schistosomiasis epidemic situation and control works could be collected, and can be used for the evaluation of the national control programmes at different levels through these four modules.

6.3 Schistosomiasis surveillance system

The revised Chinese Law of Prevention and Treatment of Infectious Diseases (version 2004) up-graded schistosomiasis from C catalogue to B catalogue in the national notifiable infectious disease list of China. Accordingly, schistosomiasis cases based on the national criteria of schistosomiasis diagnosis must be reported within 24 h after diagnosis. In order to push forward the internet-based health information system and strengthen management on infectious diseases, China CDC has launched the the National Notifiable Disease Reporting System (NNDRS) since 2004 (Dai et al., 2019b; Li et al., 2010). In contrast to the data collected from the departments of schistosomiasis control as described in the part of 6.2 of this Chapter, the human cases are normally reported by clinician through the NNDRS. The NNDRS is currently mainly used for response to submit the information of acute infections and parasitologically diagnosed cases. The message on case reporting is automatically sent to mobile phone of users once information of an acute case was submitted into this system. NIPD-CTDR staff is responsible to clarifying the cases and accomplish the annual report based on the information downloading from the NNDRS.

Community-based sentinel surveillance is one key part of national schistosomiasis surveillance system. The community-based sentinel surveillance for a purpose of epidemiological observation commenced since early 1990s. There were only 13 communities in the sentinel surveillance system in the beginning of setting up those communities as the national sentinel surveillance sites in early 1990s. The First Shanghai Medicine University (now the Medicine School of Fudan University) was responsible for running the sentinel surveillance system (Zhou et al., 2005b). After the China CDC was set up, and the central government transfer payment system for the disease control programme was initiated, the NIPD-CTDR took the responsibility to manage the community-based sentinel surveillance since 2005. The sentinel surveillance sites increased to 80 communities in 2005, and extended to the potential endemic areas for the sentinel surveillance since 2013.

Therefore, a meeting on surveillance of potential endemic areas of schistosomiasis was held in 2013, with the outcome documents of Surveillance Plan from 2013 to 2018 in Potential Endemic Areas of Schistosomiasis and its implementation protocols was formulated also. At the same time ， the quality control and evaluation programme for schistosomiasis sentinel surveillance was issued. With the significant reduction of schistosomiasis prevalence, all 450 endemic counties in P.R. China are involved in the sentinel surveillance system by selecting one community in each county as the sentinel sites since 2014. In the sentinel community each county, the active surveillance was conducted based on the standard protocol, including infections in human, domestic animals and snails. The NIPD-CTDR prepared the guideline and control the quality for the community-based sentinel surveillance.

Risk assessment was initiated in the face of declining infection rate and density and normally conducted in the highly potential transmission environment along the lakes or rivers. In contrast to the community-based sentinel surveillance, the risk assessment focussed on identifying the potential contamination of *S. japonicum* cercariae in water body by sentinel mice, looking for the infective animal faeces in snail habitats, and screening the infected snails by molecular technology, such as LAMP. The NIPD-CTDR collected the data under the help of local departments of schistosomiasis control and conducted the risk assessment annually.

6.4 Early warning and modelling

In order to eliminate schistosomiasis, it is crucial to improve the sensitivity of schistosomiasis surveillance and early warning and to improve the effectiveness of on-site intervention measures in low-prevalence areas. In order to meet the important needs in initiation of the national elimination programme of schistosomiasis, establishment of the surveillance and response system with highly sensitivity and real-time monitoring in the lower prevalence of the disease is necessary. The NIPD-CTDR focussed on three parts, i.e., development of key detecting technologies, upgrading of surveillance technologies, and development of innovative technologies on forecasting transmission risks of schistosomiasis. In addition, those technologies in the surveillance and response system to mitigate the transmission risks of schistosomiasis were applied in the field. Three major innovative achievements were gained as follows: (i) An assessment system on sero-diagnostics for schistosoma infections were established and quality control systems for diagnosis with networking laboratories were formed (Xu et al., 2011a). An innovative estimation methodology for human infection rate with *S. japonicum* was developed based on the Bayesian model (Wang et al., 2006, 2008), and the optimized strategy for diagnosis and treatment evaluation were developed for the different control stages in national schistosomiasis elimination programme; (ii) A series of surveillance indicators relevant to biological, natural and social factors that affecting the schistosomiasis transmission were developed and optimized (Cao et al., 2013a,b). The spatiotemporal distribution scenarios and transmission patterns of schistosomiasis were explored, and the transmission threshold of transmission under various snail densities were found, which have provided the technical support the formulation of the national schistosomiasis control and elimination programme; (iii) The forecast models along with various indicators relevant to the prediction and analysis of schistosomiasis transmission patterns were developed based on different impact factors and indicators in various endemic areas of schistosomiasis, and networking platform for surveillance and response in highly risk areas of schistosomiasis transmission were established in order to support the quick response activities in mitigation of the outbreaks and special status in schistosomiasis control programme (Wang et al., 2008; Xu et al., 2013).

Therefore, the impact factors related to biological, natural and social determinants in the low transmission areas in the national schistosomiasis elimination programme have been systematically investigated, with the defined key factors relevant to the transmission of schistosomiasis. Surveillance indicating system and innovative technologies for forecast models and predicted response has developed, which provided the technical support to the national schistosomiasis elimination programme. The efficiency and suitability of those developed technologies in surveillance, forecast, and rapid response in the field trials has been verified, after the NIPD-CTDR has provided the tools and updated information to the prevention of schistosomiasis infections in advance, and formulated the goals for the next stage of the national schistosomiasis elimination programme (Lei and Zhou, 2015; Zhang et al., 2016b).

With the more progress on the national schistosomiasis elimination programme which launched in 2014, the prevalence of schistosomiasis in P.R. China is at a very low level. For the purpose of effectively monitor the prevalence of schistosomiasis, the NIPD-CTDR took the leadership in carrying out schistosomiasis surveillance in P.R. China. Based on the National Infectious Disease Surveillance Network System, a direct network reporting system for schistosomiasis at the national level has been established, which fills in the scope of the national schistosomiasis elimination programme. The advanced surveillance techniques have applied in the monitoring transmission risk factors and implementing of the rapidly assessment of the epidemic risk of schistosomiasis. Accurate prediction and early warning of the transmission trend of schistosomiasis have provided a scientific reference for timely response to the potential schistosomiasis epidemics (Cao et al., 2013a; Xu et al., 2012; Zhou et al., 2007b).

7. Technical guidelines and expertise support to the control programme

Great achievements have been made in the national schistosomiasis control programme in P.R. China, due to profiting from the strong leadership of the government and the formulation and implementation of various government decisions, programmes and plans (Chen and Feng, 1999; Department of Disease Control and Prevention, 2008; Lei and Zhou, 2015; Zhou et al., 2005d, 2007a, 2018). Administrative decision-making is a process of analysis and synthesis from experience to reality and then to the future. It needs sufficient, credible, professional and creative information and suggestions (Pautz, 2011). From this view, the NIPD-CTDR played the role of professional think-tank in the national control programme. As a "thought laboratory" of decision-making organization, the NIPD-CTDR not only provides decision-making consultation for the government, but also provides human resources and intellectual support for field control activities (Kui et al., 2017a,b; Zhang, 2017). The NIPD-CTDR has been committed to comprehensive research of schistosomiasis control, with solid theoretical rationale and high scientific research capability (Tang, 2005). Meanwhile, experts and professionals of NIPD-CTDR have been stationed in the serious endemic areas for a long time, to carry out operational research on multiple pilots in various epidemic types of endemic areas in P.R. China. With familiar with the social and economic development as well as local customs and practices of most endemic areas of schistosomiasis in P.R. China, those scientists from NIPD-CTDR also grasp the real demand and current situation of schistosomiasis control in P.R. China (Cao et al., 2010; Guo and Yu, 2005; Zheng et al., 1996, 2000; Zhou et al., 2008).

7.1 Providing technical support as major member agency of the Expert Committee on Schistosomiasis Control

In 1955, the Central Research Committee on Schistosomiasis Control was established in Shanghai (Jia, 1995; Mao, 1960; Zhou et al., 2008), and then NIPD-CTDR fully supports the national comprehensive schistosomiasis survey and pilot programme of control from 1956 to 1957. In May 1987, the Ministry of Health set up the first Consultative Committee of Schistosomiasis Experts (the office located in NIPD-CTDR) (Zheng, 1987), which has been held for the fifth time, the specialists of NIPD-CTDR have served as chairman, committee member and secretary. Since its establishment, the committee has played momentous roles of technical support and policy consultation in coordinating the national key scientific research projects, providing technical guidance and formulating national schistosomiasis control plans, standards, regulations and laws.

7.2 Involving in the design and implementation of the national control strategy and work plan

In 2006, the NIPD-CTDR organized experts to assist Legal Affairs Office of the State Council and Ministry of Health in formulating the Regulations on Schistosomiasis Control, which promulgated on April 1, and put into effect on May 1 by The State Council of P.R. China, making schistosomiasis control activities having a law to follow (Anonymous, 2006; Er, 2006; Wang et al., 2009a). NIPD-CTDR has given pertinent opinions or suggestions in the drafting or formulation of national planning in the light of topographical features and current situation of schistosomiasis in P.R. China, especially in core technology steps of goal setting, strategies and measures finalization. The comprehensive strategy to reduce the roles of humans and cattle as sources of *S. japonicum* infection in snails proposed in researches along Poyang Lake has adopted by government as the national strategy for the control of schistosomiasis and rapidly expanded to other endemic provinces (Cao et al., 2014; Wang et al., 2009b). Moreover, the Thirteenth Five-year National Plan has emphasized it for Schistosomiasis Control that this strategy should be adhered to at this stage of national schistosomiasis control (Anonymous, 2017). In 2018, NIPD-CTDR assisted in drafting the part of schistosomiasis control in the Three-year Plan and Action for Endemic Disease Control and Prevention following the important instructions gave by President Xi. The short-term goal of current schistosomiasis control in P.R. China is to effectively control and eliminate the harm of schistosomiasis and a series of corresponding measures around the goal were put forward (Gao, 2019). Furthermore, experts from NIPD-CTDR attended and provided technical supports in drafting or formulating of several national plans or programmes, such as the Seventh Five-Year Plan for Schistosomiasis Control (1986), the Eighth Five-Year Plan for the Comprehensive Control of Schistosomiasis (1991), the National Medium-and Long-term Strategic Plan for Prevention and Control of Schistosomiasis (2004–2015). NIPD-CTDR took the responsibility to detail the work plan, draft technical guidance, monitor and evaluate the progress of programmes, etc.

7.3 Technical support to the response systems

In order to effectively prevent, control and standardize the emergency treatment the outbreak of schistosomiasis, the NIPD-CTDR drafted Emergency Response Trial Plan for Major Outbreak of Schistosomiasis which implemented by the Ministry of Health in 2003. Henceforth, perfected Emergency Plan for Outbreaks of Schistosomiasis for Ministry of Health, complicated of work manual of grass-roots schistosomiasis control personnel for department of Disease Control and provided advice about Trail Plan of Transmission Control and Blocking of Epidemic Relapse for the Schistosomiasis Control Office of the State Council in 2005. With the advancement of schistosomiasis control in P.R. China, the endemic areas of schistosomiasis were facing the demand of achieving standard of disease control. In 2007, NIPD-CTDR participated in the formulation of the Schistosomiasis Prevention and Control Area Achievement Assessment Programme (revised and issued in 2010) for Ministry of Health, and assisted the Schistosomiasis Control Office of the State Council in formulating the Schistosomiasis Control Achievement Assessment Pre-assessment Programme. In 2008, the NIPD-CTDR composed the Handbook for Assessment and Evaluation Achieving Infection Control of Schistosomiasis and then completed of the National schistosomiasis epidemic situation validation programme in 2011. In 2015 and 2018, participated in the compilation of Assessment Draft for the Elimination of Schistosomiasis and Code of Practice for the Elimination of Schistosomiasis. In addition, involved Programme of Joint Control of Schistosomiasis between Ministry of Health, Ministry of Agriculture and Hubei and Hunan province (2009, 2010), Evaluation of Criteria of Transmission Control Scheme in Hubei (2013). Following above files, NIPD-CTDR cooperated with Ministry of Health/NHFPC accomplished transmission control situations evaluation of Jiangsu (2011), Hubei (2013), Hunan, Jiangxi and Anhui (2015), carried out the review of schistosomiasis elimination in Shanghai, Guangdong, Fujian, Guangxi and Zhejiang in 2016, as well as the field assessment on the standard schistosomiasis interruption in Sichuan (2017) and Hubei province (2018).

7.4 Formulating national criteria related to schistosomiasis control

The NIPD-CTDR experts were devoted them to drafting and formulating national and health criteria for schistosomiasis control, and revising them with the progress of schistosomiasis control in P.R. China. To improve and standardize the diagnosis work for human schistosomiasis especially for control stations and hospitals staff, the NIPD-CTDR organized famous and skilled experts related schistosomiasis diagnosis and control formulating schistosomiasis diagnosis criteria in 1995 and then revised in 2006. The criteria for the Control and elimination of Schistosomiasis in P.R. China evolution seven editions, experts of the NIPD-CTDR chaired each version's discussion and revision. In fact, the latest three versions were led by the experts of NIPD-CTDR (GB15976-1995, GB15976-2006, GB15976-2015). To provide guidance for snails survey for schistosomiasis control stations or institutes, NIPD-CTDR chaired the draft of the criteria for snail survey and the criteria was issued in 2017. In 2018, the experts of the NIPD-CTDR participated in the preparation of "In-direct Erythrocyte Agglutination Test for Schistosomiasis Antibody Detection in Japan (WS/T630-2018).

7.5 Attending field investigation to provide accurate information for policy makers

From 2004, the NIPD-CTDR dispatch technicians to participate in the investigation and guidance of the State Council's spring investigation of schistosomiasis prevention every year. To understand the real situation of schistosomiasis endemicity and control activities, the NIPD-CTDR organized the activities on rapid risk assessment of the disease transmission in seven major endemic provinces of the lake and mountainous regions since 2012, and also coordinated the field investigation in secret to master the real situation of the disease transmission since 2016, all of which provided the evidence to monitor the programme at all level through the State Council's spring investigation (Zhou et al., 2018). In addition, the NIPD-CTDR conducted rapid assessments on extreme drought effects on Dongting and Poyang Lake at 2011, and a survey on imported schistosomiasis to assess the potential risks of imported schistosomiasis in 2012 etc.

It is honoured for the NIPD-CTDR to bring the top best experts in the country together making efforts on schistosomiasis control and research, and provide theoretical and practical evidences for the policy making on schistosomiasis control in P.R. China (Zhou et al., 2011). The NIPD-CTDR not only take the synthesis of thought from the expert consultation and demonstration, also provides suggestions for the Chinese government to formulate control plan to design the objectives, policies, measures, and national control standards for the national schistosomiasis control and elimination programmes (Zhou et al., 2010).

8. Conclusion and Perspectives

Schistosomiasis japonica has been prevalent in P.R. China for at least 2000 years. The disease mainly distributes in 12 provinces (autonomous regions and municipalities) along and to the south of the Yangtze River. After 70 years of activities in control and research, the NIPD-CTDR has mastered the epidemic situation of schistosomiasis in P.R. China, the distribution of snails, and explored the strategies and measures suitable for different environmental areas (Chen et al., 2020; Zhen and Bin, 2017). Life cycle and transmission patterns of schistosomiasis have been understood well by Chinese scientists, advanced research based on genomics conducted by the NIPD-CTDR made it possible to explore high-efficiency and low-toxicity therapeutic drugs and molluscicidal drugs, laid foundations for developing new diagnostic tools and vaccine candidates. Lessons and experiences were accumulated through the 70 years' control practices for the NIPD-CTDR to be as the only national institute specific for parasitic diseases in the country (Zhou et al., 2005a). The needs of the control programme in development of strategies, intervention measures, diagnostic techniques, monitoring or early warning have changed with the development of social-economic systems and health systems in the country. These changes require our more efforts on the research to develop new tools used in the control programme (Zhen and Bin, 2017). At the same time, the lessons and experience accumulated need to be distilled for guiding further control activities, through reviewing archived documents, publishing more articles or books, and interviewing the senior scholars who attended the national schistosomiasis control activities in early stage.

The strategic plan for the Healthy China 2030 set the goal to eliminate schistosomiasis in all endemic counties in P.R. China by 2030. To realize this final goal, the NIPD-CTDR sustaines the capacity to provide technical support to the control and research activities as follows: Firstly, sensitive and rapid surveillance-response system should strengthen to understand the endemic tendency of schistosomiasis timely and provide evidence for elimination (Feng et al., 2016). Secondly, regulations and guidelines related to schistosomiasis modify to adapt to current situation and requirements and schistosomiasis has moving to very low infection rate and infection intensity (Zhou et al., 2007c). Thirdly, basic and applied research to meet the requirements of schistosomiasis elimination programme need to continuously perform at advanced level (Zhou et al., 2015), such as advance of sensitive molecular diagnostic tools for environmental risk assessments, establishment minimized surveillance database or indicators for early warning, development of more environmental friendly molluscicides for snail control etc. Fourthly, capacity building for schistosomiasis control and research not only for Chinese but also other countries' professionals need to be strengthened, to keep a workforce involving in schistosomiasis elimination campaign in P.R. China and contribute to the China-African cooperation programme for schistosomiasis control along the "One Belt and One Road Initiative" (Chen et al., 2020).

References

Anonymous, 2006. Regulations on the Prevention and Control of Schistosomiasis. China Legal Publishing House, Beijing in Chinese.

Anonymous, 2017. Six departments including the Ministry of health and Family Planning Commission jointly issued the 13th five year plan for the prevention and control of schistosomiasis. Chin. Emerg. Manag. 11 (3), 29–31 (in Chinese).

Balen, J., Zhao, Z.Y., Williams, G.M., McManus, D.P., Raso, G., Utzinger, J., Zhou, J., Li, Y.S., 2007. Prevalence, intensity and associated morbidity of *Schistosoma japonicum* infection in the Dongting Lake region, China. Bull. World Health Organ. 85 (7), 519–526.

Bao, Z.P., Cao, C.L., Dai, J.R., Liu, J.B., Li, G.P., Guo, J.D., Guo, J.G., 2011. Evaluation on molluscicidal effect of suspension concentrate of metaldehyde and niclosamide. Chin. J. Schisto. Control 23 (1), 48–53 (in Chinese).

Cao, C.L., Zhu, H.Q., Bao, Z.P., Zheng, J., 2004. Investigation of acute schistosomiasis in Sichuan province. Chin. J. Schisto. Control 16 (5), 387–388 (in Chinese).

Cao, J.P., Xu, Y.X., Han, H.B., Song, G.C., Li, X.H., Lu, Y.Y., Shen, Y.J., Liu, S.X., 2005. Immunogenicity of thioredoxin in schistosomiasis II. Protective immunity of mice with schistosomiasis thioredoxin DNA vaccine. Chin. J. Schisto. Control 17 (4), 255–259 (in Chinese).

Cao, C.L., Bao, Z.P., Zhu, H.Q., Yu, Q., Li, S.Z., Wang, Q., Zhou, X.N., Guo, J.G., 2010. Analysis of schistosomiasis control requirements in boat fishermen in lake regions. Chin. J. Schisto. Control 22 (6), 55–59 (in Chinese).

Cao, C.L., Bao, Z.P., Zhu, H.Q., Jia, T.W., Yu, Q., Qin, Z.Q., Liang, Y.S., Zhong, B., Ren, G.H., Huang, X.B., Lin, D.D., Zhang, S.Q., Xu, J., Li, S.Z., Guo, J.G., Zhou, X.N., 2013a. Establishment and application of rapid assessment system of environment with high transmission risk of schistosomiasis in marshland and lake regions II Establishment and application of measures for field survey. Chin. J. Schisto. Control 25 (5), 451–456 (in Chinese).

Cao, C.L., Xu, J.F., Xu, J., Bao, Z.P., Jia, T.W., Yu, Q., Zhu, R., Dang, H., Zhang, L.J., Zhu, H.Q., Li, S.Z., Guo, J.G., Zhou, X.N., 2013b. Establishment and application of rapid assessment system of environment with high transmission risk of schistosomiasis in marshland and lake regions I: Establishment of an index system with Delphi method. Chin. J. Schisto. Control 25 (3), 232–236 (in Chinese).
Cao, C.L., Bao, Z.P., Yang, P.C., Chen, Z., Yan, J., Ren, G.H., Li, Y.Y., Cai, S.X., Liu, J.B., Xu, J., Li, S.Z., Guo, J.G., Zhou, X.N., 2014. Schistosomiasis control effect of measures of replacing cattle with machine for cultivation and forbidding depasturage of livestock on marshlands in marshland and lake regions. Chin. J. Schisto. Control 26 (6), 602–607 (in Chinese).
Chen, M.G., 1980. Diagnosis of schistosomiasis japonica by dry egg ring precipitation reaction. Intermediate Med. J. (12), 43–44.
Chen, M.G., Feng, Z., 1999. Schistosomiasis control in China. Parasitol. Int. 48 (1), 11–19.
Chen, J.X., Liu, S.X., Cao, J.P., Song, G.C., Xu, Y.X., 2006. The anti-disease immune effect of *Schistosoma japonicum* paramyosin gene gene vaccine in mice. Chin. J. Parasitol. Parasit. Dis. 24 (2), 81–85 (in Chinese).
Chen, H.G., Xie, S.Y., Zeng, X.J., Huang, X.B., Wang, T.P., Li, Y.S., Liang, Y.S., Zhou, X.N., 2011. Current epidemic characteristics and control strategies of schistosomiasis in lake areas of China. Chin. J. Schisto. Control 23 (1), 5–9 (in Chinese).
Chen, T.T., Zhou, X.N., Zhu, D., Li, M.Y., 2012. Preliminary screening tests of molluscicidal effect of extracts from Eucalyptus leaves. Chin. J. Schisto. Control 24 (6), 694–696 (in Chinese).
Chen, J., Ding, W., Li, Z., Zhou, D.D., Yang, P., Wang, R.B., Zheng, B., Sheng, H.F., Guan, Y.Y., Xiao, N., Li, S.Z., Zhou, X.N., 2020. From parasitic disease control to global health: new orientation of the National Institute of Parasitic Diseases, China CDC. Acta Trop. 201, 105219.
Dai, S.M., Edwards, J., Guan, Z., Lv, S., Li, S.Z., Zhang, L.J., Feng, J., Feng, N., Zhou, X.N., Xu, J., 2019a. Change patterns of oncomelanid snail burden in areas within the Yangtze River drainage after the three gorges dam operated. Infect. Dis. Poverty 8 (1), 48.
Dai, S.M., Xu, Z.M., Zhang, L.J., Xu, J., Lv, S., 2019b. Analysis of schistosomiasis cases report in National Notifiable Disease Report System in China, 2015–2017. Chin. J. Schisto. Control 31 (2), 121–125 (in Chinese).
Davis, A., Wegner, D.H., 1979. Multicentre trials of praziquantel in human schistosomiasis: design and techniques. Bull. World Health Organ. 57 (5), 767–771.
Department of Disease Control and Prevention, M.O. H, 2008. The mission and achievement of schistosomiasis control in half a century: commemoration of 50th anniversary of publishing Chairman Mao Zedong's two poems 'Farewell to the God of Plague'. Chin. J. Prev. Med. 42 (10), 711–712 (in Chinese).
Department of Diseases Control, M.O.P.H, 2006. Epidemiological Situation of Schistosomiasis in China—Results From a Nation-Wide Sampling Survey in 2004. Shanghai Science and Technology Literature Press, Shanghai in Chinese.
Department of Endemic Diseases Control, M.O.P.H, 1993. Epidemiological Situation of Schistosomiasis in China — Results from a Nation-Wide Sampling Survey in 1989. Chengdu Science and Technology Press, Beijing in Chinese.
Department of Endemic Diseases Control, M.O.P.H, 1998. Epidemiological Situation of Schistosomiasis in China—Results from a Nation-Wide Sampling Survey in 1995. Nanjing University Press, Nanjing in Chinese.
Duan, L.P., Qin, Z.Q., Li, G.P., Li, S.Z., Li, Q., Xu, D.F., 2014. Synthesis, crystal structure and biological evaluation of 4-chloro-2-(morpholine-4-carbonyl) phenyl-4¡ä-methoxybenzoate as molluscicidal agent. Struct. Chem. 33 (6), 841–845.
Er, E.T., 2006. Regulations on the prevention and control of schistosomiasis. Chin. Anim. Industry 13 (14), 4–5 (in Chinese).

Feng, Z., Yan, L.Z., Chen, M.G., Qian, Z.L., Lu, P., Zhang, Y.H., Li, H., Xue, H.Q., 1998. Analysis of the status of diagnostic reagents for schistosomiasis in China from parallel detection and monitoring results. Chin. J. Parasitol. Parasit. Dis. 16 (5), 3–7 (in Chinese).

Feng, Y., Liu, L., Xia, S., Xu, J.F., Bergquist, R., Yang, G.J., 2016. Reaching the surveillance-response stage of schistosomiasis control in the People's Republic of China: a modelling approach. Adv. Parasitol. 92, 165–196.

Gao, Y.H., 2019. Interpretation of the tasks of the special three-year program for prevention and control of endemic diseases (2018-2020). Chin. J. Endemiol. 38 (1), 1–3 (in Chinese).

Guan, X.H., Shi, Y.E., 1996. A collaborative study to evaluate the efficacy of immunological diagnostic methods for schistosomiasis in Japan. Chin. J. Schisto. Control 8 (2), 72–78 (in Chinese).

Guo, J.G., 2006. The position and role of the strategy for controlling the source of infection of schistosomiasis in China. Chin. J. Schisto. Control 18 (3), 231–233 (in Chinese).

Guo, J.G., Yu, Q., 2005. Prevalent status and trend of schistosomiasis of China in recent years. Chin. J. Schisto. Control 17 (5), 321–323 (in Chinese).

Guo, J.J., Zheng, J., 1999. Effects of water resources development on schistosomiasis prevalence. Chin. J. Parasitol. Parasit. Dis. 17, 62–65 (in Chinese).

Guo, J.G., Chen, H.G., Lin, D.D., Hu, G.H., Wu, X.H., Li, D., Liu, H.Y., Zheng, J., Chen, M.G., Tanner, M., 2002. A method of rapid identification snail habitat in marshland of Poyang Lake region by remote sensing. Chin. J. Parasit. Dis. Control 15 (5), 291–300 (in Chinese).

He, Z.Y., 2005. Effect of Returning Farmland to Lakes on Schistosomiasis Transmission in Poyang Lake Area. Chinese Center of Disease Control and Prevention, Beijing.

He, Y.K., Liu, S.X., Yu, X.L., Song, G.C., Xu, Y.X., Cao, J.P., Yang, R.Q., Hou, X.Y., Zhang, X.Y., 2002a. Preliminary observation on specific antibody level and reduction of ovulation induced by recombinant Schistosoma japonicum 26kDa GST antigen in water buffaloes. Chin. J. Parasitol. Parasit Dis 20 (3), 133–135 (in Chinese).

He, Y.K., Liu, S.X., Luo, X.S., Zhang, X.Y., Song, G.C., Xu, Y.X., Yu, X.L., Yang, R.Q., Lin, J.L., Chen, Y., 2002b. Field research on recombinant 26kDa glutathione-s-transferase (reSjc26GST) as a vaccine candidate: protection induced in bos buffelus and effectiveness on the control of Schistosomiasis japonica. Chin J Zoonoses 18 (2), 87–90 (in Chinese).

He, Y.K., Liu, S.X., Yu, X.L., Li, Y.S., Song, G.C., Xu, Y.X., Ho, X.Y., Cao, J.P., Liu, Z.S., Zhao, Y.Q., McManus, D.P., 2003. The time of no egg excretion from faeces of water buffalo infected with Schistosoma japonicum and its relevant control strategy. Pract. Prevent. Med. 10 (6), 831–834 (in Chinese).

Hu, J.Y., Huang, L.X., Zhu, D.P., Yin, J.Y., 1996. Study on the relationship between snail killing activity and structure of bisphenol compounds. Chin. J. Parasitol. Parasit. Dis. 14 (3), 81–82 (in Chinese).

Hu, W., Yan, Q., Shen, D.K., Liu, F., Zhu, Z.D., Song, H.D., Xu, X.R., Wang, Z.J., Rong, Y.P., Zeng, L.C., Wu, J., Zhang, X., Wang, J.J., Xu, X.N., Wang, S.Y., Fu, G., Zhang, X.L., Wang, Z.Q., Brindley, P.J., McManus, D.P., Xue, C.L., Feng, Z., Chen, Z., Han, Z.G., 2003. Evolutionary and biomedical implications of a *Schistosoma japonicum* complementary DNA resource. Nat. Genet. 35 (2), 139–147.

Hu, W., Brindley, P.J., McManus, D.P., Feng, Z., Han, Z.G., 2004. Schistosome transcriptomes: new insights into the parasite and schistosomiasis. Trends Mol. Med. 10 (5), 217–225.

Hu, Y., Shen, Y.J., Cao, J.P., Xu, Y.X., Lu, W.Y., Zhou, H.J., Zhang, J., Li, X.H., Quan, H., Liu, S.X., 2010. Study on the immunoprotection of Schistosoma japonicum HGPRT recombinant antigen combined with different adjuvants. J. Jiangsu Univ. (Med. Sci.) 20 (2), 93–97.

Huang, A.S., Zhou, B., Chen, W.Y., Sheng, Q.H., Jiang, G.Y., Xiao, C.Y., Xiao, S.H., Chen, M.G., Zheng, J., Liu, H.X., 1999. Primary study on effect of artemether on preventing schistosome infection in water-contacted workers. Chin. J. Schisto. Control 11 (3), 161 (in Chinese).

Hunan Medical College, 1980. Study on the Mummy From Han Dynasty Tomb No. 1 in Mawangdui in Changsha City. Wenwu Publishing House, Beijing in Chinese.

Hunter, J.M., Rey, L., Scott, D., 1982. Man-made lakes and man-made diseases. Towards a policy resolution. Soc. Sci. Med. 16 (11), 1127–1145.

Jia, Y.D., 1995. Glorious course—commemorating the 40th anniversary of schistosomiasis control in China. Chin. J. Schisto. Control 7 (6), 321–322 (in Chinese).

Kan, H.C., 1936. Further note on the intracutaneous test with *Schistosoma japonicum* antigen. Chin Med J (Engl) 50 (11), 1649–1652.

Katz, N., Chaves, A., Pellegrino, J., 1972. A simple device for quantitative stool thick-smear technique in schistosomiasis mansoni. Rev. Inst. Med. Trop. Sao Paulo 14 (6), 397–400.

Kui, Y., Cao, C.L., Tian, S., Tao, B.Y., Zhou, X.N., 2017a. Evaluation on human resource allocation in certain national institute of parasitic diseases in ten years. Chin. J. Schisto. Control 29 (2), 197–201 (in Chinese).

Kui, Y., Cao, C.L., Tian, S., Xue, J.B., Wu, J.N., Chu, Y.F., Li, S.Z., Zhou, X.N., 2017b. Analysis on the current situation of personnel constitution in nine provincial institutions of parasitic disease control and prevention. Chin. J. Parasitol. Parasit. Dis. 35 (3), 313–316 (in Chinese).

Lei, Z.L., Zhou, X.N., 2015. Elimination of schistosomiasis: a new target and a new task for the National Schistosomiasis Control Porgramme in the People's Republic of China. Chin. J. Schisto. Control 27 (1), 1–4 (in Chinese).

Li, Y.H., 1976. Schistosomiasis ring egg precipitation reaction. Foreign Med. Ref. Mater. (Parasit. Dis.) (1), 7–12 (in Chinese).

Li, H., Wang, W., 2017. Apropos: critical analysis of molluscicide application in schistosomiasis control programs in Brazil. Infect. Dis. Poverty 6 (1), 54.

Li, Y.H., Yan, G.Q., 1956. Discussion on several problems of intradermal reaction of schistosomiasis. Intermediate Med. J. (4), 9–11.

Li, X.H., Liu, S.X., Song, G.C., Xu, Y.X., Cao, J.P., Chen, J.X., 2005. Construction of a nucleic acid vaccine for myosin partial heavy chain gene of *Schistosoma japonicum* and protective study in mice. Chin. J. Schisto. Control 17 (1), 9–12 (in Chinese).

Li, H.Z., Wang, Q., Zheng, C.J., Zhang, Q., Qian, Y.J., Li, S.Z., Wu, X.H., Dong, Y.F., Zhou, X.N., 2010. Development and application of Information System for Parasitic Diseases Control and Prevention. Chin. J. Schisto. Control 22 (4), 301–303 (in Chinese).

Li, S.Z., Qian, Y.J., Yang, K., Wang, Q., Zhang, H.M., Liu, J., Chen, M.H., Huang, X.B., Xu, Y.L., Bergquist, R., Zhou, X.N., 2012. Successful outcome of an integrated strategy for the reduction of schistosomiasis transmission in an endemically complex area. Geospat. Health 6 (2), 215–220.

Li, Z.J., Ge, J., Dai, J.R., Wen, L.Y., Lin, D.D., Madsen, H., Zhou, X.N., Lv, S., 2016. Biology and control of snail intermediate host of *Schistosoma japonicum* in the People's Republic of China. Adv. Parasitol. 92, 197–236.

Lin, D.D., Zhou, X.N., Liu, Y.M., Sun, L.P., Hu, F., Yang, G.J., Hong, Q.B., 2002. Prediction of snail distribution after 1998 flood by using satellite remote sensing data in Poyang Lake area. Chin. J. Schisto. Control 14 (2), 119–121 (in Chinese).

Liu, S.X., Song, G.C., Ding, L.Y., Xu, Y.X., Cai, Z.H., Wu, G.Z., 1993. Comparative study on antigenicity and immunogenicity of 26-28 kDa antigen and recombinant Sj26 (RSj26) of *Schistosoma japonicum*. Southeast Asian J. Trop. Med. Public Health 24 (1), 65–69.

Liu, S.X., Song, G.C., Xu, Y.X., Yang, W., McManus, D.P., 1995. Anti-fecundity immunity induced in pigs vaccinated with recombinant Schistosoma japonicum 26kDa glutathione-S-transferase. Parasite Immunol. 17 (7), 355–40.

Liu, S.X., He, Y.K., Song, G.C., Luo, X.S., Xu, Y.X., McManus, D.P., 1997. Anti-fecundity immunity to *Schistosoma japonicum* induced in Chinese water buffaloes (*Bos buffelus*) after vaccination with recombinant 26 kDa glutathione-S-transferase (reSjc26GST). Vet. Parasitol. 69 (1), 39–47.

Lu, Y., Xu, B., Qi, C., Mo, W., Chen, Y.B., Feng, Z., Wang, X.N., Hu, W., 2012. Immunoscreening and identification of positive clones of cDNA library of *Schistosoma japonicum* eggs. Chin. J. Parasitol. Parasit. Dis. 30 (2), 109–115 (in Chinese).

Luo, B.R., Yao, J.M., Wang, M.S., Li, S.Z., Yang, J., Chen, S.R., Luo, J.J., Liu, Y.H., Duan, L.P., 2018. Molluscicidal effect and mechanism of action of niclosamidate against *Oncomelania hupensis*. Chin. J. Schisto. Control 30 (2), 202–204 (in Chinese).

Luo, B.R., Wang, M.S., Yao, J.M., Li, S.Z., Yang, H., Yang, J., Chen, S.R., Liu, Y.H., Duan, L.P., 2019. Molluscicidal activity of 25% wettable powder of pyriclobenzuron sulphate against *Oncomelania hupensis robertsoni*. Chin. J. Schisto. Control 31 (2), 115–120 (in Chinese).

Lv, S., Xu, J., Cao, C.L., Li, S.Z., Zhou, X.N., 2019. China fighting against schistosomiasis for 70 years: progress and experience. Chin. J. Schisto. Control 37 (5), 514–519 (in Chinese).

Mao, S.B., 1960. Achievements in schistosomiasis control after the founding of the People's Republic of China. Chin. Sci. Bull. 5 (12), 368.

Mao, S.P., Shao, B.R., 1982. Schistosomiasis control in the People's Republic of China. Am. J. Trop. Med. Hyg. 31 (1), 92–99.

Mao, S.B., Li, Y.H., Shao, B.R., Zhen, 1956. Preliminary summary of intradermal reaction test of schistosomiasis. Shanghai J. Trad. Chin. Med. (2), 10.

Pautz, H., 2011. Revisiting the think-tank phenomenon. Public Policy Adm. 26 (4), 419–435.

Qi, Y.X., Liu, S.X., 1996. Immunization test of *Schistosoma japonicum* triose phosphate isomerase. Chin. J. Parasitol. Parasit. Dis. 14 (4), 7–11 (in Chinese).

Qin, Z.Q., Xu, J., Feng, T., Zhu, H.Q., Li, S.Z., Xiao, N., Zhou, X.N., 2013. Construction and thinking of the national schistosomiasis diagnosis network platform. Chin. J. Schisto. Control 25 (4), 329–332 (in Chinese).

Qin, Z.Q., Xu, J., Feng, T., Lv, S., Qian, Y.J., Zhang, L.J., Li, Y.L., Lv, C., Bergquist, R., Li, S.Z., Zhou, X.N., 2018. Field evaluation of a loop-mediated isothermal amplification (LAMP) platform for the detection of Schistosoma japonicum infection in Oncomelania hupensis snails. Trop. Med. Infect. Dis. 3 (4), 124.

Qin, T-J., et al., 2002. Study on new drug of nicothiocyanide for schistosomiasis in cultivated cattle[M].

Qiu, L.S., 1975. Study on the egg-precipitation test of schistosomiasis in Japan. Foreign Med. Ref. Mater. (Parasit. Dis.) (1), 25–26.

Qiu, L.Z., Xue, H.C., 1990. Experimental diagnosis. In: Mao, S.P. (Ed.), Schistosome Biology and Control of Schistosomiasis. Publishing House for People's Health, Beijing, pp. 448–527.

Qiu, L.S., Feng, Z., Zhang, Y.H., Li, H., 2000. Circulating antigen detection in schistosomiasis japonica patients by sandwich ELISA using polyclonal and monoclonal antibodies. Chin. J. Parasitol. Parasit. Dis. 18 (2), 92–93 (in Chinese).

Sleigh, A., Jackson, S., Li, X., Huang, K., 1998. Eradication of schistosomiasis in Guangxi, China. Part 2: political economy, management strategy and costs, 1953–92. Bull. World Health Organ. 76 (5), 497–508.

Song, G.M., Wang, G.F., Becker, W., 1992. Oxidative phosphorylation of liver mitochondria in *Oncomelania* snail. Chin. J. Parasitol. Parasit. Dis. 10 (2), 113–116 (in Chinese).

Souza, C.P., 1995. Molluscicide control of snail vectors of schistosomiasis. Mem. Inst. Oswaldo Cruz 90 (2), 165–168.

Tallo, V.L., Carabin, H., Alday, P.P., Balolong, E.J., Olveda, R.M., McGarvey, S.T., 2008. Is mass treatment the appropriate schistosomiasis elimination strategy? Bull. World Health Organ. 86 (10), 765–771.

Tang, L.H., 2005. To serve the people's health and economic development—the 55th anniversary of the National Institute of Parasitic Diseases. Chin. J. Parasitol. Parasit. Dis. 23 (1), 321–322 (in Chinese).

Tao, Y., Xue, J., Jiang, B., Zhang, H.B., Xiao, S.H., 2015. Significance of higher drug concentration in erythrocytes of mice infected with *Schistosoma japonicum* and treated orally with mefloquine at single doses. Parasitol. Res. 114 (12), 4521–4530.

Taylor, M., 2008. Global trends in schistosomiasis control. Bull. World Health Organ. 86 (10), 738.

Tian, Z.Y., Xiao, S.H., Xiao, J.W., Zhou, Y.C., Liu, D.S., Zheng, J., Chen, M.G., Qu, G.S., Zhang, X.Y., Yao, X.M., Zhang, X.Z., Zhang, D.L., Wang, G.X., 1997. Reduction of *Schistosoma japonicum* infection in an endemic area in islet with embankment after prophylaxis with oral artemether throughout the transmission season. Chin. J. Parasitol. Parasit. Dis. 15 (4), 208–211.

Tong, Q.B., Liu, S.X., Li, X.H., Xu, Y.X., Shen, Y.J., Cao, J.P., 2007. Cloning and expression of the gene encoding the myosin-like protein of *Schistosoma japonicum* and its immunogenicity. Chin. J. Schisto. Control 19 (4), 247–251 (in Chinese).

Utzinger, J., Zhou, X.N., Chen, M.G., Bergquist, R., 2005. Conquering schistosomiasis in China: the long march. Acta Trop. 96 (2–3), 69–96.

van Dam, G.J., Xu, J., Bergquist, R., Claudia, J., Utzinger, J., Qin, Z.Q., Guan, W., Feng, T., Yu, X.L., Zhou, J., Zheng, M., Zhou, X.N., Paul, L.A.M., 2015. An ultra-sensitive assay targeting the circulating anodic antigen for the diagnosis of Schistosoma japonicum in a low-endemic area, People's Republic of China. Acta Trop. 141, 190–197.

Viana, M., Faust, C.L., Haydon, D.T., Webster, J.P., Lamberton, P.H.L., 2017. The effects of subcurative praziquantel treatment on life-history traits and trade-offs in drug-resistant *Schistosoma mansoni*. Evol. Appl. 11 (4), 488–500.

Wang, L.D., 2006. Commemorating the 100th anniversary of the discovery of schistosomiasis in China. Chin. Med. Eng. 14 (1), 126.

Wang, G.Z., 2011. The practice and enlightment of schistosomiasis control by Chinese Communist Party through integrating human power in 1950s. Res. Teach. Hist. Communist Party 221, 8.

Wang, G.F., Song, G.M., 1990. Preliminary studies on transaminase of *Oncomelania* snail. Chin. J. Parasitol. Parasit. Dis. 8 (2), 110–112 (in Chinese).

Wang, R.B., Zheng, J., 2003. Three Gorges Dam project and the transmission of schistosomiasis in China. Chin. J. Schisto. Control 15 (1), 71–74 (in Chinese).

Wang, H.Z., Jia, Y.D., Guo, J.P., Xia, C., 1989. A historic review or 40 years' control on schistosomiasis in P.R. China. Chin. J. Schisto. Control 1, 4. (in Chinese).

Wang, J.L., Xiao, S.H., Yang, Z., Wang, M.K., Yang, H., Liu, Y.H., Zhou, G.S., Zheng, J., Chen, M.G., 1997. Effect of oral artemether in controlling schistosomiasis in Yunnan mountainous endemic area. Chin. J. Parasitol. Parasit. Dis. 15 (3), 138–143.

Wang, X.H., Wu, X.H., Zhou, X.N., 2006. Bayesian estimation of community prevalences of *Schistosoma japonicum* infection in China. Int. J. Parasitol. 36 (8), 895–902.

Wang, X.H., Zhou, X.N., Vounatsou, P., Chen, Z., Utzinger, J., Yang, K., Steinmann, P., Wu, X.H., 2008. Bayesian spatio-temporal modeling of Schistosoma japonicum prevalence data in the absence of a diagnostic 'gold' standard. PLoS Negl. Trop. Dis. 2 (6), e250.

Wang, J., Cao, C.L., Guo, J.G., Zhao, G.M., 2009a. Surveys on implementation of National Schistosomiasis Control and Prevention Regulation in main schistosome-endemic areas of China. Chin. J. Schisto. Control 21 (6), 522–524 (in Chinese).

Wang, L.D., Chen, H.G., Guo, J.G., Zeng, X.J., Hong, X.L., Xiong, J.J., Wu, X.H., Wang, X.H., Wang, L.Y., Xia, G., Hao, Y., Chin, D.P., Zhou, X.N., 2009b. A strategy to control transmission of *Schistosoma japonicum* in China. New Engl. J. Med. 360, 121–128.

Wang, T.P., Zhang, S.Q., He, J.C., Zhang, G.H., Tao, C.G., 2009c. A brief explanation of comprehensive schistosomiasis control strategy based on infection source control (Continued). Trop. Dis. Parasitol. 7 (4), 244–247.

Wang, W.S., Qin, Z.Q., Zhu, D., Wei, Y.F., Li, S.Z., Duan, L.P., 2015. Synthesis, bioactivity evaluation, and toxicity assessment of novel salicylanilide ester derivatives as cercaricides against *Schistosoma japonicum* and molluscicides against *Oncomelania hupensis*. Antimicrob. Agents Chemother. 60, 323–331.

William, S., Botros, S., Ismail, M., Farghally, A., Day, T.A., Bennett, J.L., 2001. Praziquantel-induced tegumental damage in vitro is diminished in schistosomes derived from praziquantel-resistant infections. Parasitology 122 (Pt. 1), 63–66.

Wu, G., 2002. A historical perspective on the immunodiagnosis of schistosomiasis in China. Acta Trop. 82 (2), 193–198.

Wu, G.L., 2005. Review and prospect of the development of immunological diagnosis of schistosomiasis in China. Chin. J. Parasitol. Parasit. Dis. 23 (S1), 323–328 (in Chinese).

Wu, X.H., Xu, X.J., Xiao, B.Z., Wang, R.B., Dai, Y.H., Xu, J., Wu, C.G., Wei, F.H., Zhou, X.N., Zheng, J., 2007. Study on the risk factors of schistosomiasis transmission after the construction of the Three Gorges Dam II. The influence of changes in social and economic factors on the transmission of schistosomiasis in the reservoir area. Chin. J. Schisto. Control 19 (3), 183–187 (in Chinese).

Xianyi, C., Liying, W., Jiming, C., Xiaonong, Z., Jiang, Z., Jiagang, G., Xiaohua, W., Engels, D., Minggang, C., 2005. Schistosomiasis control in China: the impact of a 10-year World Bank Loan Project (1992–2001). Bull. World Health Organ. 83 (1), 43–48.

Xiao, S.H., Catto, B.A., 1989. In vitro and in vivo studies of the effect of artemether on *Schistosoma mansoni*. Antimicrob. Agents Chemother. 33 (9), 1557–1562.

Xiao, S.H., Shen, B.G., 1994. Scanning electron microscope observation on tegumental alteration of Schistosoma japonicum induced by levo- and dextro-praziquantel. Chin. J. Parasitol. Parasit. Dis. 12 (1), 46–50.

Xiao, S.H., Shao, B.R., Xu, Y.J., Pan, Q.R., 1980. Action model of praziquantel to *Schistosoma japonicum*. Acta Pharmacol. Sin. 1 (1), 51–55 (in Chinese).

Xiao, S.H., Friedman, P.A., Catto, B.A., Webster, B.L., 1984. Praziquantel-induced vesicle formation in the tegument of male *Schistosoma mansoni* is calcium dependent. J. Parasitol. 70 (1), 177–179.

Xiao, S.H., Catto, B.A., Webster, L.T.J., 1985a. Effects of praziquantel on different developmental stages of *Schistosoma mansoni* in vitro and in vivo. J Infect Dis 151 (6), 1130–1137.

Xiao, S.H., Yue, W.J., Mei, J.Y., 1985b. Analysis of the prophylactic effect of praziquantel on mice infected with *Schistosoma japonicum* cercariae. Acta Pharm. Sin. 20 (9), 641–646.

Xiao, S.H., Le, W.J., Mei, Y.Y., 1987. Effect of artemether on glyceraldehyde-3-phosphate dehydrogenase, phosphoglycerate kinase, and pyruvate kinase of *Schistosoma japonicum* harbored in mice. Acta Pharmacol. Sin. 8, 261–266.

Xiao, S.H., Yang, Y.Q., Shen, B.G., Xu, D.H., Yang, H.Z., Mei, J.Y., Yue, W.J., 1988. Effect of praziquantel on ultrastructure of Schistosoma japonicum cercariae and schistosomula within mouse skin. Acta Pharmacol. Sin. 9 (4), 360–363 (in Chinese).

Xiao, S.H., Fu, S., Ba, C., 1991a. Recent laboratory investigations by Chinese workers on antischistosomal activities of praziquantel. Chin. Med. J. (Engl.) 104 (7), 599–606.

Xiao, S.H., You, J.Q., Guo, H.F., Catto, B.A., 1991b. Uptake and effect of praziquantel and the major human oxidative metabolite, 4-hydroxypraziquantel, by *Schistosoma japonicum*. J. Parasitol. 77 (2), 241–245.

Xiao, S.H., You, J.Q., Guo, H.F., 1992. Plasma pharmacokinetics and therapeutic efficacy of praziquantel and 4-hydroxypraziquantel in *Schistosoma japonicum*-infected rabbits after oral, rectal, and intramuscular administration. Am. J. Trop. Med. Hyg. 46 (5), 582–588.

Xiao, S.H., You, J.Q., Yang, Y.Q., Wang, C.Z., 1995. Experimental studies on early treatment of schistosomal infection with artemether. Southeast Asian J. Trop. Med. Public Health 26 (2), 306–318.

Xiao, S.H., Wang, J.L., Wang, C.Z., Yang, Z., Chu, B., Yang, H., Liu, Y.H., Zheng, J., Chen, M.G., 1996. Protection of the residents from schistosome infections using oral artemether in mountainous endemic area. Chin. J. Parasitol. Parasit. Dis. 14 (1), 11–14.

Xiao, S.H., You, J.Q., Mei, J.Y., Guo, H.F., Jiao, P.Y., Sun, H.L., Yao, M.Y., Feng, Z., 1997. Effect of artemether on glucose uptake and glycogen content in Schistosoma japonicum. Acta Pharmacol. Sin. 18 (4), 363–367.

Xiao, S., Tanner, M., N'Goran, E.K., Utzinger, J., Chollet, J., Bergquist, R., Chen, M., Zheng, J., 2002a. Recent investigations of artemether, a novel agent for the prevention of schistosomiasis japonica, mansoni and haematobia. Acta Trop. 82 (2), 175–181.

Xiao, S.H., You, J.Q., Gao, H.F., Mei, J.Y., Jiao, P.Y., Chollet, J., Tanner, M., Utzinger, J., 2002b. *Schistosoma japonicum*: effect of artemether on glutathione S-transferase and superoxide dismutase. Exp. Parasitol. 102 (1), 38–45.

Xiao, S.H., Wu, Y.L., Tanner, M., Wu, W.M., Utzinger, J., Mei, J.Y., Scorneaux, B., Chollet, J., Zhai, Z., 2003. *Schistosoma japonicum*: In vitro effects of artemether combined with haemin depend on cultivation media and appraisal of artemether products appearing in the media. Parasitol. Res. 89 (6), 459–466.

Xiao, S.H., Mei, J.Y., Jiao, P.Y., 2011. *Schistosoma japonicum*-infected hamsters (Mesocricetus auratus) used as a model in experimental chemotherapy with praziquantel, artemether, and OZ compounds. Parasitol. Res. 108 (2), 431–437.

Xiao, S.H., Sun, J., Chen, M.G., 2018. Pharmacological and immunological effects of praziquantel against Schistosoma japonicum: a scoping review of experimental studies. Infect. Dis. Poverty 7 (1), 9.

Xu, J., Feng, T., Guo, J.G., Zheng, H., Wang, Q., Wu, X.H., Zhou, X.N., 2005. Comprehensive evaluation of several immunosuppressive reagents for schistosomiasis in China. Chin. J. Schisto. Control 17 (2), 116–119 (in Chinese).

Xu, J., Yan, Z., Zhang, R.J., Feng, T., Wang, Q., Qian, C.Z., Wu, X.H., Zhu, D., Guo, J.G., Zhou, X.N., 2006. Study on detection of serum antibodies against schistosomiasis by colloidal gold immunofiltration assay. Chin. J. Parasitol. Parasit. Dis. 24 (2), 146–147 155. (in Chinese).

Xu, J., Chen, N.G., Feng, T., Wang, E.M., Wu, X.H., Chen, H.G., Wang, T.P., Zhou, X.N., Zheng, J., 2007. Evaluation of the effect of on-site investigation of common diagnostic methods for schistosomiasis in Japan. Chin. J. Parasitol. Parasit. Dis. 25 (3), 175–179 (in Chinese).

Xu, J., Feng, T., Lin, D.D., Wang, Q.Z., Tang, L., Wu, X.H., Guo, J.G., Peeling, R.W., Zhou, X.N., 2011a. Performance of a dipstick dye immunoassay for rapid screening of *Schistosoma japonicum* infection in areas of low endemicity. Parasit. Vectors 4, 87.

Xu, J., Peeling, R.W., Chen, J.X., Wu, X.H., Wu, Z.D., Wang, S.P., Feng, T., Chen, S.H., Li, H., Guo, J.G., Zhou, X.N., 2011b. Evaluation of immunoassays for the diagnosis of Schistosoma japonicum infection using archived sera. PLoS Negl. Trop. Dis. 5 (1), e949.

Xu, J., Li, S.Z., Huang, Y.X., Cao, Z.G., Zhou, X.N., 2012. Risk Evaluation of Schistosomiasis japonica in Potential Endemic Areas in China. Chin. J. Parasitol. Parasit. Dis. 30 (6), 428–433 (in Chinese).

Xu, J.F., Xu, J., Li, S.Z., Jia, T.W., Huang, X.B., Zhang, H.M., Chen, M., Yang, G.J., Gao, S.J., Wang, Q.Y., Zhou, X.N., 2013. Transmission risks of schistosomiasis japonica: extraction from back-propagation artificial neural network and logistic regression model. PLoS Negl. Trop. Dis. 7 (3), e2123.

Xue, H.F., Yan, L.Z., Zhu, Z.X., Zhang, Y.H., 1990. An improved Dot-ELISA method. Chin. J. Schisto. Control 2, 35–38 (in Chinese).

Xue, J., Jiang, B., Liu, C.S., Sun, J., Xiao, S.H., 2013. Comparative observation on inhibition of hemozoin formation and their in vitro and in vivo anti-schistosome activity displayed by 7 antimalarial drugs. Chin. J. Parasitol. Parasit. Dis. 31 (3), 161–169 (in Chinese).

Xue, J., Wang, X., Dong, Y., Vennerstrom, J.L., Xiao, S.H., 2014. Effect of ozonide OZ418 against *Schistosoma japonicum* harbored in mice. Parasitol. Res. 113 (9), 3259–3266.

Yan, Z.Z., 1976. Different immunoprecipitation techniques for the immunodiagnosis of human schistosomiasis. Foreign Med. Ref. Mater. (Parasit. Dis.) (3), 126.

Yan, Z.Z., 1979. Enzyme-linked immunosorbent assay for detection of antibodies to human Schistosoma mansoni infection. Foreign Med. (Parasit. Dis.) (4), 170–171.

Yan, Z.Z., Deelder, P.R.A., 1988. Monoclonal antibody ELISA for detection of schistosomiasis circulating antigen. Chin. J. Parasitol. Parasit. Dis. 6 (S1), 67 (in Chinese).

Yan, L.Z., Xue, H.Q., Zhang, Y.H., Zhu, Z.X., 1985. Preliminary evaluation of latex antigen for diagnosis of schistosomiasis in Japan. Chin. J. Parasitol. Parasit. Dis. 3 (2), 82(in Chinese).

Yan, L.M., Xue, H.Q., Zhang, Y.H., Zhu, Z.X., Tao, Y.X., Lu, S.K., Guo, J., Tsang, V.C., Hancook, K., 1986. Experimental observation on the detection of schistosomiasis antibodies by latex agglutination test. Chin. J. Parasitol. Parasit. Dis. 4 (3), 19–22 (in Chinese).

Yan, L.Z., Xue, H.Q., Zhang, R., Zhu, Z.X., Zhang, Y.H., Tao, Y.X., Lu, S.K., Guo, J., 1987. Further study on the diagnosis of schistosomiasis by latex agglutination test. Shanghai J. Immunol. 7 (3), 150–152.

Yan, L.Z., Xue, H.Q., Zhang, Y.H., Zhu, Z.X., 1988. Application of enzyme-linked immunoelectron transfer blotting technology–antibody response to adult antigens in patients with schistosomiasis. Shanghai J. Immunol. 8 (2), 118–119,126.

Yan, Z., Lu, Z.Y., Wang, W., Wu, W., 1990a. Detection of circulating antigen in schistosomiasis by monoclonal antibody Dot-ELISA II. Chin. J. Schisto. Control 2, 39–42 (in Chinese).

Yan, Z., Lu, Z.Y., Wang, W., Wu, W., Xu, F.N., Qin, Z.Y., Xiao, X., 1990b. Detection of schistosomiasis egg antigen by monoclonal antibody spot test. Shanghai J. Immunol. 10 (2), 113–116.

Yan, Z., Wang, W., Lu, Z.Y., Wu, W., 1990c. Monoclonal Antibody Dot-ELISA for Detection of Circulating Antigens in Schistosomiasis I. Chin. J. Parasitol. Parasit. Dis. 8 (3), 3–6 (in Chinese).

Yan, Z., Lu, Z.Y., Wang, W., Wu, W., Zhu, R.S., Hu, W.F., 1992a. Field test of schistosomiasis circulating antigen detection II. Chin. J. Schisto. Control 4, 335–337 (in Chinese).

Yan, Z., Wang, W., Lu, Z.Y., Tang, W.H., Wu, W., Hu, W.F., Wang, Z.H., 1992b. Field test of schistosomiasis circulating antigen detection I. Chin. J. Schisto. Control 4, 275–277 (in Chinese).

Yan, Z., Wang, W., Lu, Z.Y., Wu, W., 1996. Rapid determination of schistosomiasis antibodies. Chin. J. Parasitol. Parasit. Dis. 14 (4), 52 (in Chinese).

Yan, Y.T., Li, X.H., Liu, S.X., Song, G.C., Xu, Y.X., 2005. Characteristics of immune response in mice with schistosomiasis glyceraldehyde triphosphate dehydrogenase nucleic acid vaccine. Chin. J. Parasitol. Parasit. Dis. 23 (5), 266–269 273. (in Chinese).

Yan, C., Feng, Z., Hu, W., 2006. Research progress in immunodiagnosis of schistosomiasis in Japan. Int. J. Med. Parasit. Dis. (5), 250–255.

Yang, G.J., Sun, L.P., Wu, F., Huang, Y.X., Yang, K., Zhou, X.N., 2010. Meta analysis of molluscicidal effect of 50% wettable powder of niclosamide ethanolamine salt in field. Chin. J. Schisto. Control 22 (6), 579–582 (in Chinese).

Yin, J.Y., Xie, F.X., Guo, Y.H., Zhang, Y., Qian, K., Zhu, D.P., Gu, J.R., Hong, G.B., Chen, P.L., Tan, H.Q., 1986. A new device for snail survey (S-1 Snail Survey Machine). Chin. J. Parasitol. Parasit. Dis. 4 (2), 90–92 (in Chinese).

Yin, J.Y., Zhu, D.P., Bao, Z.P., Cai, L.Q., Gao, S.H., 1994. Observation of the toxicity of bromoacetamide on clam. Chin. J. Parasitol. Parasit. Dis. 12 (3), 236 (in Chinese).

Yin, M., Tao, Y., Zhu, D., Zhang, H.B., 2014. GC/MS-based dynamic analysis of *Eucalyptus camaldulensis* extract on correlation of chromatography-activity against *Oncomelania hupensis*. Chin. J. Schisto. Control 26 (2), 141 (in Chinese).

Yu, S.H., Wang, W., Xu, C., Chen, M.G., Song, G.C., Yan, Z., Zhang, Y.Q., Liu, S.Q., Wang, X.F., 1979. Research on the experimental diagnosis of Schistosoma mansoni. J. Chin. Acad. Med. Sci. (2), 131–135.

Yue, W.J., Xiao, S.H., Mei, J.Y., 1989. Relationship between the effects of praziquantel on mice and rabbits with different intensities of Schistosoma japonicum infection and humoral immunity levels of the host. Acta Pharmacol. Sin. 10 (5), 476–479.

Zhai, Z.L., Jiao, P.Y., Mei, J.Y., Xiao, S.H., 2002. Glutathione inhibits the antischistosomal activity of artemether. Chin. J. Parasitol. Parasit. Dis. 20 (4), 212–215 (in Chinese).

Zhang, K.Y., 2009. Evolution of schistosomiasis control strategies in China. Southeast Natl. Def. Med. 11 (4), 340–342.

Zhang, M.Q., 2017. Reflection on the cultivation of parasitology control professionals. Chin. J. Public Health Mang. 35 (3), 313–316 (in Chinese).

Zhang, Y., Gao, Y.H., 1992. Study on the effect of bromoacetamide upon the development of snail eggs. Chin. J. Parasitol. Parasit. Dis. 10 (4), 258–261 (in Chinese).

Zhang, J.F., Xu, J., Bergquist, R., Yu, L.L., Yan, X.L., Zhu, H.Q., Wen, L.Y., 2016a. Development and application of diagnostics in the national schistosomiasis control programme in the People's Republic of China. Adv. Parasitol. 92, 409–434.

Zhang, L.J., Li, S.Z., Wen, L.Y., Lin, D.D., Abe, E.M., Zhu, R., Du, Y., Lv, S., Xu, J., Webster, B.L., Rollinson, D., Xn, Z., 2016b. The establishment and function of schistosomiasis surveillance system towards elimination in the People's Republic of China. Adv. Parasitol. 92, 117–141.

Zhen, L., Bin, Z., 2017. Analysis of academic impact of publications from National Institute of Parasitic Diseases, Chinese Center for Disease Control and Prevention based on SCIE database in recent five years. Zhongguo Xue Xi Chong Bing Fang Zhi Za Zhi 29 (2), 235–240 (in Chinese).

Zheng, J., 1987. Establishment of schistosomiasis Expert Advisory Committee of the Ministry of Health. Chin. J. Parasitol. Parasit. Dis. 5 (3), 226 (in Chinese).

Zheng, J., 2000. Relationship between schistosomiasis distribution and elevation in the mountainous plateau of Yunnan Province. Strait J. Prev. Med. 6 (3), 17–18.

Zheng, J., Gu, X.G., Qiu, Z.L., Li, X.J., Chen, J.X., 1996. Study on the control strategy of schistosomiasis in plateau mountainous area. Chin. J. Schisto. Control 8 (2), 65–71 (in Chinese).

Zheng, J., Guo, J.G., Wang, X.F., Zhu, H.Q., 2000. Relationship of the livestock trade to schistosomiasis transmission in mountainous area. Chin. J. Parasitol. Parasit. Dis. 18 (3), 146–148 (in Chinese).

Zheng, J., Gu, X.G., Qiu, Z.L., Li, X.J., Chen, J.X., Xu, F.S., Yang, S.D., Wen, S., Yin, G.L., Liang, S., Yang, W.S., Zhao, W.X., Zuo, X.F., Zhao, L.G., 2001. Control and interruption of schistosomiasis transmission in mountain area. Med. Res. Commun. 30 (5), 26–27.

Zheng, J., Gu, X.G., Xu, Y.L., Ge, J.H., Yang, X.X., He, C.H., Tang, C., Cai, K.P., Jiang, Q.W., Liang, Y.S., Wang, T.P., Xu, X.J., Zhong, J.H., Yuan, H.C., Zhou, X.N., 2002. Relationship between the transmission of schistosomiasis japonica and the construction of the Three Gorge Reservoir. Acta Trop. 82 (2), 147–156.

Zheng, J., Gu, X.G., Xu, C.L., Wang, T.P., Xu, X.J., He, C.H., Tang, C., Cai, K.P., Jiang, Q.W., Liang, Y.S., Ge, J.H., Yang, X.X., Zheng, Q.S., Yuan, H.C., Han, J.J., Liang, S., Wen, S., Zuo, J.Z., Huang, Y.X., Hu, W.Z., Wang, Y.Z., 2003. Study on the relationship between ecological environment change and schistosomiasis transmission in Three Gorges Dam. Bull. Med. Res. 32 (5), 7–10.

Zhou, X.N., 2005. Science on Oncomelania Snail. Science Press, Beijing in Chinese.

Zhou, X.N., Wang, T.P., Wang, L.Y., Guo, J.G., Yu, Q., Xu, J., Wang, R.B., Chen, Z., Jia, T.W., 2004. The current status of schistosomiasis epidemics in P.R. China. Chin. J. Epidemol. 25, 555–558 (in Chinese).

Zhou, X.N., Chen, J.X., Chen, M.G., Bergquist, R., 2005a. The National Institute of Parasitic Diseases, Chinese Center for Disease Control and Prevention: a new administrative structure for schistosomiasis control. Acta Trop. 96 (2–3), 296–302.

Zhou, X.N., Jiang, Q.W., Sun, L.P., Wang, T.P., Hong, Q.B., Zhao, G.M., Wen, L.Y., Yin, Z.C., Wu, X.H., Lin, D.D., 2005b. Schistosomiasis control and surveillance in China. Chin. J. Schisto. Control 17 (3), 161–165 (in Chinese).

Zhou, X.N., Wang, L.Y., Chen, M.G., Wang, T.P., Guo, J.G., Wu, X.H., Jiang, Q.W., Zheng, J., Chen, X.Y., 2005c. An economic evaluation of the national schistosomiasis control programme in China from 1992 to 2000. Acta Trop. 96 (2–3), 255–265.

Zhou, X.N., Wang, L.Y., Chen, M.G., Wu, X.H., Jiang, Q.W., Chen, X.Y., Zheng, J., Utzinger, J., 2005d. The public health significance and control of schistosomiasis in China—then and now. Acta Trop. 96 (2), 97–105.

Zhou, X.N., Guo, J.G., Wu, X.H., Jiang, Q.W., Zheng, J., Dang, H., Wang, X.H., Xu, J., Zhu, H.Q., Wu, G.L., Li, Y.S., Xu, X.J., Chen, H.G., Wang, T.P., Zhu, Y.C., Qiu, D.C., Dong, X.Q., Zhao, G.M., Zhang, S.J., Zhao, N.Q., Xia, G., Wang, L.Y., Zhang, S.Q., Lin, D.D., Chen, M.G., Hao, Y., 2007a. Epidemiology of schistosomiasis in the People's Republic of China, 2004. Emerg. Infect. Dis. 13 (10), 1470–1476.

Zhou, X.N., Cai, L., Zhang, X.P., Sheng, H.F., Dai, Q., 2007b. Potential risks for transmission of schistosomiasis caused by mobile population in Shanghai. Chin. J. Parasitol. Parasit. Dis. 25 (3), 180–184 (in Chinese).

Zhou, X.N., Jiang, Q.W., Wu, X.H., Zhao, G.M., Lin, D.D., Zhang, S.Q., Wang, T.P., Guo, J.G., Xu, J., Wang, S.P., Yin, Z.C., Wang, X.F., Wang, L.Y., 2007c. The function and evolution of the criteria for control and elimination of schistosomiasis in China. Chin. J. Schisto. Control 19, 1–4 (in Chinese).

Zhou, X.N., Yang, G.J., Yang, K., Wang, X.H., Hong, Q.B., Sun, L.P., Malone, J.B., Kristensen, T.K., Bergquist, R., Utzinger, J., 2008. Potential impact of climate change on schistosomiasis transmission in China. Am. J. Trop. Med. Hyg. 78 (2), 188–194.

Zhou, X.N., Jia, T.W., Guo, J.G., Wang, L.Y., 2010. Project management model and evolution of schistosomiasis control in China. Chin. J. Schisto. Control 22 (1), 1–4 (in Chinese).

Zhou, X.N., Xu, J., Chen, H.G., Wang, T.P., Huang, X.B., Lin, D.D., Wang, Q.Z., Tang, L., Guo, J.G., Wu, X.H., Feng, T., Chen, J.X., Guo, J., Chen, S.H., Li, H., Wu, Z.D., Peeling, R.W., 2011. Tools to support policy decisions related to treatment strategies and surveillance of schistosomiasis japonica towards elimination. PLoS Negl. Trop. Dis. 5 (12), e1408.

Zhou, X.N., Jiang, Q.W., Guo, J.G., Lin, D.D., Xu, J., 2012. Road map for transmission interruption of schistosomiasis in China. Chin. J. Schisto. Control 24 (1), 1–4 (in Chinese).

Zhou, X.J., Zheng, B., Yi, F.Y., Xiong, Y.H., Zhang, M.Q., 2015. Analysis on Research Projects Supported by the National Natural Science Foundation of China at the National Institute of Parasitic Diseases during 2003–2013. Zhongguo Ji Sheng Chong Xue Yu Ji Sheng Chong Bing Za Zhi 33 (2), 151–153 (in Chinese).

Zhou, X.N., Li, S.Z., Hong, Q.B., Yang, K., Lv, S., Xu, J., 2018. Remain true to our original aspiration for farewell to the God of Plague, compose the new chapter for the national schistosomiasis control programme scientifically-Commemoration of 60th anniversary of publishing Chairman Mao Zedong's two poems 'Farewell to the God of Plague. Chin. J. Schisto. Control 30 (1), 1–4 (in Chinese).
Zhu, Y.C., 2005. Immunodiagnosis and its role in schistosomiasis control in China: a review. Acta Trop. 96 (2–3), 130–136.
Zhu, D.P., Yin, J.Y., 1986. Enlarged field trials of bromoacetamide against *Oncomelania* snails. Chin. J. Parasitol. Parasit. Dis. 4 (2), 90–92 (in Chinese).
Zhu, D.P., Guo, J.R., Yin, J.Y., Hong, G.B., Chen, F.L., Guo, Y.H., 1984. Studies on a new molluscicide, bromoacetamide. Chin. J. Parasitol. Parasit. Dis. 2 (1), 17–20 (in Chinese).
Zhu, D., Yin, J., Cai, L., Gao, S., Li, Y.L., 1995. Molluscicidal effects of bromoacetamide and impact to the weeds and fish by spraying in marshland of Fang Lake. Chin J Parasitol Parasit Dis 13 (1), 75–76 (in Chinese).
Zhu, D., Yin, J., Bao, Z., Cai, L., 1998. Studies on extensive application of bromoacetamide in various types of schistosomiasis endemic area. Chin. J. Parasitol. Parasit. Dis. 16 (1), 53–57 (in Chinese).
Zhu, D., Xie, F.X., Bao, Z.P., Zhu, D.B., Ni, C.H., 2001. Climbing-upward inhibition of *Oncomelania hupensis* by niclosamide combination. Chin. J. Parasitol. Parasit. Dis. 19 (6), 348–350 (in Chinese).
Zhu, H.Q., Cao, C.L., Gao, F.H., Guo, J.G., Bao, Z.P., Wang, X.H., Dang, H., Zhou, X.N., 2005. The evaluation of modified Kato-Katz in large-scale survey on schistosomiasis. Chin. J. Schisto. Control 17 (4), 273–277 (in Chinese).
Zhu, D., Zhou, X.N., Zhang, S.Q., Zhang, G., Liu, H.X., Lu, D.B., Cai, G.Y., Ni, Q.Z., Cao, Z.G., Wu, W.D., 2006. Study on the molluscicidal effect of META-Li against *Oncomelania hupensis*. Chin. J. Parasitol. Parasit. Dis. 24 (3), 200–203 (in Chinese).
Zhu, D., Li, H.J., Liu, H.X., Zhang, Y., Guo, J., Liang, Y.S., Zhou, X.N., 2007. Enzyme histochemistry: the effect of META-Li on *Oncomelania hupensis*. Chin. J. Parasitol. Parasit. Dis. 25 (3), 198–201 (in Chinese).
Zhu, D., Liu, H.X., Xu, G.C., Gu, R.B., Zhou, X.N., 2010. Study on toxicities of Meta-li against freshwater fish. Chin. J. Schisto. Control 22 (3), 284–285 (in Chinese).

CHAPTER TWO

The contributions and achievements on malaria control and forthcoming elimination in China over the past 70 years by NIPD-CTDR

Xinyu Feng[a,b,c,d,e], **Zhi-Gui Xia**[a,b,c,d,e], **Jun Feng**[a,b,c,d,e], **Li Zhang**[a,b,c,d,e], **He Yan**[a,b,c,d,e], **Linhua Tang**[a,b,c,d,e], **Xiao-Nong Zhou**[a,b,c,d,e], **Shuisen Zhou**[a,b,c,d,e,*]

[a]National Institute of Parasitic Diseases, Chinese Center for Disease Control and Prevention, Shanghai, People's Republic of China
[b]Chinese Center for Tropical Diseases Research, Shanghai, People's Republic of China
[c]WHO Collaborating Centre for Tropical Diseases, Shanghai, People's Republic of China
[d]National Center for International Research on Tropical Diseases, Ministry of Science and Technology, Shanghai, People's Republic of China
[e]Key Laboratory of Parasite and Vector Biology, Ministry of Health, Shanghai, People's Republic of China
*Corresponding author: e-mail address: zhouss@nipd.chinacdc.cn

Contents

1. Introduction 64
2. Epidemic situation of malaria in China 66
   2.1 Key investigation and research stage (1949–59) 66
   2.2 Control of the pandemic transmission stage (1960–79) 69
   2.3 Intensive prevention and control stage (1980–99) 69
   2.4 Consolidation of the achievements of prevention and control (2000–09) 71
   2.5 Approaching the malaria elimination stage (2010–20) 72
3. Monitoring and information management 74
4. Technical support to the malaria control and elimination programmes 76
5. Experimental research in the biology of the malaria pathogen 79
   5.1 Investigation into the *Plasmodium* species 79
   5.2 Research on *Plasmodium falciparum* 80
   5.3 Research on *P. vivax* biology 82
6. Biological investigations on malaria vectors 84
7. Research and development of control tools, including diagnostics, drugs, and insecticides 88
   7.1 Diagnosis 88
   7.2 Serological diagnosis 89
   7.3 Drugs 92
   7.4 Insecticide resistance 97

Advances in Parasitology, Volume 110
ISSN 0065-308X
https://doi.org/10.1016/bs.apar.2020.03.005

8. Future perspectives 98
8.1 Malaria retransmission risk 99
8.2 Maintenance of post-malaria elimination surveillance and response capacity 99
8.3 Long-term mechanism for managing imported malaria cases and mobile population 100
8.4 Malaria diagnosis, drug resistance, and insecticide resistance issues 101
9. Conclusion 101
Acknowledgements 102
Competing interests 102
Authors' contributions 102
References 102

Abstract

Although the past decades have seen a remarkable decrease in malaria-caused mortality and morbidity, the infection remains a significant challenge to global health. In the battle against malaria, China has gained notable feat and achievement since the 1940s through the efforts of several generations. Notably, China has not recorded a single indigenous malaria case since August 2016. The National Institute of Parasitic Diseases of the Chinese Center for Disease Control and Prevention (NIPD), as the only specialized institution for parasitic disease at the national level, has played a significant role in the malaria control, prevention, and elimination in China in the different historical periods. In order to transfer Chinese experiences on malaria control and elimination to other Low and Middle Income Countries (LMICs) and to improve global health collaboration, we have summarized and reviewed the contributions and achievements by the NIPD over the past 70 years, covering the epidemic situation; field investigation and laboratory experimental research on both parasite and vector; research and development on diagnostics, drugs, and insecticides; surveillance and response; technical and international. Support and cooperation. In addition, we also focus in particular on malaria retransmission risk, strategies on management of imported malaria cases and mobile populations, surveillance and response capacity to be maintained in post-elimination stage, challenges on diagnosis, drug resistance, and insecticide resistance as future concerns.

1. Introduction

Malaria is one of most serious vector-borne diseases transmitted by the *Anopheles* mosquito and is one of the major infectious diseases that has historically affected social development in China (Qian and Tang, 2000). Malaria has been prevalent in most parts of China, except for Gansu, Ningxia, Qinghai, Jilin, Inner Mongolia, and some other provinces, in which malaria cases were occasionally reported or were malaria-free areas. In the early days after the founding of the People's Republic of China, malaria was epidemic in nearly 80% of the counties in the country, and 70% of the population was under threat (Tang, 2000). In 1954, according to an incomplete statistical estimate, there were 6.97 million malaria patients nationwide (Taiwan not included), with an annual incidence rate of 12.29 per 1000. Malaria was most serious in the southwestern border areas, such as Hainan Island and Yunnan, and was mainly falciparum malaria. Even in the 1970s, there were still more than 24 million cases. By 2010, after several decades of active prevention and control, only 7855 cases of malaria were reported nationwide, and the incidence rate dropped to 0.06 per one million (Yin et al., 2014). In 2010, in response to the global malaria eradication initiative proposed by the United Nations (UN) high-level meeting on the Millennium Development Goals, 13 ministries and commissions, including the Ministry of Health, jointly issued the China Malaria Elimination Action Plan (2010–20), which was proposed to eliminate malaria in the majority of the country by 2015, with the goal of complete malaria elimination by 2020 (Diouf et al., 2014; Zhou et al., 2013). Most strikingly, in 2017, China reported no indigenous malaria case reports for the first time (Feng et al., 2018).

China's malaria prevention and treatment work has achieved great progress in the past decades, resulting in effective control of the malaria epidemic and ensuring the health of its people along with the promotion of social and economic development (Shou-Pai, 1983). The National Institute of Parasitic Diseases of the Chinese Center for Disease Control and Prevention (NIPD), the national parasitic disease prevention and control work guidance centre, has played a central role in the prevention and treatment of malaria in China. In 2017, the State Commission Office of Public Sectors Reform formally approved a new title of the "Chinese Center for Tropical Diseases Research" and affiliated it with the NIPD, China CDC (NIPD-CTDR).

In the past 70 years, the NIPD-CTDR has played an important role in the demonstration, on-site prevention, organization, and implementation of parasitic disease control programmes and handling of emerging diseases outbreaks. The NIPD-CTDR has established a national malaria diagnostic reference laboratory network, parasitic disease prevention, and research base, as well as support platforms for monitoring and early warning, diagnostic testing, media control, and information services. National malaria surveillance, prevention and control of malaria epidemic within the central region, and prevention and control of non-malaria cases have been carried out under the guidance of the NIPD-CTDR. The research field of NIPD involves global disciplines, applied epidemiology, pathogenic and vector biology, immunology, molecular biology, pharmacy, and other multi-disciplinary research methods and modern technologies, and the institution also analyzes popular epidemic characteristics and transmission of important parasitic diseases and tropical diseases. The NIPD-CTDR has consistently focused on scientific research based on disease control and prevention, including laboratory work and field needs, and has cooperated closely with disease prevention and control institutions at all levels across the country to fight side-by-side, ensuring the completion of prevention and research tasks. The NIPD-CTDR has led the national parasitic disease control and prevention programme and achieved a number of scientific and technological achievements (Table 1). This review discusses the key role of the NIPD-CTDR in the control of parasitic disease epidemics within China, progress of parasite biology, vector research, prevention and control strategies, case management monitoring, and related technical support for the national malaria control and elimination programme.

2. Epidemic situation of malaria in China

Since the founding of the People's Republic of China, malaria has presented with different epidemiological characteristics during different historical periods.

2.1 Key investigation and research stage (1949–59)

Before the founding of the People's Republic of China in 1949, historical documents on malaria were very scarce. Endemic instances of malaria were only reported and recorded in a few provinces, such as Yunnan, Guangxi, Guangdong, and Guizhou. From 1935 to 1938, Lanzhou Feng published "China's Malaria Epidemiology" and other scientific papers, which outlined the regional and seasonal distribution of malaria within China, as well as the types of malaria parasites and vectors, and summarized the main characteristics of the malaria epidemics within China (Yin et al., 2014). From 1947 to 1949, the former Central Health Laboratory conducted a dichloro-diphenyl-tricgloroethane (DDT) indoor residual spraying and mosquito killing experiment in Dongshan Town, Nanjing. From 1951 to 1952, the Central Health Research Institute Huadong Branch worked on *Anopheles sinensis* in Pengxingzhou and Nanjing South Suburb of Dangtu County,

Table 1 Primary research focus of the NIPD during different time periods.

Period (year)	Research area of NIPD		
	Parasite	**Vector**	**Control and elimination strategies**
1949–59	• Plasmodium species investigation • Establishment of animal models of malaria • Development of new antimalarial drugs	• Investigations on malaria vector distribution • Investigations on malaria vector species • Investigation on vector competence	• Investigations on malaria epidemic situation • Classification on epidemic areas • Pilot study on malaria control prevention • Guidelines and comprehensive control strategies
1960–79	• In vitro culture of *Plasmodium falciparum* • Establishment and improvement of anti-malaria drug sensitivity • Multi-drug resistance testing • Indirect fluorescent antibody test (IFAT) • Enzyme-linked immunosorbent assay	• Investigations on malaria vector distribution • Research on *Anopheles sinensis* • Investigations on insecticide resistance	• Radical treatment on vivax malaria • Malaria vector control • Investigation on the degree and geographical distribution of drug resistance to chloroquine
1980–99	• The establishment of a single epitope monoclonal antibody against P. falciparum • Biological research on *Plasmodium vivax*	• Investigations on insecticide resistance • Research on vector competence of *An. sinensis* and other species	• Epidemiological investigation and surveillance strategies • Control strategies on malaria outbreak • Vector control
2000–09	• Identification of Plasmodium species by PCR method • Drug resistance surveillance	• Research on *Anopheles pseudovespa* in Tibet • Susceptibility study on different vector species	• Malaria control strategies • Optimization and adjustment on surveillance and control strategies
2010–20	• Molecular marker for drug resistance • Development of new antimalarial drugs • Drug resistance surveillance and related mechanism research • Development of novel antimalarial drugs	• Surveillance on insecticide resistance • Molecular studies on main malaria vector developmental immunity • Development of novel vector control strategy	• Research on malaria elimination strategy • Development on the 1-3-7 strategy • Malaria surveillance post-elimination • Risk prediction and assessment post-elimination • Pilot study in African countries

Anhui Province, performing some minor antimalarial test interventions, such as residual spraying or larval killing of mosquitoes in wintering places. At that time, 239 people were surveyed in the urban area of Quanjiao County, Anhui Province, and were found to have an infection rate of 41.4% (of which *Plasmodium falciparum* accounted for 74.7%) and a splenomegaly rate of 74.8% (Tang, 2009; Zu-Jie, 1981).

The number of malaria cases was the highest of all infectious diseases at the beginning of the founding of the People's Republic of China. Malaria was prevalent in 1829 counties (cities), accounting for 70%–80% of the counties at that time. In 1954, the Ministry of Health conducted surveys in Yunnan, Guizhou, Guangdong, Guangxi, Hubei, Sichuan, Jiangxi, Henan, and other provinces and found that the number of malaria cases accounted for about 60% of 19 infectious diseases surveyed. Malaria was more prevalent in Yunnan, Guizhou, Guangdong, and Guangxi than in other provinces, especially in the southwestern border areas of Hainan Island and Yunnan. There were local outbreaks in some places in north and central China, such as Yongzhou Hunan, Xinyang Henan, Quanjiao Anhui, and Xingcheng Liaoning. The epidemic pattern of malaria seriously jeopardized the health of the people. Anaemia and hepatosplenomegaly caused by malaria was very common in these areas. South of 25° north latitude, malaria was highly endemic in most areas and only a few areas were moderately endemic. Generally, plains areas were low or moderately endemic with an unstable pattern, while mountainous areas were highly prevalent with a relatively stable pattern. From 25 to 32° north latitude, malaria was very prevalent and mainly caused by vivax malaria with high population density where the epidemic was unstable. There were often low or moderate malaria outbreaks in this region. The malaria situation in the mountainous area was more serious than in the plains area. To the north of 32° latitude, the malaria epidemic was relatively mild. The transmission period was between 3 and 6 months, and the splenomegaly rate was generally below 10%. Malaria was mostly distributed in the low-lying areas along rivers and lakes. The immunity of the residents was very low, which often led to outbreaks. In the northwestern region, annual malaria cases ranged from 9000 to 18,000, and the annual incidence rate was 179–364/100,000. The incidence rate north of Xinjiang was higher than that south of Xinjiang. The epidemic areas were mainly distributed in swamps, low-lying areas, and some rice-growing areas on both sides of the river. Malaria maintained a low endemic profile near the Kashgar, Yarkant, and Tianhe rivers in southern Xinjiang. After 1958, malaria prevention and treatment was also affected by the "the Great Cultural Revolution," which derailed the situation at that time. In addition, after severe natural disasters in 1959, malaria prevention and control work was stopped and stagnated for many areas. In 1960, malaria outbreaks occurred in many areas. The incidence rate in 1960 was 1.8 to 600 times higher than that in 1959 in Hebei, Shandong, Henan, Jiangsu, and Anhui provinces alone, and the number of malaria cases was as high as 9.53 million.

2.2 Control of the pandemic transmission stage (1960–79)

From 1960 to 1979, comprehensive prevention and control strategies were planned and developed based on the investigation, prevention, and control of malaria in China during the 1950s. However, the malaria prevention and control work during this period was disrupted by the impact of natural disasters, the country's economic difficulties, shortage of funds for prevention and treatment, tight supply of antimalarial drugs, and the impact of social unrest, resulting in repeated epidemics and poor control effects (Zu-Jie, 1981). In the early 1960s and 1970s, vivax malaria was prevalent in the Huanghuai Plain and Jianghan Plain within the central region. According to reported data, this period had the most serious malaria epidemic in China after the founding of the People's Republic of China. In the 20 years from 1960 to 1979, more than 180 million malaria cases were reported nationwide, and the highest case number was reported in 1970 at about 24 million cases, with an incidence rate of 296 per million. The highest number of mortalities was 2,049 in 1963. During this time, the number of malaria cases in the country accounted for 12.40%–63.15% of the total number of infectious diseases reported in the epidemic, indicating that malaria was the most serious infectious disease in China. According to statistics of counties (cities, districts), the prevalence in some regions was very high. For example, during 1973–75, 46.92%–63.85% of the counties (cities, districts) in the central region, including Jiangsu, Henan, Anhui, Hubei, and Shandong provinces, had an annual incidence rate of more than 1% (Kai-Jie et al., 2016). Moreover, the incidence in 14.39%–26.61% of the counties (cities, districts) was above 10%. By 1979, many historical endemic areas still had an incidence rate above 1% (Fig. 1A and B).

2.3 Intensive prevention and control stage (1980–99)

Since the implementation of comprehensive prevention and control programmes starting in 1971, the epidemic situation of malaria in the

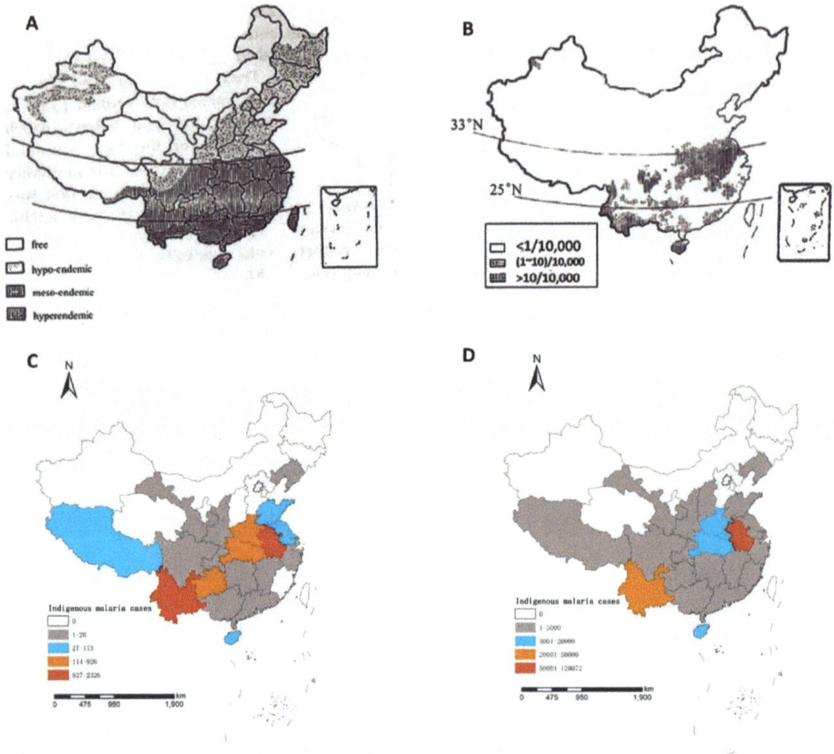

Fig. 1 The incidence and prevalence of malaria in China during different periods. (A) Malaria epidemics during the 1970s. (B) Malaria incidence rates in the 1980s. (C) Malaria epidemics in China from 2002 to 2009. (D) Malaria epidemics in China during 2010–16. Since the beginning of 2017, there have been no local malaria cases reported in China.

country has changed significantly (Huang, 1980). The severe epidemics that raged during the early 1970s are now effectively under control. The incidence of malaria in the country has dropped to its lowest level (257.54/100,000), except for the rate in 1958 and 1959. Between 1980 and 1999, the malaria epidemic in the country was in a stable downward trend, and some areas also experienced spot-like or patchy outbreaks (Goodman, 1986). Except for the rebound in 1989 and 1994 (Sleigh et al., 1998), the efforts of continuous implementation of various prevention and control measures has resulted in the gradual decline in the incidence of malaria in the country every year, as compared to the previous year. After 1980, the prevention and control work was strengthened throughout the country. Through continuous prevention and control, some counties (cities) gradually eliminated malaria, and the incidence of malaria still continued to decline every year in other counties (cities). Due to the strengthening of the prevention and treatment of falciparum malaria, the number of endemic provinces (districts) in China have been decreasing every year. After 1990, falciparum malaria was prevalent in only Hainan and Yunnan provinces, and the morbidity and mortality rates reported in those regions were significantly reduced (Xiao et al., 2012). In 2010, there were only 17 counties in Yunnan Province with falciparum malaria. In 1980, China was at the height of a 10-year cycle of malaria epidemics. According to the epidemic report, the annual national incidence rate was 337.87/100,000, which was 31.17% higher than the previous year.

From 1980 to 1999, the prevalence of malaria in the country was gradually reduced, and the counties (cities, districts) with high incidence rates have continued to decrease every year, while those with low or previously malaria-free areas have slowly reported increased malaria cases. However, the public health system was established under the planned economic system during this stage. The grassroots health organizations have been greatly affected by the formation of the market economic system. The under-reporting rate for malaria might have been high according to the monitoring and special surveys at the province level. As such, the actual number of malaria cases in the country should have been greater, and the actual prevalence might be higher than the situation reflected in the epidemic report.

2.4 Consolidation of the achievements of prevention and control (2000–09)

After 50 years of prevention and control, the malaria epidemic was controlled in most parts of the country, and the malaria cases were mainly concentrated in areas where episodes of malaria were typically more common. Since 2000, although the incidence of morbidity has increased in some areas, the overall epidemic has been stable and maintained at a low level. According to the epidemic report, the national incidence was in a dynamic trend from 2000 to 2009 (Fig. 1) (Expert Advisory Committee on Malaria, Ministry of Health, 2001). The epidemic situation was mainly caused by the malaria outbreak in the Huaibei District of Anhui Province. The prevalence of malaria was still relatively stable in other parts of the country. From 2000 to 2009, the prevalence of malaria in most areas of

China remained stable at an annual incidence rate of about 1/100,000 (Feng et al., 2015a). However, due to the rebound or outbreak of malaria in the Huaibei area of Anhui Province, the annual incidence rate was over 10/10,000. The number of high incidence rates in counties (cities, districts) increased (Fig. 1C). At this stage, the under-reporting rate of malaria in primary health care institutions was still serious, especially in high-endemic and unstable transmission areas. In 2002, the Global Fund-China Malaria Project verified the actual malaria incidence in 10 project provinces (Wang et al., 2014a). The results showed that the actual number of malaria cases in each province in 2002 was 3.95–100.09 times the number in epidemic reports, and the actual number of cases in 10 provinces was 20.55 times higher than the number reported in epidemic reports. In 2003, the Malaria Expert Advisory Committee of the Ministry of Health conducted a serious discussion and analysis on the actual incidence of malaria in the country. According to the survey results, the experts estimated that the actual number of malaria cases in the country in 2002 was about 740,000, which was 20.96 times the number of epidemic reports (35,300). The under-reporting of cases was mainly concentrated in areas where social and economic development was relatively lagging, grassroots health organizations were weak, and malaria was more common as an infectious disease. By contrast, the incidence of epidemic situation in most regions with prevalence of counties (cities, districts) below 1/10,000 was comparable to the actual incidence in these regions and areas, basically consistent with the situation reflected in the epidemic report. However, the actual prevalence of the county (city, district) with an incidence rate above 10/10,000 was much higher than that in epidemic report.

2.5 Approaching the malaria elimination stage (2010–20)

In order to construct more efficient malaria elimination strategies in China, 2858 counties in China were divided into four categories according to the local malaria epidemic situation in 2006–08, with 75 counties in tier 1, 678 counties in tier 2, 1432 in tier 3, and 664 in tier 4. These counties were divided as such to help explore and accelerate the goal of eliminating malaria, cope with the challenge of imported cases, and promote work in all areas by drawing upon the experience gained on key points (Sheng et al., 2016). In 2010, the national malaria elimination programme (NMEP) decided to choose malaria endemic counties and selected a tier of counties (districts) as pilot counties and initiated pilot work to eliminate malaria in six provinces (cities), including Shanghai, Zhejiang, Fujian, Shandong, Hebei, and Guangdong. Since the launch of the action plan, the national malaria epidemic has declined overall. The number of cases and prevalence of indigenous infections have continued to decrease (Hu et al., 2016). In 2016, only three indigenous cases were reported in Yingjiang County of Yunnan Province and Chayu County of Tibet District (Fig. 1D), which was a 99.9% decrease from 2011 (4262 cases) (Li et al., 2016). Furthermore, the distribution range also decreased from 155 counties (districts) in 13 provinces (municipalities and autonomous regions) in 2011 to 2 counties (districts) in 2 provinces (autonomous regions) in 2016. In terms of parasite species classification, 10,589 cases (31.4%) of vivax malaria, 16,211 cases (48.1%) of falciparum malaria, 1669 cases (4.9%) were oval malaria, 383 cases (1.1%) were malaria cases, 396 cases were of mixed species (1.2%), 2 cases were *Plasmodium knowlesi* infection, and 503 cases (1.5%) were not classified. There were also 3884 clinically diagnosed cases (11.5%). It is particularly noteworthy that in 2017, for the first time, China did not report any indigenous cases of malaria (Fig. 2) (Feng et al., 2018).

Fig. 2 National report on the incidence of malaria during 1993–2018.

3. Monitoring and information management

In terms of the current malaria status and malaria epidemic characteristics and trends in China, it is necessary to carry out planned, continuous, and systematic monitoring of malaria cases, vector, and drug resistance, including control, elimination, and post-elimination (Fig. 3) (Feng et al., 2014b). The monitoring objectives, monitoring priorities, and blood tests of fever patients are different at each stage. According to the National Malaria Elimination Surveillance Programme (2015 Edition), malaria surveillance includes an elimination and post-elimination phase. Counties (cities, districts) that have not yet passed the provincial malaria elimination assessment work should follow the monitoring requirements of the elimination phase. Counties (cities, districts) that have passed the provincial malaria elimination and non-malaria endemic counties (cities, districts) should carry out work in accordance with the post-elimination monitoring requirements. Elimination phase monitoring includes: case discovery and reporting, case review and epidemiological case investigation, epidemic investigation and classification, active screening of returnees, sentinel hospital monitoring,

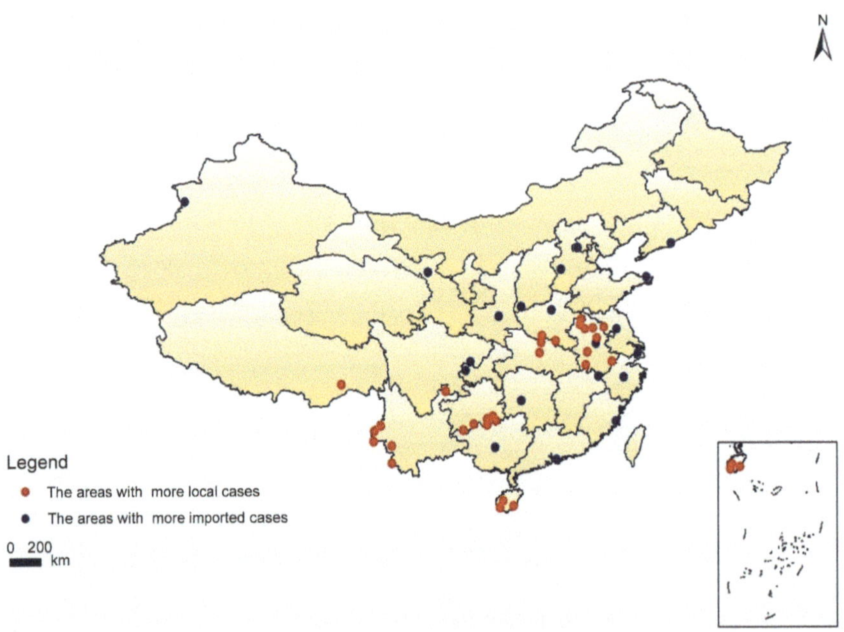

Fig. 3 Distribution map of the national malaria surveillance sites in 2014.

antimalarial drug sensitivity monitoring, and insecticide sensitivity. Post-elimination monitoring should include case monitoring, vector monitoring, and anti-malaria drug resistance.

The malaria control stage aimed to understand the current status of malaria transmission and its influencing factors and to master the prevalence and trends of malaria in order to provide a scientific basis for evaluating the prevention, treatment, and development of malaria control strategies. In the early 1960s, research at the NIPD regarding malaria epidemiology and its prevention and control in China mainly focused on two tasks. First, it was about radically curing vivax malaria. Most of China's mainland was in the northern temperate zone, where the spread of malaria began in the spring and stopped in the winter. This was conducive to the implementation of a radical treatment for vivax malaria. Therefore, it was proposed to carry out anti-relapse treatment during the malaria transmission period as one of the key measures to eliminate vivax malaria and to formulate a technical plan for malaria control. Monitoring during the elimination stage mainly focused on the timely removal of infection sources and blocking possible transmission via the timely detection of each source of infection and possible epidemic of transmission. Post-elimination monitoring is a key measure to maintain work capacity, consolidate and eliminate outcomes, and prevent secondary transmission caused by imported malaria.

The malaria control stage focuses on suspected cases, clinically diagnosed cases, and confirmed cases. The NIPD has tried to identify malaria outbreaks likely to occur, but laboratory detection and diagnosis were not emphasized in this stage. Monitoring during the elimination phase is based on the improvement of overall malaria prevention and control ability, focusing on confirmed cases, strengthening the importance of laboratory diagnosis, reducing clinical diagnosis cases and suspected cases through the establishment of malaria diagnostic reference laboratories, and improving the laboratory detection rate and proportion of confirmed cases. The monitoring in post-elimination stage focuses on the risk assessment of reintroduction, emphasizing the definition of imported and local infections and assessing transmission risk through case screening and vector investigations. For example, in 2013–18, a total of 18,658 malaria cases were reported nationwide, with a 24-h case report rate of 100.0%; 18,525 cases were detected in laboratory with a detection rate of 99.3%; and 18,336 cases were confirmed in laboratory with a coincidence rate of 98.3%. The epidemiological investigation was conducted among 17,773 cases with rate of 95.3%, and 12,509 related foci were disposed within seven days.

During the malaria control stage, the cases with suspected malaria and unexplained fever were included in the blood test. The annual blood test rate should be not less than 5% in the area with high incidence (1/1000 or more, the township as the unit), while the annual blood test rate should be not less than 2% in the unstable area (1/1000 to 10,000). In the areas with lower incidence (less than 1/10,000), the annual blood test rate should be not less than 1%. The blood test for fever patients should be carried out throughout the year. In the elimination stage, patients with fever for unknown reasons are included in the blood test. In first and second class counties (the township as the unit), the number of blood tests per year should be not less than 1% of the total population of the jurisdiction. In the third class of counties, the number of blood tests per year should be not less than 0.5‰ of the total population of the jurisdiction. The number of patients with unexplained fever during the spread season should be not less than 80% of the total number of blood tests. In the post-elimination stage, surveillance is conducted in sentinel hospitals in areas with high retransmission risk rates for patients with unexplained fever.

4. Technical support to the malaria control and elimination programmes

The control and elimination of malaria is not only an improvement of disease prevention and control goals, but also a major shift in disease prevention and control strategies. The transformation from the traditional "comprehensive prevention" mode to the new "eliminate foci and infection source" model was based on new theories, techniques and methods, and innovations and breakthroughs in key technologies. Over the years, at the technical level, in order to ensure the effective implementation of the elimination of malaria strategy measures and the timely realization of the elimination of targets, the NIPD has fully guaranteed the planning documents, funding, human resources investment, supervision and inspection, and international cooperation.

In order to strengthen the technical guidance for national malaria research and prevention work, the Ministry of Health established the Malaria Special Committee, consisting of the experts and secretaries of the NIPD, in March 1981. Since the establishment of the special committee, substantial work has been done in coordinating national malaria research projects, organizing key scientific research projects, conducting prevention and treatment technical guidance, and formulating national malaria prevention and control plans, regulations, and regulations. Among them, in the early 1960s, the NIPD's research on malaria epidemiology and its prevention and treatment in China mainly focused on some key points. For instance, it aimed to eliminate vivax malaria in most of China's mainland since most of the areas were located in the northern temperate zone, which created the opportunity to block the spread of malaria in the winter and spring and was conducive to the implementation of elimination. Therefore, it was proposed to carry out anti-relapse treatment outside the malaria transmission period as one of the measures to cure and eliminate vivax malaria. This finally formulated a technical plan for malaria control. In the early 1960s, tens of millions of mass prevention campaigns were also promoted by the technical plan. The "eight-day therapy" programme was first proposed. In 1963, the "four-day therapy" with less adverse reactions and easier application was proposed. In the same period, technical groups from the NIPD were sent to the four provinces to conduct on-the-spot investigations, write survey reports, and select Jiangsu Binhai County to track the investigation of outbreak trends.

From 1991 to 1994, the NIPD led a collaboration group with 23 counties (cities) in 11 provinces (cities), such as Henan and Hunan, and conducted malaria prevention and control in nearly 15 million population areas (*An. sinensis* as main vector) to study new features of epidemiology and to monitor programme research (Liu et al., 1996). A multidisciplinary approach, such as parasitology, serology, vector biology, and health economics, was used to investigate the new epidemiological characteristics of malaria prevention and control in this area and to explore economic and feasible simplified monitoring strategies. The transmission potential of *An. sinensis* was assessed by vector capacity, and the transmission intensity of malaria transmission index was calculated by the ratio of secondary cases induced by imported cases to imported cases. The cost and effect of the monitoring programmes were evaluated by a comprehensive effect index. The study found that in the later stage of malaria control in China, the malaria transmission was low, and the epidemic situation was stable. The main cases were imported, but rarely spread. New features of epidemiology included the epidemic foci, which were scattered, and the proportion of typical clinical symptoms of patients, which increased. The human blood index and vector competence of *An. sinensis* were significantly reduced, indicating that the potential of malaria transmission in this region has been at a low level. According to the results of the implementation of the experimental area and the new characteristics of epidemiology, the monitoring measures in this area are proposed as the blood test for fever patients with malaria and suspected malaria. The annual blood test rate is 0.3%, which can simplify the management of malaria in the mobile population and monitor the programmes for epidemic treatment, thereby saving manpower and anti-malaria funds. The research results won third prize for the National Science and Technology Progress Award in 1998.

After the national malaria elimination programme launched in 2010, more intensive work on the surveillance and response in all 24 malaria endemic areas. With strengthening surveillance and response system, particularly to identify the local and imported malaria cases, the implementational works on surveillance and response were standardized as "1-3-7" surveillance approach, which means case reporting within one day, case investigation within three days and focus investigation and action within seven days (Cao et al., 2014). In recent years, in order to ensure the quality of the "1-3-7" work, the National Anti-Malaria Technical Expert Group conducts an audit analysis of each reported case every quarter (2017) or semi-annual (from 2018) to ensure that each case is confirmed, and the information is accurate and reliable. Also, the National Critical Malaria Rescue Expert Group conducts an annual review of each death case to find out the cause of death and improve the ability of each province to diagnose and reduce the number of death. In 2017, the expert group reviewed 2651 cases from 24 malaria endemic provinces, of which 58.2% were qualified in the initial examination, and 92.5% passed the expert review. In 2018, the expert group reviewed 2546 cases from 24 malaria epidemic provinces, of which the rate of the preliminary examination was 74.27%, and the confirmation rate of the expert review was 98.08%. From 2016 to 2018, the NIPD trained 85,016 laboratory inspectors in the nationwide institutions. In addition, 256,694 clinical doctors were trained, including 44,977 ICU doctors, 20,489 doctors from communicable diseases department, and 191,228 doctors from other departments. At the same time, the National Health and Health Committee holds the National Parasitic Disease Prevention and Control Technology Competition every year. Through selection and competition, it has received good results in promoting competition. The NIPD was responsible for training and providing technical support and guidance to provincial laboratories and regularly assessing the microscopic examination and genetic testing capabilities of provincial laboratories. The NIPD also carried out parallel detection and confirmation of specimens from malaria cases with controversial results in provincial laboratories, established a national malaria case information database for the exchange and sharing of relevant information, and researched new technologies and methods related to malaria parasite diagnosis. In addition, the NIPD actively conducts international exchanges and cooperation, introduces and promotes the application of advanced and applicable technologies, and makes full use of internationally funded projects, such as the Global Fund to support malaria elimination operations. For example, the NIPD has established a cross-border malaria prevention and control cooperation mechanism to accelerate the control and elimination of malaria in China's border areas. Since 2015, in order to promote cross-border cooperation on malaria elimination in the China-Burma border region, the China CDC and relevant departments in Myanmar have held annual seminars and invited the National Health and Health Commission, WHO, Health Poverty Alleviation, and other departments and international organizations to participate. The strategic plan for malaria elimination in the border areas of China and Myanmar was drafted, and the "same strategy in the same region" was taken as the guiding ideology to jointly promote the joint prevention and control of malaria in the border areas. The NIPD has actively contacted Southeast Asia and relevant African institutions and has launched and implemented the China-British-Tanzania Malaria on-site prevention and control project and the China-Australia- Papua New Guinea malaria prevention and control project, which has effectively reduced local malaria morbidity and mortality (Wang et al., 2019; Xia et al., 2014).

5. Experimental research in the biology of the malaria pathogen

5.1 Investigation into the *Plasmodium* species

Based on survey data from 1951 to 1961, the composition of malaria parasites in China was as follows: *Plasmodium vivax* accounted for 52.1%, *Plasmodium falciparum* accounted for 36.9%, *Plasmodium malariae* accounted for 7.5%, and the remaining 3.5% was *P. vivax* mixed infection with *P. falciparum*. The composition of the *Plasmodium* species in different latitudes also differed. There are four *Plasmodium* species distributed south of 25° north latitude. *P. falciparum* and *P. vivax* were the main species, of which *P. falciparum* accounted for 45.9%, *P. vivax* accounted for 42.0%, and *P. malariae* and mixed infection accounted for 7.4% and 4.7%, respectively. Only *P. vivax* was found north of 33° latitude. Between 25° latitude and 33° latitude, three kinds of *Plasmodium* existed, but *P. vivax* dominated and accounted for 74.3%, while *P. falciparum* accounted for 17.0%, and *P. malariae* and mixed infection accounted for 7.9% and 0.8%, respectively.

The distribution and composition of malaria parasites varied in each province and region. *P. vivax* had a very wide distribution range and was distributed in various kinds of malaria endemic areas, such as the tropical, subtropical, and temperate regions of China. The distribution of *P. falciparum* had certain limitations. It was widely distributed south of Nanling Mountains, including almost all areas in the mountains, hills, and plains. The infection rate of *P. falciparum* was around 20%, which was the dominant species in these areas. Between Nanling and Qinling, *P. falciparum* was mostly distributed in mountain basins with a high temperature and low terrain. It was difficult to find its distribution north of the Qinling Mountains. *P. malariae* was scattered within a certain range, and it was mainly distributed south of the Qinling Mountains, but it was not a dominant species. *P. ovale* was only found minimally disturbed south-southwest of Yunnan Province, Guizhou Province, and Hainan Island (Xiao et al., 2010). After more than 60 years of active prevention and control, the geographical distribution of the four Plasmodium species changed greatly in China. In recent years, *P. ovale* has not been rediscovered in any place, and *P. malariae* has been rarely found. Only 17 counties (cities) in Yunnan Province have reported falciparum malaria. The proportion of falciparum malaria has been greatly reduced, while the number of all types of malaria cases has dropped significantly. In recent years, only vivax and falciparum malaria have been distributed within the border areas of Yunnan Province (Wang et al., 2014b).

5.2 Research on *Plasmodium falciparum*

5.2.1 Continuous in vitro *culture of* P. falciparum *and its application*

In 1979, the NIPD successfully established the *in vitro* culture of *P. falciparum* by using the candle cylinder method. The culture method was greatly improved. The relatively expensive cost of the foetal bovine serum in the original method was changed to rabbit serum, and the serum ratio in the medium was reduced from 10%–12% to 8%–10%. Thus, the improved method facilitated the continuous culture of *P. falciparum in vitro* and at the same time, reduced the expenditure. In addition, some progress with the *in vitro* culture of *P. falciparum* gametophytes was also made. This progress created conditions for the screening of antimalarials *in vitro*, testing and comparing the insecticidal effects of antimalarial drugs, and the development of vaccines, in addition to the preparation of applicator plates and freeze-dried culture medium for the *in vitro* microassay to determine drug resistance. Further, it also provided possibilities to compared the effects of chloroquine

(CQ) and its two enantiomers [(+) -CQ, (−) -CQ], desethylchloroquine (D-CQ), and pyronaridine (PY) drugs. The results showed that the five drugs had similar inhibitory effects on CQ-sensitive strains, and D-CQ showed a poorer effect on the anti-CQ strain. PY was significantly better than the other four drugs. In the conventional method for screening antimalarial drugs *in vitro*, the *in vitro* metabolic system consisted of rat liver microsomes and oxidized coenzyme I, which was added to detect common antimalarial drugs (CQ, PY, TCL cycloguanil, and Proguanil) activity. The results confirmed that the introduction of liver microsomes into the *in vitro* metabolic system could avoid the detection of compounds that have antimalarial effects. In addition, a method for the detection of sensitivity of *P. falciparum* to PY/sulfadoxine/pyrimethamine *in vitro* was also established based on microassay, which subsequently demonstrated a synergistic effect between PY and sulfadoxine/pyrimethamine.

5.2.2 The establishment and improvement of the sensitivity of the testing method for malaria in vitro

Due to the spread of CQ resistance among *P. falciparum* and poor treatment effect among falciparum malaria patients, which resulted in a rise in the number of cases and an increase in mortality, antimalarial drug resistance has become the first priority for malaria control. The above-mentioned *in vitro* microassay for testing the sensitivity of malaria parasites was a new technology with simple, fast, accurate, and patient-friendly advantages, but required preformed applicators and culture medium for on-site use. In 1979, the imitative CQ plate was successfully developed by the NIPD, and the applicator plate was improved to make it more suitable for field application in China. In 1980, a freeze-dried medium was developed for the convenience of on-site investigation, thereby establishing the sensitivity test for malaria parasites against drugs using the *in vitro* microassay method. In the same year, a resistance investigation workshop was held in Wanning County, Hainan Island, and the technology was trained to local health workers, which laid a foundation for the national drug resistance investigation.

5.2.3 Cloning and expression of the three repeats of a P. falciparum *gene*

In order to clone and express the three repeats 3R, 6R, and 9R of the *P. falciparum* (Pf) 11.1 gene product, the NIPD designed primers to amplify the three repeats from the genomic DNA of cultured *P. falciparum* 3D7 strain. The PCR product was cloned into the pT7 vector for bidirectional sequencing. The sequencing results were analyzed by GENETYX-MAC software. The amplified fragment was subcloned into pET32a(+) or pET32b(+) and induced by isopropyl-β-D-thiogalactoside (IPTG) to express recombinant proteins in *Escherichia coli* BL21. The 3R, 6R, and 9R fragments were successfully amplified by PCR, and the sizes were 552, 630, and 444 bp, respectively. The sequencing results showed that the Pf11.1 gene of 3D7 strain had four 3AA and one 6AA repeat units more than that of Palo Alto strain. The identities of the 3R and 6R fragments of the two *Plasmodium* strains were 92.8% and 95.1%, respectively. The amplified 9R fragment contained 13 9AA repeat units. Three recombinant proteins were expressed in the BL21 strain with relative molecular masses (Mr) of 45,000, 60,000, and 42,000, respectively.

5.2.4 *Preparation of monoclonal antibodies specific to the lactate dehydrogenase of* P. falciparum

By cloning and expressing the lactate dehydrogenase gene of *P. falciparum* and immunizing BALB/c mice with the expressed recombinant protein, the NIPD was able to prepare monoclonal antibodies via a hybridoma technique. Subsequently, the prepared monoclonal antibodies were determined by their subclasses and titres, and their specificity was analyzed by western blotting. The lactate dehydrogenase gene of *P. falciparum* was successfully cloned and expressed. The monoclonal antibody was prepared by using recombinant lactate dehydrogenase protein as the immunogen source. A total of 15 strains were screened for effective secretion titre at 1:6400–1:51,200 with the antibody subclass being IgG1 or IgG2. All antibodies recognized the *P. falciparum* protein Mr 33000 component and did not cross-react with the red blood cell components of non-malarial fever patients in the non-endemic areas.

5.3 Research on *P. vivax* biology

Researchers at NIPD built long-term collaborations with the staff in certain regions of China, including Henan, Yunnan, and Hunan, in order to investigate the relationship between the incubation and recurrence of *P. vivax*, the latency and latent period of *P. vivax*, and perform experimental research and on-site observation of the viability of gametophytes. The results further proved that the incubation period of *P. vivax* in China was of two types—the short incubation period was 10–28 days, and the long incubation period was 228–443 days. The length of the incubation period was related to the amount of sporozoite inoculation, which can be artificially controlled from long-to-short or from short-to-long. In the temperate regions of China with four distinct seasons, there were no long-latency or short-latency *P. vivax* strains with stable biological characteristics. By observing the treatment of vivax malaria patients with CQ in southern Anhui for one year, 95.2% of the cases relapsed, but the length of the latent period was different, ranging from 50 to 100 days in the short period and 160 to 300 days. The time limit for recent recurrence and long-term recurrence was about 3 months, instead of 8–10 weeks as determined by foreign scholars. For the long incubation period vivax malaria, there was a recent recurrence but no long-term recurrence, and the number of recurrences was shorter during the incubation period. The short-term was only found during the long-latency period of malaria and recurrence, which can roughly distinguish between two different types of vivax malaria and its composition ratio. It is more credible to use the median latency to estimate the length of the incubation period.

The periodic study of gametophytic viability based on the observation of *P. vivax* cases revealed that *P. vivax* needs to pass through several schizontial proliferation cycles after completing tissue development into the bloodstream in order to meet the infection threshold in mosquitoes for infection transmission. Therefore, the primary infection case with a longer incubation period is more likely to cause infection after being bitten by a mosquito vector. In the study of the growth and decline of gametophytic infectivity during the whole course of vivax malaria infection, it was found that artificially infected cases of vivax malaria, if temporarily not treated after clinical onset, were infectious to mosquito vectors throughout the course of the disease, but the duration of high infection status was very short. The reason is that the infection period of the *P. vivax* gametophyte has periodic fluctuations. The peak of mosquito infection occurs with large trophozoites in the peripheral blood of patients, while low infection occurs with mature schizonts and immature trophozoites in the blood. These findings provided a basis for future research on how to select blood donor patients and the amount of mosquitoes needed for infection experiments.

In addition, the mechanism of vivax malaria recurrence is a longstanding problem that needs to be solved. The NIPD used the non-human primate malaria model of *P. cynomolgi–Anopheles stephensi–Macaca rhesus* system to conduct experimental studies regarding malaria recurrence. After *R. macaques* were infected with *P. cynomolgus* sporozoites, serial sections of tissues were taken from different parts of the liver on day 8 post-infection. A large number of intrahepatic schizonts at different stages of development were found via microscopy, indicating that the development of *P. cynomolgus* in the liver tissue was asynchronous with the schizont matured and released merozoites into the blood circulation. The intracranial merozoites disappeared when treated with PY and M8132; however, the ring stage was present in the blood, but without other *Plasmodium* periods. This is because the mature intrahepatic schizonts released merozoites into the blood, and this time period could last for more than 50 days. After the disappearance of the drug effect, the delayed release of intrahepatic merozoites were able to repeatedly proliferate in the blood, causing a recent recurrence. When intrahepatic merozoites are released into the blood circulation to form parasitemia, administration of PY (6 mg/kg) combined with artemether (10 mg/kg) and CQ (20 mg/kg) could effectively remove the parasite and prevent recrudescence, seeing of recent recurrences indeed reliable. Rhesus monkeys were infected with $(50–55) \times 10^4$ *P. cynomolgi* sporozoites, and the pre-existing period was 8–9 days, which was not affected by the number of spores inoculated. However, the recurrence status varied. The monkeys inoculated with 55×10^4 sporozoites relapsed 3 times, while the monkeys inoculated with 11×10^3 sporozoites relapsed only once, and monkeys inoculated with 50 and 5500 sporozoites had no recurrence within 200 days. The number of sporozoites was related to the frequency of recurrence. The experimental results showed that the real recent recurrence was observed when ruled out the relapse. The results of this study provided a biological experimental basis for the recent recurrence mechanism.

6. Biological investigations on malaria vectors

Anopheles mosquitoes are the main vector for malaria. There are 3 genera and more than 400 species in the world, and about 35 of them have medical importance. More than 60 species of *Anopheles* mosquitoes are known in China. It is generally believed that the main malaria vectors include four species, namely *Anopheles minimus*, *Anopheles dirus*, *Anopheles sinensis*, and *Anopheles anthropophagus* (Fig. 4) (Feng et al., 2017).

An. sinensis is distributed in all provinces (autonomous regions and municipalities directly under the Central Government), except Qinghai and Xinjiang. North of 34° north latitude, *An. sinensis* is an important vector for malaria and filariasis in the vast plain areas, especially in the rice-planting areas, and is the main vector for *vivax* malaria. *An. sinensis* is a semi-domestic mosquito species, and 40%–75% can be found indoors. This mosquito species can feed on bovine and human blood, and it has a strong outside-feeding

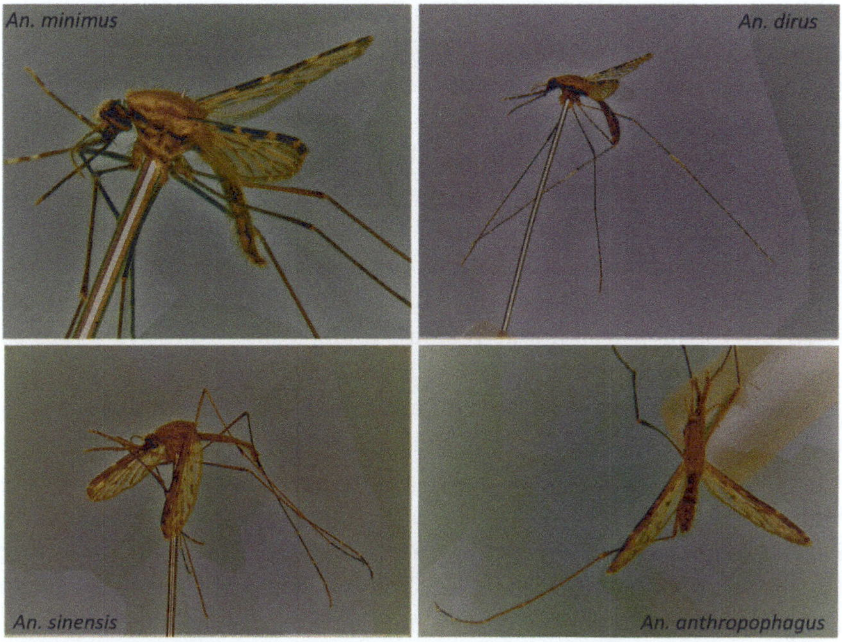

Fig. 4 The four main malaria vectors, e.g., *An. sinensis*, *An. anthropophagus* (or *An. lesteri*), *An. minimus*, *An. dirus*, in China.

habit in the plain rice areas, mountains, and hills. By contrast, it prefers inside-feeding habits in rice-planting areas. The peak of blood-sucking activity at night is one hour before and after midnight. In most parts of China, population breeding season lasts from July to September, and the peak of human-biting is from July to August. Female mosquitoes mostly lay eggs before midnight, and mosquito eggs cannot tolerate extreme low or high temperatures and will die at freezing temperatures or temperatures above 40°C. The development of the entire larval stage is 8.3 days in July and August, and 9–10 days before June or after September. Larvae have a wide adaptability to the environment, but tend to prefer a water body around a shady place. Since the 1950s, many irrigation rivers in the low-lying areas of the Huanghuai Plain tend to be the main breeding ground for *An. sinensis* because these rivers usually have a very slow water flow and are overgrown with vegetation. The larvae can survive at low temperatures, and a few larvae can even survive at 3°C. Although the infection rate of *An. sinensis* to *Plasmodium* is relatively low, it has a large population. Thus, there are many opportunities for mosquito exposure, and it plays an important role in malaria transmission, especially in some areas with poor mosquito-repellent conditions and with outside-sleeping habits during the summer. Therefore, the mosquito is the recognized as the main or only vector for maintaining low prevalence of malaria in most plain areas of mainland China.

An. anthropophagus, which used to be a subspecies of *Anopheles lesteri*, is a relative species of *An. sinensis*. It used to be recognized as a subspecies of *An. sinensis* or *An. anthropophagus* around 1975. In 2008, it was revised by an entomologist and was regarded as a synonym of *An. lesteri*. So far, it has been confirmed that *An. Anthropophagus* is distributed in more than 300 counties in 18 provinces (autonomous regions, municipalities directly under the Central Government) and the Hong Kong Special Administrative Region (Liu, 1990). *An. anthropophagus* has been found with a distribution ranging from 102°20′ to 121°30′ east longitude and 20°00′ to 42°10 north latitude. It is the main vector for falciparum malaria. *An. anthropophagus* is a semi-domestic mosquito species and also a facultative feeder. The peak season is from July to September in most areas of China. In some areas, there may be a second density peak in late October. This species is often distributed in the same area as *An. sinensis* (sympatric distribution). The population density of *An. anthropophagus* is generally not as high as *An. sinensis*, but it can be act as the dominant species in human houses in some areas. Nighttime blood-sucking activities begin around half an hour to two hours after sunset, and the peak of activity lasts from midnight until dawn (Xia et al., 2003). Larvae are found in rice fields, but prefer to be born in water bodies with plant or water bodies with water seepage. While its larvae are symbiotic with *An. sinensis*, the density of *An. sinensis* is higher in the early and middle stages of rice growth, and the density of *An. anthropophagus* larvae increases with the growth of rice. The vector competence of *An. anthropophagus*, especially for falciparum malaria, is far stronger than that of *An. sinensis*. Its salivary gland infection rate is quite high, generally above 0.5%, and its distribution is closely related to the prevalence of malaria. In 1984, through a quantitative survey of entomology and parasitology in Mu'ai Township, Qilian County, Sichuan Province, the vector competence of *An. sinensis* and *An. anthropophagus* was compared. The local malaria cases estimated to be transmitted by *An. anthropophagus* were about 94.3%, while transmission by *An. sinensis* was about 5.7%, about 20 times the former. In recent years, researchers from the NIPD have confirmed that there is *An. anthropophagus* in Faku County, Liaoning Province at 42.5° north latitude, but its ecological habits are obviously different from those south of 33°, which has caused widespread concern and research interest.

An. minimus is distributed in the hilly areas of China south of 33° north latitude, especially south of 25° north latitude. In recent years, classification studies have shown that *An. minimus* has multiple types, such as A, B, and C (Lu, 1997). This mosquito is the main vector of malaria in the mountainous and hilly areas in the south China, and is also one of the most important vectors of malaria in southeast Asia. It is also the vector of filariasis in Hainan Province. *An. minimus* prefers indoor habitats, and the habitats during the day are mainly human houses and barns. The proportion of *An. minimus* in Hainan Island during the day or night is about 60% in the mountainous area and about 40% in the hilly area. The proportion of *An. minimus* inhabiting inside of the net during the day is about 42%–66%, and at night it is about 42%–88%. The density of *A. minimus* begins to rise sharply around March, peaking from April to June before the rainy season and then gradually decreases. A small density peak appears again in some areas from September to October. The larvae are mostly found in ditches where the water quality is good, and the flow of water on the shore is slow. During winter in southern China, larvae can survive in the breeding sites. In an environment of 22–25°C, it takes 18–21 days to develop from an egg to an adult. In Hainan Island, the mosquito tends to prefer human blood (85%), while in Yunnan, Guizhou, Guangxi, and other provinces (regions), the local *An. minimus* sucks both human and bovine blood. According to the geographical observation, the farther north, the smaller the proportion of *An. minimus* that prefers human blood. The blood-sucking activity of female mosquitoes at night starts 1 h after sunset and lasts all night with the peak around midnight. However, in the special ecological environment of the beach in Wenchang County, Hainan, the peak of blood-sucking activity can occur 1–2 h before dawn. *An. minimus* is the main malaria vector in the southern malaria area of China. Its key role in malaria transmission is not only reflected in the high natural infection rate of salivary glands, but also in the area distribution of salivary gland positive infection rate and its duration in the malaria transmission season. There have been positive salivary gland infections in Hainan Island every month in the past (50 years ago).

An. dirus is a tropical jungle *Anopheles* mosquito mainly distributed in Hainan Province, west and south of Yunnan, and at the southwestern border of Guangxi (Xu and Feng, 1975). This mosquito species is only found in the forest and hilly areas of tropical rain forests or secondary mixed forests. It is a typical outdoor habitat mosquito species. It is generally believed that this mosquito species is an important vector for malaria transmission in the mountainous areas of central and southern Hainan Province.

However, the role of malaria transmission in other areas is still unclear. Historical studies have shown *An. dirus* prefers human blood (96.2%). Blood-seeking activity begins one hour after sunset, and while it peaks at around midnight, there is blood-sucking activity all night. The population peaks from July to October, and the average human-biting rate in undeveloped forest areas is 14.8%, and 2.03% in semi-development zones. The proportion of the composition of *An. dirus* is about 2%–10% in the hilly area, more than 10% in the mountainous area, and 50% in the isolated and isolated small villages in a few undeveloped forest areas. The main habitat of *An. dirus* larvae is small natural water that is shaded by trees or bushes. Due to strict requirements for the environment, the population is usually small. *An. dirus* is the main malaria vector in the mountainous forest area and mountainous area of Hainan Island. The infection rate of salivary gland fluctuates from 2.1% to 8.3%, and there are natural infections from April to November throughout the year. Because the *An. dirus* larvae must have a certain degree of shading, once the surrounding jungle is cut down and shade conditions disappeared, the *A. dirus* larvae will quickly disappear and subsequently reduce the spread of malaria.

7. Research and development of control tools, including diagnostics, drugs, and insecticides

7.1 Diagnosis

7.1.1 Etiological diagnosis

When compared with traditional methods, such as Wright's and Giemsa staining, both the fluorescent pigment acridine orange staining method and fluorescence microscopy are able to quickly detect *Plasmodium*, even at a low-density. Fluorescence microscopy is more efficient than ordinary microscopy. In a study conducted by the NIPD, the researchers first stained the blood film with fluorescein acridine orange and then examined it with fluorescence microscopy and then counterstained it with Gill's stain. The results of 179 primary school students showed that the detection rate of acridine orange method was 42.0% higher than that of Giemsa method. The acridine orange method was able to detect 11 blood films per hour on average, which is 1.8 times faster than the Giemsa method, which was able to detect four blood films per hour. In 1978, the detection efficiency of *P. falciparum* by the acridine orange method was compared in Hainan Island, where 119 people were investigated and 34 cases were detected (including 32 cases of falciparum malaria). The detection rate was 97% when compared to Giemsa method (85%). Among the falciparum patients, 7 cases were detected with 5–10 parasites in the blood smears, and 9 cases with less than 5 parasites, indicating that the acridine orange method has a high accuracy for detection of the thin ring stage and fewer parasites. Fluorescein acridine orange was used to examine *Plasmodium gallinaceum* sporozoites in the salivary gland of *Aedes* with expected results. The nucleus of these sporozoites showed a punctate yellow-green fluorescence. Most of the sporozoites were mononuclear, but a few had dual nuclei. The cytoplasm was sickle-shaped with the sharp ends showing orange-red fluorescence, and the sporozoites had clear morphological structure, sporozoite activity was still visible after 2 h.

Researchers at the NIPD have established a polymerase chain reaction (PCR) to detect *P. falciparum*. The amplified products were analyzed by agarose gel electrophoresis to amplify a specific 492 bp DNA fragment. By contrast, DNA from *P. vivax*, *P. cynomolgus*, *P. yoelii*, *Plasmodium berghei*, and healthy human leukocytes did not amplify this product. This method can detect the lower limit of parasite density at 10 parasites/20 μL with a high sensitivity and specificity. PCR was used to detect 53 blood samples of falciparum malaria patients in Yunnan. The results were all positive, and the coincidence rate with microscopy was 100%. At the same time, 8 samples of vivax malaria and 12 healthy blood samples were tested, all of which were negative, further indicating that this method has a high sensitivity and specificity. A rapid diagnosis using a malaria colloidal gold immunochromatographic strip method was established, and 54 blood samples of patients with falciparum malaria were tested, of which 51 were positive. The detection rate with this method was 94.4%. There were also 110 samples of vivax malaria tested, among which 102 were positive, and the detection rate was 92.7%, which also presented with high sensitivity and specificity.

7.2 Serological diagnosis

7.2.1 Indirect fluorescent antibody test (IFAT)

In 1976, serum antigens were detected in 126 cases of vivax malaria patients in the northern malaria area of northern Fujian Province using *Plasmodium cynomolgus* monkey (*P.c*) antigen. The antibody titres $\geq$1:10 and $\geq$1:20 were 73.8% and 60.7%, respectively, for those under 15 years old and 81.0% and 69.0%, respectively, for those over 16 years old. In 1978, *P.c* antigen was used to detect serum antibodies in 50 patients with falciparum malaria in Suining County, Jiangsu Province. The positivity rate of the antibody titre $\geq$1:20 was 98.0%. Also, 52 cases of falciparum malaria in Hainan Island were detected using the serum antibodies from each patient, and the positivity rate of the antibody titre ≥1:20 was 100%. In 1983, the serum antibodies of 35 patients with falciparum malaria were detected by *in vitro* cultured *P. falciparum* and *P.c.* antigen. The positive rates were 100% and 97.1%, respectively. However, the geometric mean titre reciprocal (GMRT) of positive individuals detected by *falciparum malaria* (*P.f*) antigen was 1445.4, which was significantly higher than 186.5 of *P.c* antigen. The positive rate of the serum antibodies in 38 cases of vivax malaria was 89.5% and 100%, respectively. The positive GMRT of P.f antigen was 40.6, which was significantly lower than 215.9 by *P.c* antigen, indicating that the same antigen was superior to the heterologous antigen, and the *P.c* antigen was superior to the P.f antigen for detecting the vivax malaria antibody. In Hainan Island and Yunnan Xishuangbanna, where *An. minimus* almost disappeared, serological investigation confirmed that malaria has been controlled for many years. In places where *An. minimus* and *An. sinensis* do not exist, the antibody level of the population indicated that malaria was still in transmission. Five different types of populations were investigated horizontally using *P.f and P.c* antigens. The results of the serological assay were consistent with the blood smear test, indicating that the two antigens are able to be used in the IFAT to investigate the prevalence of *falciparum* and *vivax* malaria. Subsequently, a three-year longitudinal survey was conducted on three mixed epidemic areas of falciparum and vivax malaria in Anhui Province, where different antimalarial measures were applied. As a result, the serum antibody positive rate, antibody titre, and malaria parasite rate showed a parallel relationship in reflecting malaria prevalence levels. The incidence and recovery rate of malaria parasitemia in the malaria transmission season in three regions, the positive rate and recovery rate of antibodies detected by *P.f* and *P.c* antigen, and their magnitude and trends were associated with the intensity of falciparum and vivax malaria transmission and the measures taken against malaria. Two years of horizontal surveys were conducted in 1983–84 in three malaria areas with different antimalarial measures and compared with the longitudinal surveys. The results of the two surveys were roughly the same. Both of them can be used to investigate the prevalence and changes of falciparum and vivax malaria in the region, and in turn, to evaluate the effects of antimalarial measures. In recent years, malaria epidemic in the Yangtze River Three Gorges reservoir area was accessed by the IFAT, and result showed a certain extent of malaria infection existed in this area (Duo-Quan et al., 2009).

7.2.2 Enzyme-linked immunosorbent assay (ELISA) for malaria epidemiology survey

In 1978, by using *P.c* antigen to prepare antigen, and labelling it with alkaline phosphatase anti-human IgG, researchers from the NIPD detected 77 dry blood drops from patients with falciparum malaria and found that the positive rate was 96.1%. Moreover, 85 cases of dry blood drops from patients with vivax malaria were analyzed by this method, and the positivity rate was 98.8%. Subsequently, by using *in vitro* cultured *P.f* was used as antigen, 307 malaria positive dry blood drops were collected and analyzed via this method. The positivity rate was 90.23%, and the positive coincidence rate of falciparum malaria was 94.29%, and the positive coincidence rate of *vivax* malaria was 84.85% (Gao et al., 1991; Zhang et al., 1998). The false positive rate of healthy people in non-malarial areas was only 1.2%. A subsequent ELISA was used to conduct a sero-epidemiological survey of residents in two types of areas—one area was a malaria endemic area and the another area was where a malaria outbreak occurred that was under control in Hainan Island. The researchers took fresh blood samples or dry blood drops for ELISA. The antigen was a soluble antigen cultured *in vitro* from *P. falciparum*, and the enzyme conjugate was anti-human IgG-HRP. The results showed that the infective rate of the malaria parasite in the former category area was 9.8%, and the ELISA positive rate was 35.2%. The antibody level of the residents increased with age, while the infection rate decreased due to the increase in age. In the latter area, the ELISA positive rate of residents was 72.2% during the second year after the outbreak, which was significantly higher than that of the former category. In the fourth year, although the infection rate in the residents did not change significantly, the average antibody level and ELISA positive rate significantly decreased. The average antibody level in all age groups under 25 recovered, indicating that malaria re-emergence may occur once there is a source of infection.

7.2.3 The establishment of a single epitope monoclonal antibody against P. falciparum

Researchers in the NIPD selected HRP-II peptides with 6–9 amino acids, which represents the antigen trait of *P. falciparum* protein. After artificial synthesis and determination of its purity, BALB/c mice were immunized with such peptides. Four hybridoma cells against a single epitope of *P. falciparum* HRP-II were obtained by spleen implantation and hybridoma technology. In this test, the carrier protein (KLH) was first adsorbed on the ELISA plate, and the nine peptides were covalently cross-linked onto the carrier protein by the cross-linking agent in order to increase the adsorption and reduce the steric hindrance, which improves the detection sensitivity. This experiment also solved the adsorption problem of small molecule peptides in ELISA assays. This has potential benefits for an in-depth study of malaria diagnosis, in addition to the structure and function of HRP-II.

7.2.4 DNA probes for the diagnosis of P. vivax

Blood samples from patients with falciparum malaria were positive for dot-hybridization with a 0.24 kb DNA probe specific for *P. vivax*, and those with negative on-site microscopy were screened by Southern blotting. The results showed that the positive reaction of the dot-hybridization reaction was also positive. The experiments confirmed and concluded that some patients with falciparum malaria and some fever patients with negative microscopy may also infected be with *P. vivax*. It is further proved that the *P. vivax* DNA probe could not only detect the presence of *P. vivax* infection in patients with falciparum malaria, but could also detect *P. vivax* in fever patients with negative microscopy. This is important for identifying whether a patient is a mixed infection, as well as for screening asymptomatic carriers in an endemic area.

7.3 Drugs

In the field of antimalarial drug research, there was only one chicken malaria model in China when the People's Republic of China was founded. Due to the need of improved research work, the NIPD established a variety of animal models to study malaria pathogenesis, in addition to animal models for antimalarial screening and pharmacological research for the supply of antimalarial drugs and other research Institutes.

7.3.1 Established malaria animal model

Avian malaria parasites were introduced from India during the Anti-Japanese War. *Aedes albopictus* is highly sensitive to chicken *Plasmodium* (*P. gallinaceum*) and can be used to study antimalarial drugs that inhibit sporozoite development in mosquitoes. The extraerythroid stage of *P. gallinaceum* can be found in tissues such as the brain, lung, spleen, liver, heart, and kidney, and can be used to study the development of drugs aiming to eradicate vivax malaria. In 1954, the Indian strain of *P. berghei* was introduced from India, and the Soviet strain of this parasite was later introduced from the Soviet Union. It can be used to screen antimalarial drugs for *erythrocytic stage* when *P. berghei* is infected in the blood of mice. Since drug resistance research was needed, a 52-fold resistant strain of *P. berghei* was developed. In this way, researchers in the NIPD induced the anti-artemisinin *P. berghei* strain with 13-fold resistance, the anti-pyronaridine with 300-fold resistance, and the resistance index of 197.2 multi-resistant to chloroquine and pyronaridine *P. berghei* strains. In order to establish a murine malaria model for screening etiological and curative drugs, *Anopheles stephensi* was introduced from London in 1973, and the Yes265 strain of *P. yoelii* was introduced from Paris in 1976, thereby establishing the *P. yoelii–An. stephensi* system. In 1973, *Plasmodium knowlesi* was introduced from the United Kingdom to establish a rhesus monkey model to test the effects of various compounds on the erythrocytic schizont. Later, one kind of monkey *Plasmodium* strain was introduced from Vietnam and identified as *P. cynomolgus* by experimental observation. The NIPD has thus established a comprehensive non-human primate malaria model consisting of *P. cynomolgus–An. stephensi*–rhesus monkey. *P. cynomolgus* is similar to *P. vivax* and is easier to maintain in the laboratory. This model can be used for the development of drugs to kill schizonts during the blood stage, as well as for etiological prophylaxis and radical drug development. Rhesus monkeys are sensitive to *P. cynomolgus* in Guangxi and other places. In monkeys infected with *P. cynomolgus*, the parasitemia was temporarily negative when treated with CQ (5 mg/kg) via the intranasal route, but recrudescence occurred soon. The parasitemia totally disappeared with a single dose of 40 mg/kg CQ. Primaquine (6 mg/(kg*d)) with CQ (20 mg/(kg*d)) for 33 days cured the infected monkeys inoculated with sporozoites. Based on these data, it was possible to test if other compounds were better than CQ in killing erythrocytic schizonts. Any drug having exo-erythrocytic effects can be tested and compared with the efficacy of primaquine in the monkey malaria model. Thus, a murine malaria model and monkey malaria model have been established, and a system for screening antimalarials has been formed. Chicken *Plasmodium*, murine *Plasmodium,* and monkey *Plasmodium* were preserved in the laboratory for a long time by means of blood transfusion and sporozoite seed transfer. At the beginning, a cryopreservation method was used to protect the blood containing *Plasmodium* with 15% glycerol phosphate buffer and stored in dry ice and acetone. In recent years, 24% dimethyl sulfoxide has been used to protect the activity of the parasites, and the stock is stored in liquid nitrogen to maintain the infection activity of parasite during the erythrocytic stage. In 1980, a cryopreservation method for the sporozoites of *P. berghei* in whole mosquitos and liver tissue without adding any protective agent was established. In addition, laboratory breeding and breeding methods for *Ae. albopictus*, *Ae. Aegypti, An. sinensis*, *An. stephensi*, and *Culex pipiens pallens* have been established to supply mosquitoes for experimental study.

7.3.2 The development of new antimalarial drugs

CQ resistance has become increasingly widespread in the world, and there is an urgent need for high-efficiency, low-toxic therapeutic new drugs without cross-resistance to CQ both at home and abroad. According to the requirements of the National Leading Group for Malaria Prevention and Treatment Drugs, the NIPD undertook research on new antimalarial drugs since the NIPD was founded. In 1964, a new compound, bispyroquine was synthesized. Animal studies proved that it had an antimalarial effect, and the clinical trial was carried out. As a result, the new drug was effective for the treatment of *P. vivax*, but the trial was stopped due to certain cross-resistance with CQ. The new compound, pyracrine, was synthesized in 1967, but it was eventually discontinued for toxicity reasons. At the end of 1969, another new compound, pyracrine phosphate, was designed and synthesized. It was found that pyrrolidine was better than CQ in animal experiments, and the adverse reactions were lower than CQ, and there was no cross-resistance with CQ. Pyracrine phosphate can be taken orally, intramuscularly, or intravenously. It entered clinical trials in 1971 and passed the production process identification in 1979. It was registered in 1980 and received three approval numbers for pyrrolidine and its tablets and injections dosages were officially put on the market. Since 1990, these tablets and injections have been included in the Pharmacopoeia of the People's Republic of China. It is the only chemical synthetic antimalarial drug created in China and included in the Pharmacopoeia. The related research work won the National Science Conference Award (1978) and the National Invention Third Prize (1985).

For the application research of malaria treatment and preventive drugs, it was necessary to explore a plan to cure children with *P. vivax* and to facilitate mass drug administration (MDA). From 1961 to 1963, patients with parasitemia under 15 years of age were selected for treatment during malaria transmission using primaquine 210 mg (7 days), 180 mg (8 days), and 120 mg (4 days) with a double course of treatment and primaquine 210 mg (7 days divided) with single course of treatment compared with CQ 600 mg double course and a no medication group as a control. The results demonstrated that all single-course regimens were not satisfactory, while the three groups treated with the dual-treatment regimen achieved good results. The length of the two treatment intervals had no significant effect on the efficacy. From the perspective of efficacy, safety, economy, and simplicity, primaquine 180 mg (8 days) and primaquine 120 mg (4 days) were suitable for field MDA application. In addition, the NIPD also organized a special study to evaluate the efficacy of CQ and chlorhexidine, chlorpyridinium chloride, and pyronaridine for the treatment of vivax malaria. For the evaluation of preventative drugs, 50 mg of pyrimethamine combined with 22.5 mg of primaquine 22.5 was tested in 1965 in the Daishan Township, Wuxing County, Zhejiang Province, and the drug was administered once every two weeks for a total of nine times. During the period of preventative medication, the incidence of malaria among residents decreased by 81.8% as compared with the same period of the previous year, and the incidence of vivax and falciparum malaria decreased by 91.8% and 70.1%, respectively. During the medication, 32 cases of falciparum malaria and 10 cases of vivax malaria occurred. The average interval time between falciparum malaria and the last preventive medication was 9.8 days, and the average interval for vivax malaria was 12 days. These results indicated that regardless of the epidemic area of vivax malaria or the mixed epidemic area of falciparum and vivax malaria, the use of pyrimethamine 50 mg for 10–15 days was effective. When chlorpyrifos used to prevent malaria, the results were poor.

The studies on CQ resistance and its geographical distribution in China showed that CQ resistance was first confirmed in Yunnan Province in 1972 and then in Hainan Island and Guangdong Province in 1974 and then Ningming County in Guangxi in 1977. In 1980, the NIPD collaborated with eight provinces (regions), including Guangdong, Guangxi, Guizhou, Guizhou, Guizhou, Jiangsu, Henan, and Anhui to carry out a large-scale investigation on the CQ resistance to *P. falciparum*. By 1984, the baseline of geographical distribution and CQ resistance status in the above areas were identified. The results showed that there was high CQ resistance in Hainan Island and southern Yunnan Province; more than 80% of cases were CQ-resistant to *P. falciparum* in Hainan Island epidemic areas, and more than 85% of cases were CQ-resistant to *P. falciparum* in Yunnan Province. The researchers also identified CQ resistance in Guangxi and Anhui, but the degree of resistance was low. The CQ resistance to *P. falciparum* in Henan, Guizhou, and Jiangsu was at the initial stage.

In order to reduce CQ resistance in falciparum malaria, from 1981 to 2003 the sensitivity of *P. falciparum* to CQ in Hainan and Yunnan was determined by an *in vitro* four-week micro-method. The results showed that the resistance rate of the *in vitro* assay in Ledong County of Hainan decreased from 97.9% in 1981 to 26.7% in 1997 ($P<0.01$), and the average concentration for the complete inhibition of schizont formation was reduced from 10.46 ± 7.14 to 1.63 ± 1.47 pmol/μL ($P<0.01$) in the blood. Moreover, the *in vivo* method determined that the resistance rate decreased from 84.2% in 1981 to 18.4% in 1997 ($P<0.01$), and the RIII (Resistance 1) in the resistant cases decreased from 53.1% to 14.3% ($P<0.01$). In 2001, the resistance rate as determined by the *in vitro* method in Yaliang Township of Sanya City was 59.8%, and the average concentration of needed to inhibit schizont formation was 3.56 ± 2.12 pmol/μL in the blood. In 2003, the resistance rate in Fuhu Township of Ledong County was 62.5%, and the RI, RII, and RIII (Resistance 1–3, respectively) accounted for 50%, 30%, and 20% of the resistant cases, respectively. The *in vitro* resistance rate of CQ in Mengla County of Yunnan Province decreased from 97.4% in 1981 to 77.8% in 1999 ($P<0.01$), and the average drug concentration that completely inhibited the formation of schizonts was reduced from 17.2 ± 12.6 to 4.4 ± 3.1 pmol/μL ($P<0.01$). In 2002, Jinghong County, Yunnan Province, demonstrated that the resistance rate was 70.4% as determined by the *in vitro* assay, and the average concentration of inhibition of schizont formation was 4.0 ± 3.3 pmol/μL. The results showed that *P. falciparum* presented with some degree of reversible resistance to chloroquine. After CQ resistance was produced, if CQ was stopped in time, *P. falciparum* was able to gradually restore sensitivity to CQ without drug pressure. However, the recovery rate was slow. For example, in Hainan Province, 20%–60% of the cases of CQ resistance have been reduced for more than 20 years; however, there is still some resistance to CQ.

In order to clarify the sensitivity of *P. falciparum* to commonly used antimalarial drugs in China, various malaria drug applicator plates were successfully developed to monitor the sensitivity of the malaria parasite to antimalarial drugs, such as piperaquine, pyronaridine, artemether, and artesunate, on the basis of successful imitation and improvement of chloroquine assay plate From 1986 to 1992, the sensitivity of *P. falciparum* to nine antimalarials was determined in Hainan and Yunnan provinces. As a result, no resistant cases were found for mefloquine and quinine, but resistant cases were found for all seven other drugs. Of note, high resistances to CQ, piperaquine, and acetaminophen were found, and 8.3% of the cases were resistant to pyronaridine. Moreover, a few cases had decreased sensitivity to artemisinin. After CQ was stopped or reduced in Hainan and Yunnan provinces, the sensitivity of *P. falciparum* to CQ recovered, while the resistance to piperaquine increased. The above results provided a scientific basis for the development of anti-malaria planning and rational selection of antimalarial drugs. In order to rationally apply the existing antimalarial drugs to delay the emergence of resistance, the Ministry of Health commissioned the NIPD to establish a principle for antimalarial use in China in 2000, and divided the antimalarial drugs used in China into first, second, and third-line drugs. CQ and piperaquine were the first-line drugs for vivax and falciparum malaria endemic areas sensitive to CQ and piperazine. Pyronaridine and artemisinin were second-line drugs for areas with moderate-to-high CQ and piperaquine resistance. Pyronaridine or artemisinin drugs and other antimalarial drugs were used as third-line drugs in areas where the efficacy of second-line drugs was not obvious.

7.4 Insecticide resistance

In the early 1960s, *An. sinensis* was very sensitive to organochlorine insecticides before the widespread use of pesticides in various places. The NIPD performed research on the sensitivity of *An. sinensis* to DDT, benzene hexachloride (BHC), organophosphorus pesticides and other insecticides (Cheng et al., 1995; Li et al., 1989; Wang et al., 2013; Wang, 1999; Wu et al., 1993). From 1978 to 1982, a survey of 62 points in 15 provinces (cities) showed that, except for Luodian, Guizhou Province, *An. sinensis* had different degrees of resistance to DDT across all areas, 34 of which (54.8%) had significant resistance, and the LC50 was >2%. Moreover, 27 points had initial resistance. In the late 1960s, *An. sinensis* was generally resistant to BHC. According to a survey of 57 points in 11 provinces (cities), *An. sinensis* in 49 points (86%) showed resistance to BHC, mostly initial resistance. With the restriction on the use of BHC, and eventual total ban on its production and use, the mosquito resistance to BHC has demonstrated a significant downward trend. In the suburbs of Shanghai, the LC50 was 1.87% in 1973, an increase of 86 times from 1961, but it fell to 1.20% and 0.3% in 1977 and 1979, respectively. The LC50 of BHC was at 0.99% in 1977 and fell to 0.17% in 1982 in Zhengzhou and Henan Province, which was close to the sensitive level. Since the 1960s, organophosphate insecticides have been widely used in the prevention and control of agricultural pests in China, but they are rarely used as a malaria vector control. According to the results of the 1979–83 survey, *An. sinensis* resistance to organophosphate insecticides was widespread and very severe in the southeastern region, while resistance to organophosphate was limited and relatively slight in the western region. Resistance to chlorpyrifos occurred only in the east, and resistance to malathion was mostly in the east and central China. This may be related to the widespread use of organophosphorus pesticides in the local area to control agricultural pests. In 1963, the sensitivity investigation of *An. minimus* to DDT and BHC in Taishan and Shangxiachuan of Guangdong showed that the LC50 of DDT was 0.044%, and the LC50 of BHC was 0.005%. This was similar to the reported LC50 of DDT in Taiwan, which was 0.09%, and the LC50 of BHC was 0.0066% in Indonesia, which can be regarded as the normal sensitivity of the mosquito to DDT and BHC. In southern China, organochlorine insecticides were mostly used for residual indoor spraying to prevent *An. minimus*. Since 1977, the results of various surveys have shown that this species of mosquito has reduced sensitivity to DDT and BHC. The LC50 of *An. minimus* in Guangdong was 0.077%–1.46%, while it was 0.3633%–0.7619% and 0.495% in Yunnan and Guizhou, respectively. By contrast, the LC50 was 0.05%–0.155%, 0.0043%–0.2172%, and 0082% in the above three regions, respectively. Compared with the surveillance data in 1963, the sensitivity to DDT and BHC decreased by 33 times and 43 times, respectively. However, this is not enough to influence the effect of residual spraying.

8. Future perspectives

Since the founding of the People's Republic of China, malaria has been regarded as a critical infectious disease requiring prevention and control. Different prevention strategies have been proposed based on malaria epidemic characteristics and degree of severity in different periods, and as such, remarkable achievements have been made. The annual incidence of malaria fell from 122.9/10,000 (6.97 million cases) in 1954 to 0.06/10,000 (7855 cases) in 2010. A solid foundation has been built such that China could move from "malaria prevention" to "malaria elimination." In the past 30 years, China's economy has developed rapidly, society has been stable, the government has paid attention to the people's livelihood, and a sound grassroots public health service system has been established, all of which provides favourable conditions for the elimination of malaria in rural areas. China was once a country with a serious epidemic of falciparum malaria. However, after a vigorous prevention and control programme, there have been no new indigenous cases of falciparum malaria in provinces since 2010, except for Yunnan Province. In addition, the island-type Hainan Province and central Jiangsu, Anhui, and Henan provinces have eliminated malaria and accumulated successful experience. Since the launch of the national malaria elimination programme (NMEP), the national work to eliminate malaria has progressed smoothly. China has been listed by the World Health Organization (WHO) as one of the 21 countries in the world to achieve the goal of eliminating malaria by 2020. The "1-3-7" work model for malaria elimination has also been officially written into the WHO's technical documents for global application. Since May 2010, the former Ministry of Health and other 13 ministries and commissions have issued the "China Malaria Elimination Action Plan (2010–20)." Since then, various local departments have carefully organized and implemented malaria prevention and control work. Of note, in 2017, China reported no indigenous malaria cases for the first time nationwide. Although the malaria prevention and treatment work in China has made significant progress and possesses the conditions and foundation for malaria elimination, there is still a long way to go to eliminate malaria eternally. At present, the challenges for eliminating malaria and consolidating results are mainly reflected in the aspects discussed below.

8.1 Malaria retransmission risk

At present, China is in a critical period to eliminate malaria. Given the fact that the source of infection (imported cases from abroad) has not been completely eliminated and that the malaria vector still exists, in addition to the Yunnan border area being a hot spot and focus for malaria elimination in China, malaria transmission conditions still exist in these areas (Zhang et al., 2016). If malaria transmission conditions cannot be discovered and removed in time, then China will still face the risk of retransmission and recovery of the epidemic. In addition, the larger number of imported cases from abroad will pose a huge challenge and threat to the prevention and elimination of malaria in China, which also faces the risk of local retransmission caused by imported malaria. Therefore, it is necessary to maintain the continuity of malaria surveillance, ensure adequate professionals, sustain investment on prevention and treatment, provide policy support, and manage support to effectively prevent the resurgence of local infections and transmission in China (Shi et al., 2017).

8.2 Maintenance of post-malaria elimination surveillance and response capacity

China has built a nationwide network epidemic reporting system and comprehensive monitoring network and has developed feasible monitoring indicators. In recent years, the malaria epidemic has dropped to a low level in

China, and as a result, the number of professionals engaged in malaria prevention and treatment has been seriously deficient. In some areas, there are no departments specializing in malaria prevention and treatment. The building capacity of the disease control institutions and construction of the talent team are still very weak, and there is not enough financial support. The relevant applied research and development of early and sensitive diagnostic tools to discover and dispose of local active epidemics and respond to unexpected imported malaria cases are also seriously lagging behind. In order to achieve the goal of eliminating malaria as scheduled and to ensure the implementation of various technical measures to eliminate malaria, capacity building is a top priority in the elimination of malaria. Therefore, it is necessary to further strengthen institutional construction and technical training, and build a professional team that can implement various measures in the process of eliminating malaria and long-term monitoring after malaria elimination. Only after these foundations are consolidated can the goal of eliminating malaria be achieved as scheduled.

8.3 Long-term mechanism for managing imported malaria cases and mobile population

With the advancement of global integration, the number of cases imported by migrants and foreign workers has been rising (Feng et al., 2014a). Since 2017, all the malaria cases have been imported malaria cases from abroad. However, at present, there are many weak points in the prevention and treatment of imported malaria. First, because the management of imported malaria involves many departments, effective management mechanisms have not been established or improved, and as such, various management measures cannot be implemented well. Second, the mobile population generally lacks the awareness, knowledge, and conditions of malaria protection, active consultation and standardized treatment. It is easy to get malaria when working or traveling in high malaria areas abroad, and after infection, these infected patients cannot seek timely treatment and obtain effective antimalarial treatment. A large number of imported malaria cases will likely occur in non-malaria endemic areas in China, where medical staff generally lack the awareness, diagnosis, and treatment skills of malaria, as well as the necessary diagnostic equipment and effective drug treatment, which may often lead to misdiagnosis or missed diagnosis and cannot effectively deal with the occurrence of falciparum malaria cases or epidemics, resulting in increased disease burden and even death of patients (Feng et al., 2014a).

8.4 Malaria diagnosis, drug resistance, and insecticide resistance issues

The conventional blood smear for fever patients has low sensitivity and is prone to missed diagnosis (Zhao et al., 2018). The rapid diagnosis test (RDT) for malaria has difficulty in identifying parasite species and provides false negatives in patients with a low-density parasite burden. These shortcomings cannot meet the requirements for the identification of malaria patients. Therefore, it is urgent to develop a new, sensitive, and specific malaria detection technology that can efficiently detect malaria infection sources in a timely manner. It has been reported that the sensitivity of *P. falciparum* to artemisinin has begun to decrease, and this resistance on emerged on the Thai-Myanmar border. A CQ-resistant strain of *P. vivax* has already evolved and begun to spread within Southeast Asia. Once these drug-resistant strains disseminate into China, they will inevitably affect the implementation and completion of our malaria elimination (Chen et al., 2010). However, the conventional in vivo drug sensitivity surveillance method is difficult to implement in China due to a few, scattered cases and cannot meet the higher requirements of the resistance test. The research on new technology for the detection of malaria parasite resistance based on resistance gene detection has become one of the key technical problems in malaria elimination in China. Also, with the long-term use of pesticides, it has been found that a variety of Anopheles mosquitoes have developed resistance worldwide (Dai et al., 2015; Feng et al., 2015b). This is an increasing trend, which poses a serious threat to the current stage of malaria prevention and control. Therefore, other new vector control strategies should be developed to mitigate the impact of pesticide resistance on global health.

9. Conclusion

Malaria, a disease that seriously endangers the health of the people, is currently at the stage of elimination. It is another disease that will be eliminated, following the eradication of smallpox and filariasis in China. It will be a milestone in the history of disease prevention in China, as well as the world. According to current work progress and overall plan to eliminate malaria, the goal of eliminating malaria nationwide by 2020 is fully achievable. However, the elimination of malaria is a complex and social system related project involving a wide range of factors with many influencing elements. Therefore, we should conscientiously implement the principles of scientific prevention, adaptation to local conditions and providing guidance to different strata, and adherence to the working mechanism of government taking the leadership, in cooperation with different departments, and participation of the whole society. We should also continuously improve science and technology on malaria control and surveillance tools, strengthen international cooperation and information exchanges on malaria elimination, make full use of domestic resources to consolidate the achievements gained in the national malaria elimination programme, and strive to achieve the national goal of eliminating malaria on schedule. At the same time, after the elimination, we should continue to focus on migrated people who return from malaria endemic areas of Africa and Southeast Asia, strengthening the surveillance and response systems to prevent introduction of imported malaria cases or foci, and maintaining malaria surveillance and response capabilities both in laboratory and epidemiological works. Through closely multi-sectoral cooperation between health and other departments, such as departments of commerce, tourism, customs and so one, we will effectively perform joint surveillance and control work to prevent secondary transmission caused by imported malaria.

Acknowledgements

The work in this chapter was supported by the Major Project on Vector-borne Pathogens in the Belt and Road (2018ZX10101002-002; 2018ZX10101002-004).

Competing interests

The authors declare that they do not have competing interests.

Authors' contributions

X.F., Z.X., X.N.Z., S.S.Z. conceived the study; X.F., Z.X., J.F., L.Z., H.Y., wrote the first version of the chapter and revised the chapter; X.F., Z.X., L.T., X.N.Z., S.S.Z. revised the chapter and gave approval of the version to be published. All the authors read and approved the final version of the chapter.

References

Cao, J., Sturrock, H.J.W., Cotter, C., Zhou, S., Zhou, H., Liu, Y., Tang, L., Gosling, R.D., Feachem, R.G.A., Gao, Q., 2014. Communicating and monitoring surveillance and response activities for malaria elimination: China's "1-3-7" Strategy. PLoS Med. 11, e1001642.

Chen, N., Chavchich, M., Peters, J.M., Kyle, D.E., Gatton, M.L., Cheng, Q., 2010. Deamplification of pfmdr1-containing amplicon on chromosome 5 in Plasmodium falciparum is associated with reduced resistance to artelinic acid in vitro. Antimicrob Agents CH 54, 3395–3401.

Cheng, H., Yang, W., Kang, W., Liu, C., 1995. Large-scale spraying of bednets to control mosquito vectors and malaria in Sichuan, China. Bull. World Health Organ. 73, 321–328.

Dai, Y., Huang, X., Cheng, P., Liu, L., Wang, H., Wang, H., Kou, J., 2015. Development of insecticide resistance in malaria vector Anopheles sinensis populations from Shandong province in China. Malar. J. 14, 62.

Diouf, G., Kpanyen, P.N., Tokpa, A.F., Nie, S., 2014. Changing landscape of malaria in China: progress and feasibility of malaria elimination. Asia Pac. J. Public Health 26, 93–100.

Duo-Quan, W., Lin-Hua, T., Zhen-Cheng, G., Xiang, Z., Man-Ni, Y., 2009. Application of the indirect fluorescent antibody assay in the study of malaria infection in the Yangtze River Three Gorges Reservoir, China. Malar. J. 8, 199.

Expert Advisory Committee on Malaria, Ministry of Health, 2001. Malaria situation in the People's Republic of China in 2000. Chin. J. Parasitol. Parasit. Dis. 19, 257–259.

Feng, J., Yan, H., Feng, X.Y., Zhang, L., Li, M., Xia, Z.G., Xiao, N., 2014a. Imported malaria in China, 2012. Emerg. Infect. Dis. 20, 1778–1780.

Feng, X.Y., Xia, Z.G., Vong, S., Yang, W.Z., Zhou, S.S., 2014b. Surveillance and response to drive the national malaria elimination program. Adv. Parasitol. 86, 81–108.

Feng, J., Xiao, H., Xia, Z., Zhang, L., Xiao, N., 2015a. Analysis of malaria epidemiological characteristics in the People's Republic of China, 2004-2013. Am. J. Trop. Med. Hyg. 93, 293–299.

Feng, X., Yang, C., Yang, Y., Li, J., Lin, K., Li, M., Qiu, X., 2015b. Distribution and frequency of G119S mutation in ace-1 gene within Anopheles sinensis populations from Guangxi, China. Malar. J. 14, 470.

Feng, X., Zhang, S., Huang, F., Zhang, L., Feng, J., Xia, Z., Zhou, H., Hu, W., Zhou, S., 2017. Biology, bionomics and molecular biology of Anopheles sinensis Wiedemann 1828 (Diptera: Culicidae), main malaria vector in China. Front. Microbiol. 8, 1473.

Feng, J., Zhang, L., Huang, F., Yin, J.H., Tu, H., Xia, Z.G., Zhou, S.S., Xiao, N., Zhou, X.N., 2018. Ready for malaria elimination: zero indigenous case reported in the People's Republic of China. Malar. J. 17, 315.

Gao, Q., Yang, C.X., Zhang, S.Y., Yang, Z.Y., Zhang, W.Q., Li, J.L., 1991. Detection of blood stage antigens of Plasmodium vivax by sandwich ELISA using pan-species monoclonal antibodies and polyclonal antibodies. Southeast Asian J. Trop. Med. Public Health 22, 393–396.

Goodman, D.S., 1986. Centre and Province in the People's Republic of China: Sichuan and Guizhou, 1955-1965. CUP Archive.

Hu, T., Liu, Y.B., Zhang, S.S., Xia, Z.G., Zhou, S.S., Yan, J., Cao, J., Feng, Z.C., 2016. Shrinking the malaria map in China: measuring the progress of the National Malaria Elimination Programme. Infect. Dis. Poverty 5, 52.

Huang, S., 1980. The present status of malaria control in the People's Republic of China. Jpn J. Trop. Med. Hyg. 8, 159–172.

Kai-Jie, L., Shun-Xiang, C., Wen, L., Jing, X., Su-Jian, P., Hua-Xun, Z., 2016. Analysis of malaria epidemic situation and control in Hubei Province from 1974 to 2015. Chin. J. Parasitol. Parasit. Dis. 28, 393–396.

Li, Z.Z., Zhang, M.C., Wus, Y.G., Zhong, B.L., Lin, G.Y., Huang, H., 1989. Trial of deltamethrin impregnated bed nets for the control of malaria transmitted by Anopheles sinensis and Anopheles anthropophagus. Am. J. Trop. Med. Hyg. 40, 356–359.

Li, S., Yin, S., Wang, J., Li, X., Feng, J., 2016. Shifting from control to elimination: analysis of malaria epidemiological characteristics in Tengchong County around China-Myanmar border, 2005-2014. Malar. J. 15, 45.

Liu, C., 1990. Comparative studies on the role of Anopheles anthropophagus and Anopheles sinensis in malaria transmission in China. Chin. J. Epidemiol. 11, 360–363.

Liu, X., Jackson, S., Song, J., Sleigh, A.C., 1996. Malaria control and fever management in Henan Province, China, 1992. Trop. Med. Int. Health 1, 112–116.
Lu, B.L., Xu, J.J., Dong, X.S., 1997. Fauna Sinica, Insecta, vol. 9, Diptera: Culicidae II. Science Press.
Qian, H., Tang, L., 2000. Achievements and prospects of malaria control in China in the past 50 years. Chin. J. Epidemiol. 21, 225–227.
Sheng, Z., Zhongjie, L., Chris, C., Canjun, Z., Qian, Z., Huazhong, L., Shuisen, Z., Xiaonong, Z., Hongjie, Y., 2016. Trends of imported malaria in China 2010–2014: analysis of surveillance data. Malar. J. 15, 39.
Shi, B., Zheng, J., Qiu, H., Yang, G.J., Xia, S., Zhou, X.N., 2017. Risk assessment of malaria transmission at the border area of China and Myanmar. Infect. Dis. Poverty 6, 108.
Shou-Pai, M., 1983. Parasitological research in institutes in China. In: Warren, K.S., Bowers, J.Z. (Eds.), Parasitology. Springer, New York, NY.
Sleigh, A.C., Liu, X.L., Jackson, S., Li, P., Shang, L.Y., 1998. Resurgence of vivax malaria in Henan Province, China. Bull. World Health Organ. 76, 265–270.
Tang, L., 2000. Progress in malaria control in China. Natl. Med. J. Chin. 113, 89–92.
Tang, L., 2009. Malaria in China: from control to elimination. Int. J. Med. Parasit. Dis. 36, 258–265.
Wang, J., 1999. Resistance to two pyrethroids in Anopheles sinensis from Zhejiang, China. J. Am. Mosq. Control Assoc. 15, 308–311.
Wang, D.Q., Xia, Z.G., Zhou, S.S., Zhou, X.N., Wang, R.B., Zhang, Q.F., 2013. A potential threat to malaria elimination: extensive deltamethrin and DDT resistance to Anopheles sinensis from the malaria-endemic areas in China. Malar. J. 12, 164.
Wang, R.B., Zhang, Q.F., Zheng, B., Xia, Z.G., Zhou, S.S., Tang, L.H., Gao, Q., Wang, L.Y., Wang, R.R., 2014a. Transition from control to elimination: impact of the 10-year global fund project on malaria control and elimination in China. Adv. Parasitol. 86, 289–318.
Wang, Y., Ma, A., Chen, S.B., Yang, Y.C., Chen, J.H., Yin, M.B., 2014b. Genetic diversity and natural selection of three blood-stage 6-Cys proteins in Plasmodium vivax populations from the China-Myanmar endemic border. Infect. Genet. Evol. 28, 167–174.
Wang, D., Chaki, P., Mlacha, Y., Gavana, T., Michael, M.G., Khatibu, R., Feng, J., Zhou, Z.B., Lin, K.M., Xia, S., Yan, H., Ishengoma, D., Rumisha, S., Mkude, S., Mandike, R., Chacky, F., Dismasi, C., Abdulla, S., Masanja, H., Xiao, N., Zhou, X.N., 2019. Application of community-based and integrated strategy to reduce malaria disease burden in southern Tanzania: the study protocol of China-UK-Tanzania pilot project on malaria control. Infect. Dis. Poverty 8, 4.
Wu, N., Qin, L., Liao, G., Zhou, W., Geng, W., Shi, Y., Tan, Y., Zhao, K., 1993. Field evaluation of bednets impregnated with deltamethrin for malaria control. Southeast Asian J. Trop. Med. Public Health 24, 664–671.
Xia, Z.G., Tang, L.H., Gu, Z.C., Huang, G.Q., Zheng, X., Wang, Y., Huang, X.P., 2003. Study on the thresholds of malaria transmission by Anopheles anthropophagus in Hubei Province. Chin. J. Parasitol. Parasit. Dis. 21, 224–226.
Xia, Z.G., Wang, R.B., Wang, D.Q., Feng, J., Zheng, Q., Deng, C.S., Abdulla, S., Guan, Y.Y., Ding, W., Yao, J.W., Qian, Y.J., Bosman, A., Newman, R.D., Ernest, T., O'Leary, M., Xiao, N., 2014. China-Africa cooperation initiatives in malaria control and elimination. Adv. Parasitol. 86, 319–337.
Xiao, D., Long, Y., Wang, S., Fang, L., Xu, D., Wang, G., Li, L., Cao, W., Yan, Y., 2010. Spatiotemporal distribution of malaria and the association between its epidemic and climate factors in Hainan, China. Malar. J. 9, 185.

Xiao, D., Long, Y., Wang, S., Wu, K., Xu, D., Li, H., Wang, G., Yan, Y., 2012. Epidemic distribution and variation of plasmodium falciparum and plasmodium vivax malaria in Hainan, China during 1995-2008. Am. J. Trop. Med. Hyg. 87, 646–654.
Xu, J.J., Feng, L.C., 1975. Studies on the Anopheles hyrcanus group of mosquitoes in China. Acta Entomol. Sin.
Yin, J.H., Zhou, S.S., Xia, Z.G., Wang, R.B., Qian, Y.J., Yang, W.Z., Zhou, X.N., 2014. Historical patterns of malaria transmission in China. Adv. Parasitol. 86, 1–19.
Zhang, L., Tang, L., Feng, X., Wang, J., 1998. Study on detection of malaria parasite DNA by PCR-ELISA. Chin. J. Parasitol. Parasit. Dis. 16, 11–15.
Zhang, Q., Sun, J., Zhang, Z., Geng, Q., Lai, S., Hu, W., Clements, A.C., Li, Z., 2016. Risk assessment of malaria in land border regions of China in the context of malaria elimination. Malar. J. 15, 546.
Zhao, Y., Zeng, J., Zhao, Y., Liu, Q., He, Y., Zhang, J., Yang, Z., Fan, Q., Wang, Q., Cui, L., Cao, Y., 2018. Risk factors for asymptomatic malaria infections from seasonal cross-sectional surveys along the China-Myanmar border. Malar. J. 17, 247.
Zhou, X.N., Bergquist, R., Tanner, M., 2013. Elimination of tropical disease through surveillance and response. Infect. Dis. Poverty 2, 1.
Zu-Jie, Z., 1981. The malaria situation in the People's Republic of China. Bull. World Health Organ. 59, 931.

CHAPTER THREE

Contribution to the echinococcosis control programme in China by NIPD-CTDR

Meihua Fu[a,b,c,d,e], Shuai Han[a,b,c,d,e], Chuizhao Xue[a,b,c,d,e], Xu Wang[a,b,c,d,e], Baixue Liu[a,b,c,d,e], Ying Wang[a,b,c,d,e], Liying Wang[a,b,c,d,e], Sihui Wei[a,b,c,d,e], Xiaoyu Cui[a,b,c,d,e], Ting Zhang[a,b,c,d,e], Hao-Bing Zhang[a,b,c,d,e], Bin Zheng[a,b,c,d,e,f], Tian Tian[a,b,c,d,e], Shijie Yang[a,b,c,d,e], Chun-Hua Gao[a,b,c,d,e], Zhisheng Dang[a,b,c,d,e], Bin Xu[a,b,c,d,e], Qing Yu[a,b,c,d,e], Weiping Wu[a,b,c,d,e,]*

[a]National Institute of Parasitic Diseases, Chinese Center for Disease Control and Prevention, Shanghai, People's Republic of China
[b]Chinese Center for Tropical Diseases Research, Shanghai, People's Republic of China
[c]WHO Collaborating Centre for Tropical Diseases, Shanghai, People's Republic of China
[d]National Center for International Research on Tropical Diseases, Ministry of Science and Technology, Shanghai, People's Republic of China
[e]Key Laboratory of Parasite and Vector Biology, Ministry of Health, Shanghai, People's Republic of China
[f]School of Global Health, Chinese Center for Tropical Diseases Research, Shanghai Jiao Tong University School of Medicine, Shanghai, People's Republic of China
*Corresponding author: e-mail address: wuwp@nipd.chinacdc.cn

Contents

1. Introduction 108
2. Active participation in the national echinococcosis control programme 110
   2.1 Echinococcosis programme in China 110
   2.2 Contribution to the national echinococcosis control programme 112
3. Promoting the pilot project on echinococcosis control 113
   3.1 Establishment of the pilot programme on echinococcosis control 113
   3.2 Contribution to the pilot project on echinococcosis control 115
4. Role in the surveillance of echinococcosis 121
   4.1 The national surveillance network on echinococcosis 121
   4.2 Contribution to surveillance 122
5. Contribution to scientific research and personnel training 123
   5.1 Progress in scientific research 123
   5.2 Control strategies 137
   5.3 Training senior professionals 138
6. Conclusions 139
References 141

Advances in Parasitology, Volume 110
ISSN 0065-308X
https://doi.org/10.1016/bs.apar.2020.04.010

Abstract

As a zoonotic parasitosis caused by the parasitism of *Echinococcus* larvae, echinococcosis imposes serious disease and economic burdens on human beings and society, and is thus a global public health issue. Its complex life history, wide distribution, the combined influence of various epidemic factors, coupled with the unique natural environment, customs, and religious beliefs in endemic areas, pose a huge challenge to the national echinococcosis control programme in China. Accurate early detection and confirmation of diagnosis of echinococcosis, the use of effective drugs, real-time surveillance of the infection status of populations and various hosts, controlling the source of infection, and blocking the route of transmission are of enormous significance for control. In this paper, the work by NIPD-CTDR on the prevention and control of echinococcosis in China is reviewed, with a view to providing reference for the further promotion of the national echinococcosis control programme.

1. Introduction

Echinococcosis, also known as hydatid disease, is a zoonotic parasitic disease caused by the larvae of *Echinococcus* in the genus of *Echinococcus* and the family of Taeniidae parasitizing humans or animals. It seriously endangers human health and life safety and affects social and economic development, and is thus a global public health issue (Chen et al., 2018; Wu et al., 2018; Zhang et al., 2015a). Up to now, there are four universally recognized independent species of *Echinococcus* infecting humans, including *Echinococcus granulosus*, *E. multilocularis*, *E. oligarthrus*, and *E. vogeli* (Liu et al., 2017). The larvae of these four species of *Echinococcus* can cause three types of human echinococcosis, among which, *E. granulosus* and *E. multilocularis* larvae infection can cause cystic echinococcosis (CE) and alveolar echinococcosis (AE), respectively, which are the most harmful to human beings. Two other species, namely *E. vogeli* and *E. oligarthrus*, are responsible for polycystic echinococcosis. In addition, *E. shiquicus* has been found in Shiqu County, Sichuan Province and some areas in neighbouring Qinghai Province, China. It has been reported that dogs, Tibetan foxes, pikas, and voles can be infected with *E. shiquicus*. However, no human infection with *E. shiquicus* has been reported (Qi and Wu, 2013).

Humans are accidental intermediate hosts of *Echinococcus* spp., which is usually caused by ingestion of *Echinococcus* spp. eggs in the faeces of dogs or other canines (Zhang et al., 2015b). Echinococcosis is a chronic disease with slow progression and long incubation period. Its lesions after infection with larvae of *Echinococcus* can involve multiple organs, mainly liver and lung. Patients with long-term echinococcosis will lose their capacity for physical labour or even die (Liu et al., 2014, 2015). The pathogenicity of echinococcosis is not only related to prolonged chronic inflammation and the toxicity of hydatid cysts to the host, but also to immunosuppressants secreted by hydatid cysts, which cause a variety of complications in the human body and result in shortened life-span and reduced quality of life of patients (Khachatryan, 2017). In particular, AE is a highly fatal disease. For patients who missed early diagnosis and treatment, the 10-year fatality rate is as high as 94% (Wu et al., 2018).

Echinococcosis, one of the neglected tropical diseases, not only endangers human health, but also brings serious economic losses and disease burden to the society. It is estimated that over 1 million people worldwide are at risk of developing AE and CE, and the economic cost incurred is approximately US $ 3 billion per year (Yu et al., 2018). The global CE-related economic loss amounts to US $ 1.92 billion per year, and the CE-related livestock production loss is up to US $ 2.19 billion per year (Qian et al., 2017), while the treatment cost and livestock loss related to echinococcosis are estimated to be US 3$ billion (Chen et al., 2018). In addition, the global disease burden of patients with echinococcosis in 2010 was about 871,000 disability-adjusted life-years (DALYs) (Li et al., 2019). The annual disease burden of CE patients worldwide is at least 1 million DALYs, and that of AE patients is 670,000 DALYs (Chen et al., 2018; Zhang and Wu, 2017). The prevalence of echinococcosis in China is also very serious. According to the survey data of China in 2004, there were about 380,000 cases of echinococcosis in China (Wang et al., 2016), and the annual CE-caused economic loss was US $ 660 million (Qian et al., 2017). It has been reported that about 30 million livestock are infected in China and 7 million livestock are infected every year, among which, over 70% are sheep, resulting in an economic loss of 1 billion Yuan (RMB) (Yang et al., 2015). The disease burden of CE patients in China is nearly 400,000 DALYs, accounting for 40% of the global CE disease burden, and the number of new AE cases accounts for 91% of the world's total every year (Chen et al., 2018; Wang et al., 2014). Echinococcosis has become a major public health problem in China. In the face of the severe situation of high prevalence of echinococcosis in China, the National Institute of Parasitic Diseases at China CDC & Chinese Center for Tropical Diseases Research (NIPD-CTDR) has effectively led, promoted, and made outstanding contributions to the prevention and control of echinococcosis in China. Based on this, this study summarized the work of echinococcosis prevention and control in China performed or coordinated by NIPD-CTDR.

2. Active participation in the national echinococcosis control programme

2.1 Echinococcosis programme in China

Echinococcosis is highly endemic in China and primarily prevalent in 368 counties in pastoral areas of western China, with a threatened population of approximately 50 million and an average prevalence rate of 0.51% (Wu et al., 2018), that can be as high as 14.3% in some areas (Gongsang et al., 2017). It is one of the main reasons for people in agricultural and pastoral areas in western China to become poor or return to poverty (Fig. 1). Since 2005, the central and local governments at all levels have stepped up efforts to prevent and control echinococcosis. In the same year, China

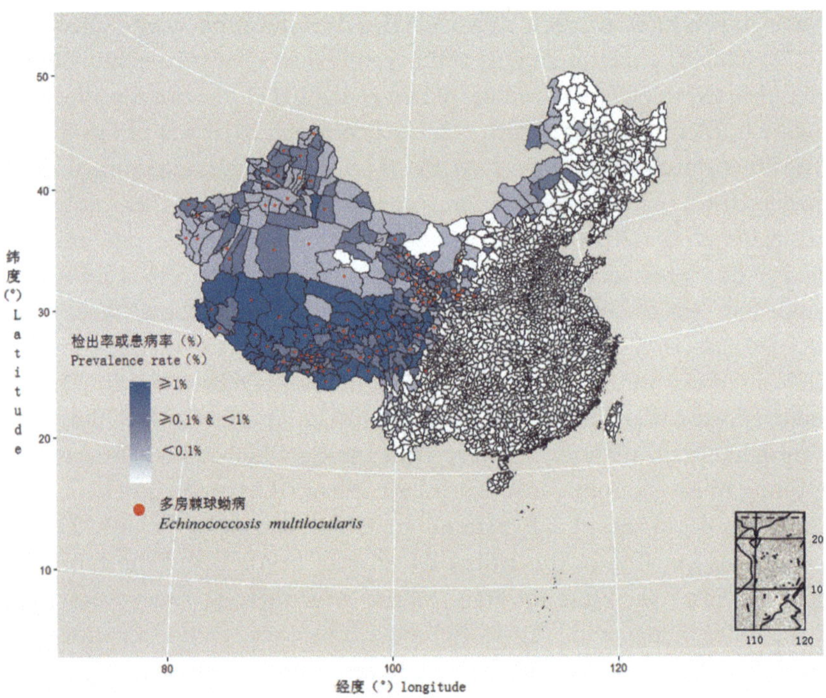

Fig. 1 Nationwide epidemic map of echinococcosis. This map showing the prevalence of echinococcosis in China, with the red dots representing the areas where human AE (county level) distributed, and the blue areas (light blue to dark blue) showing the detection rate or prevalence rate of echinococcosis in the corresponding areas.

launched the transfer payment project of central finance for the prevention and control of local echinococcosis (Zhu et al., 2016). Subsequently, echinococcosis was listed as one of the six infectious diseases benefitting of free treatment in China (Qian et al., 2019). The attention of the national government has effectively promoted the prevention and control of echinococcosis and achieved good preliminary results. However, the prevention and control of echinococcosis in China is still faced with many difficulties and challenges. Therefore, in order to comprehensively promote the prevention and control of echinococcosis, the NIPD-CTDR carried out the research project of "Research on Comprehensive Echinococcosis Control Technology" under the "National 11th Five-Year Plan" during 2008–2010, in collaboration with Shanghai Jiaotong University, Sun Yat-Sen University, Sichuan Center for Disease Control and Prevention, Sichuan Agricultural University, and Institute of Science of Ganzi Prefecture. Furthermore, the NIPD-CTDR has organized and formulated the Action Plan for the Prevention and Control of Echinococcosis (2010–2015) (referred to as the "Action Plan") and the National Plan for the Prevention and Control of Echinococcosis and other Key Parasitic Diseases (2016–2020) (referred to as "Control Plan"), and is currently participating in the formulation of the echinococcosis control programme under the "National 14th Five-Year Plan". The Action Plan required that in townships in the Qinghai-Tibet Plateau, the canine infection rate and the serological positive rate of children aged 6–12 should all be reduced to below 8%, and the prevalence rate of livestock under 2 years old should be reduced to below 15%. While in other epidemic areas, the canine infection rate and the serological positive rate of children aged 6–12 should all be reduced to below 5%, and the prevalence rate of livestock under 2 years old should be reduced to below 10%. In 2015, the final evaluation of the implementation of the Action Plan was carried out in the endemic counties of China. The evaluation results showed that the up-to-standard prevalence rate of children aged 6–12, the up-to-standard canine infection rate, and the up-to-standard prevalence rate of livestock in national endemic areas were 96.64%, 92.91%, and 92.77%, respectively. And the three rates in the Qinghai-Tibet Plateau area were 97.66%, 88.89%, and 90.91%, respectively, while in the non-Qinghai-Tibet Plateau area were 93.56%, 94.79%, and 93.60%, respectively (Wang, 2017). The overall goal of the Control Plan was to establish and improve the surveillance system for key parasitic diseases, and basically control the prevalence of echinococcosis, namely controlling the prevalence of echinococcosis below 1% and the infection rate of domestic dogs below 5%

in over 70% endemic counties. The formulation and implementation of these programmes can reflect the specific progress in echinococcosis control in China, thereby providing a basis for promoting the prevention and effective control of echinococcosis.

2.2 Contribution to the national echinococcosis control programme

NIPD-CTDR, as the core centre to provide guidance for the prevention and control of echinococcosis in China, organizes and holds a technical seminar on echinococcosis control every year, training courses on echinococcosis control technology, and academic meetings related to echinococcosis on an regular basis, as well as formulates the echinococcosis control programme and organizes the work of echinococcosis prevention and control throughout the country to identify and resolve problems in time. For instance, NIPD has coordinated the formulation of the national standards on "Echinococcosis Diagnosis" and "Echinococcosis Control", the medium-term evaluation scheme on the echinococcosis control programme under "National Program for the Prevention and Control of Echinococcosis and Other Key Parasitic Diseases (2016–2020)", and has held a series of training courses on control capacity of echinococcosis at grassroots level in Ganzi Prefecture of Western China. In addition, NIPD-CTDR actively participated in various meetings and training courses organized by the National Health Commission and the China CDC, such as China CDC Training Course on Scientific Integrity and Medical Ethics, Tibet Health Training Session of the China Health Education Center, and National Training Course of the National 13th Five-Year Mid-Term Assessment Programme for Echinococcosis.

NIPD has also provided technical support for echinococcosis control at provincial and national levels. For instance, it has actively provided technical support, guidance and supervision, and daily consultations for disease endemic provinces, regions, and counties, such as formulating the mid-term evaluation scheme for the echinococcosis control programme under "National Program for the Prevention and Control of Echinococcosis and Other Key Parasitic Diseases (2016–2020)", executing the technical and evaluation works on "Technical Plan for the Prevention and Control of Echinococcosis", and formulating and monitoring the "Echinococcosis Surveillance Programme". Focusing on the prevention and control of echinococcosis in Tibetan areas in Qinghai, Sichuan, and Tibet, NIPD has completed relevant investigations and provided technical guidance at community level.

3. Promoting the pilot project on echinococcosis control

3.1 Establishment of the pilot programme on echinococcosis control

In 2005, the Ministry of Health of China began to launch the pilot programme of echinococcosis control in key endemic provinces (regions) (Yu, 2008). In this Control Plan, NIPD-CTDR was actively involved in the promotion of the pilot project. The aim of the project was to explore the control strategy and measures for echinococcosis control in Tibetan populations in Sichuan Province, that could be scaled up in other Tibetan areas in the country. Through the promotion of pilot works, many echinococcosis endemic provinces have started to perform control activities which has allowed dramatic progress in the national echinococcosis control programme. Taking Shiqu County of Sichuan Province as an example where the prevalence of echinococcosis was among the highest in the Province, the pilot project for comprehensive control of echinococcosis was launched in November 2015, with the following activities being comprehensively promoted: screening and treatment of echinococcosis patients, standardization of management procedures for echinococcosis control in dogs, control of infectious sources among livestock and rodents, health education for behavioural change, guarantee of safe drinking water supply, and comprehensive prevention and control among monks and nuns in temples.

Thanks to the joint efforts of all partners, the pilot project in Shiqu County progressed smoothly and generated important findings in 2017. The mid-term evaluation and technical report showed that the eight output indicators and three outcome targets were achieved to meet the targets of 2017 (Table 1). The later comprehensive evaluation indicated that Shiqu County had established an effective prevention and control mechanism, that the technical control measures had been fully implemented, that all the indicators for evaluation had been achieved or had exceeded the expected targets, that the pilot project had gained remarkable outcomes, such as effective containment of transmission of echinococcosis and a suitable working model for echinococcosis control in plateau pastoral areas, laying a foundation for further prevention and control of echinococcosis in China. As such, the implementation of the pilot project on echinococcosis control in Shiqu Country is of substantial significance for controlling echinococcosis in China.

Table 1 Evaluation of pilot targets and indicators for comprehensive echinococcosis control in Shiqu County.

Targets and indicators of pilot programme

Category	Baseline data in 2014 (%)	End of 2017 (%)	End of 2020 (%)	Results of self-assessment in August 2017 (%)	On-site technical evaluation results (%)
New patient detection rate in children aged 6–12	2.91	Below 2	Below 1	0.28	0.28
Canine infection rate	6	Below 4	Below 2	3.91	3.08
Livestock infection rate	45	Below 35	Below 25	32.58	32.58
Domestic dog registration rate	72	80	Above 85	100	100
Canine deworming coverage	73	80	90	97.05	100
Treatment rate of diseased organs in slaughterhouse	–	100	100	100	100
Safe drinking water coverage in settlements	80	96	100	100	100
Rodent control rate in settlements and within 1 km radius	–	75	80	100	100
Awareness rate of prevention and control knowledge	30	80	Above 85	87.94	87.94
Standardized treatment rate of patients	50	75	Above 85	91.69	97.5
Qualification rate of professional skills	60	90	95	95	100

3.2 Contribution to the pilot project on echinococcosis control

In order to explore the replicable control model in Tibetan areas and effectively boost the smooth implementation of the pilot project on comprehensive control of echinococcosis, the NIPD-CTDR has established a workstation for echinococcosis control in Ganzi Prefecture of Sichuan Province, under the guidance of the Bureau of Disease Control and Prevention of the National Health Commission of China. As a joint venture among various departments of the NIPD-CTDR, this workstation was launched in 2016, and has been actively involved in coordinating sub-projects. A total of 25 sub-projects were implemented in the workstation from 2016 to 2019 (Table 2).

The workstation in Ganzi played a role to bridge the benches to the fields, and link to the sub-projects by performing a series of field works on the control of echinococcosis, coordinating and providing rear services supports to the work by staff from NIPD-CTDR, collecting information on the progress of each sub-project, and compiling the monthly bulletin for the workstations. The main outputs of those sub-projects are as follows: (i) After 4 years of continuous surveillance, the findings from the sub-project on "Surveillance of three townships in Shiqu County" revealed that the population's awareness rate of echinococcosis increased from 41.2% (247/600) in 2016 to 91.3% (210/230) in 2019, the number of domestic dogs decreased significantly in 2019, and the positive rate of dog faeces decreased to 1.66% (1/60), and the positive rate of field intermediate hosts decreased by 23.42% compared with 2018. (ii) A number of training courses on echinococcosis control to training professionals at county level from Ganzi Prefecture, and training courses on control technology for parasitic diseases were held to train professionals from Ganzi Prefecture and in the Center for Disease Control and Prevention in Ganzi Prefecture were arranged every year. Among them, 8 training sessions and 10 seminars were held for professionals on field echinococcosis control technology and laboratory testing technology for echinococcosis. The training of echinococcosis operators in Ganzi Prefecture has reached more than 1000 person-times, the coverage rate of the technical guidance for the prevention and control for health workers from village, township and county levels in Shiqu County has reached more than 85%. (iii) The evaluation and demonstration of new technology on detection and diagnosis of echinococcosis in humans and canines have been completed. The progress of these works on diagnostics has considerably advanced the process of echinococcosis control in Ganzi Prefecture.

Table 2 List of sub-projects implemented by the NIPD-CTDR during 2016–2019.

No.	Person in charge	2016	2017	2018	2019
1	Wu Weiping	Establishment of echinococcosis prevention and control model based on animal husbandry group	–	–	–
2	Yu Qing	Application of a new deworming bait in the control of field infection source of echinococcosis	Extended observation on the effect of applying unmanned aerial vehicles (UAVs) to release deworming bait to control canine infection sources in the field	Study on intervention and related factors of field infection source of echinococcosis centred on human settlements in Qinghai-Tibet Plateau	Comparison of the changes in host infection status and the application of identification technology after termination of field infection source intervention in echinococcosis
3	Yang Shijie	Surveillance of intermediate host of *E. multilocularis*	Development and evaluation of automatic timing canine deworming collar based on "Internet +"	Development and field evaluation of intelligent canine deworming collar for echinococcosis prevention and control	Field evaluation of intelligent collar and feeder for echinococcosis prevention and control in canines
4	Han Shuai	–	–	Study on the infection route of echinococcosis in Qinghai-Tibet Plateau	Pilot project of echinococcosis prevention and control model based on grid management

5	Liu Baixue	–	–	–	Pilot health education model of echinococcosis in Tibetan area based on WeChat public platform
6	Wang Ying	–	Optimization and application of faecal sample detection technology for dogs infected with *Echinococcus*	–	Effect evaluation of rapid test strip for *Echinococcus* antigen
7	Ai Lin	–	–	Application of quantitative fluorescence PCR in detection of *Echinococcus* eggs in canine faeces in echinococcosis endemic areas	Establishment of recombined polymerase amplification detection technology for dogs infected with *Echinococcus*
8	Tian Liguang	Construction of serum/biological sample bank for patients with echinococcosis	Construction of serum bank for patients with echinococcosis	Construction of biological sample/serum bank of echinococcosis	
9	Kui Yan	–	Study on sustainable construction of workstation capacity for echinococcosis prevention and control in Ganzi Prefecture	–	Evaluation on the effect of intervention measures of echinococcosis health education based on strengthening the training of primary school teachers

Continued

Table 2 List of sub-projects implemented by the NIPD-CTDR during 2016–2019.—cont'd

No.	Person in charge	2016	2017	2018	2019
10	Jiang Bin	Evaluation of in situ solidified long-acting praziquantel injection against canine *Echinococcus*	Study on the effect evaluation of new slow-release praziquantel for controlling canine *Echinococcus* transmission	Laboratory and field effects of new slow-release praziquantel preparation on the control of *Echinococcus* infection in dogs	Formulation optimization of slow-release praziquantel preparation and establishment of quality standards for industrial products
11	Liu Congshan	–	–	–	Study on new target-based anti-hydatid drugs
12	Zhang Ting	Field evaluation of omics-based diagnostic technology for human echinococcosis and faecal antigen detection technology for dogs infected with *Echinococcus*	Study on the demonstration, popularization, and application of new technology for rapid diagnosis of echinococcosis	Pilot trial of the application of new technology for rapid echinococcosis diagnosis in population surveillance	Laboratory promotion of new technology for real-time diagnosis of human echinococcosis
13	Xu Bin	–	–	–	Evaluation of the efficacy of vaccine against canine echinococcosis based on recombinant proteins (AgB, GP50)
14	Dang Zhisheng	Field test method and effect evaluation of echinococcosis vaccine	Effect evaluation test of polypeptide vaccine against echinococcosis		–

15	Jiang Yanyan	–	–	Rapid on-site molecular diagnosis of echinococcosis based on honeycomb chip	–
16	Liu Hua	–	–	Study on genotype characteristics and genetic polymorphism of human *Echinococcus* infection in China	Study on species distribution and genetic characteristics of human and animal *Echinococcus* infection in some areas of China
17	Cao Shengkui	–	Application of rapid detection technology for canine faecal nucleic acid in *Echinococcus* infection	–	Study on the popularization and application of detection technology of *Echinococcus* faecal and the distribution of species in the definitive host
18	Wang Liying	Effect evaluation of drug and surgical treatment in patients with echinococcosis	Evaluation of therapeutic effect in patients with echinococcosis	–	–
19	Xue Chuizhao	–	On-site epidemiological survey and output capacity training of echinococcosis prevention and control in Ganzi Prefecture	Study on the demonstration, popularization, and application of new technology for rapid diagnosis of echinococcosis	–

Continued

Table 2 List of sub-projects implemented by the NIPD-CTDR during 2016–2019.—cont'd

No.	Person in charge	2016	2017	2018	2019
20	Xia Shang	–	–	Study on the mechanism and environmental factors of echinococcosis transmission and spread in China	–
21	Wang Junyun	Application and evaluation of new technology for CE/AE diagnosis	–	–	–
22	Mo Xiaojin	–	Training on laboratory testing technology for human echinococcosis and dogs infected with *Echinococcus*	–	–
23	Duan Liping	Preclinical evaluation of anti-*Echinococcus* candidate drug "H1451"	Field application of anti-sheep *Echinococcus* candidate drug "H1451" based on DNA topoisomerase	Study on new drugs against *Echinococcus*	–
24	Tian Tian	Investigation into infection factors of echinococcosis	–	–	–
25	Chen Jiaxu	Promotional film "The Second Cancer-Echinococcosis" and health education products	–	–	–

4. Role in the surveillance of echinococcosis

4.1 The national surveillance network on echinococcosis

Echinococcosis is prevalent in nine provinces (autonomous regions) of China, including Inner Mongolia, Sichuan, Yunnan, Tibet, Shaanxi, Gansu, Qinghai, Ningxia, and Xinjiang. In order to continuously understand the dynamic changes in the epidemic situation of echinococcosis in endemic areas and evaluate the effectiveness of prevention and control, China set up 20 fixed echinococcosis surveillance sites in 2015, and the number of surveillance sites rose to 140 after 2016. At present, a total of 140 echinococcosis surveillance sites have been established nationwide, and all 370 endemic counties will be covered in 2020. The surveillance items and methods include following two aspects:

1. *Routine surveillance*. The routine surveillance is performed through the case report by the medical institutions and disease control institutions at all levels, in accordance with relevant provisions of the "Law of the People's Republic of China on the Prevention and Control of Infectious Diseases" and "the Standard for the Administration of Information Reporting of Infectious Diseases". The report on diagnosed cases of echinococcosis covers the cases of echinococcosis detected in active visits or in screening projects for echinococcosis. Medical and health institutions that implement direct case reporting upload the data to the "China Disease Prevention and Control Information system" within 24 h through the internet, and those institutions that could not submit case reports directly through the internet have to send case reporting cards to institutions at county level within 24 h. County-level disease prevention and control agencies and qualified township health centres have to report directly through the internet once they receive the case report card. On top of that, disease control agencies at all levels must timely fill in and examine the information on echinococcosis control in the web-based information management system for control of parasitic diseases.
2. *Surveillance at sentinel sites*. Sentinel sites selection in provinces, need to reflect the different epidemic types and burden of echinococcosis, with strong capacity to undertake the surveillance tasks in the county centre for disease control. Based on the above requirements, a total of 140 sentinel sites were set up in 10 epidemic provinces and autonomous regions (corps), including Inner Mongolia, Sichuan, Yunnan, Tibet, Gansu,

Table 3 Problems in surveillance and solutions.

No.	Problems in surveillance	Solutions
1	The surveillance quality carried out by the provincial professionals and the reporting of data are not satisfactory	To clarify the surveillance tasks and the time of data reporting, and to strengthen the supervision over surveillance work
2	The transmission of echinococcosis in some sentinel villages remains uncontrolled, and the transmission between the definitive host canine and the intermediate host continues to exist	To actively carry out deworming of domestic dogs, and it is necessary to control field infection sources in areas with high prevalence
3	Some of the surveillance data are inaccurate, and those data recording without information of the prevalence in the surveillance population	It is necessary to conduct further investigation and arrange more data entry training, and to improve the sensitivity and specificity of detection reagents

Qinghai, Ningxia, Shaanxi, Xinjiang, and Xinjiang Production and Construction Corps. The major tasks for the sentinel surveillance include human population surveillance, intermediate host surveillance, and definitive host surveillance. Professionals from NIPD-CTDR have taken the following responsibilities: reviewing the reported data, completing the national annual report on echinococcosis surveillance, identifying and tackling problems in surveillance, and seeking solutions through the seminar on echinococcosis surveillance and patient management (Table 3).

4.2 Contribution to surveillance

For the purpose of further improving the surveillance of echinococcosis in China, based on the pilot surveillance summarized in 2015, the "National Echinococcosis Surveillance Program (2016)" was formulated. Thus, NIPD-CTDR took the responsibility for the following tasks: (i) monitoring and implementing of the national echinococcosis surveillance programme, by providing technical guidance on quality control, and data collection and analysis in the surveillance programme; (ii) data clearance at national level, and compiling the annual report for national surveillance on echinococcosis, and timely feedback those surveillance results to each province; (iii) training of professional and technical personnel at the provincial level who are participating in surveillance at national echinococcosis surveillance sites, and holding annual national training courses on echinococcosis surveillance; and (iv) supervising the surveillance work in each province, including work plans, personnel capacity and funding arrangements, on-site implementation, ultrasonic testing technology and laboratory operating procedures, and information data management, etc.

According to the infectious disease reporting system of the China Disease Epidemic Reporting Information Network, a total of 4327 cases of echinococcosis were reported in 26 provinces and autonomous regions in 2018, and no deaths were reported. The established sentinel surveillance sites promote the quality of the control programme regularly. For examples, 1504 new cases were found in 105 county hospitals in 2018 where the sentinel surveillance sites were located, with the largest in Seda County in Sichuan Province, followed by Maqin County in Qinghai Province. Screening of human echinococcosis was also carried out in sentinel villages in 133 counties at 140 sentinel surveillance sites nationwide, in which 173 cases of echinococcosis were found out of 267,478 residents screened, with an average population incidence of 0.06%. A survey of canine infections in 125 sentinel villages detected a total of 331 dogs with faecal antigens positive for *Echinococcus* out of 14,112 domestic dog faecal samples, with the average positive rate of faecal antigens of dogs being 2.35%. A total of 41,853 cattle and sheep were investigated at 114 surveillance sites, among which 888 sick livestock were detected, with a detection rate of 2.12%. The surveillance results from 2015 to 2018 are shown in Table 4.

5. Contribution to scientific research and personnel training

5.1 Progress in scientific research

In addition to adopting a comprehensive control strategy of "controlling the source of infection mainly, in combination of the controlling intermediate host and patient investigation and treatment", the national echinococcosis control programme in China attaches great importance to the diagnosis, drug development, and control of infection sources of echinococcosis in the field. During the 34 years from 1986 to 2019, a great number of research projects on the immunodiagnosis and drug development of echinococcosis were undertaken.

Table 4 Prevalence of population and intermediate host/definitive host infection at surveillance sites during 2015–2018.

	Population			Intermediate host (livestock)			Intermediate host (rodent)			Definitive host (dog)		
Year	Number of people examined	Number of positive people	Prevalence rate (%)	Number of livestock examined	Number of positive livestock	Infection rate (%)	Examined number	Positive number	Infection rate (%)	Number of dogs examined	Number of positive dogs	Infection rate (%)
2015	12,993	180	1.39	7288	114	1.56	–	–	–	2997	22	0.73
2016	101,240	100	0.10	16,773	499	2.98	9084	730	8.04	6117	150	2.45
2017	262,699	1473	0.56	42,363	883	2.08	14,846	297	2.00	13,473	433	3.21
2018	267,478	173	0.06	41,853	888	2.12	22,292	320	1.44	14,112	331	2.35

5.1.1 Immunodiagnostics development

Several projects on immunodiagnostic development were carried out in NIPD-CTDR with a focus on antigen screening. Wang et al. (2012) obtained a recombinant protein *Eg*Eno with a molecular weight about 50 kDa by cloning and expressing the *Eg*Eno gene of *E. granulosus*. The sensitivity of this antigen diagnosis by ELISA for CE was 81.25%, which might be used as a candidate antigen of immunodiagnosis for CE, but the protein could not distinguish CE from AE. Gao et al. (2018) and Wang et al. (2013) used purified antigen of hydatid cyst fluid and recombinant *E. multilocularis* 18 antigen as coasted antigen, and anti-IgG antibody labelled colloidal gold as detection reagent, and developed the immunochromatographic strip to detect the serum of CE patients, with a sensitivity of 91.0% and a specificity of 96.9%, while the sensitivity and specificity of the immunochromatographic strip for AE were 98.0% and 99.3%, respectively. The immunochromatographic strips have been tested and evaluated in the field of in Ganzi Tibetan Autonomous Prefecture of Sichuan Province, China, a region with a high prevalence for both forms of human echinococcosis. The sensitivity and specificity, and the diagnostic accuracy of the strip were determined to be greater than 94% for both CE and AE by using the ultrasonography as the gold standard, which indicated that the strip is suitable for large-scale screening of echinococcosis cases in endemic areas. Jiao et al. (2014) held that using *Eg*PS-ELISA and *Eg*AgB-ELISA as a pair of diagnostic combination was superior to other tests for antigens detection in clinical diagnosis and efficacy evaluation of CE. These identified antigens with better diagnostic value significantly increased the early diagnosis rate of echinococcosis.

Currently, most of the human echinococcosis antibody surveillance reagents in China employ natural antigens, but there are some problems like difficulties in obtaining raw materials and unstable quality. Based on the genomic data of *Echinococcus*, Mo et al. (2018), Qian et al. (2018), and Zhang et al. (2014, 2016) using proteomic high-throughput screening technology, found highly effective new diagnostic antigens in over 200 *Echinococcus* proteins, and developed a "new technology for real-time diagnosis of echinococcosis" based on recombinant antigens. The technology included a colloidal gold rapid detection kit for echinococcosis (Fig. 2) and a mobile traceable system (Fig. 3). The characteristics and innovation of this technology are as follows: (1) The detection reagent has high diagnostic efficiency (the sensitivity of the on-site large-scale sample evaluation in the epidemic area is above 70%, and the specificity is above 90%, which

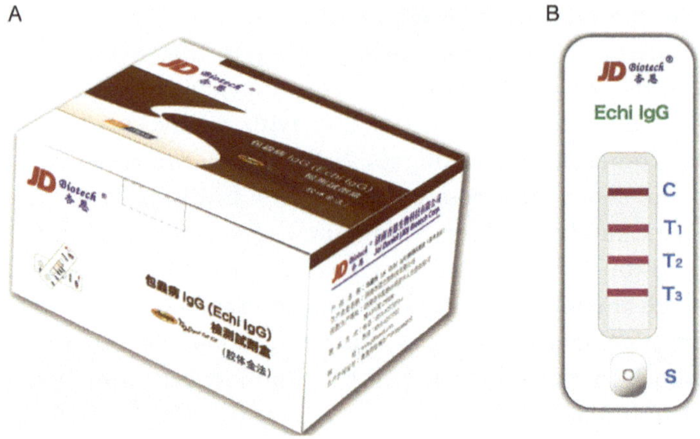

Fig. 2 Rapid diagnosis kit for echinococcosis.

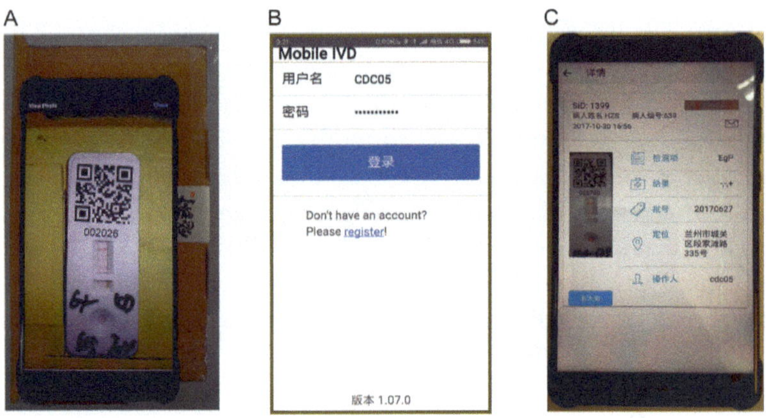

Fig. 3 Mobile traceable system for echinococcosis detection.

are higher than those of the commercial kit), it is easy and fast to operate, and the test results can be obtained in 15 min. Besides, due to the use of genetically engineered recombinant antigen, it has the advantages of simple production, easy quality control and mass production, low costs, and low cross reaction with other parasitic diseases. (2) The mobile traceable system can identify the QR code on the detection card through mobile application software and transmit the test results to the cloud expert system in real time for analysis, feedback, and storage. This technology not only maintains the traditional advantage of rapid diagnosis, but also reduces the naked eye misjudgement and the limitation on detection time and space, greatly facilitating the management and traceability of test results, especially suitable for using in remote plateau areas where there is a shortage of technicians. At present, the detection kit technology has been transferred to biological companies for industrialization. The studies on the immunodiagnostic technology for echinococcosis by NIPD-CTDR are summarized in Table 5.

5.1.2 Drug development

At present, benzimidazoles—albendazole (ABZ) and mebendazole (MBZ)—are the only two drugs approved for the treatment of echinococcosis, which mainly exert anti-parasitostatic effects through highly dosed, long-term treatment (Dang et al., 2018; Liu et al., 2013). However, due to the poor oral absorption of ABZ and MBZ, the cure rate of CE is merely about 30%, and long-term application of benzimidazoles (ABZ and MBZ) results in severe adverse side effects (Liu et al., 2015; Wang et al., 2017). Therefore, scientists from NIPD-CTDR have explored and studied new preparations of ABZ for many years. As lipid diet can significantly increase the blood concentration of ABZ, after observing the effect of different oils on the absorption of ABZ, Chai et al. (2006) developed a new preparation of albendazole emulsion containing 30% soybean oil (w/o) through a series of experiments, with a cure rate of 70–80% for CE patients, much higher than that of tablet albendazole and suspensions. And the albendazole emulsion is being used in the treatment of echinococcosis in the field. Nevertheless, in view of the short shelf life of the emulsion and insufficient compliance, the research group led by Zhang Haobing developed albendazole and mebendazole soft capsules (Liu et al., 2012a, b, 2013, 2014; Zhang et al., 2011a, b), in which the new preparation of mebendazole can improve the bioavailability of the drug, and reduce the dosage of the drug, thus reducing the side effects of drug. It is a promising new anti-echinococcosis drug (Liu et al., 2012b, 2014). For both soft capsules, cooperative development agreements have been reached with pharmaceutical companies.

In animal experiments, the research group also discovered that a class of amino alcohol structures had a good killing and inhibitory effect on the protoscoleces and germinal cells of echinococcus (Liu et al., 2016a, 2018a; Zhang et al., 2015c). Among the 130 amino alcohol compounds evaluated, 16 compounds were effective against both *E. granulosus* protoscoleces and germinal cells at a concentration of 20 μg/mL, and the compounds with best activity were also effective on the cysts cultured in vitro. The above study provided an important basis for the discovery of new structures of anti-echinococcosis drugs. Moreover, Liu et al. (2016b, 2018b, c) found that

Table 5 Studies on immunodiagnostic technology for echinococcosis by NIPD-CTDR.

Detection methods	Diagnostic principle	Technical characteristics	Technology status	References
Enzyme-linked immunosorbent assay (ELISA)	Detection of serum antibody	The recombinant antigen encoded by the *Eg*Eno gene was used for diagnosis, it has good sensitivity and specificity, with diagnostic value, but unable to distinguish CE from AE	Still in research	Wang et al. (2012)
Colloidal gold immunochrom-atography	Detection of whole blood or serum antibody	The recombinant antigen encoded by *Em*18 gene fragment and the purified crude hydatid cyst fluid antigen were used for diagnosis. The sensitivity and specificity for the diagnosis of CE were 91.0% and 96.9%, respectively, and for AE were 98.0% and 99.3%, respectively. The strip test is fast and simple to operate and needs no devices, and the results are objective, which is especially suitable for field and grassroots applications	Product development and technology transfer have been completed, and the national device registration certificate has been obtained. (National Instrument registration 2016340065)	Gao et al. (2018) and Wang et al. (2013)

Enzyme-linked immunosorbent assay (ELISA)	Detection of serum antibody	The diagnostic technology of natural antigen of *Eg*PS-ELISA and *Eg*AgB-ELISA was used as a pair of diagnostic combination for the clinical diagnosis and efficacy evaluation of CE, which demonstrated better efficacy than other natural and recombinant antigens, and thus the best choice at present	Still in research	Jiao et al. (2014)
Colloidal gold immunochromatography	Detection of whole blood or serum antibody	Three recombinant *Echinococcus* antigens were used for combined diagnosis. The technology includes a colloidal gold rapid detection kit for echinococcosis and a mobile traceable system, with high diagnostic efficiency, simple production, and low costs, real-time analysis, feedback and storage of test results, easy and fast to use, but cannot distinguish CE from AE	Technology development and transfer have been completed	Mo et al. (2018), Qian et al. (2018), and Zhang et al. (2014, 2016)

tacrine had a good killing effect on protoscoleces and germinal cells. After 3 days of exposure to 20 and 40 μg/mL tacrine in vitro, the percentage of dead protoscoleces of *E. granulosus* was 100%. The dead protoscoleces showed slight swelling in some region and the outline of tegument dispersed became evident. In the meanwhile, the ultrastructure changed significantly, with a large number of vacuoles and lipid droplets appearing in the tegument (Liu et al., 2018b). Furthermore, tacrine can also inhibit the activities of *E. multilocularis* protoscoleces in vitro but had little effect on the activity of normal host cells. After treatment with 15 mg/kg tacrine, the survival rate of mice infected with *E. multilocularis* increased (Liu et al., 2018c). In addition, our researcher Yin et al. (2018) disclosed that ursolic acid (UA) could kill around 45.95% of the protoscoleces in vitro at 40 μg/mL and inhibited the growth of over 90% of germinal cells, thus implicating UA as a potential anti-echinococcosis drug. The study carried out by Dang et al. (2017a) suggested that the anticancer drug sorafenib had a stronger inhibitory effect on *E. multilocularis* cysts than ABZ, which can block the function of vascular endothelial growth factor receptors and then reduces the supply of oxygen and nutrients, thereby causing the cysts to shrink or stop growing. Consequently, sorafenib is a potential drug for the treatment of echinococcosis. In addition, Dang et al. (2018) and Xu et al. (2019) evaluated the effects of the novel carbazole amino alcohol drugs (propylamine and its optical isomers R-propylamine and S-propylamine) on anti-*Echinococcus* in vitro and in vivo. In the experiment of anti-protoscoleces in vitro, the treatment time of different concentration of propylamine and R-propylamine was positively correlated with the mortality of the protoscoleces, while the killing effect of S-propylamine was weak. The team also introduced an epoxy group to MBZ by a reaction with epichlorohydrin, which not only improved the solubility of MBZ, but also increased its parasiticidal effect on *E. multilocularis* and reduced its cytotoxicity in rat hepatoma cells. These studies provide references for screening and developing high solubility and high efficiency drugs for echinococcosis. The studies on drug treatment of echinococcosis by NIPD-CTDR are summarized in Table 6.

5.1.3 Innovative technique development for interrupting transmission

The main transmission sources of both forms of human echinococcosis are the synanthropic cycles canines (domestic dog) and the sylvatic cycles canines (stray dog, fox, wolf, jackal, etc.), and the key to interrupt the transmission risk of echinococcosis is to focus on control of the two transmission sources.

Table 6 Studies on drug treatment of echinococcosis by NIPD-CTDR.

Drug name	Action mechanism	Characteristics	Stage of research	References
Benzimidazoles (ABZ and MBZ)	By inhibiting the uptake and utilization of glucose by *Echinococcus*, MBZ can block the transport of nutrients, especially glucose, and exhausted the endogenous glycogen in *Echinococcus*, it can also inhibit enzymes required for energy metabolism, such as phosphoenolpyruvate carboxylase, so as to inhibit or kill *Echinococcus*. The action principle of ABZ is basically similar to that of MBZ: inhibiting glucose uptake and fumarase system, hindering the production of ATP, and killing the *Echinococcus*	It can effectively kill and inhibit the metacestodes of echinococcosis. ABZ is better than MBZ in the treatment of echinococcosis, but due to the poor oral absorption of ABZ and MBZ, the cure rate of CE merely about 30% and long-term application of it resulted in severe adverse side effects	The first choice for the treatment of echinococcosis, but because of its unsafety and side effects, many researchers have done a great deal of work in the development of new preparations and new drugs	Dang et al. (2018), Liu et al. (2013, 2015), and Wang et al. (2017)
Albendazole emulsion	Soybean oil and fatty foods can delay gastric emptying, prolong the residence time of drugs in the stomach, enhance absorption, and increase the blood concentration and bioavailability of ABZ	The blood concentration of ABZ is significantly increased compared with that of tablet albendazole and suspensions but the emulsion has problems such as short shelf life and insufficient compliance	Has obtained invention patents and new drug certificates, has been transferred to Hanjiang Pharmaceutical Company, and new drugs have come into the market	Chai et al. (2006)

Continued

Table 6 Studies on drug treatment of echinococcosis by NIPD-CTDR.—cont'd

Drug name	Action mechanism	Characteristics	Stage of research	References
ABZ/MBZ soft capsules	ABZ/MBZ and appropriate auxiliary materials are fully dispersed in different liquid oils to prepare oil suspension, so as to improve the solubility and bioavailability of the drug	Improved the bioavailability of MBZ and ABZ, and can reduce the dosage and course for treatment, improve compliances, and reduce treatment costs	Cooperative development agreements have been reached with pharmaceutical companies for both soft capsules	Liu et al. (2012a, b, 2013, 2014) and Zhang et al. (2011a, b)
Amino alcohol compounds	The compounds with amino alcohol structures have a good killing and inhibitory effect on the protoscoleces and germinal cells of *Echinococcus* in vitro	These compounds can significantly kill *Echinococcus* in vitro	Patent enters the real trial stages	Liu et al. (2016a, 2018a) and Zhang et al. (2015c)
Tacrine	It has been experimentally verified that tacrine can directly affect the activity of protoscoleces and germinal cells of *E. granulosus* cultured in vitro, as well as the activity of protoscoleces of *E. multilocularis* cultured in vitro	Tacrine can kill protoscoleces and inhibit the growth of germinal cells, with a strong anti-echinococcosis effect, it is a highly effective new anti-echinococcosis drug	Has obtained invention patents	Liu et al. (2016b, 2018b, c)

Ursolic acid	It may be due to the direct effect of UA on *Echinococcus* and its impact on the host in the mechanism of anticancer. For example, UA can induce apoptosis by through suppression of the components of ERK/MAPK signalling pathway, which plays an important role in the growth and development of *Echinococcus*	UA has strong insecticidal activity of *E. granulosus* protoscoleces, germinal cells, and metacestodes in vitro and in vivo, but poor solubility of UA limits its application	Has obtained invention patents	Yin et al. (2018)
Sorafenib	Sorafenib can block the function of vascular endothelial growth factor receptors and then reduces the supply of oxygen and nutrients, thereby causing the cysts to shrink or stop growing	Sorafenib had a stronger inhibitory effect on *E. multilocularis* cysts than ABZ, with a good cystic inhibition effect, thus it's a potential drug for the treatment of echinococcosis	Patent enters the real trial stages	Dang et al. (2017a, b) and Xu et al. (2019)

For the study of intervention in domestic dogs, Jiang et al. (2017a, b) developed in situ solidified long-acting praziquantel injection that has been used in the experimental study of dog infection with *E. granulosus*. The experimental results showed that the average deworming rate of *E. granulosus* in dogs could be over 90% within five and a half months after a single administration. Considering the spawning cycle of *Echinococcus*, this preparation can protect dogs and control the spread of echinococcosis for half a year or longer. So far, two small-scale field tests have been carried out, and a joint development agreement has been reached with Sichuan Hengtong Animal Protection Company. Yang et al. (2017, 2019) designed and developed automatic timing canine deworming collars and feeders, realizing timed drug delivery, a wearing, 1 year deworming effect and obtaining relevant patents, and the field preliminary evaluation was good. Currently, field evaluations and demonstrations are being carried out in Qinghai, Tibet, and Gansu in order to obtain a more scientific assessment data.

At the same time, for the wild canine intervention study, our institute researchers Yu et al. (2017, 2018) and Han et al. (2018) carried out the field investigation of the wild transmission hosts based on the grid design concept, and developed a new type of bait, which was launched by UAVs in Shiqu, Sichuan and Yushu, Qinghai, and carried out a 4 years field epidemiology experiment. It was found that the grid design could be better used to investigate the distribution of intermediate and definitive hosts of *Echinococcus* in the field and to evaluate the intervention measures of field infection sources in a certain range. It was also revealed that the positive rate of dog faeces in the field area after the delivery of new type baits using UAVs decreased from 14.00% in 2017 to 4.98% in 2019, showing a yearly decline for three consecutive years. Compared with manual bait delivery, the comprehensive intervention measures can significantly reduce the infection risk of *Echinococcus* in the wild canines, in addition to saving a lot of cost and time. The research results are shown in Table 7.

5.1.4 Vaccine candidate development

Dogs are the definitive hosts of *Echinococcus*, and the eggs which are excreted with faeces are important source of infection. It is of great significance for the prevention and control of echinococcosis to develop an effective vaccine against echinococcosis, to block the source of infection (dog, definitive host) and to prevent the disease before it occurs. According to the transcriptome information of *E. granulosus* (Huang et al., 2016, 2017), our research group selected AgB2 and GP50 as candidate vaccine targets. AgB2 polypeptide,

Table 7 Studies on infection source control technology.

Infection source control technology	Application principle	Characteristics	Stage of research	References
In situ solidified long-acting praziquantel injection	The drugs and polymers were dissolved in suitable solvents and injected subcutaneously locally. At the site of drug administration, the polymers solidified under physiological conditions to form a semi-solid or solid drug depot, and played an efficient and long-term role in anti-*Echinococcus*	One administration can protect animals from *Echinococcus* infection in infection season or even longer, avoid the pain of surgical implantation, simple implantation preparation, good stability, and convenient dose control, reduce control cost, high security of materials	Patent enters the real trial stages. Transformed into cooperation	Jiang et al. (2017a, b)
Automatic timing canine deworming collar	The collar is equipped with a control device, which includes a timing component, a positioning component, a power supply component, a drug administration component, and a communication component. According to the time requirements for regular deworming, set the time of drug administration, realize regular drug administration in sequence, and perform remote management	Can overcome the dilemma of improper implementation such as easy to forget and large randomness, or of difficulties in deworming due to natural factors such as weather and traffic, can automatically, regularly, and intelligently administer drugs, and realize accurate deworming, can improve the coverage and effect of deworming in canines, can save time, manpower, and material and financial resources, particularly suitable for dogs in the transition between winter and summer pastures	Obtained patents, transformed into cooperation, demonstration application	Yang et al. (2017)

Continued

Table 7 Studies on infection source control technology.—cont'd

Infection source control technology	Application principle	Characteristics	Stage of research	References
Automatic timing canine deworming drug feeder	The feeder comprises a shell, the internal systems such as drug administration system, camera system, positioning system, warning and alarm system, power supply system, and communication system, as well as a cloud server and a user mobile terminal adapted to the feeder. The pre-set programme can help realize the functions such as regular drug administration for deworming and surveillance of canines eating baits	The feeder can be controlled remotely, and manual bait delivery deworming can be improved to automatic, timing, and intelligent delivery, which can improve the deworming coverage and effect in canines, and reduce the time and cost of deworming	Obtained patents, transformed into cooperation, demonstration application	Yang et al. (2019)
The technique of bait delivery with new praziquantel using UAVs for canine deworming based on the grid design	The grid-based design carry out the length-width proportion grid design and planning in the investigation or intervention area. The UAVs is light and small, and can be easily and effectively operated within the prescribed range. The modified UAVs is equipped with a new praziquantel deworming bait, carrying out fixed-point delivery drug intervention and observe intervention effect of the source of infection in the field, which can effectively act on wild canines	This technology and method could be used to investigate the distribution of intermediate and definitive hosts of *Echinococcus* in the field and to evaluate the intervention measures of field infection sources in a certain range. Compared with manual bait delivery, delivery by UAVs saved a lot of cost and time, and the comprehensive intervention measures can significantly reduce the infection risk of *Echinococcus* in the wild canines	Published articles, obtained patents, cooperative development	Han et al. (2018) and Yu et al. (2017, 2018)

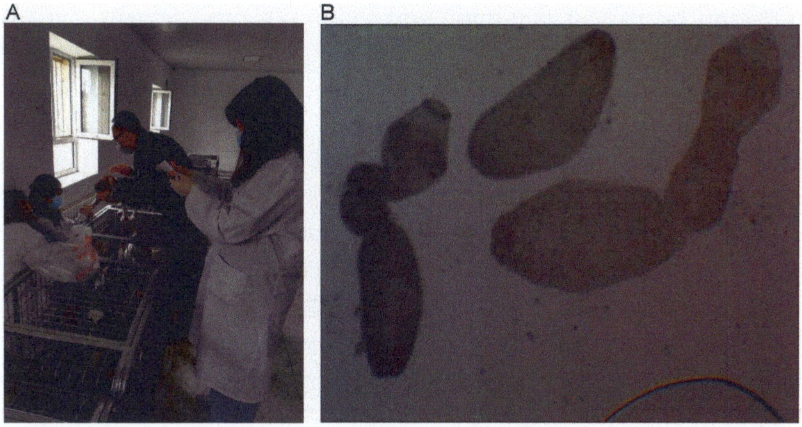

Fig. 4 The evaluation experiments for canine echinococcosis vaccine candidates. (A) Dog vaccination experiments with blood collection. (B) The 27-day adult worm of *E. granulosus* was observed under microscope, which was eluted from the cecal mucosa of dogs.

prokaryotic recombinant protein AgB2 and eukaryotic expressed protein GP50 were used as vaccine intervention protein group, combined with aluminium hydroxide and Qill A adjuvant, respectively, to evaluate the protective effect of the vaccine. The results showed that the deworming rate of AgB2 polypeptide and GP50 protein were 84.8% and 40%, respectively. Meanwhile, the research group established a semi-open vaccine evaluation system, including the construction of biological-grade experimental animal enclosures, and the development of a series of SOPs, such as experimental animal dog breeding, vaccine immunity, protoscoleces infection and anatomy (Fig. 4). It laid a good foundation for the further development of canine echinococcosis vaccine.

5.2 Control strategies

In line with the national echinococcosis control programme, some innovative control strategies that are of great significance to promote the process of echinococcosis control have been developed in the pilot activities in Shiqu County of Sichuan Province, in the following aspects. Firstly, the screening approach of "iris recording + fingerprint + ID card" was innovatively applied to fast recording the personnel information correctly in archiving the screening information, it has been used for 95.20% of the county's resident population with the screening information recorded when screening undertaking, so that it make easy to track screening population with identity information on patient condition and follow-up management. Secondly, development and application of management system and electronic chip collar as "ID card" for canines with echinococcosis was applied to information management of canines with a domestic dog registration rate of 100%. Thirdly, combined with the characteristic of scattered living of herdsmen, innovative design of small-scale non-hazardous treatment ponds for dog faeces was performed, with unified construction. The way of unified management, unified collection and unified profess was adopted for dog faeces, and the collected dog faeces were collected by the township veterinary station, then processed centrally under the supervision of the municipal station. A special person was appointed to take charge of the maintenance of non-hazardous treatment ponds for dog faeces, which effectively ensured the timely and non-hazardous treatment of dog faeces. A total of 4200 ponds were constructed with the non-hazardous treatment rate of dog faeces reached 100% during deworming.

In order to clearly understand the epidemic situation of echinococcosis in Tibet Autonomous region, an epidemiological survey on echinococcosis was carried out in Tibet Autonomous region. A coordinating office of epidemiological survey was set up in Lhasa city to help devise an investigation plan and assist Tibet in conducting this epidemiological survey, in which National Health Commission of China designated 500-odd B-ultrasound doctors from 17 provinces and cities to assist this epidemiological survey. At the same time, in order to ensure the standardization of the epidemiological survey and strengthen quality control, in collaboration with professionals from different levels in Tibet, experts from the NIPD-CTDR supervised and inspected the patient screening in seven cities of Tibet Autonomous region, including supervising the quality of population screening and the implementation of echinococcosis control projects by means of on-site discussion, data analysis, on-site inspection, and resident interviews. By 2018, the epidemiological survey was completed and the results suggested that echinococcosis is seriously transmitted in Tibet Autonomous region with 0.27% (95% CI: 0.25–0.30%) of overall prevalence in the population of six administrative divisions. It is still necessary to continue to step up screening efforts in the future in order to increase screening coverage further.

5.3 Training senior professionals

The NIPD-CTDR also plays an important role in the training professionals in the field of echinococcosis control nationwide. About 17 postgraduates

Table 8 Training list of senior professionals in the field of echinococcosis (2010–2019).

Year	Province (autonomous region)	No. of trained at doctoral level	No. of trained at master level	No. of trained as professionals	Subtotal
2010	Shanghai, Sichuan	0	0	47	47
2011	Shanghai, Sichuan	0	3	49	52
2012	Shanghai, Sichuan, Tibet	0	2	58	60
2013	Shanghai, Sichuan	0	3	46	49
2014	Shanghai, Sichuan	1	1	46	48
2015	Shanghai, Sichuan	0	3	45	48
2016	Gansu, Qinghai, Shandong, Yunnan, Shanghai, Tibet	0	0	450	450
2017	Shanghai, Tibet, Sichuan	0	3	114	117
2018	Shanghai, Tibet, Sichuan	0	1	107	108
2019	Shanghai, Sichuan, Qinghai, Tibet	0	1	119	120

have been trained in the field of echinococcosis. Since 2010, two training courses on echinococcosis control technology have been held each year, and a total of 1086 person-times from the echinococcosis endemic provinces have been trained. Additionally, more than 10 trainings have been provided to Tibetan professionals, and a group of talents on research and development of echinococcosis drugs have been trained as well. One doctoral student engaged in anti-echinococcosis drug development has graduated and now working on anti-echinococcosis drug development in the Department of Drug Development of the NIPD-CTDR. Details are shown in Table 8.

6. Conclusions

Echinococcosis has been seriously prevalent in China, the number of threatened populations and patients across the country still ranks first worldwide. The prevalence of echinococcosis is often affected by natural, biological, and social factors, coupled with some unique natural factors, such as customs, and religious beliefs, which make the transmission complex, thus posing challenges to the national echinococcosis control programme in China. For a long time, China's control strategy on echinococcosis is based on "controlling the source of infection mainly, combining the control of intermediate host with patient investigation and treatment", which has achieved certain outcomes, for instance, the establishment and improvement of the epidemic network of echinococcosis has enabled us to grasp the epidemic situation of echinococcosis more comprehensively, quickly, and accurately, which provided the evidence-based decision in count-measures taken in the programme. Application of new technologies including imaging diagnosis and immunodiagnosis, and research on new drugs development have also achieved optimistic results for the programme. Besides, some new research and development on the interruption of echinococcosis transmission in field canines has also sprung up. However, the control of echinococcosis in China still has a long way to go, and there exist many challenges. First, the ability to control echinococcosis needs to be further improved. The endemic area of echinococcosis in China covers 40% of the land area. The number of dogs is huge, and there are numerous uncontrollable ownerless dogs, which make it more difficult to implement the strategy of "deworming month by month and administering drugs dog by dog". Hence, in the later stage, efforts should be made to prevent and control field infection sources. Second, studies have shown that in Inner Mongolia, the average infection rate of *E. multilocularis* of *Microtus brandti*, overwintering adult *Microtus brandti* and Corsac fox was 2.43%, 5.23–14.29% and 33.3%, respectively (Tang et al., 1988). The infection rate in Inner Mongolia is similar to other endemic counties, but no AE patients have been reported so far. Therefore, the epidemic range and degree of AE in Inner Mongolia are worth an in-depth study. Third, the facilities and capacity of echinococcosis control programme is weak, the capital investment is difficult to sustain for a long time, and the control efforts in livestock is tough. Moreover, the number of echinococcosis control professionals at grand level is scarce with low level technically, and the control capacity is insufficient, making it hard to maintain and consolidate the existing programme and achievements gained. Finally, the tools of control cannot meet the needs of work. Most of the diagnostic methods currently used are suitable for the patients in the middle and late stage of the disease. Most of the drugs are for long-term use and have severe side effects. As a result, there is a lack of sensitivity tests for early diagnosis and effective drugs for short-term treatment. In addition, although there are many reports on animal echinococcosis vaccines, but effective vaccines have not been available for humans yet.

In conclusion, in the future control programme on echinococcosis, we need to continue to explore effective methods to control field infection sources, and to conduct a preliminary investigation on the epidemic degree of AE in Inner Mongolia in order to find out the actual epidemic status of echinococcosis. Besides, we need to strengthen the training of talents, as well as the technical training and guidance in heavily echinococcosis endemic areas. Furthermore, great efforts should be made to carry out education to disseminate the right information to the public, and improve the surveillance network so as to facilitate the national echinococcosis control programme with targeted policies and measures in China. Finally, it is necessary to strengthen the research on the key technologies of echinococcosis, strengthen the research and development of early diagnosis technology, new effective drugs and human echinococcosis vaccines, thus providing better technical support to the national echinococcosis control programme in China.

References

Chai, J.J., Xiao, S.H., Gao, F.H., et al., 2006. Study on drug treatment of cystic echinococcosis. J. Med. Res. 35 (4), 29–30 (in Chinese).

Chen, W.Q., Zhang, Y.L., Gongsang, Q.Z., et al., 2018. Analysis of hydatid disease cases in Tibet Autonomous Region. Chin. J. Parasitol. Parasit. Dis. 36 (1), 43–53 (in Chinese).

Dang, Z.S., OKU, Y., Zhou, X.N., et al., 2017a. In vitro inhibitive effect of the anticancer drug sorafenib on *Echinococcus multilocularis* larvae. Chin. J. Parasitol. Parasit. Dis. 35 (5), 417–422 (in Chinese).

Dang, Z.S., Xiao, N., Zhou, X.N., et al., 2017b. Application of anticancer small molecule compound sorafenib in the treatment of liver hydatid disease: CN106474126A. (in Chinese).

Dang, Z., Xu, S., Zhang, H., et al., 2018. In vitro and in vivo efficacies of carbazole aminoalcohols in the treatment of alveolar echinococcosis. Acta Trop. 185, 138–143.

Gao, C.H., Wang, J.Y., Shi, F., et al., 2018. Field evaluation of an immunochromatographic test for diagnosis of cystic and alveolar echinococcosis. Parasit. Vectors 11 (1), 311.

Gongsang, Q.Z., Wu, W.P., Yan, X.L., et al., 2017. Investigation on distribution of alveolar echinococcosis in human and infection of intermediate host in Tibet Autonomous Region. J. Pathogen Biol. 12 (12), 1175–1179 (in Chinese).

Han, S., Yu, Q., Yang, S.J., et al., 2018. Evaluation on the measures by a grid-based design for wildlife control in hyper-endemic areas for echinococcosis. Chin. J. Parasitol. Parasit. Dis. 36 (5), 495–498 (in Chinese).

Huang, F., Dang, Z., Suzuki, Y., et al., 2016. Analysis on gene expression profile in oncospheres and early stage metacestodes from Echinococcus multilocularis. PLoS Negl. Trop. Dis. 10 (4), e0004634.

Huang, F., Dang, Z., Zhang, H., et al., 2017. Comparative study on secretome and transmembranome of immature and mature metacestodes of Echinococcus multilocularis. Vet. Parasitol. 245, 153–159.

Jiang, B., Zhou, X.N., Zhang, H.B., et al., 2017a. Slow-release praziquantel for dogs: presentation of a new formulation for echinococcosis control. Infect. Dis. Poverty 6 (1), 140.

Jiang, B., Zhang, H.B., Zhou, X.N., et al., 2017b. In-situ solidified slow-release injection of anti-parasite drug and its preparation method: CN107811967A. (in Chinese).

Jiao, W., Fu, C., Liu, W.L., et al., 2014. Diagnostic potential of five natural antigens from Echinococcus granulosus in the patients of cystic echinococcosis with different clinical status. Chin. J. Parasitol. Parasit. Dis. 32 (2), 116–122 (in Chinese).

Khachatryan, A.S., 2017. Analysis of lethality in echinococcal disease. Korean J. Parasitol. 55 (5), 549–553.

Li, B., Quzhen, G., Xue, C.Z., et al., 2019. Epidemiological survey of echinococcosis in Tibet Autonomous Region of China. Infect. Dis. Poverty 8 (1), 29.

Liu, C.S., Zhang, H.B., Jiang, B., et al., 2012a. Enhanced bioavailability and cysticidal effect of three mebendazole-oil preparations in mice infected with secondary cysts of Echinococcus granulosus. Parasitol. Res. 111 (3), 1205–1211.

Liu, C.S., Tao, Y., Zhang, H.B., 2012b. Study on the solubility of mebendazole in different solvents. Int. J. Med. Parasit. Dis. 39 (5), 292–295 (in Chinese).

Liu, C.S., Zhang, H.B., Yin, J.H., et al., 2013. Echinococcus granulosus: suitable in vitro protoscolices culture density. Biomed. Environ. Sci. 26 (11), 912–915.

Liu, C.S., Zhang, H.B., Lei, W., et al., 2014. An alternative mebendazole formulation for cystic echinococcosis the treatment efficacy, pharmacokinetics and safety in mice. Parasit. Vectors 7, 589.

Liu, C., Zhang, H., Yin, J., et al., 2015. In vivo and in vitro efficacies of mebendazole, mefloquine and nitazoxanide against cyst echinococcosis. Parasitol. Res. 114 (6), 2213–2222.

Liu, C.S., Zhang, H.B., Xue, J., et al., 2016a. In vitro effects of aminoalcohol-carbazole compound BTB3 against Echinococcus granulosus. Chin. J. Parasitol. Parasit. Dis. 34 (3), 245–248 (in Chinese).

Liu, C.S., Zhang, H.B., Yin, J.H., et al., 2016b. Application of tacrine in preparing drugs for the treatment of echinococcosis: CN106176747A. (in Chinese).

Liu, H., Xiao, N., Yang, S.J., et al., 2017. Epidemiological characteristics of canine Echinococcus infection in Qinghai-Tibet Plateau of China. Chin. J. Schisto. Control 29 (2), 129–138 (in Chinese).

Liu, C., Yin, J., Xue, J., et al., 2018a. In vitro effects of amino alcohols on Echinococcus granulosus. Acta Trop. 182, 285–290.

Liu, C.S., Yin, J.H., Yao, J.Q., et al., 2018b. Efficacy of tacrine on Echinococcus granulosus in vitro. Chin. J. Zoonoses 34 (7), 608–612 (in Chinese).

Liu, C.S., Yin, J.H., Yao, J.Q., et al., 2018c. Inhibitory efficacy of tacrine on Echinococcus multilocularis. Chin. J. Parasitol. Parasit. Dis. 36 (3), 238–242 (in Chinese).

Mo, X.J., Qin, B., Zhang, T., et al., 2018. A cassette for rapid imaging analysis of echinococcosis test strip: CN209372673U. (in Chinese).

Qi, Y.F., Wu, W.P., 2013. Progress on the epidemiology of echinococcosis. Chin. J. Parasitol. Parasit. Dis. 31 (2), 143–148 (in Chinese).

Qian, M.B., Abela-Ridder, B., Wu, W.P., et al., 2017. Combating echinococcosis in China: strengthening the research and development. Infect. Dis. Poverty 6 (1), 161.

Qian, S., Zhang, T., Qin, B., et al., 2018. A mobile traceable system for detection of echinococcosis: CN201811563981. (in Chinese).

Qian, Y.J., Ding, W., Wu, W.P., et al., 2019. A path to cooperation between China and Mongolia towards the control of echinococcosis under the Belt and Road Initiative. Acta Trop. 195, 62–67.

Tang, C.T., Cui, G.W., Qian, Y.C., et al., 1988. Investigation on the pathogens of *Echinococcus multilocularis* in Hulunbuir Grassland, Inner Mongolia. Acta Zool. Sin. 34 (2), 172–178 (in Chinese).

Wang, L.Y., 2017. Final evaluation of the 12th five-year action plan for echinococcosis control and the 13th five-year plan. China Anim. Health 19 (7), 13–18 (in Chinese).

Wang, Y., Zhang, J., Yuan, Z.Y., et al., 2012. Cloning, expression and immunodiagnostic evaluation of enolase from Echinococcus granulosus. Chin. J. Schisto. Control 24 (5), 549–556 (in Chinese).

Wang, J.Y., Gao, C.H., Steverding, D., et al., 2013. Differential diagnosis of cystic and alveolar echinococcosis using an immunochromatographic test based on the detection of specific antibodies. Parasitol. Res. 112 (10), 3627–3633.

Wang, Q., Huang, Y., Huang, L., et al., 2014. Review of risk factors for human echinococcosis prevalence on the Qinghai-Tibet Plateau, China a prospective for control options. Infect. Dis. Poverty 3 (1), 3.

Wang, Q., Yu, W.J., Zhong, B., et al., 2016. Seasonal pattern of Echinococcus re-infection in owned dogs in Tibetan communities of Sichuan, China and its implications for control. Infect. Dis. Poverty 5 (1), 60.

Wang, W., Li, J., Yao, J., et al., 2017. In vitro and in vivo efficacies of novel carbazole aminoalcohols in the treatment of cystic echinococcosis. J. Antimicrob. Chemother. 72 (11), 3122–3130.

Wu, W.P., Wang, H., Wang, Q., et al., 2018. A nationwide sampling survey on echinococcosis in China during 2012–2016. Chin. J. Parasitol. Parasit. Dis. 36 (1), 1–14 (in Chinese).

Xu, S., Duan, L., Zhang, H., et al., 2019. In vitro efficacies of solubility-improved mebendazole derivatives against Echinococcus multilocularis. Parasitology 146 (10), 1256–1262.

Yang, S., Wu, W., Tian, T., et al., 2015. Prevalence of cystic echinococcosis in slaughtered sheep as an indicator to assess control progress in Emin County, Xinjiang, China. Korean J. Parasitol. 53 (3), 355–359.

Yang, S.J., Zhou, X.N., Xiao, N., et al., 2017. An automatic timing dog deworming collar: CN107374773A. (in Chinese).

Yang, S.J., Yu, Q., Zhou, X.N., et al., 2019. An automatic timing dog deworming drug administration feeder and its implementation: 201811061935.5. (in Chinese).

Yin, J., Liu, C., Shen, Y., et al., 2018. Efficacy of ursolic acid against Echinococcus granulosus in vitro and in a murine infection model. Parasit. Vectors 11 (1), 58.

Yu, S.H., 2008. Global progress of echinococcosis control and an insight to the National Control Program. Chin. J. Parasitol. Parasit. Dis. 26 (4), 241–244 (in Chinese).

Yu, Q., Xiao, N., Yang, S.J., et al., 2017. Deworming of stray dogs and wild canines with praziquantel-laced baits delivered by an unmanned aerial vehicle in areas highly endemic for echinococcosis in China. Infect. Dis. Poverty 6 (1), 117.

Yu, Q., Liu, H., Xiao, N., 2018. Unmanned aerial vehicles: potential tools for use in zoonosis control. Infect. Dis. Poverty 7 (1), 49.

Zhang, M.Y., Wu, W.P., 2017. Advances in study of the burden of echinococcosis in China and elsewhere around the world. J. Pathogen Biol. 12 (5), 473–475 (in Chinese).

Zhang, H.B., Liu, C.S., Jiang, B., Wu, W.P., Wang, H., et al., 2011a. Mebendazole soft capsules: CN102813639A. (in Chinese).

Zhang, H.B., Wu, W.P., Liu, C.S., et al., 2011b. Benzimidazole oil suspensions: CN102274174A. (in Chinese).

Zhang, T., Hu, W., Chen, Y., et al., 2014. A new *Echinococcus granulosus* positive antigen and its application: CN104651328A. (in Chinese).

Zhang, W., Zhang, Z., Wu, W., et al., 2015a. Epidemiology and control of echinococcosis in central Asia, with particular reference to the People's Republic of China. Acta Trop. 141 (Pt. B), 235–243.
Zhang, T., Zhao, W., Yang, D., et al., 2015b. Human cystic echinococcosis in Heilongjiang Province, China: a retrospective study. BMC Gastroenterol. 15, 29.
Zhang, H.B., Liu, C.S., Xue, J., et al., 2015c. Anti-parasite application of a class of amino alcohol compounds: CN106727581A. (in Chinese).
Zhang, T., Chen, Y., Hu, W., et al., 2016. A new diagnostic antigen for hydatid disease and its application: CN107033240A. (in Chinese).
Zhu, Y.Y., Wu, W.P., Guan, Y.Y., et al., 2016. Cost analysis of dog deworming in echinococcosis control program in Daofu, Sichuan Province. Chin. J. Parasitol. Parasit. Dis. 34 (2), 144–149 (in Chinese).

CHAPTER FOUR

Contributions to the lymphatic filariasis elimination programme and post-elimination surveillance in China by NIPD-CTDR

De-Jian Sun†, Yuan Fang†, Yun Huang, Yi Zhang*
National Institute of Parasitic Diseases, Chinese Center for Disease Control and Prevention; Chinese Center for Tropical Diseases Research; WHO Collaborating Centre for Tropical Diseases; National Center for International Research on Tropical Diseases, Ministry of Science and Technology; Key Laboratory of Parasite and Vector Biology, Ministry of Health, Shanghai, People's Republic of China
*Corresponding author: e-mail address: zhangyi@nipd.chinacdc.cn

Contents

1. Introduction 146
2. The prevalence patterns of lymphatic filariasis in China 147
3. Epidemiological and experimental investigations in China 151
4. Research and development of control tools, including diagnostics and MDA drugs 155
   4.1 Study on the transmission threshold of filariasis 155
   4.2 Pathogen biological research 158
   4.3 Research of the pathogenesis of filariasis 159
   4.4 Research of diagnostic methods 162
   4.5 Study on anti-filarial drugs and treatment methods 165
   4.6 The project 'Research achievements on the strategy and technical measures for interrupting the transmission of LF in China' provide technical support for LF elimination 171
5. Monitoring and evaluation supported by information management both in elimination and post-elimination stages 172
   5.1 The surveillance system in elimination stages 172
   5.2 Special monitoring and investigation in key areas after the elimination of filariasis 174
6. Technical supports to the elimination and post-elimination programmes 175
7. Conclusion 176
Ethical approval and consent to participate 177
Consent for publication 177
Availability of data and materials 177

† De-Jian Sun and Yuan Fang contributed equally to this work.

Advances in Parasitology, Volume 110
ISSN 0065-308X
https://doi.org/10.1016/bs.apar.2020.04.007

Competing interests 177
Author's contributions 177
References 178

Abstract

China was once one of the most heavily endemic for lymphatic filariasis (LF), with a heavy disease burden. Due to decades of sustained efforts, LF was eliminated from China in 2007. The historical tales in the control and elimination of LF in China and current post-elimination surveillance are reviewed and concluded. In the course of LF control and elimination, National Technical Steering Group for Filariasis Control and Research, and Collaborating Research Group on the Transmission Threshold of Filariasis, led by the National Institute of Parasitic Diseases, carried out researches, developed control measures, and applied in practical control and surveillance. The main of the results include: in theoretical study, (1) formulating the microfilaria rate below 1% as the transmission threshold of filariasis, (2) developing anti-filarial drugs and establishing animal model for drug screening, (3) studying the vectorial capacity of mosquitoes transmitted filariasis; in clinical diagnostic and therapy, (1) developing antigens for serological diagnostics, (2) studying different administration regimes for alleviating the side-response of diethylcarbamazine (DEC), (3) studying on the recurrence of acute adenolymphangitis and lymphadenitis of malayan filariasis, (4) observing lymphoedema of the lower extremity using lymphangiography; in practical control and surveillance, (1) establishing the control strategy taking elimination of infection source as the major focus, (2) formulating safe, feasible and effective DEC-mass drug administration, (3) revealing the transmission pattern of filariasis at the post-control phase, (4) establishing comprehensive active surveillance system, (5) formulating the criteria of basic elimination and elimination of filariasis. In the post-elimination surveillance phase, special monitoring and investigation in key areas were conducted for clearing way potential residual infection sources. Two schemes on LF surveillance in the post-elimination phase and caring for chronic filariasis patients were issued to maintain the diagnosis and treatment skills of filariasis all levels of disease control and prevention workers in future. Also, the Notifiable Diseases Reporting System, which included LF in 2004, plays an important role in LF post-elimination surveillance to prevent LF resurgence in China. The strategies and measures of LF control, elimination and post-elimination surveillance in China provide valuable practical experience and promote the progress of Global Programme of Elimination of Lymphatic Filariasis in worldwide.

1. Introduction

Lymphatic filariasis (LF) is a parasitic disease that seriously endangers human health, listed as the second cause of disability in the world by World Health Organization (WHO) (WHO, 1995). The Chinese government paid great effort on LF control and elimination, and included it in the long-term work plan for control of key infectious diseases in the country. Therefore, the investigation, control and elimination, and scientific research on LF were organized systematically and comprehensively in the country. After more than 50 years of intensive and sustained efforts, there were three major stages for the national LF control and elimination programme, including preparation, control, and post-control stages (Fang and Zhang, 2019). Finally, the goal of elimination of LF in nationwide was achieved by 2006, which effectively guaranteed the health of the people and promoted the development of the social economy. In 2006, Chinese Ministry of Public Health (MOPH) submitted *the National Report on the Elimination of Lymphatic Filariasis in China* to the fourth global alliance for the elimination of LF, which was approved by WHO on May 9th, 2007.

After the goal of elimination of LF in nationwide achieved in 2007, several documents including *Working Program for the Morbidity Management and Disability Prevention of Chronic Filariasis Patients* and *National Monitoring Program for the Post-elimination of LF* was issued in 2010, to prevent the resurgent of LF, and provide basic personal care for morbidity in local clinical systems in the post-elimination phase (Fig. 1). These sustainable works are very important to keep our capacity in the LF surveillance and response in future. This paper aims to review the historical tales in the control and elimination stages of the LF control and elimination programmes, and synthesize the lessons learnt from the previous efforts and current post-elimination surveillance.

2. The prevalence patterns of lymphatic filariasis in China

Before 1949, according to the limited survey, LF due to *Wuchereria bancrofti*, and *Brugia malayi* has been detected in Guangdong, Fujian, Taiwan, Zhejiang, Jiangsu, Shandong, Hunan and Hubei Provinces at that time, with severe disease burden in some places (Feng, 1936). Two large-scale of LF investigation and control programmes in nationwide began in the late 1950s and early 1970s, respectively. Scientists in our Institute, and many medical workers from health institutions, medical colleges and military medical units participated in field studies in endemic areas with great enthusiasm (Fig. 2).

In the early 1980s, entrusted by former Department of Disease Control, Ministry of Health (MOPH), the Institute drew up a detailed outline of the survey on LF, in order to collect results obtained in the two large-scale programmes from the provinces/autonomous regions/municipalities (P/A/M). In further, retrieved data was carefully checked by the

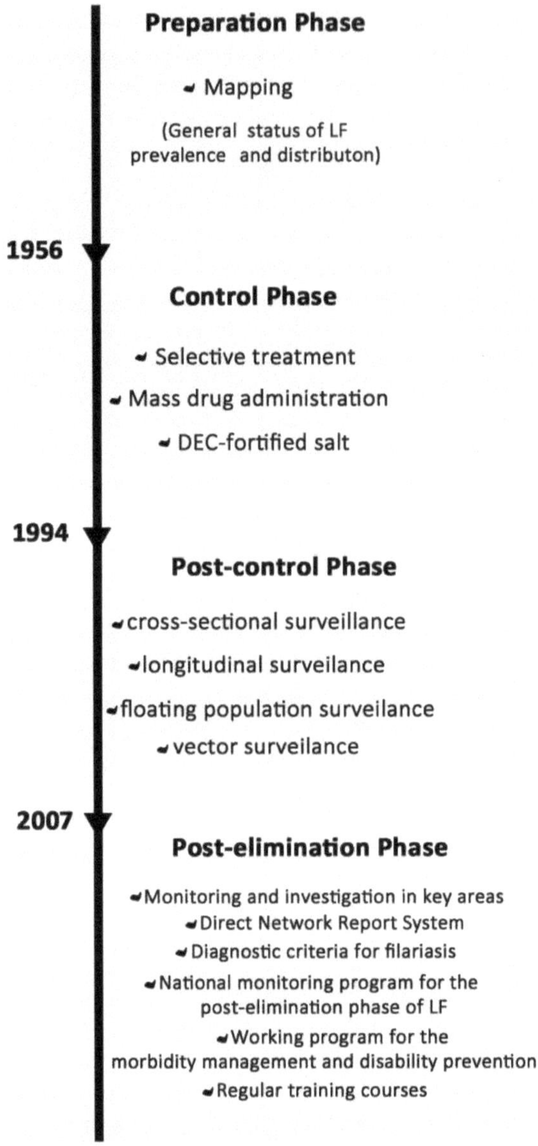

Fig. 1 The strategies implemented in each phase in the course of LF control and elimination in China. Abbreviations: DEC, diethylcarbamazine; LF, lymphatic filariasis.

collaboration between our Institute and provincial sanitary and anti-epidemic stations, and institutes of parasitic disease. Since then, the distribution and prevalence of LF in China were relative comprehensive understanding. The disease was spreading in the temperate, subtropical,

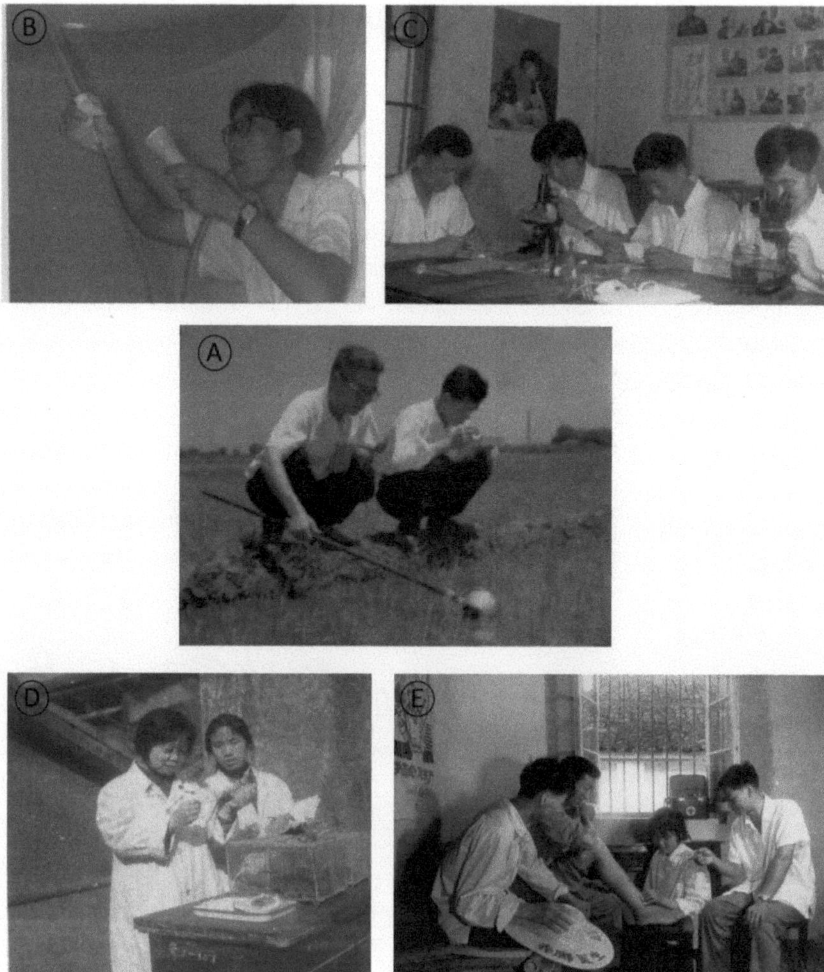

Fig. 2 Epidemiological investigations, vector control and establishment of cotton rat-filariasis animal conducted by NIPD members. (A) Prof. Lanchou Feng was catching larvae of *An. sinensis* in breeding sites (1960). (B) Catching mosquitoes inside the bed net (1965). (C) Mosquito dissection for filaria larvae detection (1965). (D) Infecting cotton rats with filaria larvae (ca. 1972). (E) Epidemiological survey on filariasis patients (ca. 1970).

and tropical zone of China from Lelin County of Shandong Province in the north (37°48′ N), to Sanya City of Hainan Province in the south (24°10′ N), and from Zhoushan Archipelago of Zhejiang Province in the east (122°30′ E), to Ya'an City of Sichuan Province (103°E) in the west (Wang et al., 2003). It was endemic in 864 counties/cities in 16 P/A/M, including Shandong, Henan, Hubei, Anhui, Jiangsu, Zhejiang, Jiangxi, Fujian, Guangdong, Hunan, Guizhou, Hainan and Sichuan provinces, Shanghai Municipality, Chongqing Municipality, and Guangxi Zhuang Autonomous Region (Fig. 3). The total population at risk of LF in endemic areas across the country was about 342 million (National report on elimination of lymphatic filariasis in China, 2007). Among the 864 counties/cities endemic for filariasis in China, 501 were hypo-endemic areas (microfilaria rate <5%) and 317 were meso-endemic areas (microfilaria rate ranging from 5.1% to 20.0%), while 44 were hyper-endemic areas (microfilaria rate ranging from 20.1% to 30.0%) and two were super-endemic areas (the microfilaria rate >30.1%).

According to epidemiological data, it was estimated that before the implementation of the national filariasis control programme in mainland China, the total number of filariasis patients (including bancroftian and malayan filariasis) was 30.994 million, while 25.594 million microfilaremia patients can be as infectious source attributed to the disease transmission,

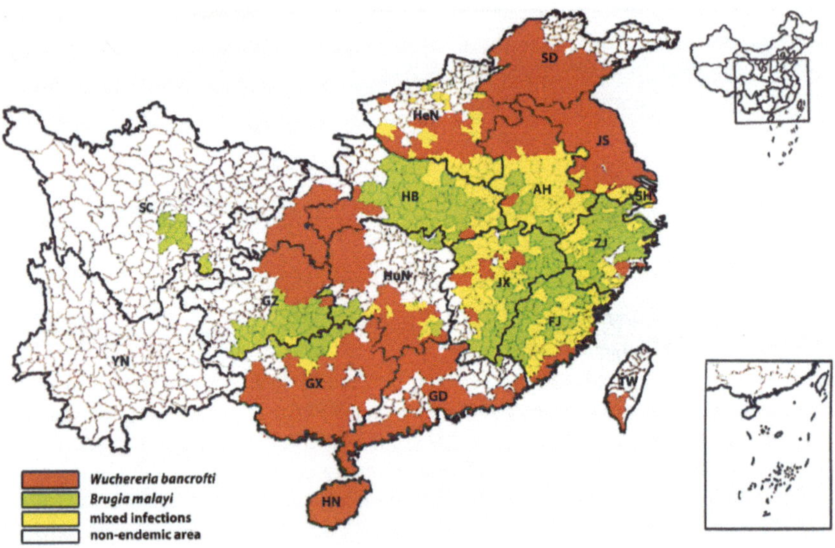

Fig. 3 The distribution of endemic areas of lymphatic filariasis in China before control was cited from a previous study (Wang et al., 2003). Abbreviations: AH, Anhui Province; FJ, Fujian Province; GD, Guangdong Province; GX, Guangxi Zhuang Autonomous Region; GZ, Guizhou Province; HB, Hubei Province; HN, Hainan Province; HeN, Henan Province; HuN, Hunan Province; JS, Jiangsu Province; JX, Jiangxi Province; SC, Sichuan Province; SD, Shandong Province; SH, Shanghai Municipality; TW, Taiwan Province; YN, Yunnan Province; ZJ, Zhejiang Province.

5.4 million people have recurrent acute adenolymphangitis, lymphangitis, lymphedema, elephantiasis, chyluria, hydrocele, and other clinical manifestations, caused by LF (National Technical Steering Group for Filariasis Control and Research, 1983). In mainland China, there were 21.962 million patients with bancroftian filariasis (17.398 million with microfilaremia and 4.564 million patients with clinical manifestations) and 9.032 million patients with malayan filariasis (8.196 million patients with microfilaremia and 836,000 patients with clinical manifestations) (National Technical Steering Group for Filariasis Control and Research, 1991a).

The above survey results and data provided evidence-based information for the formulation of the national LF control programme, and evaluation and certification of LF elimination in China.

3. Epidemiological and experimental investigations in China

Professor Lanchou Feng, a famous parasitologist and member of the Chinese Academy of Sciences, was the director of the Institute and chief of the Department of Filariasis Research. Based on the biological and entomological features of filariasis and prevalent characteristics, a control strategy based on elimination of infection source as the major focus was proposed by the research group led by Prof. Feng (Feng, 1933, 1936). Using the administration of diethylcarbamazine (DEC) to eliminate the source of infection as the control measure, could interrupt the LF transmission, and thereby achieve the goal of LF elimination (Feng, 1962).

Meanwhile, other members from Institute of Parasitic Diseases, Chinese Academy of Preventive (the predecessor of our Institute), Shandong, Guangxi and other Provincial Institute of Parasitic Diseases, went down to the field sites in hyper- and super-endemic areas. With the effort of several years, the experience of the strategy and measures for filariasis control obtained in a number of pilots indicated that repeated diagnosis and treatment of microfilaremia patients (selective treatment), and mass drug administration (MDA) in entire populations at-risk could reduce the microfilaria rate significantly in the human population in an economic way. It has a similar control effect to that of integrated measures, which is combined elimination of the source of infection and vector control. The integrative strategy required a major investment in human and material resources, and the effect of controlling mosquito vector was not ideal (Shi and Sun, 1983).

The suburb of Wuhan, was an originally hypro- and meso-malayan filariasis endemic area. In 1950s, after successive years (3–5 years) of extensive field investigations, and retreatment in target populations, the microfilaria rate reduced below 1%. The programme was interrupted since 1959. In 1970, surveying group entrusted by the office of National Technical Steering Leading Group for Schistosomiasis Control and Research, and MOPH, carried out blood survey in 1420 persons in local area. The results showed that after a lapse of 11 years, six microfilaremia patients were found, with the microfilaria rate of 0.4%, indicating that the control efficiency is sustained. Thus, the strategy of eliminating infection sources as the principle measure for LF control in China, supported by the practical evidence, was established.

From 1979 to 1981, as a technical personnel organized by the MOPH, members of our Institute divided into three groups to provide technology and professional training for three consecutive years, in order to verify the effect of LF control in 14 P/A/M (Hainan Province, and Chongqing Municipality have not been set up at that time) with a uniform approach. Under the direction of MOPH, national training courses on filariasis prevention and control techniques have been held by our Institute for 10 periods. In further, the Institute edited and published the *Manual of Filariasis Control*; coordinated the manufacture and supply of DEC and microscopes for filariasis control; and mobilize resources. Since 1982, under the organization of the MOPH, the Collaborating Research Group on the Transmission Threshold of Filariasis was founded. It was led by the Institute of Parasitic Diseases, Chinese Academy of Preventive, with the collaboration of relative provincial LF control organizations. According to species of filariasis, mosquito species, microfilaria rate, and microfilarial density, 21 sentinel sites in 11 P/A/M were set for longitudinal surveillance after intervention. The results showed that in endemic areas, where the microfilarial density was low in residual microfilaremia cases, when the microfilariae rate was $\leq$1.71% in areas with bancroftian filariasis prevalence and $\leq$1.5% in areas with malayan filariasis prevalence, which meant that transmission could no longer be sustained. The microfilaria rate had been declined to below 1% in the human population in most administrative villages after LF control. Obviously, filariasis transmission has already dropped down below the transmission threshold of filariasis, adopted in the LF control strategy. This study provided important theoretical basis for formulating technical strategies for LF surveillance in areas where filariasis had been basically eliminated, and developing strategies and criteria for LF elimination (Collaborating Research Group on the Transmission Threshold of Filariasis, 1994).

In further, the transmission threshold of filariasis has been evidenced by the results of surveillance obtained from the sentinel sites. Though several years of study it was found that in areas where filariasis had been basically eliminated, the microfilaria density in most residual microfilaremia patients and natural filarial infection in mosquitoes would decrease year by year. Majority of the residual microfilaremia patients, even do not receive drug treatment, could gradually turn to negative in 10 years. By 1994, all of the 864 endemic counties/cities had reached the criterion for the basic elimination of filariasis (Table 1). Since then, the strategy and criteria for the final elimination of filariasis were formulated. A comprehensive and active surveillance system, combining with longitudinal and cross-sectional surveillance was conducted in areas achieved basic filariasis elimination (National Technical Steering Group for Filariasis Control and Research, MOPH, 2002). Assessment of the elimination of filariasis in originally endemic areas, was strictly performed according to the *Criteria for Filariasis Elimination*, in order to achieve the goal of LF elimination in national wide.

The *Criteria for Verification of Elimination of Filariasis* (trial) and *Verification of Elimination of Filariasis* (trial) was issued by MOPH in 1994 and implemented in 1996. The *Criteria for Verification of elimination of filariasis* (trial) included the following major points. By county or equivalent administrative region as a unit, through provincial level assessment, filariasis transmission has been interrupted (basic elimination) for more than 10 years: (1) parasitological surveillance covers 3% of the total population in 30% of the endemic towns or townships, no microfilaremia case was found; and (2) no human filarial larva infection was detected in mosquitoes. The content about the survey, treatment, and caring of chronic patients of filariasis has been added in the *Criteria for Elimination of Filariasis* issued by State Administration for Market Regulation and State Standardization Administration.

As the major drafters, scientists in our Institute participated in research and formulation of three national standards, *Criteria for Elimination of Filariasis* (GB20048–2006), *Diagnostic Criteria and Principles of Management of Filariasis* (GB15985–1995), and *Diagnostic Criteria for Filariasis* (WS260–2006).

Table 1 Time of each previous endemic province (autonomous regions/municipalities) achieved basic elimination and elimination of filariasis through national assessment.

Province	Year achieved basic elimination	Year achieved elimination	References
Guangxi	1985	1995	Pan et al. (1992), Pan et al. (1995)
Guizhou	1984	1996	Liu et al. (2002), Tao et al. (1987)
Shanghai	1984	1996	Ji et al. (1991)
Sichuan	1986	1997	Wei et al. (1994), Zhang (2004), Zhang et al. (1993), Zhang et al. (2001), Zhang et al. (2002b), Zhu (1990)
Chongqing	1986	1997	Li et al. (2002)
Hunan	1986	1997	Deng et al. (2008), Duan (2001, 2004), Duan et al. (2001)
Jiangsu	1989	1998	Liu (1990), Sun et al. (2001), Xu et al. (2008)
Guangdong	1987	2000	Huang et al. (1995), Huang et al. (2003), Huang et al. (2004b), Ou (1990)
Hubei	1988	2001	Pei and Zhang (1996), Peng et al. (2000), Zhan et al. (2012), Zhang (1990)
Zhejiang	1989	2001	Wang et al. (1989), Yao et al. (2002)
Fujian	1988	2002	Lin et al. (1996), Liu et al. (1989), Liu et al. (1990), Liu and Liu (1984)
Shandong	1983	2004	Cui et al. (1984), Deng and Chen (2002), Sheng et al. (1995), Wang (2005)
Henan	1987	2004	Chang et al. (1992), Lin et al. (2005), Ru et al. (1991), Xu et al. (2002), Zhao et al. (1992)
Jiangxi	1990	2005	Liu et al. (1991), Zhang et al. (2002a), Zhang et al. (2006a,b)
Hainan	1987	2005	Hu et al. (2008), Wu et al. (1990), Wu et al. (1997), Wu et al. (2003a)
Anhui	1994	2006	Chen et al. (2007), Hu et al. (2009), Zou et al. (1996), Zou et al. (2002)

4. Research and development of control tools, including diagnostics and MDA drugs

4.1 Study on the transmission threshold of filariasis

In 1982, the MOPH approved the study of the transmission threshold of filariasis as a key project. A unified research method was designed and formulated by Collaborating Research Group on the Transmission Threshold of Filariasis. Twenty-one representative sentinel sites were selected in 10 provinces, and one autonomous region. All the sentinel sites were in previously hyper-endemic areas where the microfilaria rate in the population had decreased to a low level through LF control with DEC administration. Four levels were set according to the microfilaria rate in population: 0.5%, 1.0%, 1.5%, and 2.5%. Three levels were set by the mean microfilaria density per 60 μL peripheral blood: <5, 5–10, and >10. The dynamics of the microfilaria rate and residual microfilaremia cases in populations from different levels were systematically observed after interruption.

The research results from the 21 sentinel sites were collected, analysed and concluded by the Collaborating Research Group on the Transmission Threshold of Filariasis in 1994. It was demonstrated that after intervention, where the man-biting rate of mosquito vectors was within 20–50 mosquito/person/night, and about 40% of the population used bed-nets seasonally, the microfilaria rate in the 20 sentinel sites was found to be below 1.71% in bancroftian filariasis endemic areas, and it was below 1.55% in endemic areas of malayan filariasis. The mean microfilaria density in residual microfilaremia cases in most sites was below 10/60 μL blood. The microfilaria rate in population and the positive rate of filarial larvae in mosquito vectors would decrease year by year, and the filariasis transmission tended to be interrupted. The above results confirmed that the microfilaria rate below 1% in the population of an administrative village, which is the criterion for the basic elimination of filariasis, is less than the transmission threshold of filariasis in field.

The achievement obtained in the study on the transmission threshold of filariasis has been proved in the large scale filariasis control campaign (Table 2). Through strict evaluation, the results of the systematic surveillance showed that all the 16 LF endemic P/A/M had reached the criteria of filariasis elimination until January 2006.

Table 2 Studies on the role of low density microfilaremia cases in the transmission of filariasis in late stage of filariasis control in China.

Time	Endemic area	County/ village	Province	Microfilaria rate	Microfilaria density	No. people	Trans-negative rates	References
1988–1992	Bancroftian filariasis	Jiaoshang Village	Jiangxi	5.33% (70/1257, 1986)–0% (0/1144, 1989)–0% (0/1244, 1992)	≥50/60 μL	3	33.33% (4th year)	Fang and Zheng (2002)
					6–50/60 μL	3	100% (4th year)	
					≤5/60 μL	3	100% (4th year)	
1981–1986	Malayan filariasis	Deqing County	Zhejiang	0.53% (5/945, 1981)	4.2/60 μL (1–8)	5	100% (4th year)	Shi et al. (1988)
	Bancroftian filariasis	Queshan County	Henan	14.96% (1982)–1.04% (87/8329, 1984)–0.20% (15/7320)	0.5–29.5/60 μL	87	82.9% (3rd year)	
1982–1985	Malayan filariasis	Linhai City	Zhejiang	<0.5%	1–10/60 μL	11	90.9% (5th year)	Li et al. (1987)
1984–1988	Bancroftian filariasis	Pingle Village, Yining County	Guangxi	0.5% (1983)–0.34% (1986)–0% (1988)	34.4/60 μL (1–62)	5	100% (5th year)	Pan et al. (1990)
1982–1996	Bancroftian filariasis	Xiangxi Autonomous Prefecture	Hunan	0.396% (1982)–0.086% (1986)–0.005% (1991)–0% (1996)	27.5/60 μL	1	0% (5th year)	Zhang et al. (1999)
					43/60 μL	1	0% (5th year)	
					≤7/60 μL	7	100% (5th year)	

1984–1993	bancroftian filariasis	Tan County	Shandong	0.56% (9/1611, 1984)–0.64% (11/1707, 1986)–0.33% (5/1538, 1989)–0.12% (2/1680, 1993)	6–50/60 μL	4	75% (5th year)	Wang et al. (1997)
					≤5/60 μL	5	60% (5th year)	
1986–1989	Malayan filariasis	Xiantao City	Hunan	1.55% (1986)–0.84% (1987)–0.42% (1988)–0.31% (1989)	≥50/60 μL	1	100% (3rd year)	Guo et al. (1989)
					6–50/60 μL	1	0% (3rd year)	
					≤5/60 μL	19	89.47% (3rd year)	
1984–1993	Bancroftian filariasis	Erketuo Village	Hunan	19% (19/100, 1984)–1.3% (1/79, 1986)–1.8% (2/112, 1990)–0% (1991)–0% (1993)	6–50/60 μL	2	50% (7th year)	Luo et al. (1996)
1982–1985	Bancroftian filariasis	Zou County	Shandong	0.08% (4/4597, 1982)–0% (1984)	≤5/60 μL	4	100% (3rd year)	Song and Ma (1988)
1984–1996	Bancroftian filariasis	Leshan City	Sichuan	1.29% (46/3554, 1984)–1.01% (36/3569, 1985)–0.33% (12/3599, 1987)–0.14% (5/3621, 1988)	≥25/60 μL	3	66.7% (3rd year)	Zhang et al. (1996)
					6–25/60 μL	12	100% (7th year)	
					≤5/60 μL	55	98.2% (7th year)	

4.2 Pathogen biological research

4.2.1 *Comparative studies on the morphology of larva stage of four species of* Setaria *and* Dirofilaria repens

The filaria larvae in mosquito vectors in China are mainly from cattle, horse and dog. During 1986–1987, to distinguish the species of filarial larvae in vectors in filariasis epidemiological investigation and surveillance, our Institute cooperated with Hunan provincial sanitary and anti-epidemic station, and Xiangxi Tujia and Miao autonomous prefectural sanitary and anti-epidemic station, used the infective larvae of *Setaria* (*S. labiatopapillosa*, *S. erschovi*, *S. equina*, and *S. leichungwingi*), and *D. repens* to infect laboratory strains of *Anopheles sinensis* and *Culex pipiens pallens* (Li et al., 1989). The morphology of different larval stages of the above filaria was described. In further, with the addition of *W. bancrofti*, *B. malayi*, *D. immitis*, and *S. digitate,* the identification key of nine common filariae in the infective stage in mosquitoes was compiled.

4.2.2 *Establishment of the animal model of cotton rat-filariasis*

For screening anti-filariasis drugs, cotton rat-filaria model consisting of cotton rats (*Sigmodon hispidus*) infected with *Litomosoides carinii*, and rat mites (*Ornithonyssus bacoti*) were introduced in 1972 through the courtesy of the Swiss scholar, Dr. Friedheim (Xi et al., 1979). Since then, the model has been well maintained in NIPD to screen known or newly synthesized compounds and herbs. The effective drug DEC for filariasis treatment was found subsequently.

4.2.3 *Studies on the transmission of malayan filariasis via* An. sinensis *and* An. lesteri

In the late 1950s and early 1960s, two types of the primary malayan filariasis vector, *An. sinensis* were found in thorough studies of the classification of the *An. hyrcanus* group and the relationship between some species of this group and filariasis (Zhang et al., 1964). Two types of *An. sinensis* are significantly different in their role in the transmission of malayan filariasis (Feng and Ma, 1956). The survey (Zhang et al., 1964) in the two malayan filariasis endemic areas, Wujiang County in Jiangsu Province and Yihuang County in Jiangxi Province showed that the natural infection rates of filarial larvae in *An. sinensis* were 28.0%, and 22.1% respectively in the above two areas. And the positive rates of natural infection with L3 *B. malayi* were 1.5%, and 1.9%, respectively. While the natural infection rates to filarial larvae of *An. lesteri* was

47.6%, with natural infection of L3 larvae was 12.9% in Wujiang; and the natural infection rates to filarial larvae of *An. lesteri* was 40.7%, with natural infection of infective larvae L3 was 5.8% in Yihuang. In addition, the positive rates of experimental infection with L3 in *An. sinensis* and *An. lesteri* were close, which were 83.7%–95.7%, and 94.3%–95.5%, respectively. *An. lesteri* is an indoor-resting species, prefers human blood, while *An. sinensis* is a semi-indoor resting species, fed on humans, poultry and other animals, but prefers animal blood (Wang and Zheng, 1989). Since *An. lesteri* is much more closely related to the human being, it is more important than *An. sinensis* in the transmission of malayan filariasis.

4.2.4 *Study on the susceptibility of* An. lesteri *to experimental infection with* W. bancrofti

In the literatures on the relationship between *An. sinensis* (*s.l.*) and filariasis published before 1950s, *An. lesteri* was mixed in the mosquito samples in some sympatric places of both species. To illuminate the role of *An. lesteri* in the transmission of bancroftian filariasis, study on the susceptibility of *An. lesteri* to experimental infection with *W. bancrofti*, comparing with that of *An. sinensis*, *Cx. p. quinquefasciatus*, and *Aedes togoi* was conducted (Xu et al., 1993). Of 188 *An. lesteri*, 202 *An. sinensis*, 280 *Cx. p. quiquefasciatus* and 129 *Ae. togoi* infected by the blood of a bancroftian microfilaremia patient with microfilarial density of 190/20 μL, the infective rates were 35.64%, 9.41%, 30.00%, and 65.98%, respectively; while of 188 *An. lesteri*, 134 *An. sinensis*, 289 *Cx. p. quinquefasciatus* and 176 *Ae. togoi* infected by the blood of another bancroftian microfilaremia with microfilarial density of 83/20 μL, the infective rates were 18.09%, 3.73%, 13.84% and 39.77%, respectively. It is concluded that the susceptibility of *An. lesteri* to experimental infection with *W. bancrofti* is significantly higher than that of *An. sinensis*, significantly lower than that of *Ae. togoi*, and at the same level with that of *Cx. p. quinquefasciatus*.

4.3 Research of the pathogenesis of filariasis

4.3.1 *Study of the recurrence of acute adenolymphangitis and lymphadenitis of malayan filariasis*

To clarify the impact of filarial and bacterial infection on the acute adenolymphangitis and repeated lymphadenitis, our Institute carried out relative studies in malayan filariasis endemic areas, Zhenze commune of Wujiang County in Jiangsu Province, and Miaoxi commune of Wuxing County in Zhejiang Province, during 1961 to 1973. The results showed that (1) the acute attacks had a clear seasonal fluctuation. The transmission season of filariasis coincided with the vector activity season. In Zhenze, a patient with historical lymphadenitis, stopped attacking of lymphadenitis for up to 35 years after leaving the endemic area, but re-attacked every year when back to the original living place. It indicates that the seasons of onset of lymphadenitis probably associate with the repeated infection of infective filarial larvae, transmitted by mosquitoes. (2) In the majority of patients with acute adenolymphangitis, except in the abscess, bacteria were absent in inflammatory site (Xie and Yuan, 1987). In the serum antistreptolysin 'O' titer assay on target patients, the antistreptolysin 'O' titer only increased 27.2% in cases of elephantiasis complicated with acute adenolymphangitis. Whereas cases with primary adenolymphangitis and lymphadenitis, and simple adenolymphangitis and lymphadenitis without the complication of elephantiasis, the antistreptolysin 'O' titer hardly increased. This suggests that bacterial infections, especially hemolytic streptococcal infections, were not the cause of acute adenolymphangitis. In a small number of patients with elephantiasis, it is one of the factors in complex aetiology. (3) After the filariasis transmission has been effectively controlled, not only in cases with primary adenolymphangitis and lymphadenitis, and simple adenolymphangitis, but also in cases with elephantiasis complicated with acute adenolymphangitis, the inflammatory attack was decreased significantly, indicating that repeated inflammatory attack is related to the filariasis transmission. (4) The positive effect of treatment against adult filaria to reducing inflammatory attack has not been found. It is included that in the endemic areas of bancroftian filariasis, the recurrent acute adenolymphangitis and lymphadenitis is mainly caused by the existence of filariasis transmission, induced by repeated infection of infective larvae injected in the human body.

4.3.2 Observation of lymphoedema of the lower extremity using lymphangiography

To understand the mechanism of leg lymphoedema caused by malayan filariasis, in endemic areas of malayan filariasis, located at the border of Wuxing County in Zhejiang Province and Wujiang County in Jiangsu Province, our Institute, collaborating with the Chinese People's Liberation Army 99 Hospital, started to observe lymphoedema of the lower extremity using lymphangiography, in selected elephantiasis cases caused by malayan filariasis (Sun et al., 1964). Twelve lower extremities of nine patients were observed, including three main types (1) one with no experience of acute adenolymphangitis or lymphadenitis (the other one with experience of acute adenolymphangitis or lymphadenitis, and lymphedema), (2) three with experience of acute adenolymphangitis or lymphadenitis, and no lymphedema, (3) eight with both lymphedema, and experience of acute adenolymphangitis or lymphadenitis (disease of duration ranging from 2 to 16 years). The technology of direct lymphangiography on superficial lymphatic vessels was performed. First, 3–5 mL 0.4% indigo carmine was injected subcutaneously in toe-webs. Then, find subcutaneously dyed blue peripheral lymphatic vessels after small transverse incision on the middle of dorsum pedis, and inject 5–18 mL sodium acetrizoate as contrast media. X-ray anterior, lateral views were performed on each patient from foot to groin.

In the case of type 1, contrast agent leaked to surrounding subcutaneous tissue in a lumpy way. In cases of type 2, lymphatic contortion and mild collateral hyperplasia were visualized in the lower limb of one lower extremity. Whereas, contrast agent only raised to 1/3 of the low limb in the other two lower extremities. In cases of type 3, seven legs showed lymphatic varicose in different degree, with following characteristics: the obvious phenomenon of lymphatic contortion collateral hyperplasia, enlarged and unevenly thicken lymphatic spaces, indistinguishable trunk and the lateral branch of lymphatic vessels, unorganized distribution of lymphatic vessels. In severe cases, the lymphatic vessels representing as spider network with the disappearance of valve shadow. Contrast media back flowed to the distal lymphatic vessels and leaked in subcutaneous tissue. The degree of lymphatic varices was consistent with the patient's clinical manifestation. The distal portion of the extremity was the most severe part. Femoral lymphatic vessels were observed in four lower extremities. Inguinal lymph nodes, linking with the afferent lymphatic vessel, were visualized in two of them, with clear contour. Thickening fibrosis in the subcutaneous tissue of foot was observed in one lower extremity with severe elephantiasis.

The above results suggest that there are two main reasons, which may cause lymphoedema of the lower extremity in patients with malayan filariasis. One is that repeated attacks of adenolymphangitis and lymphadenitis would induce different degrees of obstruction of the lymphatic vessel and a change in the circulation dynamics of lymph, resulting in backflow obstruction of lymph, followed by the appearance of lymphedema in the related extremity. The other is that granulation tissue proliferation, and fibrosis inside the lymphatic vessels would cause lymphatic obstruction, and result in lymphoedema of the lower extremity.

4.4 Research of diagnostic methods

4.4.1 Four antigens were used to diagnose filariasis by intradermal test

Since 1973, the study on the serological diagnostic techniques was started to carry out. Four prepared polypeptide antigens, including adult *D. immitis* polypeptide antigen (DFPT), adult *Setaria* sp. polypeptide antigen (SFPT), and other two adult *Setaria* spp. polypeptide antigens ($SFSC_1$ and $SFSC_4$) isolated and purified to evaluate the effect of purified antigen from animals to filariasis diagnostic by intradermal tests. The study was performed in the malayan filariasis endemic area (Wuxing County of Zhejiang Province), non-endemic area (Chongming County of Shanghai Municipality, and Fushan County of Shandong Province), and bancroftian endemic area (Teng County of Shandong Province). The intradermal test results were compared with that performed in local microfilaremia patients.

Serum samples for malayan microfilaremia detection were diagnostic by the four antigens, DFPT, SFPT, $SFSC_1$, and $SFSC_4$, with 94.1% (176/187), 87.5% (35/40), 74.0% (37/50), and 78.3% agreement to results of the blood test, respectively. While in samples for bancroftian microfilaremia detection, the results of the serological method using DFPT, SFPT, and $SFSC_1$ antigens were 88.3% (136/154), 73.6% (39/53), and 44.9% (22/49) agreed to that of the blood test. In microfilaremia negative populations from non-endemic areas, the false positive rate of DFPT antigen was 4.8%, and 9.6% in Chongming County, Shanghai, and Teng County, Shandong Province, respectively. In Chongming, the false-positive rates of SFPT, $SFSC_1$, and $SFSC_4$ in microfilaremia negative populations were 33.5%, 20.8%, and 76.9%, respectively. In Teng County, the false-positive rates of SFPT, and $SFSC_1$ in microfilaremia negative populations were 26.1%, and 5.0%, respectively. In crossover test, the DFPT antigen showed 6% (3/50) false-positive cross-reactions with sera from patients infected by schistosomiasis, and 8.6% (2/23) false-positive rate with sera from patients infected by hookworm and whipworm.

4.4.2 Application of intradermal test in the evaluation of filariasis longitudinal surveillance

Intradermal test with the filariasis polypeptide antigen (FPT) purified from *D. immitis*, as a donation from Guizhou Provincial Institute of Parasitic

Disease, was applied in two villages of Queshan County in Henan Province for longitude surveillance from 1982 to 1987, in order to trace the dynamic of filarial specific anaphylaxis level in populations before and after the implementation of DEC mass chemotherapy for the whole population. The two administrative villages Zhaozhuang and Zhongdian were located in meso- bancroftian endemic area treated as two sentinel sites in the survey.

The microfilaria density in populations of Zhaozhuang and Zhongdian were 9.98% and 18.66%, respectively, in the whole population blood survey conducted in 1982. The control strategy of treatment of microfilaremia cases followed by DEC-fortified salt for the whole population was implemented since May 1983. After intervention in May 1984, the microfilaria density decreased to 0.42% and 1.75% in populations of Zhaozhuang and Zhongdian, respectively. With the residual microfilaremia cases gradually turned to negative, the microfilaria density in populations continued to decline, and down to 0.05%, and 0.42% in populations of Zhaozhuang and Zhongdian, respectively, in May 1986.

The diameter of skin papule exceeds 9 mm after injection of antigens, as a positive criterion in intradermal test. Surveys by the intradermal test in spot-check sites were conducted in October 1982 (before control), September 1983 (5 months after control), and May 1987 (4 years after interruption). In the three surveys, the positive rates in populations from Zhaozhuang were 48.9%, 35.4%, and 20.6%, respectively, and 63.2%, 43.7%, and 19.4%, respectively in Zhongdian. The results showed that the positive rates of filarial specific anaphylaxis in populations from both spot-check sites started to decline sustainably after the implementation of selective treatment and MDA.

The false-positive rate of the serological diagnostic in non-endemic areas was approximately 5%, indicating that the method may not be sensitive in the lower age group, which is reflecting in the results in the field survey. No significant change in positive rate has been observed in the group with 1–4 years old children, comparing with the results of surveys, conducted in 1982 and 1987. Whereas the mean diameter of skin papule obviously appeared declining in this group before and after the control. If setting the diameter of skin papule exceeds 13 mm as a significant positive criterion, the positive rate in populations from Zhaozhuang was declined significantly from 19.1% to 3.0%, since 1982 to 1987, and it declined from 29.5% to 2.5% in Zhongdian. The serologic test result showed negative in a few persons in the primary survey, while turned to positive in the tracing surveys from 1983 to 1987. The annual positive conversion rate can reflect the level of filariasis transmission in a period. Thus, the dynamic of annual positive conversion rate is one indicator to evaluate the control effect. In conclusion, the positive rate, significant positive rate, and annual positive conversion rate obtained from the intradermal test of FPT antigen in the population can be a valuable reference to evaluate the filariasis transmission level and tendency in duration, in a systematic and comprehensive way.

4.4.3 Studies on detection anti-filarial antibodies by dot-enzyme-linked immunoassay (Dot-ELISA)

In order to improve the accuracy and sensitivity serological diagnosis of filariasis, Dot-ELISA was carried out in the late 1980s to detect anti-filarial antibodies. In further, the comparative study among Dot-ELISA, and other two methods indirect fluorescent antibody test (IFAT) and ELISA, which had been widely used in China at that time, was conducted for detecting antibody levels in bancroftian microfilaremia patients. The conjugate, carrier, and substrate of dot-ELISA were Horseradish peroxidase Rabbit Anti-Human IgG, nitro cellulose membrane, and 4-chloro-1-neophenol, respectively. Negative, positive and TBS controls were set in each experiment. The appearance of the blue spot on the membrane was determined as a positive reaction. Serum samples were collected from patients with *B. malayi* microfilaremia in Anhui Province and patients with *W. bancrofti* microfilaremia in Jiangxi Province. In addition, sera of healthy persons from non-endemic areas in Shanghai were selected as the negative control. Serum samples of patients with schistosomiasis from non-endemic areas in Jiangxi Province and samples of co-infection with hookworm and ascaris from non-endemic areas in Henan Province were chosen for cross-reaction test. The results showed the positive rates of the sera from 62 *B. malayi* microfilaremia patients, and 49 *W. bancrofti* microfilaremia patients were 96.8%, and 85.7%, respectively; while in 50, and 39 healthy individuals the false-positive rates were 4.0%, and 5.1%, respectively. In 18 patients with schistosomiasis, and 39 patients with hookworm and ascaris, the false-positive rates were 5.6%, and 17.9%, respectively (Yuan et al., 1992). It indicates that this method was highly sensitive, simple and convenient, and might be suitable for serological surveillance of filariasis. Dot-ELISA is simple, sensitive, and saving the use amount of antigen. Additionally, the reaction results can be identified by the naked eye and don't have to use a special instrument. It could be used for monitoring the control of filariasis transmission. However, there were some cross-reactions with the sera of persons infected with hookworm and ascaris.

A comparative study was conducted for detecting antibody levels of *W. bancrofti* microfilaremia cases by using three serologic tests, IFAT, ELISA and dot-ELISA (Sun et al., 1994). IFAT was performed with the frozen sectional antigen of adult *B. malayi*. While adult *B. malayi* soluble antigen was used for ELISA. A total of 102 serum samples were collected from *W. bancrofti* microfilaremia cases in bancroftian filariasis endemic area in Guzhen County, Anhui Province. The positive rate was 89.2% with IFAT, 84.3% with ELISA, and 89.2% with dot-ELISA, respectively. While in healthy individuals from non-endemic areas, the false-positive rate was 0%, 1.8%, and 0%, respectively. The results suggested that there were no statistically significant differences among three serologic tests, and no correlation was found in this study between microfilaria density and the antibody level. Therefore, all the three methods can be used to monitor the trend of the antibody level in the population for post-control surveillance. Each sentinel site can choose the method for the serologic test, according to its own conditions and the advantages and disadvantages of each method. The results can be compared with each other.

In further, other sero-epidemiological diagnostic methods have been used in post-control (after basic elimination) and post-elimination phase in China for filariasis surveillance (Table 3). Additionally, DNA probe was applied for *W. bancrofti* identification in vectors for filariasis surveillance in the late stage of post-control phase (Huang et al., 2000; Zheng et al., 1992).

4.5 Study on anti-filarial drugs and treatment methods

4.5.1 Development of anti-filarial drugs

The National of Bureau of Medical Pharmaceutical Administration and the Shanghai Academy of Pharmaceutical Industry cooperated with the Shanghai Institute of Parasitic Diseases to study new anti-filariasis drugs. Several thousand chemical compounds were screened. Through experimental tests on animals, it was found that furapyrimidone (M170) could not only kill *L. carinii* microfilaria and adult worms, but also *B. malayi* microfilaria and adult worms in the blood and abdominal cavity in jirds, especially in killing the adult *L. carinii*. By the end of 1979, our Institute cooperated with the Shandong Institute of Parasitic Diseases in clinical trials, and then Guangdong, Hubei, Jiangsu, Henan, Zhejiang, Fujian, Jiangxi, Shandong and Shanghai provinces/cities continued further clinical

Table 3 Sero-epidemiological diagnosis for filariasis detection in post-control (after basic elimination) and post-elimination phase in China.

		Subjects						
		Sensitivity	Specificity					
Method	Test principle	Microfilaremia	Control	Patients infected by other parasites	Purpose of use in LF control and elimination in China	Technical difficulty	Cost	References
ICT card	Antigen (finger prick)	95.7% (111/116)	0% (71)	0% (12 malayan filariasis, 33 ascaris, 20 schistosomiasis, 6 trichinosis)	case diagnosis, monitoring and investigation in key areas after the elimination	+	++ +	Zheng et al. (1998)
		88.24%–93.33% (15/17, 14/15, 11/12)	0% (69)					He et al. (1999)
		94.92% (56/59)	0% (40)	0% (30 cysticercosis, 25 clonorchiosis)				Deng et al. (2000)
(McAb)-ELISA	Antibody (venous blood)	96.61% (57/59)	0% (40)	0% (30 cysticercosis, 25 clonorchiosis)	Systematic surveillance in previous endemic areas in post-control and post-elimination phase	++	+	Deng et al. (2000)
		94.1% (64/68)	0% (38)	0% (60, each 10 of schistosomiasis, ascaris, cysticercosis, clonorchiosis, hydatidosis, paragonimiasis)				Huang et al. (2004a)
		97.10% (67/69)	3.66% (3/82)	17.65% (6/34, children borne after control) 12.50% (6/48, control from original endemic areas) 18.03% (11/61, previous microfilaremia cases)				Wang and Gao (1990)

		95.91% (47/49)	0% (50)	0% (77 chronic filariasis) 0% (30 intestinal helminthes, 20 clonorchiosis, 20 cysticercosis)				Xu et al. (1996)
		96.3% (104/108)	0% (90)	0% (302, children born after control) 0.30% (10/3291, population in previous endemic area) 0% (132, chronic filariasis) 0.78% (1/129, previous microfilaremia but turn negative) 0% (30 intestinal helminthes, 45 clonorchiosis, 50 cysticercosis)				Chen et al. (2002)
ELISA	Antibody (venous blood)	92.64% (63/68, malayan filariasis) 83.33% (70/84, bancroftian filariasis)	1.24% (2/161)		Sentinel site surveillance in previous endemic areas in post-control phase	+++	++	Deng et al. (1986)
		87.8% (79/90)	1.63% (2/123)					Li et al. (1993)

Continued

Table 3 Sero-epidemiological diagnosis for filariasis detection in post-control (after basic elimination) and post-elimination phase in China.—cont'd

Method	Test principle	Subjects			Purpose of use in LF control and elimination in China	Technical difficulty	Cost	References
		Sensitivity	Specificity					
		Microfilaremia	Control	Patients infected by other parasites				
IFAT	Antibody (venous blood)	97.50% (39/40)	2.44% (2/82)	13.33% (10/75, chronic filariasis) 4.44% (2/45, children born after control) 16.31% (46/282, floating population)	Sentinel site surveillance in previous endemic areas in post-control phase	++	+	Zou et al. (1995)
		87.2% (20, malayan filariasis) 97.1% (105, bancroftian filariasis)	6.55% (62)	0% (7 hookworm, 30 ascaris, 20 suspected schistosomiasis)				Li et al. (1988)
		93.88% (92/98, malayan filariasis) 90.48% (38/42, bancroftian filariasis)	3.6% (4/111)	3.23% (1/31, children born after control)				Jiang and Zhong (1990)
		86.8% (99/114)	1.57% (3/191)	0% (54, more than one parasite of hookworm, ascaris, or trichuriasis) 0% (96, pneumonia)				Ji et al. (1988)

		89.2% (91/102)	0% (0/30)					Sun et al. (1994)
		99.1% (226/228, malayan filariasis) 92.8% (412/444, bancroftian filariasis) 96.9% (31/32, chronic filariasis)	99.3% (149/150)					Wang et al. (1987)
Dot-ELISA	Antibody (venous blood)	96.8% (60/62, malayan filariasis) 81.6% (40/49, bancroftian filariasis)	4.0% (2/50) 5.1% (2/39)	0% (20, schistosomiasis) 5.1% (2/39, schistosomiasis) 17.8% (8/45, hookworm and ascaris) 17.9% (7/39, hookworm and ascaris)	Sentinel site surveillance in previous endemic areas in post-control phase	++	++	Yuan et al. (1992)
		89.2% (91/102)	0% (54)					Sun et al. (1994)

The rating of technical difficulty is based on the steps of the test and the time required, with '+' lowest and '+++' highest. Cost ratings are based on the cost of materials, laboratory equipment is not included. The control group is healthy person from non-endemic area. Abbreviations: ELISA, enzyme-linked immunoassay; ICT, immunochromatorgraphic test; IFAT, indirect fluorescent antibody test; LF, Lymphatic filariasis.

observations. Till 1983, 981 cases had been treated. Among them, 755 cases were bancroftian filariasis and 266 cases were malayan filariasis. The results showed that using M170 in a total dose of 140 mg/kg body weight for 7 days had a good effect on bancroftian microfilaria and adult worms. The effect was better than using the DEC standard dose in long-term or short-term treatment. Therapy consisting of M170 total dosage of 120 mg/kg body weight for 6 days in malayan filariasis did not have a good effect. The general side reaction was similar to DEC, and vomiting usually occurred. When the total dosage was increased to 200 mg/kg body weight, many patients showed significantly increased serum glutamic pyruvic transaminase, which indicated that a large dosage could damage the liver (Ren and Wang, 1984; Wu, 1984). Pharmacological studies found that a large dosage was toxic on rat embryos. TA_{100} and TA_{98} could be mutations. Finally, M170 was not submitted for production.

4.5.2 Study on alleviating the therapeutic response of DEC

DEC is known to be an effective therapeutic drug for filariasis, with low toxicity. The side reactions are mainly due to the heterologous protein released by the microfilaria killed. Generally, the reaction is mainly the febrile symptom, which is a therapeutic response. It greatly hindered the promotion of DEC-MDA in filariasis control. Therefore, in 1974–1975, investigations on reducing the DEC side reaction, combing with surveillance on DEC-MDA spot-check sites were carried out in malayan filariasis endemic areas, Deqing County and Tongxiang County in Zhejiang Province, and bancroftian filariasis endemic area, Yongning County in Guangxi Province.

Practices have proved that antipyretic and analgesic drugs can reduce the febrile reaction. But antipyretic drugs maintain in the blood for a short time. Therefore, the interval time of administration should not be too long. Through observation, taking paracetamol regularly can effectively reduce and restrain the reaction caused by DEC in the treatment of microfilaremia, thus ensuring the smooth implementation of MDA.

Two regimes of the administration and dosage of taking paracetamol have been designed successively for alleviating the therapeutic response of DEC (Sun et al., 1990). The first was to take the first dose of diethylcarbamazine together with paracetamol rapid-release tablet (ordinary dosage) 0.5 g, every 4 h for 2 days. A total of 31 microfilaremia cases of *B. malayi*, with microfilaria density more than 30/60 μL, undergoing DEC chemotherapy for the first time were investigated. The dose of DEC was 0.75 g each time, once a day for 2 days, 1.5 g in total. Fifty patients in the control group were selected as the same requirement of the experimental group, but paracetamol was not given. The results showed that in the experimental group, the fever rate was 71.0% after taking DEC, with 9.7% high fever (≥39 °C) reaction. The mean cumulative duration of fever was 16 h. Whereas in the control group, the fever rate was 100% after taking DEC, with 74.0% high fever reaction. The mean cumulative duration of fever was 36 h. Compared with the control group, the fever rate was reduced by 29.0% in the experimental group, and the hyperthermia response rate was reduced 64.3% in the experimental group. The cumulative fever duration of the febrile patients was shortened 29 h. The concomitant symptom was also lessened and relieved.

In order to reduce the times of administration of paracetamol, the second regime was developed. The difference from the first regimen was that paracetamol rapid-release tablets and sustained-release tablets (enteric-coated tablets) were given at the same time, 0.5 g each, every 8 h for six times for malayan filariasis therapy, and every 8 h for eight times for bancroftian filariasis therapy. The dose of DEC for therapy of malayan filariasis patients was the same as regime 1. Whereas for patients of bancroftian filariasis, the DEC treatment regimen was 3 g for 3 days, given 0.5 g each time, twice a day. The results showed that in the experimental group of malayan filariasis patients, the fever rate was 80.9% in 110 cases, with 17.3% high fever reaction. The mean cumulative duration of fever was 14.2 h; while in the control group, the fever rate was 100% in 15 cases, with 63.8% high fever reaction. The mean cumulative duration of fever was 19.2 h. In the experimental group of bancroftian filariasis patients, the fever rate was 63.8% in 47 cases, with 6.4% high fever reaction. The mean cumulative duration of fever was 14.3 h; while in the control group, the fever rate was 94.7% in 19 cases, with 15.8% high fever reaction. The mean cumulative duration of fever was 29.3 h. The investigation results were consistent with regime 1. Since compared with regime 1, the administration interval time of paracetamol was extended to 8 h in regime 2, and the times of taking the drug were reduced by half, regime 2 is easy to be accepted by the public.

4.6 The project 'Research achievements on the strategy and technical measures for interrupting the transmission of LF in China' provide technical support for LF elimination

Based on the of pathogenic biological characteristics of filariasis, through studies on the dynamics and threshold of filariasis transmission, it was for the first time that the control strategy taking elimination of source of infection as the major focus was established, the residual low-density microfilaremia cases being of no practical importance in filariasis transmission at the post-control stage was elucidated (Shi et al., 1988), the indicator for filariasis transmission interruption was proposed, and the technical measures were worked out accordingly. The main of the results include: (1) establishing the control strategy taking elimination of infection source as the major focus, (2) formulating safe, feasible and effective MDA and DEC-fortified regimens, (3) revealing the transmission pattern of filariasis at the post-control phase, and proposing indicators for filariasis transmission interruption, (4) establishing longitudinal, cross-sectional, and floating population active surveillance system, (5) formulating the criteria of basic elimination and elimination of filariasis and corresponding technical indicators to guide the realization of the goal of basic elimination (transmission interrupting) and the elimination of filariasis in the whole country by stages (Shi and Sun, 1999).

The above research results are implemented nationwide, which plays an important and crucial role in realizing the elimination of filariasis in China (Shi et al., 2001). The project 'Research achievements on the strategy and technical measures for interrupting the transmission of LF in China' led by our Institute, with the cooperation of other provincial institutes of parasitic control, was won the first prize of National Science and Technology Advance Award in 2000.

5. Monitoring and evaluation supported by information management both in elimination and post-elimination stages

5.1 The surveillance system in elimination stages

Summing up the parasitological data in cross-sectional surveillance conducted in 16 endemic P/A/M from 1972 to 2005, a total of 3.315 million persons received blood examination, and 21,710 residual microfilaremia cases were found. The residual microfilaremia cases distributed in 456 counties/cities, occupying 53.3% sentinel counties/cities. After basic elimination of filariasis in China, microfilaremia cases have been found in Longyan city and Shaowu city, Fujian Province in the last 14 years (the longest). While the last microfilaremia case was found in Sixian County, Anhui Province in 2005. During 1976–2005, cross-sectional mosquito vector surveys were performed. A total of 4.71 million mosquitoes were dissected and 809 filarial larvae positive mosquito were found. Since almost all the original filariasis endemic counties/cities reached the level of basic elimination, mosquito vector and their natural infection rate with filaria larvae has decreased gradually. In 1976, the natural infection rate of filarial larvae in *Cx. pipiens* complex (*Cx. p. pallens*, and *Cx. p. quiquefasciatus*) was 0.19% in bancroftian filariasis endemic area. In the malayan filariasis endemic area, the infection rate of *Anopheles* spp. (*An. sinensis* and *An. lesteri*) was 0.84%. Subsequently, both the infection rates of vectors of bancroftian filariasis and malayan filariasis has been dropped to 0%. Filaria-infected mosquitoes have been detected in Dejiang County, Guizhou Province and Cangshan City, Shandong Province in the 12th year (the longest) after basic elimination. While Xiamen is the place where the last positive mosquito of filaria was found in 1992. Eighty-two longitudinal surveillance sentinel sites were established between 1980 and 2004. Among them, 45 sites of which conducted long-term and effective surveillance. After the intervention, the majority of residual microfilaremia cases turned to negative successively in 10 years. In the fifth and ninth year, no microfilaremia case was found for bancroftian and malayan filariasis sentinel sites, respectively, in the surveillance. The microfilaria rate has decreased subsequently. The positive rate of mosquitoes decreased gradually, and after the eight-year of the surveillance, no positive mosquito was found. Neither new infection, nor new chronic patient has been found in sentinel sites during filariasis surveillance. In order to consolidate the achievements of filariasis control after basic elimination, the search for residual epidemic foci has been intensified in various regions. Only in a few administrative villages in Huilai, Haikang of Guangdong Province, and Gao'an and Zhangshu of Jiangxi Province, the microfilaria rates were found higher than 1% (National Technical Steering Group for Filariasis Control and Research, MOPH, 1991b). All residual endemic foci have been treated timely and effectively. From 1984 to 2005, LF surveillance of the floating population was carried out. The total number of 363.5 thousand persons received blood test and 419 microfilaremia cases were found, all of them being detected before 1992. The results indicated that the probability of detecting microfilaremia cases in the floating population reduced greatly since all the original filariasis endemic counties/cities reached the level of basic elimination (National Technical Steering Group for Filariasis Control and Research, MOPH, 2002; Sun, 2005).

From 1995 to 2006, the MOPH organized evaluation in Guangxi (1995), Guizhou (1996), Shanghai (1996), Sichuan (1997), Chongqing (1997), Hunan (1997), Jiangsu (1998), Guangdong (2000), Hubei (2001), Zhejiang (2001), Fujian (2002), Shandong (2004), Henan (2004), Jiangxi (2005), Hainan (2005), and Anhui (2006), altogether 16 P/A/M, to assess the elimination of filariasis (Wu et al., 2003b) (Table 1). In May 2007,

China was officially confirmed as having achieved the elimination of LF as a public health problem by the WHO (Sun et al., 2013). The success of LF elimination in China provides practical evidence for the call of elimination of LF as a public health problem, which was adopted at the 50th World Health Assembly in 1997 (Fang and Zhang, 2019; Sun, 2001; WHO, 1997, 2011).

5.2 Special monitoring and investigation in key areas after the elimination of filariasis

In order to consolidate the achievements of filariasis elimination in China, further clear away the infective sources, and provide scientific evidences for filariasis surveillance at the post-elimination phase, entrusted by Disease Control Bureau, MOPH and Chinese Center for Disease Control and Prevention (CDC), our Institute, cooperating with Guangxi, Hunan, Hubei, Guizhou, Chongqing, Jiangxi and Anhui Provincial CDC carried out special monitoring and investigation in key areas, with a unified guideline, from January to June in 2008. The investigation was divided into three steps: baseline survey in previous endemic areas, spot check in key areas, and review of the results of spot checks.

The investigation was conducted previous bancroftian filariasis endemic area (including the endemic area with mixed infection areas with bancroftian filariasis as the primary). The principles of sentinel sites were as follows: (1) lack historical material on mass blood survey for filariasis; (2) remote areas with inconvenient transportation and poor sanitary conditions; (3) sites at the boundary of provinces or counties; (4) low frequency of mass blood survey and selective treatment; (5) low coverage or less than the prescribed amount of DEC-fortified salt; (6) relatively high microfilaremia rate at the end of the mass blood survey or spot check (The evaluation unit is based on the administrative village). Where, the administrative villages accord with one item of (1), (2) or in any two items of (3), (4), (5), and (6) were identified as key areas. According to the results of the baseline survey, 80 administrative villages in 38 counties/cities of 7 P/A/M were selected as the sentinel sites.

The filariasis-specific IgG4 testing kit was applied for serologic diagnostic in permanent residents over 10 years old (no less than 600 people) in sentinel sites. In the large administrative village, random cluster sampling was adopted for selecting natural villages as spot-check sites. In villages with resident population less than 600, the scope of the investigation was expanded to neighbouring villages, in order to meet the requirement of sample size.

Filter paper blood specimens were collected and examined by Shandong Provincial Institute of Parasitic Diseases. The results showed that 165 were found positive among 50,424 samples, with a rate of 0.327% (0.051–0.630). The results of the pathological examination in serologic positive persons showed that all were negative, no microfilaremia case was found.

The project of 'Special monitoring and investigation in key areas after the elimination of filariasis' indicates that the achievements of filariasis elimination in China is very consolidated. Considering the following factors: long period and difficult in filariasis control, imbalance in different endemic areas, and the impact of the floating population, the residual foci of filariasis cannot be completely excluded. It suggests that the surveillance work after elimination of filariasis should be carried out in a certain period after the declaration of LF elimination in China. Firstly, the LF should be included in the direct-network reporting of the Notifiable Disease Monitoring System continuously. Secondly, all levels of disease prevention and control agencies at the endemic areas should continue to take the initiative surveys for clearing away the potential infective source. Thirdly, imported cases from aboard should be paid serious attention.

6. Technical supports to the elimination and post-elimination programmes

Lymphatic filariasis, as a national statutory B infectious disease, has been included in the Notifiable Diseases Reporting System since 2004. Annually, approximately 450 direct-network reported cases of LF have been identified over the last 10 years. But, the majority of these cases were false alarms due to mis-clicking, and the remaining cases were chronic filariasis. Until now, almost no case was identified as microfilaremia through direct-network reporting of the Notifiable Disease Monitoring System, except that in August 2007, a clinically confirmed case of bancroftian filariasis from Fuchuan County of the Guangxi Zhuang Autonomous Region, has been found through the system. Since the case has been detected, our Institute, together with the Guangxi provincial CDC, carried out an emergency investigation in the residual focus promptly (Li et al., 2008). Nineteen microfilaremia cases (1052 persons involved) with the positive rate of 1.8%, were found by pathological examination in Changtang Village; no microfilaria-positive case (4119 residences involved) was found in three other villages in Fuchuan County. Immunochromatographic technology was conducted for those going out but retuned and those in neighbouring areas. Positive individuals were checked by thick blood smear for pathological diagnostic. Mosquitoes were collected in human dwellings. No filaria larva was found in 54 dissected *Cx. p. quinquefasciatus*. More identified patients were more than 60 years old, indicating that the residual focus was an original hypro-endemic area of bancroftian filariasis. Based on the discovery and timely disposal of the residual foci, two experiences can be concluded in the progress of filariasis elimination. In the preliminary stage of post-control, evaluation on the historical control process should be conducted with regular surveillance simultaneously, such as special monitoring and investigation in weak points, which can make more effective in the endeavour. In the later of post-control and post-elimination phases, besides health personnel of the disease control system, the medical personnel of the medical system should maintain the diagnosis and treatment skills of filariasis and keep high vigilance against filariasis, which is important to discovery of residual foci.

Since the declaration of lymphatic filariasis elimination in China, in 2007, the *Working Program for the Morbidity Management and Disability Prevention of Chronic Filariasis Patients* was issued by the MOPH. The dynamic digital management of chronic filariasis patient documents and the providing of basic personal care for morbidity in local clinical systems were proposed. In 2010, the *National Monitoring Program for the Post-elimination of Lymphatic Filariasis* was issued. The program was created for the timely discovery of possible residual and imported sources of infection, thereby preventing retransmission of filariasis. For the filariasis cases, which are reported through the Notifiable Diseases Reporting System, our Institute is in charge to inform and cooperate provincial CDC to verify the cases based on evidence from epidemiology, pathogenic diagnosis, serologic tests, and molecular biologic tests.

7. Conclusion

The achievements and experience of the elimination of LF in China are highly appraised and valued in the word. The strategies and technical measures for preventing LF transmission implemented in China have been recognized and widely recommended by WHO (Fig. 1). They are included establishing the control strategy taking elimination of infection source as the major focus; formulating the microfilaria rate below 1% as the criteria for termination of control measures; including DEC-fortified salt in the

MDA; establishing the surveillance strategy after the termination of control measures. In the post-control phase, special monitoring and investigation in key areas were conducted for clearing way potential residual foci. Two schemes and one criterion were issued to guild all levels of disease control and prevention workers on LF surveillance, epidemic disposal, and managing and caring for chronic filariasis patients in the post-elimination phase. In addition, the Notifiable Diseases Reporting System keeps an eye on the LF cases reported in national wide. These efforts prevent the LF resurgence in China, and simultaneously, promote the progress of Global Programme of Elimination of Lymphatic Filariasis in worldwide.

Ethical approval and consent to participate

Not applicable.

Consent for publication

All the authors have read and approved the submitted version of the manuscript to this journal.

Availability of data and materials

All data generated or analysed during this study are included in the published article.

Competing interests

The authors declare that they have no competing interests.

Author's contributions

D.-J.S, Y.F., Y.H., and Y.Z. designed the study; D.-J.S., and Y.F. wrote the original manuscript; D.J., Y.F., Y.H., and Y.Z. revised the manuscript and gave approval for its publication. All the authors read and approved the final version of the manuscript.

Acknowledgements

Abbreviations

CDC	Center for Disease Control and Prevention
DEC	diethylcarbamazine
ELISA	enzyme-linked immunoassay
IFAT	indirect fluorescent antibody test
LF	lymphatic filariasis
MDA	mass drug administration
MOPH	Ministry of Public Health
P/A/M	provinces/autonomous regions/municipalities
WHO	World Health Organization

References

Chang, J., Zhao, Q., Sun, J., Huang, Q., Lin, X., Zhao, X., Xu, B., 1992. Reports of filariasis longitudinal surveillance in Henan Province from 1988 to 1989. Chin. J. Parasit. Dis. Control 5 (1), 57–58.

Chen, X., Deng, X., Li, G., Xu, F., Fu, B., Liu, X., Gao, C., Zhao, Z., 2002. Study on application of McAb-ELISA to filaria surveillance by detecting filaria-specific IgG4. China Pub. Heal. 18 (5), 579–580.

Chen, G., Fan, H., Jiang, J., Li, C., 2007. The surveillance study on eliminating lymphatic filariasis in Anhui Province. Anhui J. Prev. Med. 13 (5), 324–325. 346.

Collaborating Research Group on the Transmission Threshold of Filariasis, 1994. Study on the transmission threshold of filariasis. Chin. J. Parasitol. Parasit. Dis. 12 (1), 1–6.

Cui, Z., Xu, J., Wang, Z., Li, W., Yin, S., 1984. Progress in the control and research in filariasis in Shandong Province during the past ten years. Chin. J. Parasitol. Parasit. Dis. 2 (2), 66–72.

Deng, X., Chen, X., 2002. Research progress on filariasis control and surveillance in Shandong Province. Chin. J. Parasit. Dis. Control 15 (2), 82–84.

Deng, S., Shen, L., Zhao, Y., Xu, R., Xia, L., Ren, X., 1986. Application of ELISA to diagnosis and seroepidemiological survey of filariasis. Chin. J. Parasitol. Parasit. Dis. 4 (4), 280–283.

Deng, X., Gao, C., Li, G., Chen, X., Xu, F., ZHao, Z., Miao, F., Li, S., Liu, X., Wang, P., Liu, B., 2000. Studies on etiological surveillance in filariasis-eliminated areas. Chin. J. Parasit. Dis. Control 13 (4), 266–268.

Deng, Y., Duan, J., Wang, L., Wu, W., Tan, Y., 2008. Cost-benefit analysis of lymphatic filariasis control in Hunan Province. China Trop. Med. 8 (1), 6–9.

Duan, J., 2001. Geographical distribution of filariasis and non-endemic area investigations in Hunan Province. Chin. J. Parasit. Dis. Control 14 (1), 72–73.

Duan, J., 2004. Surveillance on filariasis after elimination in Hunan Province. Chin. J. Parasit. Dis. Control 17 (6), 378–380.

Duan, J., Li, Q., Hu, G., Li, Z., Li, J., Yi, H., Liang, R., Liu, X., 2001. Longitudinal observation on filariasis after its basic elimination in Hunan Province. Chin. J. Dis. Control Pev. 5 (1), 88–89.

Fang, Y., Zhang, Y., 2019. Lessons from lymphatic filariasis elimination and the challenges of post-elimination surveillance in China. Infect. Dis. Poverty 8 (1), 66.

Fang, W., Zheng, L., 2002. Field observation on the natural conversion of microfilaremia cases with different density. Chin. J. Parasit. Dis. Control 15 (2), 95.

Feng, L., 1933. A comparative study of the anatomy of microfilaria *Brugia malayi*, 1927 and microfilaria *Wuchereria bancrofti*, 1877. Chin Med J (Engl) 47, 1214–1246.

Feng, L., 1936. The distribution and transmission of filariasis in China. Acta Cono. Ter. Trop. Malar. Morb. 1, 239–248.

Feng, L., 1962. Filariasis and its control problems. Chin Med J (Engl) 48 (4), 265–271.

Feng, L., Ma, S., 1956. The transmission of *Brugia malayi* in nature by *Anopheles hyrcanus* var. *sinensis*. Acta Microbiol. Sin. 4 (1), 137–154.

Guo, Q., Lu, Y., Liu, S., Wang, S., Zhang, Q., Gu, J., Zhang, S., 1989. Analization of the residual microfilaremias after the basic elimination of malayan filariasis in the Jianghan plans. Chin. J. Parasit. Dis. Control 2 (4), 259–263. 316.

He, S., Zheng, G., Zhang, Y., Tu, S., 1999. The evaluation of ICT detection in the diagnosis of chronic patients and microfilaremia cases of bancroftian filariasis. Strait J. Prev. Med. 5 (3), 36–37.

Hu, X.M., Wang, S.Q., Huang, J.M., Lin, S., Tong, C., Li, S., Zhen, W., 2008. The control and surveillance of filariasis in Hainan Province, China. Southeast Asian J. Trop. Med. Public Health 39 (1), 27–32.

Hu, X., Wang, S., Huang, J., Lin, S., Tong, Z., Li, S., Zeng, W., Chen, D., Liu, Y., 2009. Surveillance of filariasis after its basic elimination in Hainan Province. China Trop. Med. 9 (1), 41–42. 130.

Huang, S., Ou, Z., Zhao, X., Chen, Z., Li, Y., Tan, D., 1995. Long-term surveillance after basic elimination of filariasis in Guangdong Province. Chin. J. Parasit. Dis. Control 8 (2), 84–86.

Huang, B., Chang, J., Han, G., Zhang, Y., Liu, F., Wang, S., Chen, X., Wang, K., Zeng, T., Li, L., Jia, F., Liu, J., Cheng, Y., 2000. Detection of filarial larvae in mosquitoes with *Wuchereria bancrofti* specific DNA probe labeled with biotin/PCR system. Chin. J. Vector Biol. Control 11 (3), 204–207.

Huang, S., Luo, X., Ou, Z., Pan, B., Cui, H., Fang, Y., Zhang, Q., 2003. Surveillance and assessment for elimination of filariasis in Guangdong Province. J. Trop. Med. 3 (3), 323–326.

Huang, B., Jia, F., Fu, T., Chen, S., Hu, S., Bai, B., Deng, X., 2004a. The research of filariasis-specific IgG4 testing kit. Chin. J. Parasit. Dis. Control 17 (3), 176–178.

Huang, S., Luo, X., Ou, Z., Pan, B., Cui, H., Zhang, Q., 2004b. Prevalence and control of filariasis in Guangdong Province. Chin. J. Parasit. Dis. Control 17 (6), 346–349.

Ji, S., Wang, K., Zhang, Z., Zhao, W., Wang, L., Song, J., Chen, M., 1988. Diagnosis of filariasis by indirect fluorescent antibody test with sonicated microfilariae malayi as antigen. Chin. J. Parasitol. Parasit. Dis. 6 (4), 256–258.

Ji, S., Shi, Q., Wang, K., Gao, Z., Yao, Y., Fang, Y., 1991. The effect of filariasis control in Shanghai municipality. Chin. J. Parasit. Dis. Control 4 (4), 287–288.

Jiang, Q., Zhong, C., 1990. Studies on application of serological tests in the filariasis surveillance. Chin. J. Parasit. Dis. Control 3 (3), 206–209.

Li, Z., Shi, Z., Li, G., Wang, J., Ding, S., Yao, R., Shi, H., Luo, X., Yuan, Y., Zhou, J., Chen, Z., 1987. Field observation on role of low density microfilaremia cases in the transmission of filariasis. Chin. J. Parasitol. Parasit. Dis. 5 (1), 63–64.

Li, Y., Hu, Y., Xu, W., Wu, Z., Yang, G., Yin, S., Luo, W., Gong, W., 1988. Studies on the diagnosis of human filariasis by indirect fluorescent antibody test (IFAT) and immunoenzymatic staining technique (IEST) with homologous and heterologous adult filarial antigens. Chin. J. Parasitol. Parasit. Dis. 6 (1), 12–14.

Li, Z., Yu, L., Li, Z., Luo, H., 1989. Comparative studies on the morphology of larva stages of four species of *Setaria* and *Dirofilaria repens*. Chin. J. Parasit. Dis. Control 2 (3), 188–192.

Li, J., Li, Y., Ou, Z., Cui, H., Huang, S., 1993. Application of ELISA in seroepidemiological surveillance of bancroftian filariasis. Guangdong J. Health Epidemic Prev. 19 (4), 68–71.

Li, J., Yan, W., Wu, C., Xiao, P., Lai, J., Huang, W., 2002. Surveillance after filariasis basically eliminated in Chongqing Municipality. Chin. J. Pract. Parasit. Dis. 10 (3), 126–127.

Li, X., Yang, Y., Huang, K., Xu, H., Xie, Z., Mai, F., Shang, S., Su, A., Li, S., 2008. Discovery of a residual focus of bancroftian filariasis after declaration of its transmission interruption in Guangxi. Chin. J. Parasitol. Parasit. Dis. 26 (6), 404–408.

Lin, G., Yang, F., Chen, Z., 1996. Surveillance and epidemiological analysis after filariasis basic elimination in Fujian Province. Strait J. Prev. Med. 2 (4), 1–3.

Lin, X., Xu, B., Zhao, X., Li, H., Deng, Y., Huang, Q., Wang, H., Zheng, X., 2005. Surveillance and assessment of filariasis elimination in Henan Province, China. Chin. J. Parasit. Dis. Control 18 (4), 275–277.

Liu, H., 1990. Epidemiological features and control of filariasis of Jiangsu Province. Chin. J. Parasit. Dis. Control 3 (1), 1–3.

Liu, X., Liu, J., 1984. Epidemiological survey of filariasis in Fujian Province. Chin. J. Parasitol. Parasit. Dis. 2 (2), 73–77.

Liu, X., Liu, J., Chen, Z., 1989. Control of filariasis in Fujian Province. Chin. J. Parasit. Dis. Control 2 (2), 65–68.

Liu, X., Liu, J., Chen, Z., Tu, Z., 1990. Filariasis surveillance of floating population in Fujian Province. Chin. J. Parasit. Dis. Control 3 (4), 323.
Liu, Z., Xu, H., Peng, J., Wu, J., Yan, Y., 1991. Prevalence and control of filariasis in Jiangxi Province. Chin. J. Parasit. Dis. Control 4 (1), 1–3.
Liu, T., Zheng, H., Wang, S., Ye, H., 2002. Surveillance after eliminating filariasis in Guizhou Province. Chin. J. Parasit. Dis. Control 15 (2), 75–76.
Luo, H., Zeng, X., Li, Z., Xiang, H., Yao, B., Duan, J., 1996. Field observation on role of residual microfilaremia cases in filariasis transmission in a natural village of the mountain areas in West Hunan Province. Chin. J. Parasitol. Parasit. Dis. 14 (2), 164–165.
Ministry of Health of the People's Republic of China, 2007. National Report on Elimination of Lymphatic Filariasis in China. People's Medical Publishing House, Beijing, China.
National Technical Steering Group for Filariasis Control and Research, MOPH, 1983. Current status of prevalence and control of filariasis in China. Chin. J. Parasitol. Parasit. Dis. 1 (1), 5–10.
National Technical Steering Group for Filariasis Control and Research, MOPH, 1991a. Major achievements and experience in filariasis control in the People's Republic of China. Chin Med J (Engl) 104 (6), 446–453.
National Technical Steering Group for Filariasis Control and Research, MOPH, 1991b. Advances in filariasis control and research in China during 1986–1990. Chin. J. Parasitol. Parasit. Dis. 9 (2), 81–85.
National Technical Steering Group for Filariasis Control and Research, MOPH, 2002. Progress in surveillance and elimination filariasis in China. Chin. J. Parasit. Dis. Control 15 (2), 80–81.
Ou, Z., 1990. Control and surveillance of filariasis in Guangdong Province. Chin. J. Parasit. Dis. Control 3 (2), 97–100.
Pan, S., Xie, Z., Lv, X., He, G., Liu, C., Li, Y., 1990. On the transmission role of residual microfilaremia cases in the area with filariasis virtually eradicated. Chin. J. Parasitol. Parasit. Dis. 8 (3), 191–194.
Pan, S., Lv, X., Mai, F., Li, H., Xie, Z., 1992. Surveillance of filariasis in Guangxi Zhuang Autonomous Region. Chin. J. Parasit. Dis. Control 5 (1), 4–5.
Pan, S., Xie, Z., Mai, F., X., L., Li, H., Yang, X., 1995. Filariasis surveillance and check-up in Guangxi Zhuang Autonomous Region. Guangxi Prev. Med. 1 (4), 200–203.
Pei, S., Zhang, S., 1996. Surveillance of filariasis in Hubei Province. Chin. J. Parasit. Dis. Control 9 (2), 84–85.
Peng, G., Zhang, S., Zhang, Q., Wang, L., Huo, X., 2000. Results and analysis of filariasis surveillance in Hubei Province from 1989 to 1996. Chin. J. Parasit. Dis. Control 13 (1), 28–30.
Ren, Y., Wang, B., 1984. Clinical trials on furapyrimidone in the treatment of filariasis. Chin. J. Parasitol. Parasit. Dis. 2 (2), 123–125.
Ru, X., Chang, J., Liu, J., Zheng, X., Zhao, Q., Ma, Y., Sun, J., Cheng, Y., Zhao, X., 1991. Research on filariasis control in Henan Province. Chin. J. Parasit. Dis. Control 4 (1), 55–57.
Sheng, Q., Ren, Z., Gao, C., Cao, X., Wang, P., Deng, X., Zhao, Z., Liu, L., Gong, M., Li, D., 1995. A ten-year longitudinal surveillance of filariasis in the controlled areas of Shandong Province. Chin. J. Parasit. Dis. Control 8 (2), 87–90.
Shi, Z., Sun, D., 1983. Objective, strategy and measure for the lymphatic filariasis control in China. Chin. J. Epidemiol. 15 (Suppl. 8), 187–189.
Shi, Z., Sun, D., 1999. Achievements in the research on filariasis in China in the past 50 years. Chin. J. Parasitol. Parasit. Dis. 17 (5), 267–270.
Shi, Z., Yuan, Y., Sun, J., Li, Z., Zheng, X., 1988. Study on the role of low density microfilaremia cases in the transmission of filariasis in late stage of filariasis control. Chin. J. Parasitol. Parasit. Dis. 6 (2), 99–102.

Shi, Z., Sun, D., Wang, Z., Tao, Z., Pan, S., Liu, X., Zhang, S., Ou, Z., Zhu, S., Li, Q., Chang, J., Wu, R.Z., Deng, S.S., Zheng, X.Q., 2001. A brief introduction to the research achievement on the strategy and technical measures for interrupting the transmission of lymphatic filariasis in China. Chin. J. Parasitol. Parasit. Dis. 19 (2), 110–112.

Song, J., Ma, K., 1988. The role of low density microfilaremia cases in the transmission of filariasis in bancroftian filariasis endemic area after basic elimination. Chin. J. Parasitol. Parasit. Dis. 6 (1), 71.

Sun, D., Li, Z., Xu, S., Wang, Q., 1990. Paracetamolum in relief of febrile reaction in DEC treatment of filariasis. In: Papers of Filariasis Control and Research in China. Publishing House of Science and Technology, Fuzhou, Fujian Province, pp. 225–227.

Sun, D., 2001. The progress of the global programme to eliminate lymphatic filariasis (outline). Foreign Med. Sci. (Parasitic Dis.) 28 (2), 59–61.

Sun, D., 2005. Global significance of the elimination of lymphatic filariasis in China. Chin. J. Parasitol. Parasit. Dis. 23 (5 Suppl), 329–331.

Sun, D., Gan, Y., Xue, Z., 1964. Lymphangiography of lower limbs in malayan filariasis. Chin. J. Radiol. 9 (2), 154–155.

Sun, D., Yuan, Y., Shi, H., Luo, M., Chen, S., Shi, Z., 1994. Comparative study on three serologic tests for detecting antibodies in persons with bancroftian microfilaremia. Chin. J. Parasitol. Parasit. Dis. 12 (3), 185–187.

Sun, F., Cao, H., Xu, Z., Zhao, L., Qian, Y., Chen, J., Zhao, Y., Ge, J., He, H., Liu, H., Wu, Z., 2001. An exploration of surveillance on eliminating filariasis in Jiangsu Province. Chin. J. Schisto. Control 13 (3), 166–169.

Sun, D., Duan, J., Deng, X., 2013. The history of the elimination of lymphatic filariasis in China. Infect. Dis. Poverty 2 (1), 30.

Tao, Z., Zheng, H., Luo, Y., Gao, G., Luo, L., 1987. Longitudinal surveillance in areas with bancroftian filariasis virtually eradicated in Guizhou. Chin. J. Parasitol. Parasit. Dis. 5 (2), 81–84.

Wang, P., 2005. Study on the elimination of lymphatic filariasis in Shandong Province. China Trop. Med. 5 (7), 1447–1450.

Wang, D., Gao, J., 1990. Observations on IgG4 antibodies in bancroftian filariasis. Chin. J. Parasit. Dis. Control 3 (3), 214–215.

Wang, S.H., Zheng, H.J., 1989. Blood meals of mosquito vectors of filariasis in Guizhou Province, China. Southeast Asian J. Trop. Med. Public Health 20 (2), 175–177.

Wang, Y., Feng, Z., Jin, X., Yang, R., Mao, F., 1987. Studies on indirect fluorescent antibody test for diagnosis and epidemiological investigation of filariasis. Chin. J. Parasitol. Parasit. Dis. 5 (2), 93–95.

Wang, K., Zhou, J., Fan, L., Xia, S., Zheng, S., 1989. Prevalence and control of filariasis in Zhejiang Province. Chin. J. Parasit. Dis. Control 2 (3), 157–160.

Wang, P., Yin, S., Gao, C., Sheng, Q., Zhao, Z., Li, D., Cao, X., Liu, L., Wan, G., Liu, B., Gong, M., Liu, X., Zhang, D., Wang, H., 1997. Study on the transmission threshold of bancroftian filariasis in Shandong Province. Chin. J. Parasit. Dis. Control 10 (1), 11–15.

Wang, Z., Shi, Z., Xu, S., Zhong, C., Zhang, B., 2003. Control of Lymphatic Filariasis in China. WHO, Geneva, Switzerland.

Wei, Y., Yang, Y., Zhang, T., Wang, X., Yi, Z., Dan, M., Yan, J., Chen, Z., 1994. Cross-sectional surveillance on filariasis in Sichuan Province from 1982 to 1991. Chin. J. Parasit. Dis. Control 7 (2), 98–100.

WHO, WHO, (Eds.), 1995. The stage of world health. In: World Health Report 1995: Bridging the Gaps. World Health Organization, Geneva, Switzerland, pp. 1–49.

WHO, WHO, (Eds.), 1997. Elimination of lymphatic filariasis as a public health problem. In: Hbk Res. World Health Organization, Geneva, Switzerland1.16.3.3; 7.1.3.

WHO, 2011. Monitoring and Epidemiological Assessment of Mass Drug Administration in the Global Programme to Eliminate Lymphatic Filariasis: A Manual for National Elimination Programmes. World Health Organization, Geneva, Switzerland.

Wu, R., 1984. Observation of clinical therapeutic effect of furapyrimidone on 82 cases of bancroftian filariasis. Chin. J. New Drug Clin. Remedy 3 (4), 16–18.

Wu, R., Chen, J., Xu, F., 1990. Prevalence and control of filariasis in Hainan Province. Chin. J. Parasit. Dis. Control 3 (2), 101–104.

Wu, R., Li, S., Lin, S., Liu, R., 1997. Epidemiological surveillance after filariasis basic elimination in Hainan Province by serological diagnosis. Chin. J. Parasitol. Parasit. Dis. 15 (2), 122–123.

Wu, R., Xu, F., Xing, F., Liu, Z., 2003a. Summary of control of filariasis in Hainan Province. China Trop. Med. 3 (1), 33–37.

Wu, W., Shi, Z., Sun, D., 2003b. A sustainable effort for achieving elimination of lymphatic filariasis in China. Chin. J. Parasitol. Parasit. Dis. 21 (6), 321–322.

Xi, Y., Xu, J., Ren, Y., Yuan, Y., Tao, G., Zhao, M., Xu, Y., 1979. Drug screening with the cotton-rat model of filariasis. Acta Pharm. Sin. 14 (8), 455–460.

Xie, J.R., Yuan, Y., 1987. Studies on the relationship between bacterial infections and filarial lymphangitis. Chin. J. Guangzhou. Med. Coll. 15 (4), 36–41.

Xu, J., Zhu, H., Luo, X., 1993. Studies on the susceptibility of Anopheles anthropophagus to experimental infection with *Wuchereria bancrofti*. Chin. J. Parasitol. Parasit. Dis. 11 (2), 112–115.

Xu, F., Li, G., Li, C., Sheng, Q., Zhao, Z., 1996. Detection of specific IgG4 in the surveillance of filariasis. Chin. J. Parasit. Dis. Control 9 (2), 119–121.

Xu, B., Li, H., Han, Q., Lin, X., Chang, J., Huang, Q., Zhao, Q., 2002. Results and analysis of filariasis surveillance in Henan Province in 1988–1999. Chin. J. Parasit. Dis. Control 15 (2), 69–71.

Xu, X., Sun, F., Cao, H., Qian, Y., Shen, M., Jiang, W., 2008. Epidemiological surveillance after filariasis elimination in Jiangsu Province. J. Trop. Dis. Parasitol. 6 (1), 37–38.

Yao, L., Tu, X., Zhou, J., Fan, L., Yu, K., Chen, H., 2002. Surveillance after filariasis was basically eliminated in Zhejiang Province. Chin. J. Parasit. Dis. Control 15 (2), 66–68.

Yuan, Y., Sun, D., Chen, S., Qiu, L., Xue, H., Zhu, Z., Shi, Z., 1992. Preliminary study of dot-ELISA for the detection of filariasis antibody. Chin. J. Parasit. Dis. Con. 5 (3), 169–171.

Zhan, F., Zhang, S., Wang, L., Yuan, F., Zhang, H., Dong, X., Li, K., Yu, P., 2012. Lymphatic filariasis in Hubei Province: from prevailing to elimination. Chin. J. Parasitol. Parasit. Dis. 30 (1), 65–70.

Zhang, S., 1990. Filariasis control in Hubei Province. Chin. J. Parasit. Dis. Control 3 (4), 325–326.

Zhang, F., 2004. Analysis of surveillance results after filariasis basically eliminated in Sichuan Province. Parasitol. Infect. Dis. 2 (4), 176–177.

Zhang, B., Xu, J., Lin, J., Zhao, Y., Li, Z., Chen, L., Sun, D., Yuan, Y., 1964. Studies on bionomics of different types of *Anopheles sinensis* and its transmission in malayan filariasis endemic areas. Chin Med J (Engl) 50 (12), 776–779.

Zhang, F., Yang, Y., Rao, Z., 1993. Surveillance on key communities in late stage of filariasis control program in Sichuan Province. Chin. J. Parasit. Dis. Control. 1 (3), 32–34.

Zhang, F., Rao, Z., Hou, C., Wang, W., Wei, Y., 1996. Follow-up observation on natural conversion of residual microfilariemia of malayan filariasis at late stage of control program. Chin. J. Pract. Parasit. Dis. 4 (4), 162–163.

Zhang, K., Huang, S., Ye, M., 1999. Analysis on the results of surveillance on residual microfilaremia cases after basic elimination of filariasis. Chin. J. Pract. Prev. Med. 6 (1), 42–43.

Zhang, F., Rao, Z., Xi, Y., Lu, H., 2001. Distribution characteristics of residual infectious sources and consolidating measures after filariasis basically eliminated in Sichuan Province. Chin. J. Pract. Parasit. Dis. 9 (1), 24–26.
Zhang, K., Li, Z., Yan, Y., Gong, Y., Huang, J., 2002a. Progress of filariasis surveillance in Jiangxi Province. Chin. J. Parasit. Dis. Control 15 (2), 85–86.
Zhang, F., Rao, Z., Lu, H., 2002b. Consolidating and surveillance measures after malayan filariasis basically eliminated in Sichuan Province. Chin. J. Parasit. Dis. Control 15 (2), 72–74.
Zhang, K., Li, Z., Yan, Y., Jin, J., Gong, Y., Qiu, H., Chen, D., Hu, J., 2006a. Epidemiological characteristic and control of filariasis in Jiangxi Province. Chin. J. Pathog. Biol. 1 (1), 7–11.
Zhang, K., Li, Z., Yan, Y., Jin, J., Gong, Y., Qiu, H., Chen, D., Hu, J., 2006b. Epidemiological surveillance after filariasis transmission interrupted in Jiangxi Province. Chin. J. Parasitol. Parasit. Dis. 24 (2), 158–159.
Zhao, Q., Chang, J., Huang, Q., Lin, X., Zhao, X., Xu, B., 1992. Reports of filariasis cross-sectional surveillance in Henan Province from 1988 to 1990. Chin. J. Parasit. Dis. Control 5 (3), 220.
Zheng, H., Tao, Z., Cheng, X., Chen, X., 1992. Application of DNA probe-PCR technique for identification of *Wuchereria bancrofti* in mosquitoes. Guizhou Med. J. 16 (2), 68–70.
Zheng, H., Tao, Z., Fang, R., Chang, B., Zhang, Y., Turner, P., 1998. Application of immunochromatographic test for diagnosis and surveillance of bancroftian filariasis. Chin. J. Parasitol. Parasit. Dis. 16 (3), 168–171.
Zhu, S., 1990. An overview of technical measures on filariasis basic elimination in Sichuan Province. Chin. J. Parasit. Dis. Control. 3 (1), 74–75.
Zou, Y., Bai, X., Tang, L., Fang, Y., Ouyang, H., Duan, J., Li, Z., 1995. Application of IFAT in the sero-epidemiological surveillance of filariasis. Chin. J. Parasit. Dis. Control 8 (1), 4–6.
Zou, J., Zheng, X., Wang, J., 1996. Prevalence and control of filariasis in Anhui Province. Chin. J. Parasit. Dis. Control 9 (1), 1–3.
Zou, Z., Jia, S., Fan, H., Shen, Y., Lu, H., Xie, W., Xu, F., 2002. Analysis on surveillance results of filariasis from 1995 to 2000 in Anhui Province. Chin. J. Parasit. Dis. Control 15 (2), 77–79.

CHAPTER FIVE

Contributions of the National Institute of Parasitic Diseases to the control of visceral leishmaniasis in China

Zheng-Bin Zhou[a,b,c,d,e,f], **Jun-Yun Wang**[a,b,c,d,e,f], **Chun-Hua Gao**[a,b,c,d,e,f], **Shuai Han**[a,b,c,d,e,f], **Yuan-Yuan Li**[a,b,c,d,e,f], **Yi Zhang**[a,b,c,d,e,f], **Xiao-Nong Zhou**[a,b,c,d,e,f,*]

[a]National Institute of Parasitic Diseases, Chinese Center for Disease Control and Prevention, Shanghai, People's Republic of China
[b]Chinese Center for Tropical Diseases Research, Shanghai, People's Republic of China
[c]WHO Collaborating Centre for Tropical Diseases, Shanghai, People's Republic of China
[d]National Center for International Research on Tropical Diseases, Ministry of Science and Technology, Shanghai, People's Republic of China
[e]Key Laboratory of Parasite and Vector Biology, Ministry of Health, Shanghai, People's Republic of China
[f]School of Global Health, Chinese Center for Tropical Diseases Research, Shanghai Jiao Tong University School of Medicine, Shanghai, People's Republic of China
*Corresponding author: e-mail address: zhouxn1@chinacdc.cn

Contents

1. Introduction 186
2. Epidemiological patterns of visceral leishmaniasis 187
   2.1 The anthroponotic visceral leishmaniasis (AVL) 189
   2.2 The desert sub-type zoonotic of visceral leishmaniasis (DST-ZVL) 189
   2.3 The mountainous sub-type of zoonotic VL (MST-ZVL) 189
3. Vectors and their distribution 190
   3.1 Confirmation of VL vectors 190
   3.2 Distribution of sandflies 191
4. Visceral leishmaniasis in China: Then and now 196
   4.1 Control efforts—What was done 196
   4.2 Supportive research from NIPD-CTDR 204
   4.3 Monitoring and evaluation 208
   4.4 Training 209
5. Conclusion and way forward 210
Acknowledgements 212
Competing interest 212
Authors' contributions 212
References 212

Advances in Parasitology, Volume 110
ISSN 0065-308X
https://doi.org/10.1016/bs.apar.2020.04.003

Abstract

Visceral leishmaniasis (VL) caused by *Leishmania* spp. is an important vector-borne disease prevalent in China. VL was rampant in the vast area of China north of the Yangtze River before the founding of the People's Republic of China in 1949. As a result of strenuous interventions, the disease was basically eliminated in most of the former epidemic areas in 1958–60. At present, only sporadic cases occur in the western regions of China. In the process, National Institute of Parasitic Diseases at China CDC and the Chinese Center for Tropical Diseases Research (NIPD-CTDR) have achieved great impact in controlling the diseases as well as in research on *Leishmania* spp. This review summarized the contribution of experts from NIPD-CTDR to the control and elimination of VL in various aspects, such as understanding the epidemiological features of VL, confirmation of VL vectors and their distribution, development of control tools including diagnostics and insecticides, monitoring and evaluation supported by information management, technical supports to the control programmes, as well as analysis of the challenges faced. At the same time, it puts forward constructive suggestions for the ultimate interruption of VL transmission in China.

1. Introduction

Visceral leishmaniasis (VL), also known as Kala-azar, is caused by *Leishmania* spp. and transmitted by sandflies of the genus *Phlebotomus*. VL was one of the five most serious parasitic diseases endangering the human health in the People's Republic of China. It was prevalent in rural areas north of the Yangtze River, affecting more than 600 counties and cities in 16 provinces, municipalities and autonomous regions. Before the founding of the People's Republic of China in 1949, VL was spread continuously owing to the lack of prevention and control measures, resulting in large numbers of deaths. After the founding of the People's Republic of China, China had formulated the work plan to take the control and elimination efforts for a variety of serious diseases as the priority task of health work. This included nine diseases, among which VL were required to be eliminated within the time frame set by the "Framework of the National Agricultural Development Programme". The governments at all levels have attached great importance to the prevention and control of VL and many human and material resources were invested in the national VL control programme. This has led to remarkable achievements, such as the development of a control strategy and measures through pilot research, the establishment of working mechanism on strong governance leadership at all levels and active participation of the people in the effective implementation of control measures, the establishment of an efficient prevention and control network throughout the country, the establishment of professional institutions for prevention and research at all levels, and the understanding of the epidemic situation and epidemic factors of VL in China. Through active efforts in control of VL, the national VL control programme in the country has achieved remarkable results, for instance, VL was basically eliminated in 1958–60 in most of the epidemic areas, and at present, only scattered cases occur in the western region of China. The objectives of this article is to summarize the contribution of experts from the National Institute of Parasitic Diseases at China CDC & Chinese Center for Tropical Diseases Research (NIPD-CTDR) to the control and elimination of VL, analyse the challenges faced, and put forward the constructive suggestions for the ultimate interruption of VL transmission in China.

2. Epidemiological patterns of visceral leishmaniasis

In the early 1950s, in order to strengthen the control work on VL in China, the Eastern-China Branch of the National Institute of Health (the predecessor of the NIPD-CTDR) was authorized by the Ministry of Health to carry out VL investigation in the VL endemic provinces in cooperation with the provincial institutions working on VL control. The results from various investigations indicated that VL was epidemic in more than 680 counties and cities out of 16 provinces, autonomous regions and municipalities, north of the Yangtze River, especially in northern Jiangsu and Anhui, Shandong, Henan, southern Hebei and Shaanxi plains. There were about 530,000 VL cases in 1951 in the endemic areas of China, resulting in high mortality (Ho et al., 1959; Wang and Wu, 1956). At that time, epidemiological characteristics of VL in China were analyzed based on demographic information of cases, such as age distribution, and the relationship with canine visceral leishmaniasis (CVL). It was found that in the northern plains in China, similar to India, the majority of VL cases were in young adult humans, with rare cases of CVL. However, in Gansu and Qinghai provinces, the majority of VL cases were in infants and young children and CVL was more common. The latter picture was similar to the situation which was also consistent with the epidemiologic in the Mediterranean and central Asia.

As the national VL control programme in China progressed, the NIPD-CTDR accumulated abundant of epidemiological data from investigations conducted from 1956 to 1975 in Shanxi, mountain area of the southern

Gansu, northern Sichuan, the Loess Plateau of northern Shaanxi, and the desert zone in Xinjiang and western Inner Mongolia, where VL were still endemic in (Wang et al., 1963, 1964b, 1966; Xiong, 1972; Xiong et al., 1963a, 1963b, 1964a, 1964b, 1964c, 1965, 1974), as well as from 1967 to 1973 in northern Jiangsu, northern Anhui and Shaanxi plains, where VL had been eliminated in 1960s. In 1976, the new perspective of three epidemiological types (Fig. 1) was formulated (Guan et al., 1976; Xiong et al., 1976), providing a scientific reference to the tailoring of the implementation of the national VL control programme to local settings.

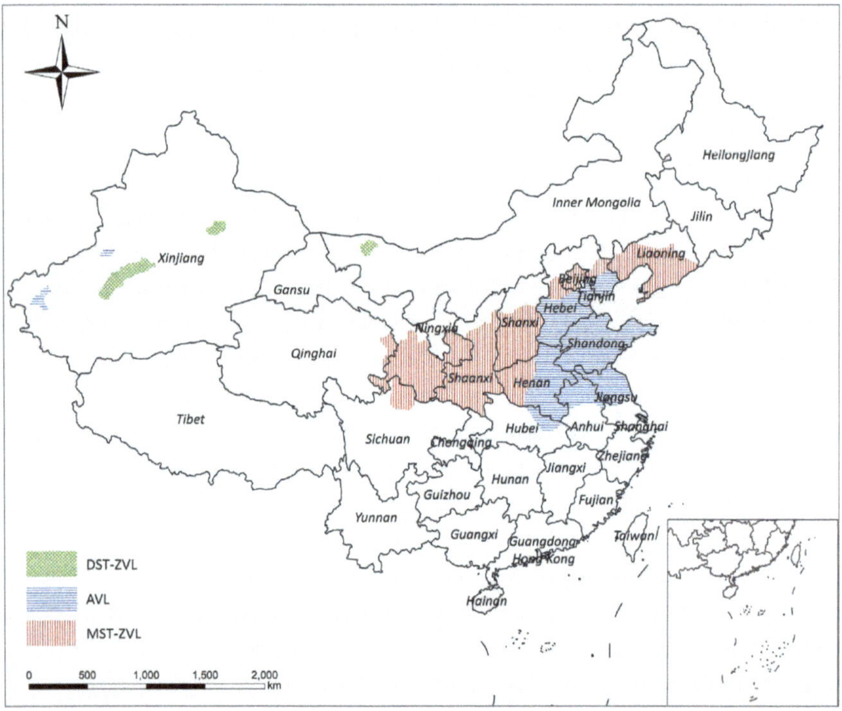

Fig. 1 Geographical distribution of the three sub-types of VL in China. The AVL endemic areas mainly distributed in south-eastern suburbs of Beijing, southern Hebei, Shandong, eastern Henan, northern Jiangsu, northern Anhui, Shaanxi plains, northern Jianghan plain of Hubei and ancient oasis area of southern Xinjiang. The DST-ZVL endemic areas mainly distributed on the edge of Tarim basin and Turpan in Xinjiang as well as Ejina banner in western Inner Mongolia. The MST-ZVL endemic areas mainly distributed in the northern suburbs of Beijing, northern Hebei, Liaoning, western Henan, Shanxi, northern Shaanxi and southern Shaanxi, southern Gansu and northwestern Sichuan, and the eastern part of Qinghai.

2.1 The anthroponotic visceral leishmaniasis (AVL)

The AVL endemic region in China included the south-eastern suburbs of Beijing, southern Hebei, Shandong, eastern Henan, northern Jiangsu, northern Anhui, Shaanxi plains, the northern Jianghan plain of Hubei and ancient oasis area of southern Xinjiang, where majority cases were juveniles, occasionally followed by post kala-azar dermal leishmaniasis (PKDL), and *Phlebotomus chinensis* (endophilic species) and *Ph. longiductus* (peridomestic species) were the vector. Because the patients were the primary reservoir host of AVL, concurrent or consecutive cases often occurred in one household with spatial clusters, sometimes leading to outbreaks. Except for Shaanxi plain, cases of CVL were rare or absent in the AVL endemic area.

2.2 The desert sub-type zoonotic of visceral leishmaniasis (DST-ZVL)

The DST-ZVL region was considered to be a natural nidus of VL, with wild animals presumed to be the source of infection. It had been uncultivated desert for a considerable period of time before it was populated by immigrants who introduced agriculture cultivation and other activities. Consequently, autochthonous infantile VL occurs after immigration. The region can be subdivided into two different sub-type, namely dry desert zone and gravel desert zone.

Dry desert zone. The dry desert zone is distributed on the edge of Tarim basin in Xinjiang and Ejina banner in western Inner Mongolia with vegetation of *Populus diversifolia-Tamarisk* as main vegetation, where *Ph. wui* (an exophilic species) was the primary vector. The incidence of the infection was sporadic and affected mainly infants under 2 years old, usually occurred in autumn and winter, rarely in summer. Lymphoglandular leishmaniasis occurred frequently in adults migrating to the dry desert region.

Gravel desert zone. The gravel desert zone is distributed in the mountain gravel desert area in Turpan, Xinjiang, where *Ph. alexandri* (an exophilic species) was the vector. The incidence of the infection was sporadic, and mainly affecting children under 5 years old, with a few adult cases accounting for about 10%. Typical VL occurs frequently in adults migrating to the gravel desert region.

2.3 The mountainous sub-type of zoonotic VL (MST-ZVL)

The MST-ZVL endemic region in China included the northern suburbs of Beijing, northern Hebei, Liaoning, western Henan, Shanxi, northern

Shaanxi and southern Shaanxi, southern Gansu and northwestern Sichuan, and the eastern part of Qinghai, where a sub-type of *Ph. Chinensis* (a semi-wild species) was the vector and dogs were reservoir with high infection rates. MST-ZVL was hypoendemic in humans and affected mainly infants under 5 years old, rarely followed by PKDL. The human incidence rate is much lower than the canine one, and no obvious correlation relationship was found between cases. In this region, it is highly probable that wild animals also were infected with *L. donovani* or *L. infantum* (Guan and Shen, 1991).

3. Vectors and their distribution

3.1 Confirmation of VL vectors

Before 1950s, investigations and experimental studies had been conducted in Beijing and North Jiangsu, confirming that *Ph. chinensis* was the vector of VL in China. Since then, the NIPD-CTDR has conducted infection experiments on dogs or hamsters using local *Ph. chinensis* in Shandong (1958, 1960), Gansu (1958) and Sichuan (1981) (Institute of Parasitic Diseases and Chinese Academy of Medical Sciences, 1958; Institute of Parasitic Diseases and Chinese Academy of Medical Sciences, 1960; Yin and Yu, 1981). The results showed that once the vector was infected, the promastigotes could multiply in the stomach and enter the anterior of the digestive tract, pharyngeal and proboscis of *Ph. chinensis*, indicating that local *Ph. chinensis* was well suited for the reproduction of *L. donovani* and *L. infantum*, and therefore the transmission of VL. In addition, a natural infection of promastigotes was found in *Ph. chinensis* captured in the wild in VL epidemic county of Jiuzhaigou of Sichuan Province (Xiong et al., 1992), confirmed to be *L. donovani* by ELISA, and further confirming the assertion that *Ph. chinensis* is the main vector of VL in China.

After 1963, investigations were extended to the western border of China, by NIPD-CTDR in cooperation with the relevant agencies of Gansu, Inner Mongolia and Xinjiang Provinces and autonomous regions. Those investigations have revealed the distribution of *Ph. longiductus*, *Ph. wui* and *Ph. alexandri* in the west of Inner Mongolia and south and north of Tianshan Mountains in Xinjiang (Guan et al., 1986; Jin et al., 2008; Xiong et al., 1974, 1976; Zhao et al., 1985; Zuo et al., 2005). In the endemic areas, the aforementioned three species of sandflies were always the main species, and their geographical distribution was consistent with that of VL. These three kinds of sandflies feed on human, and after engorging the blood of the animal infected with *L. donovani* or *L. infantum*, the promastigotes not only survived in the process of blood digestion, but also flourished and migrated to the front of the stomach, the anterior of the digestive tract, pharyngeal and proboscis. In the endemic area of VL, natural infections of the promastigotes was found in *Ph. wui* and *Ph. alexandri*, which was further confirmed by biochemical or immunological techniques to be the same kind of *Leishmania* species isolated from VL cases; after artificial inoculation with promastigotes isolated from *Ph. wui* and *Ph. alexandri*. Finally, it was found that VL occurred in healthy experimental animals after being bitten by *Ph. alexandri* infected with *L. donovani*.

Studying on the vectors of VL in Xinjiang, scientists from the NIPD-CTDR confirmed that *Leishmania* species isolated from VL cases in Beijing and Shandong could multiply in the stomach of *Ph. longiductus*, and *Leishmania* species isolated from cases in Henan and Gansu Province could flourish and migrate to the pharyngeal and proboscis of *Ph. alexander*, confirming the ability to transmit those types of VL (Guan et al., 1990). Therefore, it is necessary to prevent the entry of exotic infectious sources into Xinjiang to cause local secondary transmission, which is of great significance in epidemiology of VL in Xinjiang Uygur Autonomous Region.

All of these investigations have been of great significance to determine *Ph. longiductus*, *Ph. wui* and *Ph. alexandri* as vectors of VL in China, especially, the confirmation of *Ph. alexandri* as a vector of VL in China, which broke the view of the past that vectors of VL in the old-world were limited to species of major group of Phlebotomus.

3.2 Distribution of sandflies

Research on sandflies has shown for a long time that the species composition of VL vectors varied in different geographic regions in China. Although some sandflies were euryecious species with extensive adaptability to different environment, there werespecies that were typically representative in certain ecological environments. As of 2011, scientists from the NIPD-CTDR have conducted a series of studies on the geographical distribution of 5 important sandflies species in China, to define their geographical scope, distribution characteristics, and map them.

3.2.1 Ph. chinensis

In the 1950s, investigations on the distribution of *Ph. chinensis* have accumulated a wealth of data, which was compiled and summarized by the NIPD-CTDR. The distribution of *Ph. chinensis* sandflies covered 358 counties and cities in 21 provinces, autonomous regions and municipalities

(Guan and Yang, 2012; Ho et al., 1959). As of 2011, the geographical distribution of *Ph. chinensis* was mapped by NIPD-CTDR (Guan et al., 2016) showing that, in terms of latitude, *Ph. chinensis* was present as far north as Changchun of Jilin Province (43°90′N, 125°50′E), as far south as Hekou of Yunnan Province (23°40′N, 104°E), as far west as Zhangye of Gansu Province (38°90′N, 100°40′E), and as far east as Jilin of Jilin Province (43°80′N, 126°60′E). In terms of altitude range, the vector was present in areas varying from 10 meters above the sea level in Jiangsu coastal plain, to 2,750 meters above the sea level in mountains areas of the northwestern Sichuan and southern Gansu Provinces (Guan et al., 2016; Wu et al., 1995; Yin et al., 1990). Even though *Ph. chinensis* is widely distributed, it is mostly encountered in the plains, mountainous, and Loess Plateau area of 32°–43°N, 102°–121°E (see Fig. 2), which is consistent with the geographical distribution of VL (Guan et al., 2016).

The *Ph. chinensis*, distributed in the central and eastern plains of China, is endophilic and highly sensitive to insecticides (Wu et al., 1955), and has

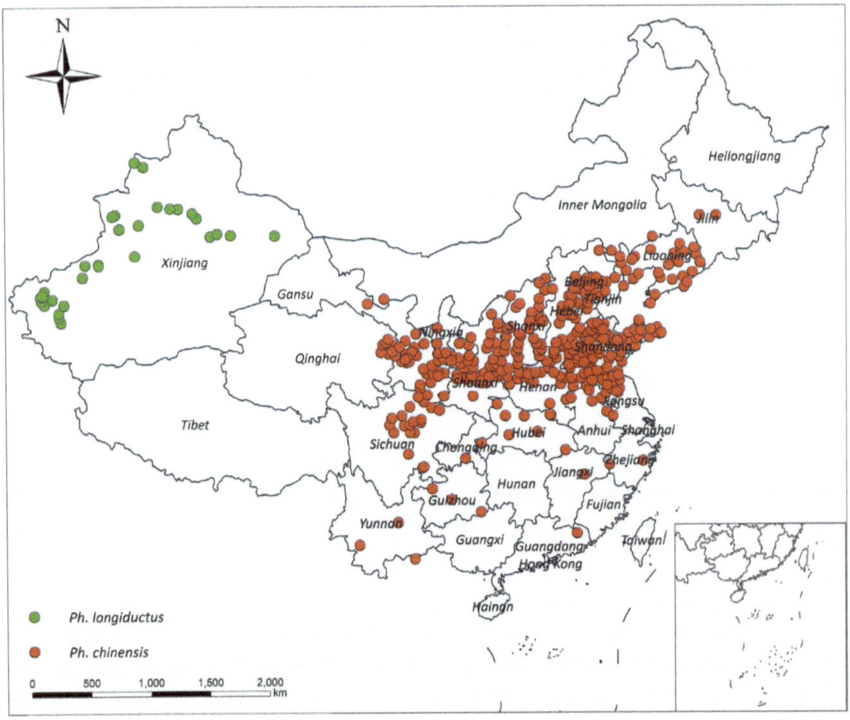

Fig. 2 Distribution of *Ph. chinensis* and *Ph. longiductus* in China.

been effectively got under control by indoor residual spraying (IRS) (Wu, 1957). In the mountainous and Loess Plateau areas, however, *Ph. chinensis* is an exophilic species. Due to a lack of similar effective control measures (Guan, 2010; Guan et al., 1980a), it is still spread widely, with new cases both of VL and CVL frequently emerging (Guan, 2009; Tian et al., 2012).

3.2.2 *Ph. mongoliensis*

Ph. mongoliensis was distributed in 240 counties/cities in 17 provinces, autonomous regions or municipalities (Chen et al., 2003; Ho et al., 1959; Zuo et al., 1998). In terms of latitude it was present as far north as Hebukesaier of Xinjiang Uygur Autonomous Region (46°80′N, 85°70′E), as far south as Jingmen of Hubei Province (31°N, 112°10′E), as far west as Huocheng of Xinjiang Uygur Autonomous Region (44°N, 80°80′E), and as far east as Suizhong of Liaoning Province (40°30′N, 120°30′E) (see Fig. 3). It was present at an altitude range of 10 meters above the sea level (in the Northern Jiangsu Coastal Plain) to 1,900 meters above the sea level (in

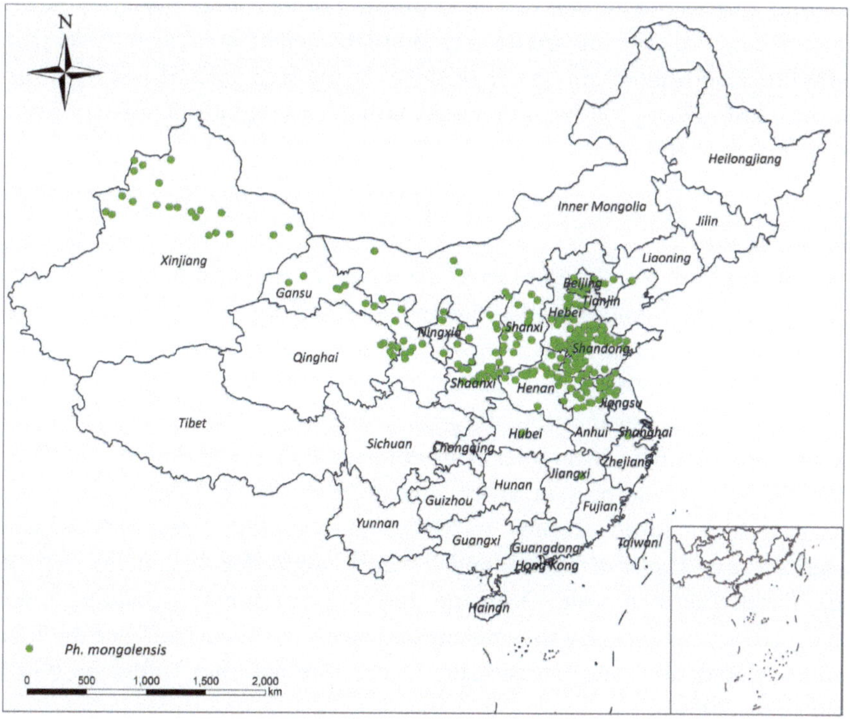

Fig. 3 Distribution of *Ph. mongoliensis* in China.

Dongxiang of Gansu Province), with the highest density observed in the northern China plain region (32°–40°N, 114°–120°′E), as well as desert areas covered with Chenopodiaceae plants in Western Inner Mongolia, the Hexi Corridor of Gansu Province, and the Zhungeer basin north of Mt. Tianshan in Xinjiang Uygur Autonomous Region (Guan et al., 2016). *Ph. mongoliensis* has been proven the vector for *L. gerbilli* and *L. turanica*, rather than for human VL (Feng, 1951), since it has been found in the subcutaneous tissue of the gerbils in some desert areas in western China (Liu et al., 1982; Wang et al., 1963, 1964a; Xu et al., 1991).

3.2.3 *Ph. longiductus*

The distribution of *Ph. longiductus* sandflies in China was limited to 30 counties and cities inside Xinjiang Uygur Autonomous Region (Zuo et al., 1998), as far north as Tacheng (46°45′N, 83°E), as far south as Yecheng (37°52′N, 77°24′E), as far west as Kashgar (37°50′N, 76°E), and as far east as Shanshan (42°90′N, 90°13′E) (see Fig. 2). Compatible altitude ranged from 90 meters (Turfan) to 2,100 meters, vertically at an altitude ranging from 90 (Turfan) to 2100 above the sea level (Wensu) (Guan et al., 2016), with the highest density found in ancient oases, at altitudes from 1,000 to 1,500 meters above the sea level in the western and northern rims of the Tarim Basin, followed by the mountainous regions. It was not found in desert areas (Guan and Yang, 2012; Guan et al., 2016).

3.2.4 *Ph. wui*

Ph. wui sandflies were present in 37 counties (banners) in Xinjiang, Gansu, and Inner Mongolia, as far north as Tacheng of Xinjiang Uygur Autonomous Region (46°45′N, 83°E), as far south as Minfeng, Xinjiang Uygur Autonomous Region (37°04′N, 82°41′E), as far west as Shufu of Xinjiang Uygur Autonomous Region (39°50′N, 75°85′E), and as far east as Eji'naqi of Inner Mongolia Autonomous Region (41°57′N, 77°50′E) (see Fig. 4). It was present at an altitude from 90 to 1,500 meters, and vertically at an altitude from 90 to 1500, mainly in desert areas covered with vegetation of *Poplar diversifolia* and *Tamarix taklamakanensis* at the edge of the Tarim Basin in south Xinjiang and western Inner Mongolia (Guan et al., 2016), followed by the ancient oasis with less than 16.8% of the local sandfly population, and occasionally in the mountainous landscape (Guan and Chai, 1996).

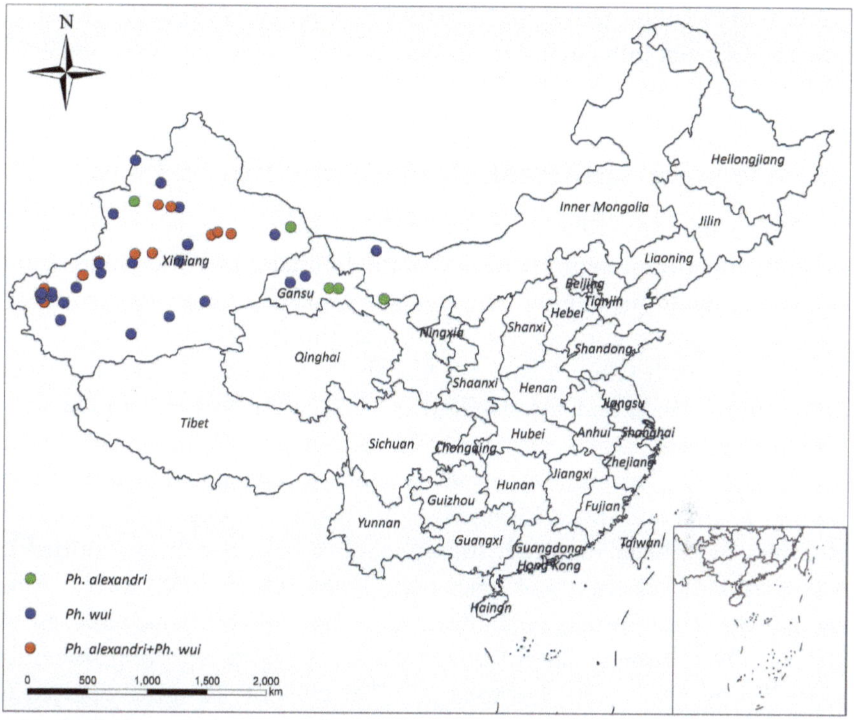

Fig. 4 Distribution of *Ph. wui* and *Ph. alexandri* in China.

3.2.5 *Ph. alexandri*

Ph. alexandri sandfly were present in 17 counties in Xinjiang, Gansu, and Inner Mongolia, as far north as Guertu in Wusu of Xinjiang Uygur Autonomous Region (44°50′N, 76°10′E), as far south as Yingjisha of Xinjiang Uygur Autonomous Region (39°10′N, 76°10′E), as far west as Atushi of Xinjiang Uygur Autonomous Region (39°70′N, 76°10′E), and as far east as Yabulai Mountains in Alxa Youqi of Inner Mongolia Autonomous Region (39°40′N, 103°10′E) (see Fig. 4) (Guan et al., 2016). *Ph. alexandri* is primarily present in areas of piedmont and stony desert, at an altitude from 500 to 1,750 meters above the sea level (Guan et al., 2016), where there is no other natural vegetation than small shrubs and various kinds of burrows can serve as the habitat of *Ph. alexandri* sandflies (Guan, 2009).

Acquired knowledge about the vector distribution and their favoured ecological environments have revealed that these four vectors, namely *Ph. chinensis*, *Ph. Longiductus*, *Ph. wui* and *Ph. alexandri*, have different requirement of ecological environment, which provides a reference for further studies on the geographical distribution of leishmaniasis and the environmental factors influencing it, as well as on the planning of national VL control programme

4. Visceral leishmaniasis in China: Then and now

Since the early 1950s, the national VL control programme has been initiated with comprehensive measures to eliminate infectious sources and control sandflies. As a result, the incidence of VL in the 16 historically endemic provinces, autonomous regions and municipalities declined rapidly. In the North China Plain and Shaanxi Plain regions, AVL had been controlled by 1958, and the standard of elimination was met in the early 1980s (Wang et al., 2000). However, VL has never totally disappeared from six provinces and autonomous regions including Xinjiang, Gansu, Sichuan, Shaanxi, Shanxi and Inner Mongolia, where sporadic cases and small-scale epidemics have been reported. Imported cases have also been notified in 51 non-endemic districts and cities nationwide (Guan et al., 2000). After entering the 21st century, case information was collected and analyzed by NIPD-CTDR through the infectious disease network reporting system in China. From 2004 to 2018, a total of 4,793 cases of VL were reported in China, with 306, 311, 270, 355, 518, 523, 375, 300, 219, 158, 293, 510, 310, 183 and 162 annual cases reported respectively, an annual average number of reported cases of 320 (Table 1), showing a decline trend. The epidemic area of VL has expanded to 82 counties and cities in 8 provinces and autonomous regions, namely Xinjiang, Gansu, Sichuan, Shaanxi, Shanxi, Inner Mongolia, Henan, Hebei Province (Table 2, Fig. 5). Generally, cases are scattered, with some concentration of cases seen in Jiashi County in Xinjiang, Zhouqu County, Dangchang County and Wudu District in Gansu Province. An annual incidence rate of over 1/10,000, and even periodic outbreaks with an interval of 7–8 years, were even reported in Jiashi County in Xinjiang in 2008 and 2015 with 202 and 356 cases reported respectively. The number of imported cases has also increased in non-endemic areas of the country (Han et al., 2019; Wang et al., 2012).

4.1 Control efforts—What was done

Establishment of professional institutions and training professionals for VL control in major endemic areas. After the founding of the People's Republic of China, the central people's government began to set up professional institutions for VL control at the provincial level in major endemic areas. Between 1950 and 1953, anti-VL stations were established in

Table 1 Distribution of VL endemic areas in China from 2004 to 2016.

Province	City (state, union)	Endemic county (district, county)
Xinjiang Autonomous Region	Bayingolin Mongol Autonomous Prefecture	Yuli, Korla, Luntai, Muchi, Ruoqiang, Yanqi hui Autonomous County, Hejing, Heshu, Bohu
	Aksu Prefecture	Wushi, Kuqa, Shaya, Aksu, Wensu, Xinhe, Baicheng, Awati, Keping
	Kizilsu Kirgiz	Acto, Artosh, Aheqi, Ucha
	Kashgar Prefecture	Kashgar, Shufu, Shule, Jiashi, Yingjisha, Megeti, Zepu, Yuepuhu, Shaka, Bachu, Yecheng
	Hotan Prefecture	Hetian city, Hetian, Moyu, Pishan, Lopu, Cele, Yutian, Minfeng
	Turpan	Gaochang, Toksun
	Hami	Yizhou
	Municipality directly under the jurisdiction of prefecture-level city	Tumushuk
Subtotal	8	45
Gansu Province	Longnan	Wudu, Wenxian, Tanchang, Kangxian, Lixian, Xihe
	Gannan state	Zhouqu, Diebu
	Jiuquan	Guazhou
	Pingliang	Kongtong, Jingning
	Dingxi	Tongwei
	Qingyang	Huanxian, Qingcheng
Subtotal	6	14
Sichuan Province	Aba state	Jiuzhaigou, Maoxian, Heishui, Wenchuan, Lixian
	Mianyang	Beichuan, Pingwu
Subtotal	2	7

Continued

Table 1 Distribution of VL endemic areas in China from 2004 to 2016.—cont'd

Province	City (state, union)	Endemic county (district, county)
Inner Mongolia Autonomous Region	Alxa League	Ejin Banner
Subtotal	1	1
Henan Province	Anyang	Linzhou
Subtotal	1	1
Shaanxi Province	Weinan	Hancheng
	Yanan	Baota, Yichuan
Subtotal	2	3
Shanxi Province	Jiuquan	Urban area, Suburb, Mengxian, Pingding
	Changzhi	Wuxiang, Suburb, Xiangyuan, Tunliu
	Jinzhong	Yushe
	Linfen	Yaodu, Daning
Subtotal	4	11
Total	24	82

Shandong, Jiangsu, Anhui, Henan, Hebei, Shaanxi and Gansu province to take responsibility to determine local epidemic situation and epidemiological factors of VL, train technical personnel at different levels and establish network from county to village levels for VL control, and provide technical guidance and supervise for VL control (Guan and Wu, 2014). In order to strengthen technical guidance, in 1951, the NIPD-CTDR was authorized to work in cooperation with Shandong provincial institutes to develop the standard manuals and embarked on 3–6 months multidisciplinary training courses for health personnel, via which, thousands of trainees were trained to become proficient in dealing with patient diagnosis, clinical treatment and case management and chemotherapy, as well as collection, identification, and control of vectors and reservoirs (Guan and Yang, 2012; Lun et al., 2015), which laid solid foundation for VL control in the endemic areas.

Table 2 Distribution of the reported VL cases in China from 2004 to 2018.

Endemic regions			Non-endemic regions		
Province	**The cumulative cases**	**Composition ratio/%**	**Province**	**The cumulative cases**	**Composition ratio/%**
Xinjiang Autonomous Region	2,347	48.97	Chongqing	19	0.40
Gansu Province	1,557	32.48	Zhejiang Province	9	0.19
Sichuan Province	583	12.16	Hubei Province	10	0.21
Shaanxi Province	97	2.02	Hunan Province	10	0.21
Shanxi Province	86	1.79	Yunnan Province	7	0.15
Inner Mongolia Autonomous Region	10	0.21	Tianjing	7	0.15
Henan Province	13	0.27	Heilongjiang Province	5	0.10
			Guangxi Zhuang Autonomous Region	5	0.10
			Beijing	4	0.08
			Hebei Province	6	0.13
			Jiangxi Province	3	0.06
			Qinghai Province	3	0.06
			Jilin Province	3	0.06
			Jiangsu Province	2	0.04
			Anhui Province	3	0.06
			Shandong Province	2	0.04
			Tibet Autonomous Region	1	0.02
			Fujian Province	1	0.02
Subtotal	4,693	97.91	Subtotal	100	2.09

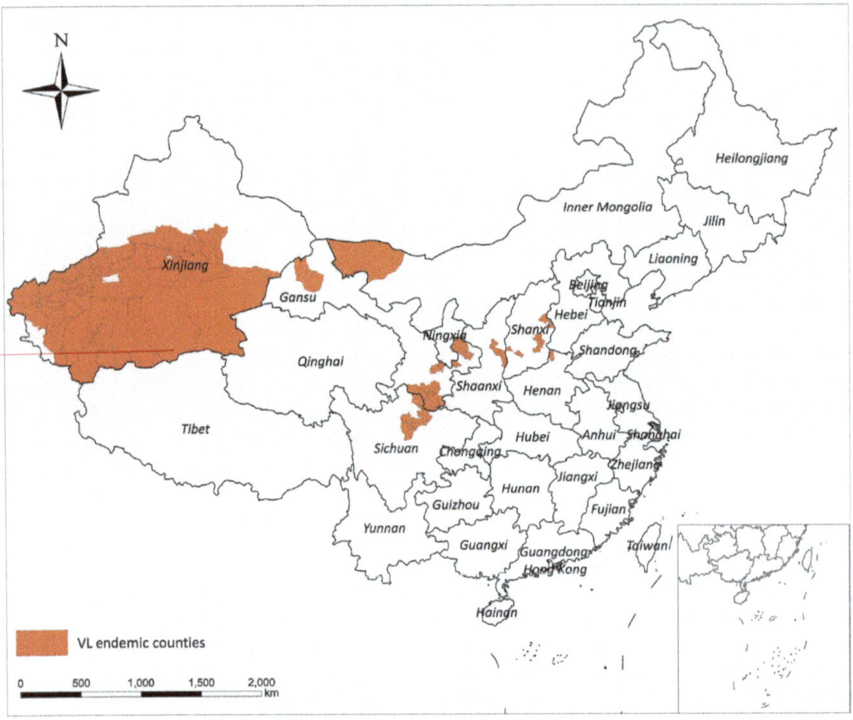

Fig. 5 Distribution of VL endemic counties and cities in China from 2004 to 2016.

Setting up control pilot sites and conducting control technology research. Although in the 1930s, pentavalent antimony, such as Neostibosan, Urea Stibamine, Solustibosan, and Pentostam have been used to treat VL in China (Guan and Wu, 2014), it was still too expensive to be used extensively for most people in China, because all of them had to be imported from abroad. In 1950, the Shandong Xinhua Pharmaceutical Factory successfully synthesized pentavalent sodium antimony gluconate which was suitable for rural medical staff to use it either for intravenous or intramuscular injection, fully meeting the needs for VL treatment in China. Before large-scale use of homemade pentavalent sodium antimony gluconate, pilot site was firstly set up in Shandong province to conduct clinical trials to determine its therapeutic effect and the method suitable for its application in rural areas. As a result, the "six-day therapy" was recommended as the therapeutic scheme of choice in the endemic areas. Besides, the "three-day therapy" was suitable for patients in good health and without complications, and "three-week therapy" was suitable for those in poor health or with serious illness (Guan and Yang, 2012).

Simultaneous efforts were devoted by the NIPD-CTDR and provincial institutes to control vectors. Between 1951 and 1953 (Wu et al., 1955), vector control pilot sites were set up in Shandong, Jiangsu, and Anhui Province, to conduct sandflies bionomics investigation and elimination approach. The results demonstrated the following bionomical characteristics of *Ph. chinensis* in the plain region: The breeding sites of the larvae were widely scattered; hence it was difficult to control effectively. However, the adult is endophilic and susceptible to insecticide, with their activity season as short as 3.5 months. The obtained information led to the implementation of vector control by indoor DDT spraying of human dwellings and animal shelters (Guan et al., 2000; Wang et al., 2000; Wu et al., 1955). The insecticides, dichlorodiphenyltrichloroethane (DDT) 1.5 g pure insecticide/m^2 and hexachlorocyclohexane (HCH) 0.15 g/m^2, were used to spray whole villages in the endemic areas. No new cases were found in the sprayed area of the whole village, while new cases were found in the control village (where patients were only treated) every year. It fully showed that the combination of killing sandflies and treatment of VL patients in the whole village had significant effects on the control of AVL in the central and eastern plain of China. According to the successful experience of vector control with insecticide, Wu and others (Wu and Wang, 1956) extended the experimental process and specific measures to the whole country, which significantly promoted the AVL control process in endemic areas of eastern and central plain in China.

The control outcomes in different types of endemic area. Since 1950, the government provided homemade pentavalent sodium antimony gluconate for free to all VL patients. If they could afford it, the patients in principle should pay for the treatment of complications. Otherwise, the government would provide relevant subsidies for them (Guan and Wu, 2014). Since 1955, free insecticides were provided for vector control by the VL control institutions, and village doctors responsible for VL treatment would get a service subsidy from governments based on the number of cases treated (Guan and Yang, 2012).The massive mobilization of the general public was conducted by a top-down approach from the provincial to the village level for education of the population about VL by trained health personnel to actively engage in the control programmes (Lun et al., 2015). In the process of VL control, NIPD-CTDR played a key role as technical guidance.

AVL endemic area. From 1950 to 1958, in the endemic areas, professional VL control institutions as the mainstay, with participation from rural health workers, carried out massive screening to identify patients through spleen palpation and inquiry of disease history. Then necessary serological or parasitological examination was performed on suspected VL cases. Later, free treatment was provided for confirmed cases in the township clinics, and treatment for severe patients in the county hospital or competent institutions at the higher level (Guan and Wu, 2014). During the period, a total of 248,427, 53,820, and 110,999 cases were treated in Shandong, Shaanxi, and Jiangsu, respectively (Guan and Yang, 2012). Since 1955, massive and thorough screening and treatment, combining vector control with insecticide spraying was conducted to eliminate the source of infection and kill the sandflies. At the end of the 1950s, the seriously epidemic areas of VL had been controlled in Southern Hebei, Northern Jiangsu, Northern Anhui, Eastern Henan, Northern Hubei and central Shaanxi and reached the elimination standard in the early 1980s (Guan and Yang, 2012).

Since 1977, annually one–two times massive screening to identify VL patients for treatment was carried out in eight AVL endemic counties and cities in the Kashgar ancient oasis in southern Xinjiang. Simultaneously, the wh1ole village with patients last year was sprayed with BHC or deltamethrin once a year. The number of VL cases in these eight counties had dropped from 222 cases in 1979 to 14 cases in 1988 (Chai and Guan, 2006). However, once the aforementioned measures were halted, the number of cases rebounded year by year, with a vengeance from 1996 to 2002 (Chai and Guan, 2006). The reason maybe that even after IRS, a large number of *Ph. Longiductus* with peridomestic bionomic characteristic still exist in AVL endemic areas. Meanwhile, asymptomatic patients cannot be identified and eliminated in campaign of massive screening (Burza et al., 2018), which could develop into new source of infection (Hasker et al., 2014). As infection source begin to accumulate, VL would rebound, even outbreak again (Medley et al., 2015).

MST-ZVL endemic area. Before 1958, comprehensive interventions were implemented in MST-ZVL endemic area, including provision of free VL treatment services for patients, IRS killing sandflies and dogs culling campaign which led to domestic dogs were once all but extinct there, thus, the prevalence of VL constantly decline. For example, the number of cases in Gansu province dropped from 3698 cases in 1953 to 33 cases in 1961 (Guan and Yang, 2012). No new autochthonic case was reported from Qinghai, Ningxia and Liaoning province after 1959, 1973 and 1975, respectively (Guan and Yang, 2012). However, different from that in plain areas, the local *Ph. chinesis* is an exophilic sandfly species, which lead to poor effectiveness of IRS. Since the mid-1960s, as the number of domestic dogs in rural areas gradually increased in the areas of southern Gansu, northwestern Sichuan, northern Shaanxi and southern Shanxi, both CVL and human VL rebounded significantly.

From 1980s, the interventions integrated IRS of patient and their neighbour's household with deltamethrin, timely treatment of patients and culling infected dogs confirmed by bone marrow aspiration have constricted MST-ZVL at a low incidence in China, whereas, the endemic area is extending gradually (Guan and Yang, 2012). In 1994, scientists from the NIPD-CTDR (Xiong et al., 1994) reported that domestic dogs bathed with 25 ppm water solution of 2.5% deltamethrin wettable powder in Jiuzhaigou County of Sichuan Province could prevent *Ph. chinensis* sandflies for the first time. The experimental dogs were exposed to wild *Ph. chinensis* sandflies in mosquito nets for five times within 30 to 104 days after a medicated bath. The blood absorption rate of sandflies was 2.7–4.9% for 8 h of exposure each time, and then the blood-sucking and non-blood-sucking sandflies were reared in the laboratory. The mortality rate of sandflies was 98% in 24 h. Under the same conditions, the blood absorption rate and blood-sucking mortality rate of sandflies to healthy dogs were 61.5% and 5.4–14.9% respectively. The results showed that the use of deltamethrin in bathing dogs could prevent the bite of *Ph. chinensis* sandflies, and subsequently can block the transmission of CVL, and thus effectively reduce the incidence of VL in the population. However, it is a challenge to promote this technology with sporadic VL cases scattered in remote mountain villages, for close corporation from dog owners are required. If the number of domesticated dogs can be contained after large-scale culling of infected dogs, it will have a more significant effectiveness on control the epidemic of VL, but it is still difficult in practice for lacking of sufficiently sensitive and specific diagnostics to identify CVL and ethical issues.

DST-ZVL endemic area. Due to the DST-ZVL cases is scattered, IRS is inefficient for controlling *Philebotomus wui*, an exophilic sandfly (Chai and Guan, 2006; Guan and Shen, 1991), and reservoir hosts are yet unconfirmed wild animals (Liao et al., 2009), the local health department has no other intervention measures except to identify VL patients for treatment timely (Chai and Guan, 2006). Evidences have shown that large-scale reclamation of desert to change the original ecological environment, or continuous spraying of insecticides in the *Populus diversifolia-Tamarisk* forest belt to reduce the density of sandflies, can greatly decrease the DST-ZVL incidence, but the former is contrary to the policy of the ecological environment conservation, and the latter is also uneconomical (Guan and Yang, 2012).

4.2 Supportive research from NIPD-CTDR

4.2.1 Research on immunoassay

4.2.1.1 Antibody detection

Indirect immunofluorescent antibody test (IFAT). IFAT was used to detect antibodies in the body infected with *Leishmania* spp. Serum of patients was once used as the test sample, but collecting blood from infants and young children was inconvenient. In 1979, scientists from the NIPD-CTDR tried to replace serum with "filter paper dried blood droplets". IFAT was performed on 23 patients with VL and 13 mice infected with *Leishmania donovani*. The results showed that the dried blood droplets and serum had equal efficiency, with similar results after two to four times of repeated tests (Guan et al., 1980b).

Counterimmunoelectrophoresis (CIEP). In 1981, the soluble antigen of leishmanial promastigotes was used to test the serum of 30 patients with VL confirmed by pathogenic examination. The positive rate of the test by the methods of counterimmunoelectrophoresis (CIEP) was 96.7%. The serum of 26 patients for post-cure 23–28 months was still able to be detected by CIEP. Only two cases showed weak positive reaction and the rest were negative. No positive reaction with sera of tuberculosis, hepatitis and healthy people was noticed (Wang et al., 1981).

Enzyme-linked immunosorbent assay (ELISA). ELISA was used to detect specific antibodies in serum of VL patients with soluble crude antigen of leishmanial promastigotes, intact leishmanial promastigotes or recombinant antigen. Intact leishmanial promastigotes was used as coating antigen in ELISA to detect 16 sera of VL patients confirmed by pathogenic examination, which showed positive results in ELISA by using leishmanial promastigotes soluble antigen as coating antigen and negative in 9 sera of healthy persons. The results showed that all sera of patients were positive and all sera of healthy persons were negative, thus proving that intact leishmanial promastigotes antigen and soluble crude antigen had the same diagnostic efficiency (Wang and Chen, 1987). The rK39 was used as coating antigen in ELISA for immunodiagnosis, and sera of 107 VL cases from Western China was detected. The positive rate was 96.2%, and 1 was found as seropositive in 31 cases of cutaneous leishmaniasis in Karamay of Xinjiang Uygur Autonomous Region (Qu et al., 1996a).

Immunochromatographic strip test. Immunochromatographic strip test is a rapid diagnostic method developed in recent years. It combines immune molecule affinity principle with Western blot and classic thin layer chromatography. Because of its high sensitivity, simple operation and fast reporting, it has been widely used in diagnosis. Scientists from the NIPD-CTDR successfully developed immunochromatographic strip test for detection of specific antibodies in VL patients using colloidal gold method with leishmania promastigotes crude antigen as coating antigen (Yang et al., 2014). The sensitivity and specificity of the strip were 96.1% and 95.6%, respectively, in 129 sera of VL patients diagnosed and confirmed by pathogen examination in the laboratory. The field evaluation in different VL endemic areas in China also showed high sensitivity and specificity, and the results had no statistical difference with parallel detection of rK39 (Feng et al., 2016). The strip is also suitable for detecting specific antibody in dog blood samples from zoonotic visceral leishmaniasis (ZVL) endemic areas (Feng et al., 2015).

4.2.1.2 Detection of circulating antigen

Monoclonal antibody labelled with peroxidase dot—ELISA. Scientists from the NIPD-CTDR used monoclonal antibody against the target antigen of *L. donovani* promastigotes labelled horseradish peroxidase in dot-ELISA test for detecting circulating antigen in serum of patients with VL, and the positive rate was 88.9–90.6% (Qu et al., 1990, 1991, 1996b). Positive correlation was exhibited between the titre of circulating antigen in the sera and the number of protozoa in the bone marrow smear. No positive result was obtained from patients who had been recovered from the disease. The sera of healthy people tested were all negative, and no cross reaction was found with the sera of patients with malaria, leprosy, brucellosis and schistosomiasis. It was suggested that McAb dot-ELISA method could be used not only for the diagnosis of current VL, but also for field evaluation of therapeutic effectiveness for VL.

Monoclonal antibody-enzyme linked immunotransfer blot technique (McAb-EITB). In 1991, scientists from the NIPD-CTDR (Wang et al., 1991) first used monoclonal antibody (HRP-L12H4) against *L. donovani* labelled horseradish peroxidase and enzyme-linked immunotransfer blot technique (EITB) to detect circulating antigen in sera of VL patients, and results showed that McAb-EITB was a sensitive, specific and practical diagnosis method for VL.

Immunochromatographic test (ICT). Scientists from the NIPD-CTDR (Gao et al., 2015c) prepared monoclonal antibodies (mAbs) against *L. donovani* promastigotes soluble crude antigen. Two secreted mAbs recognizing the same leishmanial protein were used to produce an ICT for the detection of circulating antigen in serum and blood of patients diagnosed with VL. Sensitivity, specificity and diagnostic efficiency of the new ICT was 95.8%, 98.7% and 97.3%, respectively. There was no significant difference in sensitivity or specificity between the test strips prepared in this study and the rK39 test strips. As the ICT detecting a circulating antigen, the ICT will also be useful in monitoring treatment efficiency and identifying VL patients with immunodeficiency.

4.2.2 Development of molecular diagnostics

4.2.2.1 Polymerase chain reaction (PCR)

Scientists from the NIPD-CTDR (Gao et al., 2006) selected six primer pairs for detecting Chinese strain of *L. infantum* by optimizing conditions. Their sensitivity and specificity were compared by using DNAs extracted from human blood seeded with cultured *L. infantum* promastigotes (MHOM/CN/86/GS) as a template. The results showed that the specificity of all six primer pairs reached 100%, and the sensitivity varied among the primer pairs. The primer pairs RV1-RV2 (0.1 parasite/mL blood) and K13A-K13B (1 parasite/mL blood) were most sensitive. This study suggested that RV1-RV2 and K13A-K13B primer pairs were suitable in detecting the asymptomatic infection of *L. infantum*. The PCR method was also suitable for detection of asymptomatic infection of *Leishmania* in humans, dogs and domestic mammals in the VL endemic area (Gao et al., 2015b; Han et al., 2018; Wang et al., 2006, 2007).

4.2.2.2 Loop-mediated isothermal amplification (LAMP)

Scientists from the NIPD-CTDR (Gao et al., 2015a) developed a loop-mediated isothermal amplification (LAMP) assay to detect *L. infantum* infections in dogs using conjunctival swab samples specifically. The primers used in the LAMP assay were designed to target kinetoplast DNA minicircle sequences of the *L. infantum* isolate MCAN/CN/90/SC. The *L. infantum*-positive rates obtained for field-collected samples were 61.3% by LAMP. The sensitivity and specificity of LAMP and PCR are similar, but conjunctival swab is a non-invasive method, which is easy to operate and more acceptable by dog owners. The findings indicate that the LAMP assay is a sensitive and specific method for the field surveillance of domestic dogs, particularly of asymptomatic canines, in ZVL-endemic areas in western China.

4.2.3 Control of exophilic sandflies

Individual protection by skin coating of repellent. The efficacy of five repellents were tested against *Ph. alexandri* on the southern slope of Bogda mountain, Turfan City, Xinjiang Uygur Autonomous Region during 1986–88. At the dose of 0.25 $\mu L/cm^2$, the protective durations of mosquito repellent perfume (MRP), N, N-diethyl-m-toluamide (DETA), mosquito repellent liquid (MRL), dimethyl phthalate (DMP) and di-butyl phthalate (DBP) were (6.94 ± 0.67)h, (5.19 ± 0.32)h, (4.98 ± 0.89)h, (3.77 ± 0.43)h and (1.94 ± 0.52)h, respectively (Jia et al., 1990). Another experiment showed that the protective durations of MRP, DETA and MRL against *Ph. wui* in Karamay were 6–9 h (Jia et al., 1988). All the repellents were proved to have practical value.

4.2.4 Animal infection sources in different types of endemicity

From 1964 to 1984, in collaboration with local professional institutes, the NIPD-CTDR conducted a series of examinations to detect *Leishmania* spp. infection in wild reservoir hosts in either the dry or the stony deserts in Xinjiang as well as in the Ejina Banner of Inner Mongolia. The method used was smear examination and Novy-MacNeal-Nicolle medium culture from samples of liver, spleen, bone marrow and lymph node. No infection of *Leishmania* spp. was observed in a total of 5,409 wild rodents, 114 hares, 157 carnivorous foxes, and 142 carnivorous hedgehogs. Wild rodents should therefore not be considered as VL reservoir hosts (Chai and Guan, 2006; Guan et al., 1984). In recent years, it was reported that Tarim rabbits were a suspicious reservoir host of VL in the dry desert area in Xinjiang, but further studies are required to confirm this (Liao et al., 2009; Lun et al., 2015).

In the Hexi Corridor Zone of Gansu Province, the Junggar Basin of Xinjiang and the Ejina Banner of Inner Mongolia, *L. gerbilli* and *L. turanica* were detected from subcutaneous tissues of gerbils. However, neither of those two leishmania species can cause human VL. Therefore, it was concluded that the gerbils were not a host reservoir for human VL (Guan et al., 1982; Liu et al., 1982; Wang et al., 1964b; Xiong et al., 1964b).

After the 1990s, in view of the reemergence of VL cases in the mountainous areas of western China, three methods, namely polymerase chain reaction (PCR), enzyme-linked immunosorbent assay (ELISA) and rK39 immunochromatographic strip, were used to detect leishmania antibodies in healthy dog serum in Wenxian County in Gansu province, with positive rates for the three methods being 50.6% (40/79), 22.2% (16/72) and 33.3% (19/57), respectively (Wang et al., 2006). This demonstrated that highly infectious (Molina et al., 1994), but asymptomatic infected dogs played an important role in the epidemiology in MST-ZVL.

These investigations have laid the foundation for discovering the infection sources of zoonotic VL in China and then developing targeted prevention measures.

4.2.5 Development of VL diagnosis standard and management principles

The Ministry of Health issued standard document of "Diagnostic Criteria and Principles of Management of Kala-Azar (GB 15986-1995)" authored by the NIPD-CTDR in 1995, as the last national guidelines available for the management of leishmaniasis control programme (Qu and Guan, 1995), which served as the nationwide general guidelines and contributed a lot for the national VL control programme.

4.3 Monitoring and evaluation

After 1958, the prevalence of VL in former epidemic provinces, autonomous regions and municipalities has dropped significantly to less than 1/10,000. Thus, monitoring work has begun to take the priority for VL control in all areas of endemic areas which includes three aspects, namely, tracing the current VL patients to grasp the trend of the epidemic, using leishmanin intradermal test for delayed-type hypersensitivity (DTH) to evaluate the VL epidemic trend in local population from the perspective of immunology, and the investigation of the density of vector to assess whether the transmission has been blocked (Guan and Yang, 2012).

Large-scale monitoring and evaluation in control stage. Under the guidance of the NIPD-CTDR, from 1970s to 1990s, several rounds of large-scale monitoring and evaluation works were performed in previous major AVL endemic provinces, such as Shandong, Henan, Jiangsu, Anhui, Hubei province (Guan and Yang, 2012). However, no new VL case was found, except for a few imported cases and rare relapse cases infected before 1970. Leishmanin intradermal test was carried out among rural residents of different age groups. Nevertheless, no positive result was found in the population born after 1970. For the vector survey, no *Ph. chinesis* was found in the eastern China plain region, except for a few sergentomyia species, not a vector of VL. The above results indicated that VL in previous AVL endemic area in eastern and central plain in China has been eliminated (Guan and Yang, 2012). However, the persistence of AVL is evident in southern Xinjiang and that of MST-ZVL in the western mountain areas, mainly Sichuan and Gansu provinces.

Infection Disease Reporting Management System. In 2004, the web-based National Diseases Reporting Information System (NDRIS) which is a network covers all the medical and health institutions at or above the township level, was established in China and operated by the Chinese Center for Disease Control and Prevention (China CDC). According to the Law of the Infectious Disease Control and Prevention, each of the medical and health institutions in the country have to notify the cases of VL online within 24 h after the clinical cases confirmed with the diagnostic criteria of VL in China (WS258-2006), and county CDC where the patient resides will take responsibility for conducting epidemiological investigation of individual cases in 15 days after case notification in NDRIS to determine whether the patient is an autochthonic case or not, and taking countermeasures on-site according relevant guidance. The cases information reported through NDRIS includes age, gender, occupation, current residential address, symptom onset date, diagnosis date, and therapeutic outcome and so on, through which demographic distribution, geographic distribution, temporal distribution and the lag time between symptom onset and diagnosis can be analyzed. NIPD-CTDR takes the responsibility to analyze VL epidemic trend, evaluate local intervention measure and issue the epidemic analysis report to guide the prevention and control of VL.

Setting pilot monitoring site. Since 2015, pilot monitoring sites are gradually set up in accordance with epidemiological type and epidemic magnitude of the endemic area to carry out case monitoring based on NDRIS, vector surveillance and reservoir surveillance where applicable. NIPD-CTDR takes the responsibility to develop monitoring and evaluation programmes scheme and provide technical guidance, as well as evaluation. CDC at country level of monitoring sites are in charge of surveillance activities implementation. At present, the pilot monitoring of VL in China is still at the initial stage, however it could provide a favourable reference for formulating the strategies of VL control in China, especially through analysis of VL cases information reported in the NDRIS.

4.4 Training

Training of key technical personnel is critical for VL control. Since 1951, the NIPD-CTDR has been authorized by health ministry to embark on multidisciplinary training courses for health personnel. via which, thousands of trainees were trained to become proficient in dealing with patient diagnosis, clinical treatment and case management, as well as collection, identification, and control of vectors and reservoirs (Lun et al., 2015), which laid solid foundation for VL control programme in the endemic areas. In 1953, scientists from NIPD-CTDR have developed and extended IRS technology with DDT to control the vector to the whole country through national VL control technical training courses. Later a serial of diagnosis technology, technical standards for VL diagnosis and management principles, monitoring and evaluation scheme have been developed and promoted by NIPD-CTDR through regular training course every year. It plays an important role in promoting the VL control process in endemic areas in China.

5. Conclusion and way forward

The number of VL patients in China has decreased from 530,000 in 1951 to 180 in 2019, showing a remarkable achievement has been made, which is undoubtedly attributable to the control programmes that were launched over half a century ago as a national priority at an unprecedented scale of historical proportion (Wang and Wu, 1956). In the process of the control, NIPD-CTDR has made a great contribution in various aspects of VL control. AVL rampant in eastern and central plain in China in history has been eliminated. However, the remaining problems are zoonotic VL transmission mostly in mountainous and desert areas in western China, and a few remnant AVL cases in the Kashgar ancient oasis in southern Xinjiang. Thus, more efforts are required to fill the knowledge gaps. Advances are also surprisingly limited in the areas of diagnosis, for example, on asymptomatic infection in AVL endemic areas in southern Xinjiang, and what the role it plays in local VL transmission remains to be studied (Wang et al., 2006, 2007). On infected dogs culling, it is still difficult in practice for lacking sufficiently sensitive and specific diagnosis agents to identify infected dogs (Courtenay et al., 2002). On clinical diagnosis, a point-of-care biomarker is still unavailable for diagnosis of patients with relapse VL (Burza et al., 2018). In recent years, DST-VL outbreaks have occurred in southern Xinjiang. However, the wild animal host remains to be elucidated and a suitable control strategy is still unavailable for control exophilic sandfly at present (Wang et al., 2010). In MST-ZVL endemic areas, a suitable control strategy is still unavailable for control exophilic *Ph. chinesis* and more efforts are needed to develop a canine vaccine for protection of dogs. Until now, none of the four major vector species in China has been successfully colonized for laboratory breeding, as is needed for xenodiagnoses examination and other studies (Molina et al., 1994). Surveillance in the endemic areas is as necessary as new laboratory and field investigation toward better control of VL in the western regions. Epidemiological dynamics should be introduced to model and predict epidemic trend.

Strictly, MSZ-VL is a zoonotic disease with both dogs and wild animals serving as reservoir hosts, so it is impossible to completely eliminate the disease. In order to restrict the prevalence of human VL and canine leishmaniasis to very low level, the health department in endemic areas must establish a long-term combat ideology against VL, maintain a relatively stable prevention and research team, strengthen investigations and field experiments, and propose prevention methods that can be promoted and applied as much as possible to control the spread of VL to reduce the number of cases. As the increasing national and international travel related to economic development, more and more imported VL cases were reported. It is especially necessary considering the need for managing VL cases in no-endemic areas, especially, regular training of health personnel for diagnosis skills and knowledge relevant to VL under the framework of One Health concept, for doctor there usually are unfamiliar with it and leads to misdiagnosis and mistreatment. Adequate support from government is still very necessary, not only to make advances but also to maintain the past success of keeping VL under control.

What can other countries learn from China

There are numerous sites of endemicity for VL in the world widespread in temperate, subtropical, and tropical areas. So knowledge and experiences gained from the Chinese VL control programmes will be of special value to those low and middle income endemic countries. In the process of VL control programme in China, there has been gradually formulating a working environment or mechanism, which is the government-led initiative, multisector cooperation, society participation and technical support from professional institutions. Professional institutions top-down from national to village level were established in the country forming an integrated anti-VL network, and thousands of professionals were trained who became the mainstay for VL control in endemic areas. Control pilot sites were set up to explore control technology and successful experience which was extended to the whole country later. Sufficient supply of homemade highly-effective, low toxic therapeutic drugs for VL; development of a suitable therapeutic regime; and implementation of effective sandflies control aiming to ensure achieving the goal of VL elimination. The massive and effective mobilization of the general public and health workers also great contributed to the success of the control programmes. The control programmes mainly consisted of (i) massive screening to identification patients for timely treatment, (ii) identification and disposal of infected dogs, and (iii) residual insecticide indoor spraying for vector control.

Acknowledgements

This study was supported by the National Special Science and Technology Project for Major Infectious Diseases of China (Grant No. 2017ZX10303404 and 2016ZX10004222-004) and the fourth round of Three-Year Public Health Action Plan (GWIV-29). We express our gratitude to Jingbo Xue, from the National Institute of Parasitic Diseases, Chinese Center for Disease Control and Prevention, for drawing the geographic distribution map of sandflies and VL endemic counties.

Competing interest

The authors declare that they do not have competing interests.

Authors' contributions

Z.-B. Zhou, J.-Y. Wang and X.-N. Zhou conceived the study; Z.-B. Zhou, J.-Y. Wang, C.-H. Gao, S. Han, Y.-Y. Li, Y. Zhang, X.-N. Zhou wrote and revised the manuscript; Z.-B. Zhou, J.-Y. Wang, Y. Zhang, X.-N. Zhou revised the manuscript and gave approval of the version to be published. All the authors read and approved the final version of the manuscript.

References

Burza, S., Croft, S.L., Boelaert, M., 2018. Leishmaniasis. Lancet 392 (10151), 951–970.

Chai, J.J., Guan, L.R., 2006. Leishmaniasis and Phlebotomine Sandflies in the Xinjiang Uygur Autonomous Region the People's Republic of China. Xinjiang People's Publishing House, Urumchi, pp. 293–348 (in Chinese).

Chen, S.B., Guan, L.R., Li, F., Liu, P.Z., Feng, Y., Yang, C.M., 2003. The fauna of Phlebotomus sandflies in Gansu Province. Chin. J. Parasitol. Parasit. Dis. 21 (5), 314–315 (in Chinese).

Courtenay, O., Quinnell, R.J., Garcez, L.M., Shaw, J.J., Dye, C., 2002. Infectiousness in a cohort of Brazilian dogs: why culling fails to control visceral leishmaniasis in areas of high transmission. J. Infect. Dis. 186 (9), 1314–1320.

Feng, L.C., 1951. The role of the peritrophic membrane in Leishmania and Trypanosoma infections of sandflies. Peking Nat. Hist. Bull. 19 (2–3), 327–334 (in Chinese).

Feng, X.P., Shi, F., Yang, Y.T., Gao, C.H., Wang, J.Y., 2015. Evaluation of colloid gold labeled immunochromatographic strip test for rapid detection of canine leishmanial infection in China. Int. J. Med. Parasit. Dis. 42 (4), 189–192 (in Chinese).

Feng, X.P., Gao, C.H., Shi, F., Yang, Y.T., Wang, J.Y., 2016. Evaluation of colloid gold labeled immunochromatographic strip test for rapid detection of visceral leishmaniasis patients infection in China. Chin. J. Zoonoses 32 (1), 24–27 (in Chinese).

Gao, C.H., Wang, J.Y., Yang, Y.T., Bao, Y.F., 2006. Study on PCR method for detecting the asymptomatic infection of Leishmania infantum. Chin. J. Parasitol. Parasit. Dis. 24 (2), 92–96.

Gao, C.H., Ding, D., Wang, J.Y., Steverding, D., Wang, X., Yang, Y.T., Shi, F., 2015a. Development of a LAMP assay for detection of Leishmania infantum infection in dogs using conjunctival swab samples. Parasit. Vectors 8, 370.

Gao, C.H., Wang, J.Y., Zhang, S., Yang, Y.T., Wang, Y., 2015b. Survey of wild and domestic mammals for infection with Leishmania infantum following an outbreak of desert zoonotic visceral Leishmaniasis in Jiashi, People's Republic of China. PLoS One 10 (7), e0132493.

Gao, C.H., Yang, Y.T., Shi, F., Wang, J.Y., Steverding, D., Wang, X., 2015c. Development of an immunochromatographic test for diagnosis of visceral leishmaniasis based on detection of a circulating antigen. PLoS Negl. Trop. Dis. 9 (6), e0003902.

Guan, L.R., 2009. Present situation of visceral leishmaniasis and prospect for its control in China. Chin. J. Parasitol. Parasit. Dis. 27 (5), 394–397 (in Chinese).

Guan, L.R., 2010. A supplement to the research on sandfly biology in China. Int. J. Med. Parasit. Dis. 37 (2), 65–67 (in Chinese).

Guan, L.R., Chai, J.J., 1996. Relationship between geographical distribution of sandflies and landscape features in Xinjiang Uyger Autonomous Region of China. Chin. Parasitol. Parasit. Dis. 14 (1), 26–32 (in Chinese).

Guan, L.R., Shen, W.X., 1991. Recent advances in visceral leishmaniasis in China. Southeast Asian J. Trop. Med. Public Health 22 (3), 291–298.

Guan, L.R., Wu, Z.X., 2014. Historical experience in the elimination of visceral leishmaniasis in the plain region of Eastern and Central China. Infect. Dis. Poverty 3 (1), 10.

Guan, L.R., Yang, Y.Q., 2012. Leishmaniasis. In: Tang, L.H., Xu, L.Q., Chen, Y.D. (Eds.), Parasitic Disease Control and Research in China. In: vol. 1. Beijing Science and Technology Publishing, Beijing, China, p. 161. 915 (in Chinese).

Guan, L.R., Wang, J., Hu, Y.D., Xiong, G.H., Zheng, S.C., 1976. Epidemiology of kala-azar in China and its significance in the prevention and treatment. J. Control Res. Epidemiol. Dis. 3, 225–232 (in Chinese).

Guan, L.R., Wang, J., Liu, P.Z., Zhang, Z.G., 1980a. The bionomics of Phlebotomus chinensis in the mountainous regions of southern Kansu and the loess plateau of northern Shansi. Acta Entomol. Sin. 23 (1), 25–31 (in Chinese).

Guan, L.R., Xu, Y.X., Wu, K.C., Jin, C.F., Cai, J.J., 1980b. Observation on indirect fluorescent antibody test of kala-azar with dry blood drops of filter paper. J. Epidemiol. 1 (2), 118–120 (in Chinese).

Guan, L.R., Xu, Y.X., Jin, C.F., Cao, H.X., Yu, J.G., Liu, P.Z., Chen, S.B., Geng, C.Y., Chen, X., 1982. Studies on Leishmania in big gerbils and lizards and its sandfly vectors in Ejin banner, Inner Mongolia Autonomous Region. Acta Acad. Med. Sin. 4 (5), 261–265 (in Chinese).

Guan, L.R., Cao, H.X., Yu, J.G., Xu, Y.X., Jin, C.F., Liu, P.Z., et al., 1984. Investigation of KALA-AZAR hosts and human skin carpeting insects in Ejina banner, Inner Mongolia autonomous region. J. Lanzhou Univ. Med. Sci. 2, 11–14 (in Chinese).

Guan, L.R., Xu, Y.X., Maoyiding, K., Wang, W., 1986. The sandfly fauna and its role in transmission of kala-azar in four landscape zones of Aksu region, Xinjiang. J. Parasitol. Parasit. Dis. 4 (3), 169–172 (in Chinese).

Guan, L.R., Xiang, X.Y., Wang, G., Jia, J.X., 1990. Observation on the infectivity of Leishmania donovani from different areas to Phlebotomus alexandri from Xinjiang. Chin. J. Parasitol. Parasit. Dis. 8 (2), 104–107 (in Chinese).

Guan, L.R., Qu, J.Q., Cha, J.J., 2000. Leishmaniasis in China–present status of prevalence and suggestions on its control. End. Dis. Bull. 15, 49–52 (in Chinese).

Guan, L.R., Zhou, Z.B., Jin, C.F., Fu, Q., Chai, J.J., 2016. Phlebotomine sandflies (Diptera: Psychodidae) transmitting visceral leishmaniasis and their geographical distribution in China: a review. Infect. Dis. Poverty 5, 15.

Han, S., Wu, W.P., Chen, K., Osman, I., Kiyim, K., Zhao, J., et al., 2018. Epidemiological survey of sheep as potential hosts for Leishmania in China. BMC Vet. Res. 14 (1), 378.

Han, S., Wu, W.P., Xue, C.Z., Ding, W., Hou, Y.Y., Feng, Y., et al., 2019. Endemic status analysis of visceral leishmaniasis in China from 2004 to 2016. Chin. J. Parasitol. Parasit. Dis. 37 (2), 189–195 (in Chinese).

Hasker, E., Malaviya, P., Gidwani, K., Picado, A., Ostyn, B., Kansal, S., Singh, R.P., Singh, O.P., Chourasia, A., Kumar Singh, A., Shankar, R., Wilson, M.E., Khanal, B., Rijal, S., Boelaert, M., Sundar, S., 2014. Strong association between serological status and probability of progression to clinical visceral leishmaniasis in prospective cohort studies in India and Nepal. PLoS Negl. Trop. Dis. 8 (1), e2657.

Ho, K.Z., Wang, J., Huang, C.M., 1959. Collected Papers on the Research of Visceral Leishmaniasis (1950–1959). Shanghai Publishing House of Science and Technology, Shanghai, pp. 34–42 (in Chinese).

Institute of Parasitic Diseases, Chinese Academy of Medical Sciences, 1958. Annual Report 360–376. (in Chinese).

Institute of Parasitic Diseases, Chinese Academy of Medical Sciences, 1960. Annual Report 426–436. (in Chinese).

Jia, J.X., Guan, L.R., Xu, Y.X., Wang, G., Zuo, X.P., Hao, K.F., 1988. Experiments on the efficacy of DETA, DMP, MRP and MRL against *phlebotomus wui*. Chin. J. Parasitol. Parasit. Dis. 6 (S1), 143–144 (in Chinese).

Jia, J.X., Guan, L.R., Xu, Y.X., Wang, G., Hao, K.F., 1990. Studies on the efficacy of five repellents against *phlebotomus alexandri*. Chin. J. Parasitol. Parasit. Dis. 8 (3), 203–206 (in Chinese).

Jin, C.F., Zuo, X.P., Gu, D.A., Yisilayin, O., Lan, Q.X., Zhang, Y., et al., 2008. A newly identified endemic area of visceral leishmaniasis in Minfeng county of South Xinjiang II. Investigation on phlebotomine vectors. Chin. Parasitol. Parasit. Dis. 26 (2), 132–135 (in Chinese).

Liao, L.F., Yan, S.S., Wuso, B., Wu, M., Xu, B., Zhang, Y., Hou, Y.Y., Lei, G., 2009. Leishmania infantum firstly isolated from Yarkand hare (Lepus yarkandensis). Chin. J. Vector Biol. Control 20, 45–47 (in Chinese).

Liu, P.Z., Guan, L.R., Li, S.X., Yu, J.G., Li, T.K., Cao, H.X., Chen, S.B., 1982. Leishmanis gerbilli in desert area of Hexi Zoulang of Gansu Province. Chin. J. Epidemiol. 3 (5), 304–305 (in Chinese).

Lun, Z.R., Wu, M.S., Chen, Y.F., Wang, J.Y., Zhou, X.N., Liao, L.F., Chen, J.P., Chow, L.M., Chang, K.P., 2015. Visceral Leishmaniasis in China: an endemic disease under control. Clin. Microbiol. Rev. 28 (4), 987–1004.

Medley, G.F., Hollingsworth, T.D., Olliaro, P.L., Adams, E.R., 2015. Health-seeking behaviour, diagnostics and transmission dynamics in the control of visceral leishmaniasis in the Indian subcontinent. Nature 528 (7580), S102–S108.

Molina, R., Amela, C., Nieto, J., San-Andres, M., Gonzalez, F., Castillo, J.A., Lucientes, J., Alvar, J., 1994. Infectivity of dogs naturally infected with Leishmania infantum to colonized Phlebotomus perniciosus. Trans. R. Soc. Trop. Med. Hyg. 88 (4), 491–493.

Qu, J.Q., Guan, L.R., 1995. Diagnostic Criteria and Principles of Management of Kala-Azar (GB 15986–1995). Standardization Administration of China, Beijing, China (in Chinese).

Qu, J.Q., Xu, Y.X., Bao, Y.F., Yang, Y.T., Wang, J.Y., 1990. A study of circulating antigen in visceral leishmaniasis using monoclonal antibody labeled with peroxidase. End. Dis. Bull. 5 (4), 55–57 (in Chinese).

Qu, J.Q., Xu, Y.X., Bao, Y.F., Yang, Y.T., Wang, J.Y., 1991. Application of monoclonal antibodies against *leishmania donovani* II. Detection of circulating antigen in sera of visceral leishmaniasis before and after treatment. Chin. J. Parasitol. Parasit. Dis. 9 (1), 21–23 (in Chinese).

Qu, J.Q., Bao, Y.F., Ruzhi, G.L., Reed, S.G., Chang, K.P., 1996a. Application of a recombinant antigen rK39 of visceral leishmaniasis in northwest of China. Chin. J. Parasitol. Parasit. Dis. 14 (3), 231–233 (in Chinese).

Qu, J.Q., Bao, Y.F., Xu, Y.X., Guan, L.R., Wang, J.Y., Emanuala, H., 1996b. Sandwich dot-ELISA in diagnosis of cutaneous leishmaniasis in Xinjiang. Chin. J. Pract. Parasit. Dis. 44 (4), 149–151 (in Chinese).

Tian, T., Wu, W.P., Wang, L.Y., Chen, H.T., 2012. Epidemiological analysis on visceral leishmaniasis in China during 2008–2011. Int. J. Med. Parasit. Dis. 39 (4), 223–226 (in Chinese).

Wang, J., Chen, C.Y., 1987. Intact promastigotes antigen for immunodiagnosis of kala-azar. End. Dis. Bull. 2 (2), 69–72 (in Chinese).

Wang, Z.J., Wu, Z.J., 1956. Kala-Azar. People's Health Publisher, Beijing, China, pp. 433–609 (in Chinese).
Wang, J., Xiong, G.H., Liu, P.Z., 1963. A bionomics study of Phlebotomus mongolensis in the desert area of Kansu Province, Northwest China. Acta Entomol. Sin. 12 (5–6), 679–687 (in Chinese).
Wang, J., Qu, J.Q., Guan, L.R., 1964a. A study on the Leishmania parasite of the big gerbil in Northwest China. Acta Parasitol. Sin. 1, 105–114 (in Chinese).
Wang, J., Xiong, G.H., Hu, Y.D., Liu, P.Z., Niu, H.F., Sai, S.Y., 1964b. The Leishmania infection found in Rhombomys opimus its transmission and relation to human. Acta Parasitol. Sin. 1 (1), 17–23 (in Chinese).
Wang, J., Peng, X.D., Guo, Y.K., Yang, G.Y., 1966. A preliminary survey of epidemicity of kala-azar in Xinjiang Urghur autonomous region. J. Epidemiol. 4 (1), 30–32 (in Chinese).
Wang, J., Bao, Y.F., Hu, Y.D., Qu, J.Q., Zhao, E.M., Jiao, L.E., Bai, R.L., 1981. Comparison of indirect hemagglution (IHA) with counter immuno-eletrophoresis (CIEP) to sero-diagnosis for kara azar. Acta Acad. Med. Sin. 3 (2), 74–75 (in Chinese).
Wang, J.Y., Qu, J.Q., Bao, Y.F., Yang, Y.T., Xu, Y.X., 1991. Detection of circulating antigen in kala-azar patients by monoclonal antibody-enzyme linked immunotransfer blot technique (EITB). End. Dis. Bull. 6 (3), 27–29 (in Chinese).
Wang, Z.J., Xiong, G.H., Guan, L.R., 2000. Achievement on the epidemiology and control of kala-azar in China. Chin. J. Epidemiol. 21, 51–54 (in Chinese).
Wang, J.Y., Chen, S.B., Gao, C.H., Jin, C.F., Feng, Y., Zhang, C.J., He, J.P., Yang, C.M., Yang, Y.T., Bao, Y.F., 2006. Survey on the *leishmania infantum* asymptomatic infection in dogs in Wenxian county of Gansu Province. Chin. J. Zoonoses 22 (8), 734–737 (in Chinese).
Wang, J.Y., Feng, Y., Gao, C.H., Jin, C.F., Chen, S.B., Zhang, C.J., He, J.P., Yang, C.M., Yang, Y.T., Bao, Y.F., 2007. Asymptomatic Leishmania infection in human population of Wenxian County, Gansu Province. Chin. J. Parasitol. Parasit. Dis. 25 (1), 62–64.
Wang, J.Y., Gao, C.H., Yang, Y.T., Chen, H.T., Zhu, X.H., Lv, S., Chen, S.B., Tong, S.X., Steinmann, P., Ziegelbauer, K., Zhou, X.N., 2010. An outbreak of the desert sub-type of zoonotic visceral leishmaniasis in Jiashi, Xinjiang Uygur autonomous region, People's Republic of China. Parasitol. Int. 59 (3), 331–337.
Wang, J.Y., Cui, G., Chen, H.T., Zhou, X.N., Gao, C.H., Yang, Y.T., 2012. Current epidemiological profile and features of visceral leishmaniasis in People's Republic of China. Parasit. Vectors 5, 31.
Wu, Z.J., 1957. Ecology of sandflies and control measures. Chin. J. Hyg. 5 (4), 231–239 (in Chinese).
Wu, Z.J., Wang, Z.J., 1956. Method of using dichlorodiphenyltrichloroethane (DDT) and hexachlorocyclohexane (HCH) to kill adult sandflies and its extension. Chin. J. Med. 42 (8), 713–733 (in Chinese).
Wu, Z.J., Wang, C.J., Ho, K.T., Wang, C., 1955. Observations on the bionomics of sandflies (Genus Phlebotomus) in East China from 1951 to 1953. Acta Entomol. Sin. 5 (4), 393–414 (in Chinese).
Wu, Y.X., Li, G.R., Liao, Q.Y., Chen, Y.F., Gao, B., 1995. Epidemiology and control of kala-azar in Sichuan Province. End. Dis. Bull. 10 (4), 25–29 (in Chinese).
Xiong, G.H., 1972. Current situation of kala-azar epidemic in China. End. Dis. Bull. 7 (3), 113–125 (in Chinese).
Xiong, G.H., Wang, J., Hu, Y.D., Liu, P.Z., 1963a. Experimental infection of P. alexandri and P. mongolensis with Leishmania donovani. Acta Zool. Sin. 15 (4), 607–610 (in Chinese).
Xiong, G.H., Wang, J., Hu, Y.D., Liu, P.Z., 1963b. Studies on the bionomics of Phlebotomus alexandri in Gansu Province. Acta Entomol. Sin. 12 (4), 458–462 (in Chinese).

Xiong, G.H., Guan, L.R., Chai, J.J., Guo, Y.K., 1964a. Observation on the bionomics of sandflies in Atushi, Sinkiang. Acta Parasitol. Sin. 1 (2), 195–200 (in Chinese).

Xiong, G.H., Hu, Y.D., Yang, K.Y., Huang, G.M., 1964b. Studies on gerbil Leishmania in Xinjiang: its sandfly vectors and relation to human. Acta Parasitol. Sin. 1 (2), 130–136 (in Chinese).

Xiong, G.H., Wang, J., Guan, L.R., 1964c. Two species of sandflies of subgenus Paraphlebotomus found in north-west China. Acta Entomol. Sinca 13 (1), 141–144 (in Chinese).

Xiong, G.H., Guan, L.R., Chai, J.J., 1965. The relation of Phlebotomus minutes sinkiangensis Ting & Ho 1962 to Leishmania infection in sinking, China. Acta Parasitol. Sin. 2 (1), 18–22 (in Chinese).

Xiong, G.H., Guan, L.R., Guo, Y.K., 1974. Studies on vectors of transmission of kala-azar in Xinjiang, China. J. Control Res. Epidemiol. Dis. 4, 327–334 (in Chinese).

Xiong, G.H., Hu, Y.D., Guan, L.R., Wang, J., 1976. The natural nidus of kala-azar in Northwest China. J. Control Res. Epidemic Dis. 1, 63–68 (in Chinese).

Xiong, G.H., Jin, C.F., Chen, X.Z., Hong, Y.M., Su, Z.W., Liu, P.Z., et al., 1992. Studies on the bionomics of sandfly, *Phlebotomus chinensis* Newstead, 1916, and relation to the visceral leishmaniasis in southern Gansu and northern Sichuan, China. J. Wuyi Sci. 9, 7–18 (in Chinese).

Xiong, G.H., Jin, C.F., Chen, X.Z., Su, Z.W., Hong, Y.M., 1994. Deltamethrin bath of domestic dog in the prevention of sandfly bite. End. Dis. Bull. 92 (2), 32–34 (in Chinese).

Xu, Y.X., Jia, J.X., Zuo, X.P., Wang, G., Hou, Y.Y., 1991. Leishmaniasis in Karamay V. observation on the bionomics of sandflies in Karamay, Xinjiang, China. End. Dis. Bull. 6 (2), 55–61 (in Chinese).

Yang, Y.T., Shi, F., Gao, C.H., Wang, J.Y., 2014. Establishment and evaluation of colloid gold labeled immunochromatographic strip test for rapid detection of visceral leishmaniasis. Int. J. Med. Parasit. Dis. 41 (3), 121–124 (in Chinese).

Yin, Z.C., Yu, F.Q., 1981. Experimental infection of *Phlebotomus chinensis* from mountainous area with *Leishmania donovani* from dogs. Bull. Control Res. Parasit. Dis. 9 (1), 49–50 (in Chinese).

Yin, Z.C., Wu, Y.X., Liu, P.Z., Guan, L.R., 1990. Study on the vertical distribution of canine visceral leishmaniasis in Wenchuan and Nanping counties, Sichuan Provinces. End. Dis. Bull. 5 (3), 25–40 (in Chinese).

Zhao, E.M., Chai, J.J., Li, B.S., Yima, M., Jiao, L.E., 1985. Epidemiologic survey of kala-azar in Bachu reclamation area, Xinjiang. J. Xinjiang Med. Coll. 8 (3), 259–262 (in Chinese).

Zuo, X.P., Chai, J.J., Guan, L.R., 1998. Forty-year research of Phlebotominae sandflies in Xinjiang, China (1958–1998). End. Dis. Bull. 3 (4), 97–102 (in Chinese).

Zuo, X.P., Deng, Z.B., Ali, M., Zhang, S., 2005. Study of the prevalence of visceral leishmaniasis in Wushi county of Xinjiang. End. Dis. Bull. 20 (1), 33–34 (in Chinese).

CHAPTER SIX

Engagement of the National Institute of Parasitic Diseases in control of soil-transmitted helminthiasis in China

Hui-Hui Zhu[a,b,c,d,e], Chang-Hai Zhou[a,b,c,d,e], Mi-Zhen Zhang[a,b,c,d,e], Ji-Lei Huang[a,b,c,d,e], Ting-Jun Zhu[a,b,c,d,e], Men-Bao Qian[a,b,c,d,e], Ying-Dan Chen[a,b,c,d,e,*], Shi-Zhu Li[a,b,c,d,e], Xiao-Nong Zhou[a,b,c,d,e]

[a]National Institute of Parasitic Diseases, Chinese Center for Disease Control and Prevention, Shanghai, People's Republic of China
[b]Chinese Center for Tropical Diseases Research, Shanghai, People's Republic of China
[c]WHO Collaborating Centre for Tropical Diseases, Shanghai, People's Republic of China
[d]National Center for International Research on Tropical Diseases, Ministry of Science and Technology, Shanghai, People's Republic of China
[e]Key Laboratory of Parasite and Vector Biology, Ministry of Health, Shanghai, People's Republic of China
*Corresponding author: e-mail address: chenyd@nipd.chinacdc.cn

Contents

1. Introduction 218
2. Epidemiological surveys and national surveillance system 219
   2.1 National survey on soil-transmitted helminthiasis 219
   2.2 Establishment and improvement of the national surveillance system on soil-transmitted helminthiasis 227
3. Control pilots and new strategy development 229
4. Applied research on diagnostics and drugs 233
   4.1 Applied research on diagnosis 233
   4.2 Research on drugs application 235
5. Future perspectives 236
6. Conclusions 238

Acknowledgements 241
Competing interests 241
Authors' contributions 241
References 241

Abstract

Soil-transmitted helminthiases (STHs) have been widely transmitted in China and the control of STHs was initiated by NIPD-CTDR since its foundation. Three national surveys on STHs have been carried out in China, and the infection rate has dropped from 53.58% in the first national survey (1988–92) to 4.49% in the third national survey (2014–16)

Advances in Parasitology, Volume 110
ISSN 0065-308X
https://doi.org/10.1016/bs.apar.2020.04.008

due to strong interventions including mass drug administration, health education and environment improvement. National surveillance of STHs started in 2006 and has been implemented successively until now, which allows to understand the endemic status and trends of STHs prevalence in China. Surveillance has been expanded to 30 provinces of China since 2016. Integrated pilot programmes have been implemented between 2006 and 2009, in which an integrated strategy, with health education and control of infection sources as key components, was adopted. Since 2019, new control pilots have been started, which will be continued for five successive years to further explore appropriate control strategies in the current "new era". With the decline of infection rate of STHs, China is approaching the elimination stage for STHs. In order to achieve this final target, poverty alleviation programmes should be integrated with precise control measures, according to real situations.

1. Introduction

Soil-transmitted helminths (STHs) are nematodes that do not require intermediate hosts for their development. Their eggs or larvae have the capacity to infect the human body directly from the environment (mainly soil). Soil-transmitted helminths include hookworm, *Ascaris lumbricoides* and *Trichuris trichiura* (Jourdan et al., 2018; Tang et al., 2012). Infections due to STHs cause serious harm to human health, especially in children, as it retards their growth and development, leading to malnutrition and developmental disorders (Bethony et al., 2006; Brooker et al., 2008). STHs account for the largest number of infections among neglected tropical diseases (NTDs), with a disease burden of 3.38 million disability-adjusted life years (DALYs), out of which the disease burdens of hookworm, *A. lumbricoides* and *T. trichiura* infections are 1.76 million, 1.08 million and 0.54 million, respectively (GBD 2015 DALYs and HALE Collaborators, 2016). These important parasitic diseases are mainly endemic in the developing regions of Asia, Africa and Latin America (Chammartin et al., 2013; Karagiannis-Voules et al., 2015; Silver et al., 2018). An updated national survey was implemented in China in 2015, which was coordinated by the National Institute of Parasitic Diseases at China CDC and Chinese Center for Tropical Diseases Research (NIPD-CTDR). The results of the survey estimated that about 29.12 million people were infected with STHs in the country, out of which 16.97 million, 8.83 million and 6.60 million people were infected with hookworm, *A. lumbricoides* and *T. trichiura*, respectively (Zhou, 2018).

STHs have been prevalent in China for a long time, but no detailed investigation was carried out until 1950 by the Eastern-China Branch of the Chinese Academy of Medicine, by then the main institute for parasitic disease control in China and a predecessor of NIPD-CTDR. Thereupon the central government, prioritized hookworm control as of 1956 onwards with the announcement of the "National Program for Agricultural Development from 1956 to 1967 (draft)" in which hookworm was listed as one of the five important parasitic diseases for urgent control efforts. The Hookworm Disease Research Laboratory was established by the Institute of Parasitic Diseases (IPD) at the Chinese Academy of Medical Sciences in the beginning of 1958, but research work on hookworm disease was also carried out by the Pathogen Biology Research Laboratory and the Pharmacology Research Laboratory of the IPD. Between 1983 and 2001, the IPD went through two developmental stages, i.e. the IPD at Chinese Center for Preventive Medicine and the IPD, Chinese Academy of Preventive Medicine. These were the stages before the institutes name was officially changed to the NIPD under the Chinese Center for Disease Control and Prevention in 2002. The name of Chinese Center for Tropical Diseases Research was added in 2017 (hereinafter referred to as "NIPD-CTDR"). The department of soil- and food-borne parasitic diseases was established and entrusted with the responsibility of effective prevention and control of soil-transmitted and food-borne parasitic diseases in cooperation with other departments in NIPD-CTDR (Yu and Zhou, 2013).

This paper summarizes the activities and progress made on STH prevention and control in NIPD-CTDR, so as to provide references to optimize STH control efforts in the future.

2. Epidemiological surveys and national surveillance system

2.1 National survey on soil-transmitted helminthiasis

Three national surveys have been carried out in China, in the 1990s, 2000s and 2010s, respectively, in which STHs were all included.

2.1.1 The national survey on distribution of human parasites

In 1986, the former Ministry of Health issued a "Notice on the National Survey of Human Parasites Distribution". The national survey on distribution of human parasites (hereinafter referred to as "the first national survey") were implemented between 1988 and 1992 under the leadership of the former Ministry of Health (Yu et al., 1994a). Investigations were carried out in 30 provinces, autonomous regions and municipalities (hereinafter referred to as "provinces") in mainland China and 1,477,742 persons were investigated.

Stratified cluster random sampling was employed in situations with large number of parasite species, large sample sizes, various local conditions that enhance STHs transmission, and areas with no prior survey reports. Three steps random sampling was strictly followed in the sampling process and control measures were designed to reduce sampling errors.

Great achievements were made in the first national survey. The main species, distribution and prevalence of human parasites were ascertained in China. Overall STH infection rate was 53.58% and the estimated number of infected populations was 536 million. The infection rates of *A. lumbricoides*, hookworm and *T. trichiura* were 47.00%, 17.17% and 18.80%, with estimated infected populations of 531 million, 194 million and 212 million, respectively. High endemic areas were mainly distributed in the eastern and southern parts of China (Fig. 1). The survey also revealed

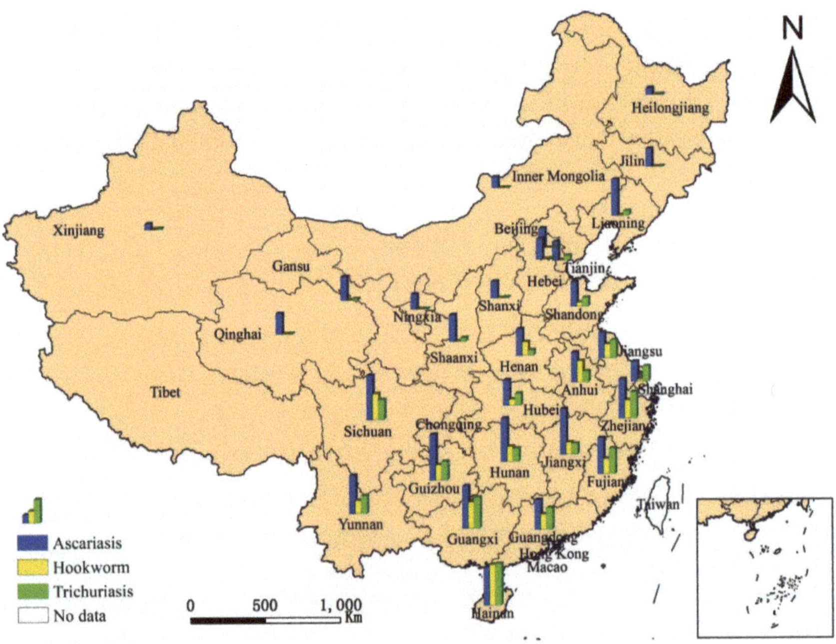

Fig. 1 Distribution of STHs in China according to the first national survey. *Data source: according to data from literature Yu S.H., Xu L.Q., Jiang Z.X., Xu S.H., Han J.J., Zhu Y.G., Chang J., Lin J.X., Xu F.N., 1994a. Report on the first nationwide survey of the distribution of human parasites in China. 1. Regional distribution of parasite species. Chin. J. Parasitol. Parasit. Dis. 12(4), 241–247.*

that mixed infections of the two hookworm species (*Necator americanus* and *Ancylostoma duodenale*) occurred in most regions of China, but with varying proportions across regions. Generally, *A. duodenale* was distributed in the northern region whereas *N. americanus* dominated the southern region (Tang et al., 2012; Xu et al., 2000). It was also discovered during the national survey that STHs infections varied in different climatic and geographical zones in China. The tropical region showed the highest infection rate, followed by subtropical zones, warm temperate zones, special temperature zones, intermediate temperate zones and cold temperate zones. Also, infection rates were highest in wet areas, followed by semi-humid areas, while the infection rate in semi-arid and arid areas was low. Finally, the STH prevalence trend in China was revealed with the implementation of the national survey. The first national survey created a significant pathway that led to effective prevention and control in China, which was reflected in the following aspects.

The results obtained from the survey, provided evidence-based references to the development of the "Strategic Target of Preventive Health Care in China (2000)", the "National Eighth Five-Years Programme (1995–2000)", the "National Ninth Five-Years Programme (2001–10)", and the "National Tenth Five-Years Programme (2011–15)" on parasitic diseases control and the National Control Programme on Key Parasitic Diseases (2016–2020). The range of control work was successively expanded and infection rates of STHs decreased from 1995 to 2016. During this period the tasks for parasitic diseases control were always listed in the national control programme by the National Health Commission of China.

Control strategies and recommendations were put forward based on the findings from the first national survey, which helped to develop the "Implementation Plan on Control of Common Helminthiasis" by the former Ministry of Health, and promote the national parasitic disease control programmes. Through the national parasitic diseases control programme initiated in 1992, a total of 120 million people received anthelmintic treatment in 1994, 80% of the patients had a reduction in intensity of infection, and more than 14 million people were cured.

Children were considered as the key targets in the national parasitic diseases control programme, according to the results from the first national survey. The number of infections with *A. lumbricoides*, hookworm and *T. trichiura* was 190 million, 70 million and 40 million, respectively, among 310 million children under 14 years old in China. In view of this situation, a document entitled "National Control Program for Student's Helminth

Infection" was jointly issued by the former Ministry of Health and the former National Education Commission, and collective deworming in the students from elementary and middle school was conducted nationwide since 1992. STHs infection rates in school-age children have decreased significantly following 10 years of effective deworming programme.

The parasitic control institutions in the country became more established with high-quality professionals. Before the first national survey, the specialized institutions for parasitic diseases control in the southern regions were tending to reduce staff, or even to abolish few of these institutions, which prompted professionals to consider changing career paths following the control of schistosomiasis, malaria, or filariasis. The timely implementation of this survey prompted local governments to retain or strengthen capacities in specialized institutions for parasitic disease control in each province in southern China. In addition, capacities of staff were enhanced by putting them through adequate training while pilot surveys were implemented. A total of 7404 professional technicians across the country were enrolled for 504 training courses to improve their knowledge on STH epidemiology and control. This strategy provided a platform to train considerable number of professionals, and many senior and middle-level professionals have updated their knowledge and mastered new technologies through training, which prompted parasitic control accordingly.

Also, in-depth development of scientific research was promoted through the execution of the survey, which provided insights to raise many research questions and led to many proposed research projects. Pharmaceutical companies and other health service providers that envisioned huge market opportunities from the survey invested heavily in developing new drugs and diagnostic tools for the prevention and treatment of parasitic diseases both domestic and abroad.

Finally, effective implementation of the survey and the national parasitic diseases control programmes have further enhanced China's reputation globally. This has also increased both domestic and international cooperation and exchanges, which prompted knowledge sharing and transferring. The article of the "National survey of human parasitic diseases in China" was published as a special report in the Southeast Asian Journal of Tropical Medicine and Public Health (Yu et al., 1994b), and was appraised highly by this journal in term of the huge sampling size and quality of the survey. The journal opined that it might be difficult to achieve similar objective elsewhere around the world. The wealth of information contained in the report have been crucial in making informed decisions and this with other notable achievements have enhanced Chinas reputation in the global health community.

In addition, the first national survey received the "First Prize of the Major Medical Science and Technology Award by Ministry of Health in 1994" and the "Second Prize of the National Science and Technology Progress Award in 1995". Also, the book entitled "Color Atlas of Human Parasitology" (Yu and Xu, 1992) was compiled and published by experts using samples obtained from the survey, and has become an important reference book for staff in disease control and teaching agencies. A large amount of data was obtained from the survey and an important review titled "Distribution and pathogenic impact of human parasites in China" (Xu et al., 2000) was published. This book was one of the most cited books in parasitology for 10 years after its publication. Moreover, a special edition of Chinese Journal of Parasitic Diseases was published based on survey of many provinces (special edition on Survey of Human Parasites, 1991).

The first national survey initiated a new era in China's STH control and research. In 2010, WHO's first Global Report on Neglected Tropical Diseases (NTDs) classified 17 types of diseases as NTDs, including 14 soil-borne, food-borne or arbo-parasitic disease (WHO, 2010). The results of the first national survey led China to prioritize "NTDs" 15 years before WHO.

2.1.2 Survey on the status of human key parasitic diseases in China

The national survey on the status of human key parasitic diseases in China (hereinafter referred to as "the second national survey") was implemented between 2001 and 2004, covering 31 provinces across China (Hong Kong, Macao and Taiwan not included) (Xu et al., 2005). NIPD-CTDR was responsible for organizing the survey and also summarizing the survey report for the second national survey, in which a total of 356,629 people were investigated.

Findings from the second national survey indicated that STHs infection rate was 19.56% and the estimated population infected was 129 million. The infection rate of hookworm, *A. lumbricoides* and *T. trichiura* was 6.12%, 12.72% and 4.63%, while the estimated population infected was 39.3 million, 85.93 million and 29.09 million, respectively (Wang, 2008). Highly endemic areas were mainly distributed in the south and southeast part of China (Fig. 2).

This survey was of great significance because STHs prevalence and distribution characteristics were updated. The survey showed that parasite

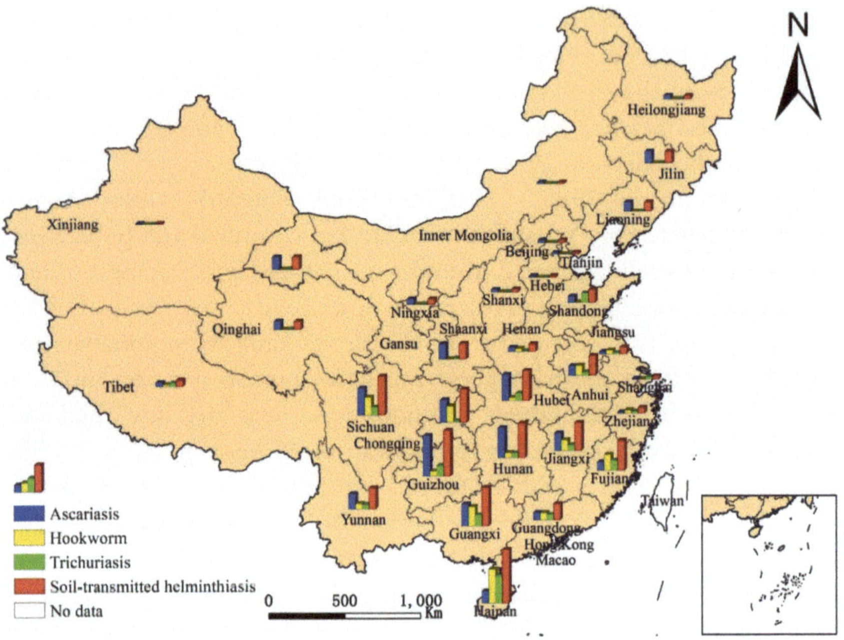

Fig. 2 Distribution of STHs in China according to the second national survey. *Data source: according to data from literature Xu, L.Q., Chen, Y.D., Sun, F.H., Cai, L., Fang, Y.Y., Wang, L.P., Liu, X., Li, L.S., Feng, Y., Li, H., 2005. A national survey on current status of the important parasitic diseases in human population. Chin. J. Parasitol. Parasit. Dis. 23, 332–340.*

infections were most severe in the western part of China, with high infection rates among women and children. Compared with the first national survey, the infection rate of hookworm, *A. lumbricoides* and *T. trichiura* has decreased significantly, and the number of infected persons also declined appreciably. The second survey also helped stakeholders to evaluate the effectiveness of STHs prevention and control in the past 10 years. Compared with the first national survey, the infection rate, number of infected persons and the number of STHs multi-species infections have all decreased significantly, indicating the success of 10 years' massive deworming in elementary and middle school in rural areas, together with the integrated control measure of health education, improved access to potable water sources and sanitary facilities. Lastly, the severity of STHs was objectively evaluated during the second survey. Findings indicated the proportion of hookworm and *A. lumbricoides* infection in fatality among hospitalized patients with parasitic diseases was 0.17% and 0.60%, respectively. Misdiagnosis of STHs remained very high, which caused high burden.

This survey provided the basis to optimize the strategies and develop effective prevention and control measures. To further promote the control on STHs in China, the establishment of a national surveillance network for STHs started. In 2006, a total of 22 national surveillance spots were established by the former Ministry of Health. Meanwhile, provincial surveillance spots were set up by many provinces, thus leading to the establishment of STHs surveillance network in China. STHs control strategies were proposed based on the prevalence of each endemic area. These included, (i) implementation of collective deworming for areas with infection rate >20% in prevalence, to reduce infection rate and intensity; (ii) adoption of counter-measures such as deworming of key populations to reduce infection rate in areas with 5–20% in prevalence; (iii) selective deworming for areas with <5% in prevalence. The National Control Plan for Major Parasitic Diseases (2006–15) was initiated by the former Ministry of Health during the 12th Five-Year Plan following the second survey. Fourthly, the infection rate of *A. lumbricoides* was first included in the evaluation index system for national hygiene in city (town) by the National Patriotic Health Campaign Committee in 2005. In addition, STHs epidemic status and control implementation were included in the National Health Statistics Survey System for the first time. Comprehensive STHs demonstration plots were established. In 2006, the "Working Plan for parasitic diseases comprehensive demonstration plots" was established by the former Ministry of Health, and eight STHs comprehensive demonstration plots were set up in highly endemic areas. Finally, this survey paved the way for the development and use of technology in control, thereby, making up for shortage of technical staff for parasitic disease control.

2.1.3 National survey on current status of major human parasitic diseases

The former National Commission for Health and Family Planning organized the third National Survey on Current Status of Major Human Parasitic Diseases (hereinafter referred to as the "the third national survey") between 2014 and 2016. The survey covered 31 provinces in mainland of China (Hongkong, Macao and Taiwan not included), and 484,210 people from rural areas were investigated.

Findings from the third national survey indicated that STHs infection rate was 4.49%, with an estimated population of 29.12 million under infection. The prevalence of hookworm was 2.62%, *A. lumbricoides* 1.36% and *T. trichiura* 1.02%. Correspondingly, the population under infection was 16.97 million, 8.82 million and 6.60 million, respectively

(Karagiannis-Voules et al., 2015). Highly endemic areas were still distributed in the south and southwest part of China (Fig. 3).

The endemic status of STHs in the new era was revealed by the third national survey. Results showed that STHs infection rates in rural areas have dropped to 4.49%, yet areas with high endemicity still existed. Low endemic or sporadic status was shown in most areas and endemic areas were evidently shrunk. Yet, STHs infection rate in some provinces or regions were still high, even up to 20%. Also, dominant species in STHs were altered due to control effort as indicated in the results obtained. The order of dominant species in this survey were hookworm, *A. lumbricoides* and *T. trichiura*, with infection rates of 2.62%, 1.36% and 1.02%, respectively. The order of dominance was *A. lumbricoides* (47.00%), *T. trichiura* (18.80%) and hookworm (17.17%), while the order in second survey was *A. lumbricoides* (12.57%), hookworm (6.08%) and *T. trichiura* (4.56%). Moreover, the regional distribution has been illustrated. Meso-endemic areas with infection rate >5% and hyper-endemic areas >20% were mainly

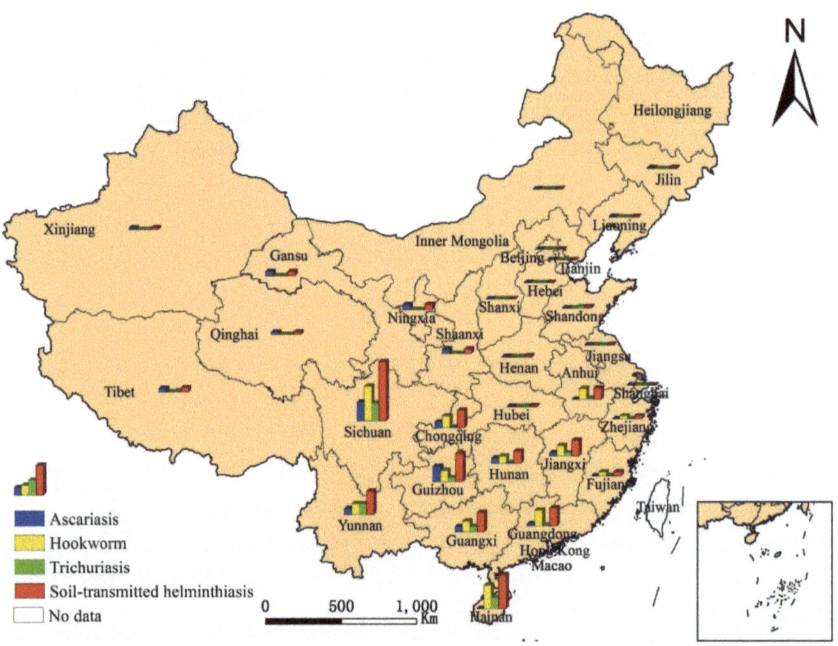

Fig. 3 Distribution of STHs in China according to the third national survey. *Data source: according to data from literature Zhou X.N., 2018. Report on the National Survey of Important Human Parasitic Diseases in China (2015). People's Medical Publishing House, Beijing.*

distributed in the southern and southwestern parts of China including Sichuan, Hainan, Guizhou, Yunnan, Chongqing, Guangxi, Guangdong and Jiangxi. The third national survey also revealed that the population infected with STHs was still huge. Hence, the disease burden could not be neglected.

Suggestions for the next STHs control stages were proposed following execution of the third national survey. It was opined that prevention and control efforts should be maintained, while the control plan for major parasitic diseases should be further implemented. Also, multi-sectoral co-operation was to be strengthened and comprehensive measures to be implemented. Furthermore, prevention and control key points should be emphasized, and precise control measures be adequately implemented. Lastly, there was a need to improve the comprehensive surveillance system and also pursue regular execution of the national survey to adequately inform decisions for control plan and implementation.

The third national survey played an important role in promoting STHs prevention and control in the new era. This survey did not only provide detailed information for the "National Control Program for Major Parasitic Diseases (2006–15)" evaluation, but also helped to lay a solid foundation for the implementation of the "National Control Program for Echinococcosis and other Important Parasitic Diseases (2016–2020)"; This also enhanced the effectiveness of the national surveillance system in the country. The STH national surveillance system was greatly expanded from 22 surveillance spots in 22 provinces to more than 200 surveillance spots in 30 provinces after the survey (Chen and Zang, 2013). Lastly, new national STHs control pilots were established based on this survey, and 12 STHs pilots have been being conducted in 9 provinces since 2019.

2.2 Establishment and improvement of the national surveillance system on soil-transmitted helminthiasis

STHs surveillance was included in the national surveillance system for key infectious diseases and vectors by China CDC in 2006, while the national surveillance programme for STHs was issued by the former Ministry of Health (Chinese Center for Disease Control and Prevention, 2006). This led to the establishment of 22 national surveillance spots of STHs, which monitors STHs endemic status and environmental transmission factors. The drafting of the national surveillance programme for STHs, staff training, annual checks, data analysis and surveillance report were all undertaken by NIPD-CTDR. In addition, an "Operation Manual for STHs

National Surveillance Program" was also provided to ensure surveillance quality (Chinese Center for Disease Control and Prevention, 2011).

In total, 1000 people were investigated at each surveillance spot, and more than 20,000 people were investigated nationally every year. The results showed that STHs infection rates from 2006 to 2010 were 20.88%, 18.93%, 16.59%, 13.30% and 11.25% respectively, which indicated a year-by-year declining pattern. The surveillance programme was revised in 2011 following the surveillance situation and the operation manual was modified accordingly (Chinese Center for Disease Control and Prevention, 2014). STHs infection rates at the national surveillance spots from 2011 to 2015 were 9.67%, 6.90%, 3.12%, 4.49%, and 4.95% respectively (Chen and Zang, 2013; Chinese Center for Disease Control and Prevention, 2015; Zhu et al., 2019), indicating continuing decline in the endemic trends. Establishment of provincial surveillance spots were promoted by the 22 national surveillance spots, and 158 provincial surveillance spots were set up by the end of 2015 (Zhu et al., 2019). People's awareness on parasitic disease control was improved through the implementation of the surveillance work.

STHs surveillance was included in the "Central Transfer Payment for control of Malaria and Other important Parasitic Disease" in 2016 by the former National Health and Family Planning Commission. All the provinces in mainland of China were required to conduct STHs surveillance, and in each P/A/M, the number of surveillance spots (counties) implemented each year should be more than 10% of all counties. STHs surveillance spots were more than 200 annually from 2016 to 2018 (Zhu et al., 2019), and the surveillance range has been continuously expanded. Surveillance results obtained between 2016 and 2018 showed that STHs infection rates were 2.46% 1.78%, 1.29% respectively, and this is the lowest point ever achieved in the history of STHs control efforts in China (Chinese Center for Disease Control and Prevention, 2016, 2018; Zhu et al., 2019).

Since 2017, data from national surveillance has been reported through the "Information Management System for Parasitic Diseases". This system is of great significance for the management of surveillance data, and it has been successively improved through usage, with faster report speed and more restrict logistic limitation. In addition, the annual surveillance together with the national competition on parasitic disease control had provided opportunities for professionals at provincial, city and county levels. Training and selection were carried out every year, and a proportion of grass-roots professionals were trained, thus improving their skills, thereby, becoming backbones of grass-root disease control agencies.

3. Control pilots and new strategy development

After the first national survey, the large-scale population-based deworming was implemented in China according to the high endemic status of STHs in China revealed, and the infection rate of STHs was significantly decreased. After the second national survey, the comprehensive control strategy of "health education as leadership, control of infection source as priority" explored by demonstration plots were carried out (Fig. 4).

In 2006, the "Work Program for Comprehensive demonstration plots of Parasite Diseases" was initiated by the former Ministry of Health to speed up control implementation in highly endemic areas, and eight comprehensive demonstration plots for STHs were established (Disease control agency of the Ministry of Health, China CDC, 2010) (Figs. 5 and 6). This control initiative was implemented for three successive years and completed in 2009. All tasks regarding control implementation including baseline investigation, selection of deworming drugs, drawing deworming plan, quality control, mid-term and final evaluation, data collection and data analysis were undertaken by NIPD-CTDR.

The control strategies of "health education as leadership, control of infection source as priority" and the control measures of "improving water,

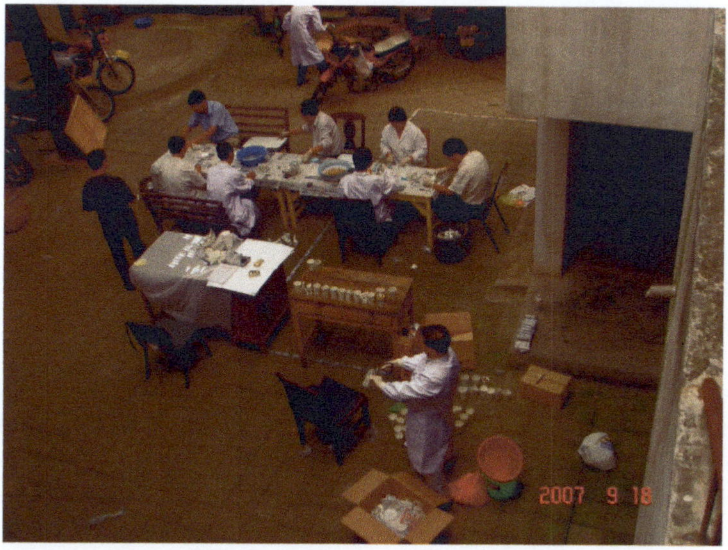

Fig. 4 Control strategies of STHs in different stages of China.

Fig. 5 Sample examination in comprehensive demonstrating zones of STHs.

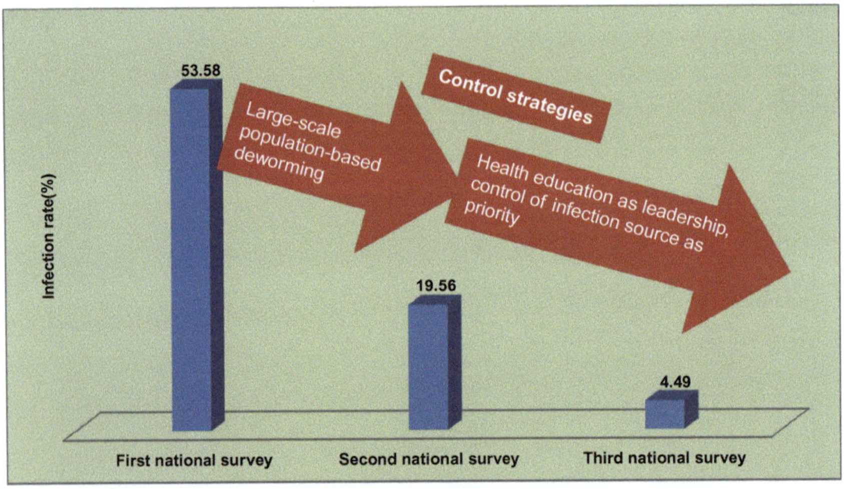

Fig. 6 Drug dispensing in comprehensive demonstrating zones of STHs.

lavatory and hygiene in conjunction with behavioural changes and drug administration" were implemented in the demonstration plots. Under the above guidelines, health education activities of various forms and contents were carried out in all comprehensive demonstrating plots. Mass media of radio, television, newspapers and internet were utilized, and seminars, training, school lectures, door-to-door propaganda, and posters in public places were also used for public education, the harm, transmission routes, and prevention measures of STHs were all publicized. Professionals at all levels were also trained in the way of cascade training. Also, drug administration was implemented according to the prevalence level in comprehensive demonstrating plots from 2006 to 2008. In towns with prevalence more than 50%, mass drug administration was implemented twice in 2006, and once in 2007 and 2008 respectively; In towns with prevalence between 20% and 50%, the masses were guided to deworm voluntarily once a year in 2006–08; In towns with prevalence between 10% and 20%, high risk population were guided to deworm voluntarily once a year in 2006–08, other people were encouraged to have voluntary examination and treatment; In towns with prevalence less than 10%, all people were encouraged to have voluntary examination and treatment. Moreover, the water, lavatory and hygiene of environment were improved in the comprehensive demonstrating plots under the leadership of government.

In 2010, the final evaluation of the 3 years control implementation on demonstration plots was assessed by NIPD-CTDR. From 2006 to 2009, deworming drug was administrated to 6.43 million person-times, and health education was given to 4.23 million person-times while the control initiative lasted. STHs infection rates in demonstration plots, Class I, II, and III were 55.85%, 31.16%, and 19.14% and was reduced by 75.63%, 79.52% and 87.08%, respectively. The outcome of control efforts exceeded the original targets. Meanwhile, living conditions of people and sanitary facilities in demonstration plots were significantly improved. Also, water and toilet reforms have been strengthened, with enhanced public's awareness to imbibe strong personal hygiene culture and improved knowledge on STHs parasites transmission routes to curtail transmission (Zhang et al., 2011).

Driven by the achievements made with the demonstration plots control initiative, a total of 99,761,380 person-times of health education were conducted nationwide, 97,658,472 person-times of STHs deworming were carried out and 390,506 person-times of professionals were trained at all levels. Decline in STHs infection rates in 23 provinces reached the target of the National Control Program for Key Parasitic Disease (2006–15) target, that is, provinces with infection rates $\geq$20%, 20% $>$ infection rate $\geq$ 5%, and $<$5% according to the second national survey decreased by more than 80%, 70% and 60%, respectively (Chen et al., 2019).

The successful experience of the comprehensive demonstration plots has explained the control strategies of "health education as leadership, control of infection source as priority" for effectively controlling STHs in China, and health education was one of the most important measures deployed to achieve this feat. A series of health education activities popular with the masses was carried out by making full use of the available resources in various places to enhance people's awareness on disease prevention and treatment, improve their compliance, thus making control efforts more effective. Experts were organized to develop a promotion plan for the propagation of the demonstrating plots control initiative experiences (Wang, 2011).

Experts from NIPD-CTDR participated in the formulation of the "Technical Plan for the Prevention and Control of STHs" in 2009, adopting "health education as leadership, control of infection source as priority" and the control measures of "improving water, lavatory and hygiene in conjunction with behavioural changes and drug administration" strategies in the demonstration plots. The endemic areas were divided into three categories in terms of STHs parasites infection rates such as hookworm, and drug administration plan was determined accordingly. In areas predominantly endemic with hookworm, knowledge of "footwears when farming and no use of fresh manure as fertilizer" was propagated in health education, and the tribendimidine, independently produced by NIPD-CTDR, was listed in the priority drugs.

The National Program for Prevention and Control of Key Parasitic Diseases (2016–2020) proposed the establishment of 1–2 control pilots in each of the key endemic provinces to achieve the planned targets. With implementation of the Poverty alleviation Project in China, the living conditions of people in poverty-stricken areas were improving, yet the prevalence of STHs in many areas were still high. So it was decided that the control of STHs be combined with the poverty alleviation project. The main purpose of the control pilots was to explore and establish sustainable control mechanisms and efforts in local endemic settings to gradually reduce parasitic diseases infection rates, so as to ensure that the programme goals are achieved, and the achievements on disease control and poverty alleviation were expanded and consolidated. The criteria for the selection of pilot areas include high STHs endemicity, poor or underdeveloped county, need for improvement on access to good water sources and lavatory, and zeal by the local government to achieve parasitic disease control objectives. Based on the result of the third national survey and the poverty alleviation process in each county, 12 counties (cities) from 9 provinces were determined as the new control pilots of STHs by NIPD-CTDR, which were all poverty-stricken areas with high prevalence of STHs.

The control pilots (counties) commenced in 2019 with 5 years completion target, and new areas for control pilots will be selected 5 years later. Following the success achieved with STHs control in demonstration plots between 2006 and 2009 with the experiences gained using combinations of control strategies. Control efforts have been strengthened. Specifically, the following aspects will be implemented in the new control pilots: in-depth health education, effective patient investigation and treatment, comprehensive improvement of water and toilet, strengthening capacity building for disease control, and enhancing information management.

4. Applied research on diagnostics and drugs

4.1 Applied research on diagnosis

4.1.1 Improved Kato-Katz thick smear method

In 2018, experts in NIPD-CTDR formulated the health industry standard (WS/T 570-2017) patent of "Detection of intestinal helminths-The Kato-Katz method for Intestinal Worm Detection", which was issued in 2017 but officially implemented on 1 February 2018. Moreover, "Diagnosis of hookworm disease" (WS 439-2013) was officially implemented on 1 December 2013. Similarly, Diagnosis of enterobiasis (WS 469-2015) and "Diagnosis of ascariasis" (WS/T 565-2017) were formally implemented on 1 September 2015 and 1 February 2018, respectively.

4.1.2 Little tube counting method

Experts from NIPD-CTDR established the little tube counting method other than the commonly used saturated saline flotation method and the modified Kato-Katz staining smear method. This method was compared with the Hong's and Stoll's method using 13 positive stools samples, and results obtained showed that there was no significant difference in the three methods in terms of total number of eggs, but the fluctuation range in the little tube counting method (33.3%~281.5%) was smaller than the others. Infection intensity was divided into light infection (number of eggs: 0–399), moderate infection (400–2999) and heavy infection (above 3000), according to the Kato-Katz method, while 30 stool samples with light infection, 58 with moderate infection, and 18, with heavy infection were counted by the little tube counting method and Hong's method respectively. The number of eggs detected by the little tube counting method was 143.5% ($P<0.05$) compared with Hong's method, and there was no significant difference between these two methods with the moderate and heavy infected faecal samples. Therefore, the little tube counting method was not only accurate in counting hookworm eggs, but also easy to carry. The quantity of saturated saline and faecal samples required are much less than those in Hong's method, saving 1/3–1/2 of the examination time (Wang and Chen, 1959).

4.1.3 Isolation of hookworm larvae and autogenous nematodes

Faecal samples collected from manure pit are often mixed with autogenous nematodes (nematodes that live in the pit), which affects the counting accuracy of hookworm larvae. Therefore, it is necessary to devise a method to kill the autogenous nematodes without affecting hatching and development of hookworm larvae. Through multiple experiments, it was found that the double-layer cotton paper is best for separating hookworm larvae, with a formaldehyde concentration of 0.7%, and the effects were best if the samples were prepared in 6% salt water or treated in 6% formaldehyde solution for 30 min before cultivation (Yu and Zhou, 2013).

4.1.4 New methods for isolation of hookworm larvae

Isolate hookworm larvae from the soil are important to understand the source of hookworm infection and, also assess the effectiveness of hookworm disease control. In the past, the method of separating hookworm larvae was difficult to apply because of the complicated operations and heavy equipment needed. To solve this problem, experts from NIPD-CTDR developed a new method. The operation steps are as follows; soil samples were place on a cotton paper in a separating device, and the separating device was loaded in a container, then salt water with temperature 42–48 °C and 4.8–5.2% concentration were added to the soil without passing the surface of the cotton paper, then, the load container was placed in an incubator and kept in 42–48 °C for 1.5–2 h. Thereafter, the separating device was taken out, and the liquid in the container was decanted. Natural sedimentation was obtained by decanting the supernatant after 15–20 min. This new method was commonly used in the national surveillance of STHs (Chinese Center for Disease Control and Prevention, 2011; Yu and Zhou, 2013).

4.1.5 Quantitative method for hookworm larvae culture with glass tube and filter

The glass tube filter paper method is often required in parasites investigation and drug efficacy assessment to culture hookworm larvae for identification and counting. Repeated tests had shown that the detection rate of glass tube filter paper method developed by NIPD-CTDR is higher than that of Hong's method. The glass tube filter paper method can determine the degree of infection and identify the parasites species with convenient operation and high accuracy (Hookworm Research Laboratory, Shanghai Institute of Parasitic Diseases, 1977). It was widely used in national epidemiological investigation and drug efficacy evaluation for a long time.

4.2 Research on drugs application

Between the 1960s and mid-1970s, pyrantel, levamisole, albendazole and mebendazole were developed abroad successively. These drugs were then synthesized in China and, after pharmacological and toxicological tests, used in clinical treatment of intestinal helminth infections (Xiao, 2009a; Xiao et al., 1990, 2002; Zhang et al., 2019). Other than that, in the mid-1980s, researchers in NIPD-CTDR independently synthesized the new drug tribendimidine and conducted a series of acute and subacute toxicity tests (Xiao et al., 2004, 2005). It was found that in treating *Nippostrongylus brasiliensis*, *Ancylostoma caninum* and *Necator americanus*, smaller dose and shorter courses of treatment were needed for tribendimidine when compared with mebendazole and pyrantel pamoate (Xiao et al., 2004). Also, through pharmacokinetics, tribendimidine was shown to have stable chemical properties, a wide deworming spectrum, a fast deworming effect, a low dose required and low toxicity for the effective dose, as advantages. It was therefore recommended for clinical use (Xiao et al., 2005, 2013). In 2004, the new drug certificates and production licences for tribendimidine was issued by State Food and Drug Administration (SFDA) (Zhang et al., 2019), and then clinical trials were conducted by the NIPD-CTDR (Xiao et al., 2007, 2013; Zhang et al., 2008, 2019). From 2005 to 2007, a clinical experiment was carried out on 1292 cases infected with intestinal nematodes in Hainan, Sichuan and Guizhou, and the effectiveness and safety of tribendimidine on hookworm disease and ascariasis were evaluated. The result showed that under the used dosage for expanding treatment in larger sample of patients with various ages and infected with *A. duodenale*, *A. lumbricoides* and other helminths, tribendimidine enteric coated tablet is safe with satisfactory efficacy (Zhang et al., 2008). Also, from 2006 to 2007, an open and multi-centre clinical trial was conducted in the provinces of Hainan, Sichuan and Guizhou to evaluate the efficacy and safety of tribendimidine in treatment of children with hookworm and

A. lumbricoides infections. A total of 899 children aged 4–14 years were enrolled in the study. The results showed that tribendimidine gave at a single dose of 200 mg exhibits lower adverse effect rate and potential efficacy in the treatment of children with hookworm and *A. lumbricoides* infections (Xiao et al., 2007).

5. Future perspectives

STH prevention and control in China has evolved over nearly 70 years and the control strategy has developed from large-scale population-based deworming to comprehensive control strategy explored by demonstration plots (Fig. 4). Through the implementation of effective control efforts STHs infection rate in China has dropped from 53.58% in the first national survey to 4.49% in the third national survey. Also, we achieved the lowest STHs infection rate in the country's history, and infected population has been greatly reduced, meanwhile, majority of the population are implicated with mild infections. Hence, STHs control has achieved a remarkable feat.

STHs epidemic also shows new features in the new period. Firstly, its epidemic is characterized by regionalization. Results obtained from national surveillances and surveys indicated that >5% infection rate are mainly distributed in the southern and southwestern parts of China. Climate condition in these areas is warm and humid, and most of the areas are underdeveloped. Vegetable farms are still common in rural areas, making them suitable for STHs (such as hookworm) transmission. Secondly, the distribution characteristics of population are significant. Data showed that STHs infection rates (hookworm, *A. lumbricoides* and *T. trichiura*) were all higher in female than that of male, and was highest in age group 60 years and above than in any other age groups. These age group are mainly engaged in agricultural production in rural areas, and their hygiene and habits while on their farms could hardly be changed through simple health education. Thirdly, the dominant STH species have changed. Currently, hookworm infection rate are higher than that of *A. lumbricoides* and *T. trichiura*, making it dominant STH species in China. There is need to prioritize targeting the elderly who are actively involved in agricultural practices in rural areas for STHs prevention and control activities.

Despite the achievements made with control efforts in China, STHs transmission risk remains. The national surveillance on STHs in 2018 showed that 11 of 24 provinces conducted soil pollution surveillance and detected hookworm larval contamination, with 2.48% average detection rate by households. Detection rate in the most polluted provinces was 25.49%. Similarly, contamination of soil with *A. lumbricoides* eggs was detected in 16 out of 24 provinces, with detection rate of 5.23%, and the detection rate of the most polluted provinces was 26.00% (Chinese Center for Disease Control and Prevention, 2018). This suggest that the use of fresh manure as fertilizer still exists in rural areas, and there are some shortcomings with the management of harmless toilets. This, therefore, indicate that STHs transmission risk still exists.

In view of the recent STH endemicity characteristics in China, strategies developed for effective disease prevention and control are as follows.

In the new era, the "National Control Program on Key Parasitic Diseases" target will be achieved by combining control efforts with the health poverty alleviation project. Findings from the third national survey indicated that STHs (hookworm, *A. lumbricoides* and *T. trichiura*) infection rates decreased with improved economic situation and standard of living because the infection rate positively correlated with the poverty level. Among the 436 counties (cities/districts) surveyed, 145 were poverty-stricken counties, and infection rates were higher than counties with good standard of living (Zhou, 2018). Therefore, STHs control efforts should be carried out by the local government authorities in conjunction with the National Health Poverty alleviation Project.

It is therefore necessary for stakeholders to improve the national surveillance mechanisms to further enhance its efficiency and quality. There is also, need to strengthen the capacities of professionals at all levels including provincial, city, and county levels to improve their performances in disease diagnosis, data analysis and reporting by engaging in workshop programmes, training exchanges and competitions. Also, in-depth application of the data management system for national surveillance should be promoted by completing the archives, making it a large platform for parasitic disease surveillance, and data sharing, thus, providing data support for formulation and implementation of control strategies.

Lastly, it is imperative to target accurate control plan and precise implementation of control strategies due to STH transmission dynamics and distribution characteristics in different areas across the country. Women, adolescents, children, middle-aged and elderly people involved in agricultural practices in rural areas should be prioritized for effective monitoring and encouraged to adhere to information passed onto them using health education materials and other control platforms. Control strategies of "health education as leadership, control of infection source as priority" and the control measures of "improving water, lavatory and hygiene in conjunction with behavioural changes and drug administration" should be implemented in hyper- and meso-endemic areas to improve people's awareness on imbibing good personal hygiene culture. While health education should be implemented areas with in low endemicity and the pursuit of improved compliance to STHs examination and treatment should be enhanced. People in endemic areas should be adequately enlightened to cooperate with the government to actively promote the construction of harmless toilets and standardize their use in rural areas targeted for control.

6. Conclusions

STHs control in China has achieved remarkable accomplishments (Table 1), and this giant stride has not gone unnoticed globally (Qian et al., 2013, 2019; Song et al., 2018; Wang et al., 2016). Owing to establishment of the "National Control Plan of Key Parasitic Disease (2016–2020)" and the initiation of the "Criteria for transmission control and interruption of soil-transmitted nematodiasis" in 2018 (National Health Commission of the People's Republic of China, 2018), efforts will be geared towards eliminating STHs in the country. To achieve the elimination goals, experiences should be summed up and shared, new strategies should be implemented accordingly, control implementation should be standardized and poverty alleviation programme should be integrated in order to lay a solid foundation for effective control and elimination of STHs in China.

Table 1 Main achievements made by the NIPD-CTDR.

Type of achievements	Name	Main accomplisher/ author	Achievements
National surveys	The national survey on distribution of human parasites	Yu Senhai, Xu Longqi, Jiang Zexiao, Xu Shuhui[a], Han Jiajun[a], Zhu Yuguang[a], Chang Jiang[a], Lin Jinxiang[a], Xu Funiu[a]	First Prize of the Major Medical Science and Technology Award by Ministry of Health in 1994 Second Prize of the National Science and Technology Progress Award in 1995

Table 1 Main achievements made by the NIPD-CTDR.—cont'd

Type of achievements	Name	Main accomplisher/ author	Achievements
	Survey on the status of human key parasitic diseases in China	Wang Longde[a], Qi Xiaoqiu[a], Wang Yu[a], Hao Yang[a], Yang Weizhong[a], Tang Linhua, Xu Longqi, Feng Zheng	Survey report by People's medical publishing house in 2008
	National survey on current status of major human parasitic diseases	Zhou Xiaonong, Li Shizhu, Li Zhongjie[a], Chen Yingdan, et al.	Survey report by People's medical publishing house in 2018
Control strategies	Control strategies and application effectiveness of soil- and food-borne parasitic diseases	Chen Yingdan, Li Huazhong[a], Xu Longqi, Yang Weizhong[a], Tian Hongchun[a], et al.	Third Prize of Science and Technology by Chinese Preventive Medicine Association in 2013
Publications	Colour atlas of human parasitology	Yu Senhai, Xu Longqi	Published by China Science and Technology Press in 1992
	Distribution and pathogenic impact of human parasites in China	Yu Senhai, Xu Longqi, Xu Shuhui	Published by People's medical publishing house in 2000
	Series books of human parasitic disease for grass-roots staff: soil-borne parasitic diseases	Zhou Xiaonong	Published by People's medical publishing house in 2011
	Illustrated parasitology and parasitoses (two volumes)	Xu Longqi	Published by China Science and Technology Press in 2016

Continued

Table 1 Main achievements made by the NIPD-CTDR.—cont'd

Type of achievements	Name	Main accomplisher/ author	Achievements
Health industrial standards	Diagnosis of ascariasis	Chen Yingdan, Zhou Changhai, Xu Longqi, Tian Hongchun[a], Yao Linong[a], Yang Yichao[a], Cai Li[a], Zhou Xiaonong, Zheng Bin, Zang Wei, Zhu Huihui	Chinese health industry standards, issued in 1 February 2018 (WS/T 565—2017)
	Detection of intestinal helminths-the Kato-Katz method	Chen Yingdan, Zhu Tingjun, Zhou Changhai, Xu Jing, Zheng Bin, Li Shizhu, Xiao Ning, Zhou Xiaonong, Zeng Xiaojun[a], Liu Hongkun[a]	Chinese health industry standards, issued in 1 February 2018 (WS/T 570—2017)
	Criteria for transmission control and interruption of soil-transmitted nematodiasis	Chen Yingdan, Zhou Xiaonong, Zhou Changhai, Li Shizhu, Zhu Tingjun, Xu Yili[a], Xu Longqi, Zheng Bin, Cai Li[a], Li Huazhong[a], Sun Fenghua[a]	Chinese health industry standard, issued in 1 April 2018 (WS/T 629—2018)
Patents	Enteric-coated tablet of tribendimidine for deworming	Yao Runhua, Gao Hongfei, Ren Hainan, Qiang Huiqin, Chen Yaoqing, Hu Ling, Sun Yi[a], Zhang Jun[a]	Invention Patent in 2002 (96116469.7) National certificate of class I new drug in 2004
	Special quantitative board of Kato-Katz thick smear method	Xu Longqi, Chen Yingdan, Zhou Changhai	Certificate of Utility Model Patent in 2011 (ZL 2011 20025026.3)

Table 1 Main achievements made by the NIPD-CTDR.—cont'd

Type of achievements	Name	Main accomplisher/ author	Achievements
	Device for separating hookworm larva from soil	Wang Jujun, Zhou Changhai, Wang Guofei, Chen Yingdan	Certificate of Utility Model Patent in 2013 (ZL 2013 20092966.3)
	Anal swab for detection of *Enterobius vermicularis*	Chen Yingdan, Zhou Changhai, Wang Jujun, Zhu Tingjun, Zang Wei, Zhu Huihui	Certificate of Utility Model Patent in 2015 (ZL 2014 20707045.8)
	Immuno-chromatographic test strip for rapid detection of hookworm disease and its preparation	Wang Junyun, Shi Feng, Yang Yuetao, Gao Chunhua, Yang Yi, Zhu Huihui, Chen Yingdan	Invention Patent in 2015 (ZL 2015 10405625.0)

[a]Accomplishers not from the NIPD-CTDR.

Acknowledgements

This study was supported by the National Special Science and Technology Project for Major Infectious Diseases of China (Grant No. 2016ZX10004222-004), The fourth round of Three-Year Public Health Action Plan (GWIV-29), and the National Natural Scientific Foundation of China (No.81973108).

Competing interests

The authors declare that they do not have competing interests.

Authors' contributions

H.-H.Z., M.-B.Q., Y.-D.C., S.-Z.L., and X.-N.Z. conceived the study; H.-H.Z., M.-B. Q., and Y.-D.C. wrote and revised the manuscript; H.-H.Z., C.-H.Z., M.-Z.Z., J.-L.H., T.-J.Z., M.-B.Q., Y.-D.C., S.-Z.L., and X.-N.Z. revised the manuscript and gave approval of the version to be published. All the authors read and approved the final version of the manuscript.

References

Bethony, J., Brooker, S., Albonico, M., Geiger, S.M., Loukas, A., Diemert, D., et al., 2006. Soil-transmitted helminth infections: ascariasis, trichuriasis, and hookworm. Lancet 367 (9521), 1521–1532.

Brooker, S., Hotez, P.J., Bundy, D.A., 2008. Hookworm-related anaemia among pregnant women: a systematic review. PLoS Negl. Trop. Dis. 2 (9), e291.
Chammartin, F., Scholte, R.G., Guimaraes, L.H., Tanner, M., Utzinger, J., Vounatsou, P., 2013. Soil-transmitted helminth infection in South America: a systematic review and geostatistical meta-analysis. Lancet Infect. Dis. 13 (6), 507–518.
Chen, Y.D., Zang, W., 2013. Current situation of soil-transmitted nematodiasis monitoring in China and working keys in future. Chin. J. Schisto. Control 27, 111–114.
Chen, Y.D., Zhu, H.H., Huang, J.L., Zhu, T.J., Zhou, C.H., Qian, M.B., Li, S.Z., Zhou, X.N., 2019. Status and working principals of soil-transmitted nematodiasis during new period in China. Chin. J. Parasitol. Parasit. Dis. 31 (1), 23–25.
Chinese Center for Disease Control and Prevention, 2006. Operation Manual of the National Surveillance on STH. Chinese Center for Disease Control and Prevention, Beijing.
Chinese Center for Disease Control and Prevention, 2011. Operation Manual of the National Surveillance on STH. (Revised in 2011), Chinese Center for Disease Control and Prevention, Beijing.
Chinese Center for Disease Control and Prevention, 2014. Annual Report on Surveillance of Selected Infectious Diseases and Vectors, China. Chinese Center for Disease Control and Prevention, Beijing.
Chinese Center for Disease Control and Prevention, 2015. Annual Report on Surveillance of Selected Infectious Diseases, China. Chinese Center for Disease Control and Prevention, Beijing.
Chinese Center for Disease Control and Prevention, 2016. Annual Report on Surveillance of Selected Infectious Diseases and Vectors, China. Chinese Center for Disease Control and Prevention, Beijing.
Chinese Center for Disease Control and Prevention, 2018. Annual Report on Surveillance of Selected Infectious Diseases and Vectors, China. Chinese Center for Disease Control and Prevention, Beijing.
Disease control agency of the Ministry of Health, China CDC, 2010. Evaluation Report on Comprehensive Demonstration Plots of Parasitic Diseases (2006–2009). vol. 5. People's Medical Publishing House, Beijing, p. 71.
GBD 2015 DALYs and HALE Collaborators, 2016. Global, regional, and national disability-adjusted life-years (DALYs) for 315 diseases and injuries and healthy life expectancy (HALE), 1990-2015: a systematic analysis for the Global Burden of Disease Study 2015. Lancet 388 (10053), 1603–1658.
Hookworm Research Laboratory, Shanghai Institute of Parasitic Diseases, 1977. Glass tube filter paper method for quantifying fecal culture of hookworm larvae. Ji Sheg Chong Bing Fang Zhi Yan Jiu Jian Bao 52–53.
Jourdan, P.M., Lamberton, P.H.L., Fenwick, A., Addiss, D.G., 2018. Soil-transmitted helminth infections. Lancet 391 (10117), 252–265.
Karagiannis-Voules, D.A., Biedermann, P., Ekpo, U.F., Garba, A., Langer, E., Mathieu, E., et al., 2015. Spatial and temporal distribution of soil-transmitted helminth infection in sub-Saharan Africa: a systematic review and geostatistical meta-analysis. Lancet Infect. Dis. 15 (1), 74–84.
National Health Commission of the People's Republic of China, 2018. Criteria for Transmission Control and Interruption of Soil-Transmitted Nematodiasis. WS/T629–2018, Chinese Specification Press, Beijing, pp. 1–6.
Qian, M.B., Chen, Y.D., Zhou, X.N., 2013. Research priorities for the control and elimination of major helminthiases. Chin. J. Parasitol. Parasit. Dis. 31 (2), 155–159.
Qian, M.B., Chen, J., Bergquist, R., Li, Z.J., Li, S.Z., Xiao, N., Utzinger, J., Zhou, X.N., 2019. Neglected tropical diseases in the People's Republic of China: progress towards elimination. Infect. Dis. Poverty 8, 86. https://doi.org/10.1186/s40249-019-0599-4.

Silver, Z.A., Kaliappan, S.P., Samuel, P., Venugopal, S., Kang, G., Sarkar, R., et al., 2018. Geographical distribution of soil transmitted helminths and the effects of community type in South Asia and South East Asia—a systematic review. PLoS Negl. Trop. Dis. 12 (1), e0006153.

Song, L.G., Zeng, X.D., Li, Y.X., Zhang, B.B., Wu, X.Y., Yuan, D.J., Ai, H., Wu, Z.D., 2018. Imported parasitic diseases in mainland China: current status and perspectives for better control and prevention. Infect. Dis. Poverty 7, 78. https://doi.org/10.1186/s40249-018-0454-z.

Special edition on survey of human parasites, 1991. Chin. J. Parasitol. Parasit. Dis.

Tang, L.H., Xu, L.Q., Chen, Y.D., 2012. Parasitic Disease Control and Research in China. Beijing Science and Technology Press, Beijing.

Wang, L.D., 2008. Report on the National Survey of Current Status of Major Human Parasitic Diseases in China, Beijing.

Wang, Y., 2011. Spreading experience of demonstration plots and strengthening control of parasitic diseases. Chin. J. Schisto. Control 23 (5), 471–472.

Wang, Z.Y., Chen, Y.Q., 1959. Achievements in hookworm prevention and treatment research. In: Chief Editor of the Ministry of Health of the People's Republic of China. Proceedings of Scientific Achievements Celebrating the 10th Anniversary of the Founding of the People's Republic of China. People's Medical Publishing House, Beijing, pp. 39–42.

Wang, J.L., Li, T.T., Huang, S.Y., Cong, W., Zhu, X.Q., 2016. Major parasitic diseases of poverty in mainland China: perspectives for better control. Infect. Dis. Poverty 5, 67. https://doi.org/10.1186/s40249-016-0159-0.

WHO, 2010. First WHO Report on Neglected Tropical Diseases: Working to Overcome the Global Impact of Neglected Tropical Diseases. .

Xiao, S.H., 2009a. Progress in anthelmintic agent study since the founding of the People's Republic of China and current challenges. Zhongguo Ji Sheng Chong Xue Yu Ji Sheng Chong Bing Za Zhi 27 (05), 383–389.

Xiao, S.H., You, J.Q., Jiao, P.Y., Guo, H.F., Huang, L.X., 1990. Studies on the effects of mebendazole, albendazole and its metabolites in an experimental therapy of mice infected with secondary cysts of Echinococcus granulosus. Endem. Dis. Bull. 5, 11–16.

Xiao, S.H., You, J.Q., Wang, M.J., Jiao, P.Y., Gao, F.H., Chai, J.J., Jiao, W., Hotez, P., 2002. Augmented bioavailability and cysticidal activity of albendazole reformulated in soybean emulsion in mice infected with Echinococcus granulosus or Echinococcus multilocularis. Acta Trop. 82, 77–84.

Xiao, S.H., Wu, H.M., Wang, C., 2004. Tribendimidine-a new broad-spectrum drug against intestinal helminths. Zhongguo Ji Sheng Chong Xue Yu Ji Sheng Chong Bing Za Zhi 22, 312–315.

Xiao, S.H., Wu, H.M., Tanner, M., Utzinger, J., Chong, W., 2005. Tribendimidine: a promising, safe and broad-spectrum anthelmintic agent from China. Acta Trop. 94 (1), 1–14.

Xiao, S.H., Wu, Z.X., Zhang, J.H., Wang, S.Q., Wang, S.H., Qiu, D.C., Wang, C., 2007. Clinical observation on 899 children infected with intestinal nematodes and treated with tribendimidine enteric coated tablets. Zhongguo Ji Sheng Chong Xue Yu Ji Sheng Chong Bing Za Zhi 05, 372–375.

Xiao, S.H., Utzinger, J., Tanner, M., Keiser, J., Xue, J., 2013. Advances with the Chinese anthelminthic drug tribendimidine in clinical trials and laboratory investigations. Acta Trop. 126 (2), 115–126.

Xu, L.Q., Yu, S.H., Xu, S.H., 2000. Distribution and Pathogenic Impact of Human Parasites in China. People's Medical Publishing House, Beijing.

Xu, L.Q., Chen, Y.D., Sun, F.H., Cai, L., Fang, Y.Y., Wang, L.P., Liu, X., Li, L.S., Feng, Y., Li, H., 2005. A national survey on current status of the important parasitic diseases in human population. Chin. J. Parasitol. Parasit. Dis. 23, 332–340.

Yu, S.H., Xu, L.Q., 1992. Color Atlas of Human Parasitology. Beijing Science and Technology Press, Beijing.

Yu, S.H., Zhou, X.N., 2013. Annals of the National Institute of Parasitic Diseases, Chinese Center for Disease Control and Prevention (1950–2010). Shanghai Science and Technology Publishing House, Shanghai.

Yu, S.H., Xu, L.Q., Jiang, Z.X., Xu, S.H., Han, J.J., Zhu, Y.G., Chang, J., Lin, J.X., Xu, F.N., 1994a. Report on the first nationwide survey of the distribution of human parasites in China. 1. Regional distribution of parasite species. Chin. J. Parasitol. Parasit. Dis. 12 (4), 241–247.

Yu, S.H., Xu, L.Q., Jiang, Z.X., et al., 1994b. National survey of human parasitic diseases in China. Southeast Asian J. Trop. Med. Public Health 25, 4–10.

Zhang, J.H., Xiao, S.H., Wu, Z.X., Qiu, D.C., Wang, S.H., Wang, S.Q., Wang, C., 2008. Tribendimidine enteric coated tablet in treatment of 1 292 cases with intestinal nematode infection a phase IV clinical trial. Zhongguo Ji Sheng Chong Xue Yu Ji Sheng Chong Bing Za Zhi 01, 6–9.

Zhang, Q., Chen, Y.D., Xu, L.Q., Zheng, C.J., Li, H.Z., 2011. Effect of control on infections of soil—transmitted helminthes in demonstration plots of China for 3 years. Chin. J. Parasitol. Parasit. Dis. 23 (5), 476–482.

Zhang, H.B., Liu, C.S., Zheng, Q., 2019. Development and application of anthelminthic drugs in China. Acta Trop. 200, 105181.

Zhou, X.N., 2018. Report on the National Survey of Important Human Parasitic Diseases in China (2015). People's Medical Publishing House, Beijing.

Zhu, H.H., Huang, J.L., Zhu, T.J., Duan, L., Zhou, C.H., Qian, M.B., Chen, Y.D., 2019. National surveillance of soil-transmitted helminth infections in 2017. Chin. J. Parasitol. Parasit. Dis. 37 (1), 12–17.

CHAPTER SEVEN

From awareness to action: NIPD's engagement in the control of food-borne clonorchiasis

Men-Bao Qian[a,b,c,d,e], Chang-Hai Zhou[a,b,c,d,e], Hui-Hui Zhu[a,b,c,d,e], Ting-Jun Zhu[a,b,c,d,e], Ji-Lei Huang[a,b,c,d,e], Ying-Dan Chen[a,b,c,d,e], Xiao-Nong Zhou[a,b,c,d,e,*]

[a]National Institute of Parasitic Diseases, Chinese Center for Disease Control and Prevention, Shanghai, People's Republic of China
[b]Chinese Center for Tropical Diseases Research, Shanghai, People's Republic of China
[c]Key Laboratory of Parasite and Vector Biology, Ministry of Health, Shanghai, People's Republic of China
[d]National Center for International Research on Tropical Diseases, Ministry of Science and Technology, Shanghai, People's Republic of China
[e]WHO Collaborating Centre for Tropical Diseases, Shanghai, People's Republic of China
*Corresponding author: e-mail address: zhouxn1@chinacdc.cn

Contents

1. Introduction 246
2. Implementation of national epidemiological surveys 247
3. Establishment of a national surveillance system 251
4. Evaluation of disease burden 252
5. Capturing social determinants 254
6. Applied research on diagnostics and drugs 256
   6.1 Diagnostics 256
   6.2 Drugs 258
7. Exploration of an optimal control strategy and interventional measures 260
8. Conclusions and perspectives 262

References 263

Abstract

Clonorchiasis is caused by ingestion of raw freshwater fish containing infective larvae of *Clonorchis sinensis*. China harbours the largest number of people with *C. sinensis* infection. During the past three decades, the National Institute of Parasitic Diseases, affiliated to the Chinese Center for Disease Control and Prevention (NIPD) conducted many studies to facilitate the control on clonorchiasis. Three national surveys have shown the updated epidemiology of clonorchiasis in China. Recently, a national surveillance system has also been established, which will enable the production of high-resolution map. The evaluation of the disease burden has enhanced the awareness on clonorchiasis. Diverse diagnosis techniques including rapid screening by questionnaire,

Advances in Parasitology, Volume 110
ISSN 0065-308X
https://doi.org/10.1016/bs.apar.2020.04.004

serological tests, faecal examination and a molecular method have been developed or evaluated. The NIPD also participated in the early evaluation of praziquantel against clonorchiasis, which enhanced its application in China. Also, the NIPD has verified the efficacy of tribendimidine against clonorchiasis. A new sustainable strategy is also being explored. However, more research is expected to further facilitate control of clonorchiasis in China, as well as international cooperation in fighting human liver fluke infections in Asia.

1. Introduction

The discovery of *Clonorchis sinensis* could data back to over 140 years ago (Qian et al., 2014b, 2016b). On 21 August 1875, the Lancet first reported the finding of a new fluke, in which Dr. McConnell described a case of a Chinese male who was admitted to the Medical College Hospital in Calcutta, India, on 8 September 1874 (McConnell, 1875). The patient died within a few hours because of severe liver disease. During the post-mortem examination, many narrow, flattened, and lanceolate flukes were detected in the bile ducts. Four weeks later, Dr. Cobbold suggested publicly to name the parasite *Distoma sinense* (Cobbold, 1875). Then, in 1907, Dr. Looss renamed the parasite *Clonorchis sinensis* because of its characteristically branched testes (Looss, 1907). However, the history of the endemicity of clonorchiasis is thought to be much longer. An ancient corpse of the Warring States Period was unearthed in Hubei province, China in 1994 (Wei et al., 1980). Many *C. sinensis* eggs were found in the corpse which was buried about 2400 years ago.

Nowadays, clonorchiasis is predominantly endemic in East Asia, including China (Chen et al., 2012), the Republic of Korea (Jeong et al., 2016), northern Vietnam (Doanh and Nawa, 2016) and part of Russia (Solodovnik et al., 2018) due to the popularity of ingesting raw freshwater fish. Globally, about 15 million people were infected with *C. sinensis*, with 13 million in China, 1 million in the Republic of Korea, another million in northern Vietnam and several thousand in Russia (Qian et al., 2012, 2013b). Adult worms of *C. sinensis* live in the hepatobiliary system and cause major damage there. Early infections usually present no symptoms, but some cases might present some symptoms in nervous and digestive systems, such as asthenia, nausea, indigestion, headache, diarrhoea, or abdominal pain (especially in the right upper quadrant) (Chen et al., 1994; Kim et al., 1982).

Chronic and long-term infections can lead to severe hepatobiliary conditions, particularly cholelithiasis, cholangitis, and cholecystitis (Choi et al., 2008; Qiao et al., 2012). Furthermore, *C. sinensis* has been ascertained as a carcinogen to human beings, leading to fatal bile cancer (cholangiocarcinoma) (Bouvard et al., 2009).

As China harbours most cases with *C. sinensis* infection, research on and control of clonorchiasis in China will determine global progress in the control of clonorchiasis (Qian et al., 2019a). Additionally, the experiences in China benefit other endemic countries. The National Institute of Parasitic Diseases, affiliated to the Chinese Center for Disease Control and Prevention (NIPD) is the only national technical organization in charge of research and control on parasitic diseases (Chen et al., 2020; Zhou et al., 2005). As such, the NIPD has also done much work in combatting clonorchiasis. This paper is summarizing the progress of research and control on clonorchiasis in China promoted by the NIPD in different stages over the previous decades.

2. Implementation of national epidemiological surveys

Since the early 1900s, many surveys were conducted, gradually showing the high endemicity of clonorchiasis in Guangdong province, southern China (Faust, 1925; Whyte, 1908). National surveys were implemented until the late 1980s, which has enabled to map the epidemiology nationally (Qian et al., 2019a). All these national surveys were organized by NIPD, including the design, implementation, supervision, data analysis and report writing.

The first national survey on important parasitic diseases was implemented between 1988 and 1992 (Xu et al., 2000). A total of 1,477,742 participants were included from 726 counties in 30 provinces, and one Kato-Katz thick smear was prepared from each faecal sample. 4606 persons were detected with *C. sinensis* infection, with a prevalence of 0.31% (Chen et al., 2012; Xu et al., 2000). The prevalence was 0.38% (2758/733,143) in males and 0.25% (1848/744,599) in females. In males, the prevalence was low in young people, increased with age peaking in the 30s and then declining in the 50s (Fig. 1). In females, the prevalence peaked in those over 15, and then persisted. Out of 30 provinces, 22 were detected with *C. sinensis* (Fig. 2). The highest prevalence was found in eastern China,

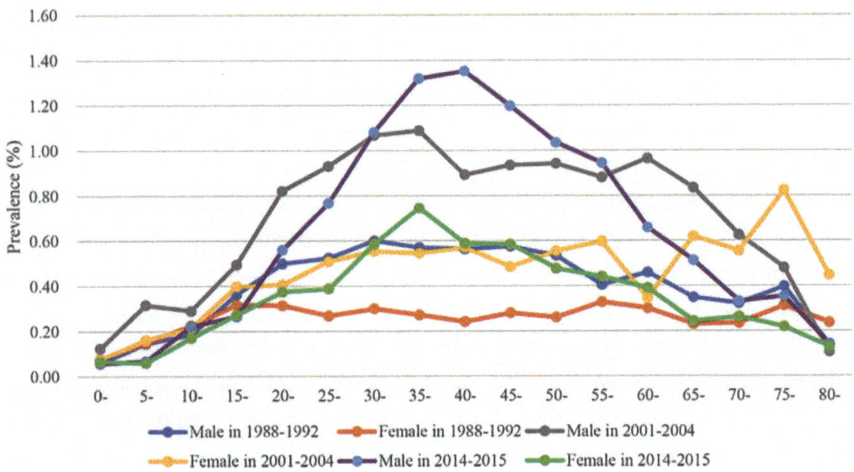

Fig. 1 Epidemiology of clonorchiasis by sexes and ages in China in three national surveys.

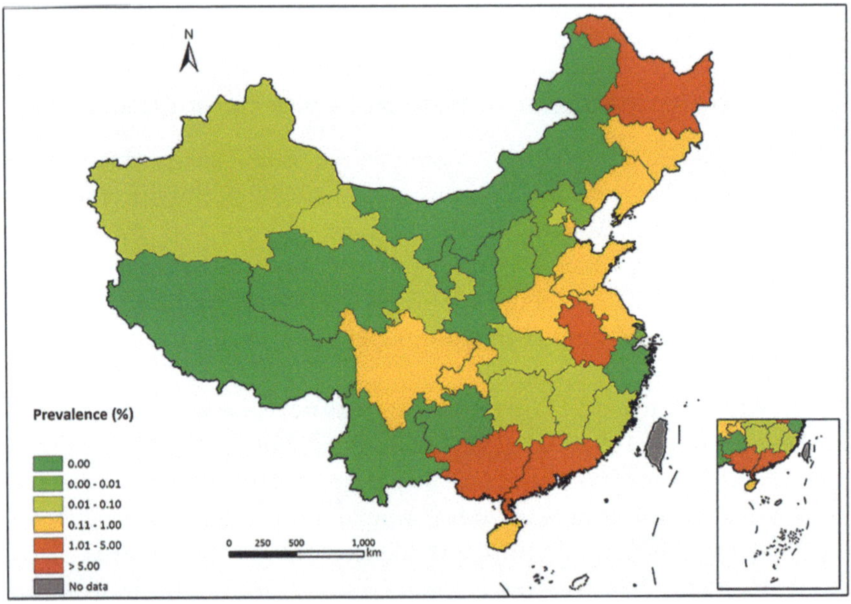

Fig. 2 Epidemiology of clonorchiasis by provincial level in China in 1988–1992.

including Guangdong (1.82%), Guangxi (1.20%) and Heilongjiang (1.18%) provinces. Anhui province in central China showed also a high prevalence (1.44%). Based on this survey, about 4.7 million people were estimated to have *C. sinensis* infection (World Health Organization, 1995).

From 2001 to 2004, another national survey and one special survey were implemented for clonorchiasis (Technical Steering Panel for National Survey of Current Status of Major Human Parasitic Diseases, 2008). *C. sinensis* was embedded in this national survey on important parasitic diseases, which covered 229 counties in 31 provinces. A total of 356,629 persons were included, out of which 2065 persons were detected with *C. sinensis*, with a prevalence of 0.58% (Chen et al., 2012; Technical Steering Panel for National Survey of Current Status of Major Human Parasitic Diseases, 2008). The prevalence was again higher in males (1290/179,341, 0.72%) than in females (775/177,288, 0.44%). In males, the prevalence was high in those aged between 20 and 70, while in females the prevalence increased with age, peaked in the 30s and then fluctuated (Fig. 1). Out of 31 provinces, *C. sinensis* infection was detected in 17 provinces (Fig. 3). High prevalence was found in Guangdong (5.35%), Guangxi (3.71%) and Jilin (4.77%) province. In addition to this, a special survey on clonorchiasis was conducted in the endemic areas in 27 endemic provinces identified on the basis of literature reports (Chen et al., 2012; Fang et al., 2008; Technical Steering Panel for National Survey of Current Status of Major Human Parasitic Diseases, 2008). Out of a total of 217,829

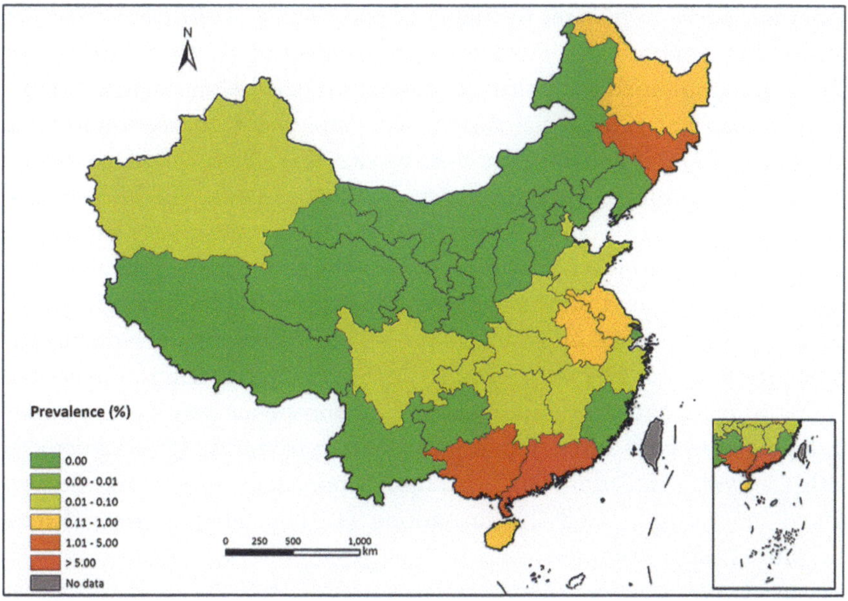

Fig. 3 Epidemiology of clonorchiasis by provincial level in China in 2001–2004.

participants, 5230 (2.40%) were found with *C. sinensis* infection. The prevalence was 2.94% in males and 1.84% in females. The epidemiological age profile was similar to the national survey. The highest prevalence was also demonstrated in eastern China including Guangdong (16.42%), Guangxi (9.76%), Heilongjiang (4.73%), Jilin (2.90%) and Hunan (1.33%) provinces. Because the distribution of clonorchiasis varied significantly by areas, and three Kato-Katz thick smears were prepared from a single faecal sample in the special survey compared to only one smear in the national survey, the findings in the special survey were believed to be more accurate (Chen et al., 2012; Fang et al., 2008; Technical Steering Panel for National Survey of Current Status of Major Human Parasitic Diseases, 2008). Based on the special survey, about 12.49 million persons were estimated to be infected with *C. sinensis*, most of which distributed in southeastern and northeastern China (Chen et al., 2012; Fang et al., 2008; Technical Steering Panel for National Survey of Current Status of Major Human Parasitic Diseases, 2008). Clonorchiasis is also endemic in Taiwan province, Hongkong Special Administrative Region and Macao Special Administrative Region (Qian et al., 2012; World Health Organization, 1995). It is estimated that about 13 million people were infected in China in the early 2000s (Qian et al., 2012).

Between 2014 and 2015, another national survey was carried out for important parasitic diseases in China (Zhou, 2018). Two parts were performed for clonorchiasis. First, the investigation of clonorchiasis in rural China was combined with that of soil-transmitted helminthiases. Second, in the urban areas, a special survey was implemented for clonorchiasis. Out of 617,441 participants examined by two Kato-Katz thick smears from a single faecal sample, 3466 were infected with *C. sinensis*, which represents a crude prevalence of 0.56% (Zhou, 2018). Unlike the surveys before, the population sampled was not proportional to the total population in different areas. Thus, weighted prevalence should be applied. The weighted prevalence was 0.47%, based on which about 5.98 million people were estimated to be infected with *C. sinensis* (Zhou, 2018). The weighted prevalence was 0.60% in males and 0.34% in females. The prevalence was high in middle aged people in both sexes (Fig. 1). Out of 31 provinces, 18 were detected with *C. sinensis* infection (Fig. 4). Highest prevalences were also demonstrated in Guangxi (6.68%), Guangdong (1.91%), Heilongjiang (1.62%) and Jilin (1.02%) provinces.

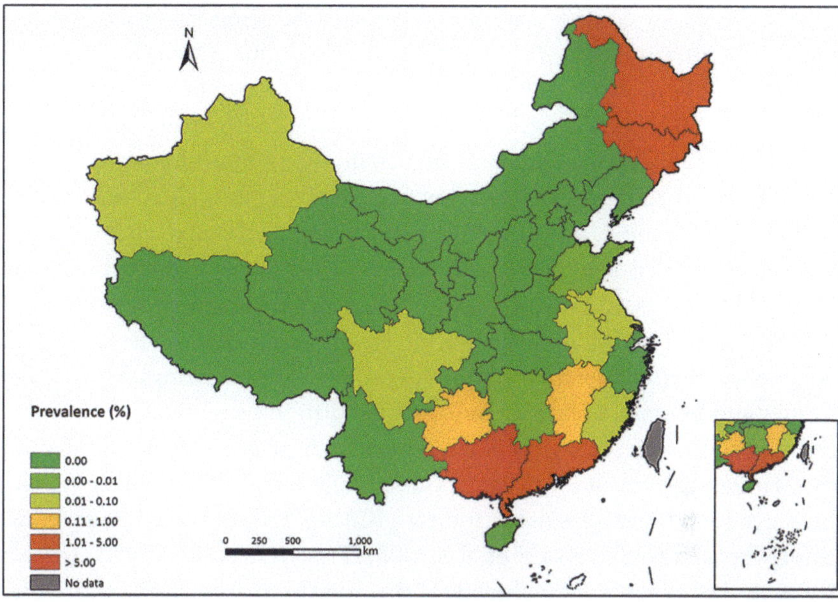

Fig. 4 Epidemiology of clonorchiasis by provincial level in China in 2014–2015.

3. Establishment of a national surveillance system

Surveillance is important in the control of parasitic diseases. Although surveillance systems had been available for many parasitic diseases in China before 2015, clonorchiasis was not included (Qian et al., 2016a, 2019a). Owing to the awareness of the importance of clonorchiasis in China, a national surveillance system for clonorchiasis has been gradually established in China since 2016 after the most recent national survey (Qian et al., 2018).

The surveillance system consists of two major parts, namely the active surveillance and passive surveillance (Qian et al., 2018). The passive surveillance involves collection of *C. sinensis* cases through the national reporting system of infectious diseases. The active surveillance system contains two components, namely fixed surveillance and mobile surveillance. In fixed surveillance, the surveillance spots are fixed for 5 years, while those in mobile surveillance vary each year. Although a national map for clonorchiasis is available owing to the three national and one special surveys, the resolution is low with a precision only down to subprovincial level, which hinders the implementation of control activities (Technical Steering Panel for National Survey of Current Status of Major Human Parasitic Diseases, 2008; Xu et al., 2000; Zhou, 2018). The fixed surveillance shows the changing trend, while the mobile surveillance aims to demonstrate the prevalence by county level which is expected to be completed within 5–10 years.

In the fixed surveillance part, seven provinces were included, namely Guangdong, Guangxi, Heilongjiang, Jilin, Hunan, Jiangxi and Guizhou, where clonorchiasis is highly endemic based on data from the national survey in 2014 and 2015 (Zhou, 2018). Several counties were selected from each province. Then, five villages are selected from each county. The surveillance system includes a survey on infection in human beings, knowledge, attitudes and practice in human beings, infection in fish, infection in animal reservoirs and other social and economic data. As to the survey on infection in human beings, about 200 villagers are included from each of five villages. The Kato-Katz method is used, and two Kato-Katz smears are prepared for a single faecal sample (Hong et al., 2003; Ting-Jun et al., 2018). In each village, 30 participants are included for the questionnaire survey, with a total of 150 villagers in each county. In each county, 100 fish are investigated by the digestive method (Li et al., 2012) and 25 animals (dogs, cats and pigs) (Chen et al., 1997; Lin et al., 2011) are included using the Kato-Katz method for examination. In mobile surveillance, only the survey on infection in human beings is conducted, in which the methodology is same to that in fixed surveillance.

In 2016, surveillance was completed in 301 counties from 30 provinces (Zhu et al., 2019). Out of 305,081 participants submitting faecal samples, 6226 were infected with *C. sinensis*, yielding a crude prevalence of 2.04%. *C. sinensis* infection was detected in 70 counties, distributed in 15 provinces. Nearly 90% persons with infection were distributed in four highly endemic provinces in southeastern and northeastern China, namely Guangdong, Guangxi, Heilongjiang and Jilin provinces. The epidemiological profiles in sex and age were similar to those in national surveys. Especially, high prevalence was demonstrated in those counties nearing to rivers.

4. Evaluation of disease burden

Compared to other helminthiases, clonorchiasis has long been neglected globally (Qian et al., 2013c; World Health Organization, 2010). It has not been listed among the global burden of disease before 2010. Thus, to fully demonstrate the burden of clonorchiasis, scholars from NIPD have implemented several studies.

First, the global epidemiology of clonorchiasis is captured in an article by Qian et al. from NIPD, who comprehensively reviewed the literature and summarized the data (Qian et al., 2012). It was found that globally the number of people with *C. sinensis* infection has increased from 7 million in the early 1990s to 15 million in the early 2000s, mainly occurring in China, the Republic of Korea, northern Vietnam and part of Russia. Regular national surveys have only been conducted in China and the Republic of Korea, every 5–10 years. Due to a lack of data, the number of people infected in the 2000s in Vietnam and Russia was assumed to be the same to those in the 1990s, as estimated by the World Health Organization. The people with *C. sinensis* infection didn't change significantly from the 1990s to the 2000s in the Republic of Korea. The increase in China contributed to the significant increase globally, with China harbouring about 85% of the global clonorchiasis cases. Even taking into consideration that opisthorchiasis is caused by *Opisthorchis viverrini* in southeast Asia and by *O. felineus* in Russia and neighbouring countries, China also shares over 50% of the global cases of human liver fluke infections (Qian et al., 2013b). The control of clonorchiasis in China is therefore likely to determine the global progress against human liver fluke infections. Furthermore, more surveys are expected in Vietnam and Russia, which will enable to capture accurate epidemiology there. Additionally, even in China and the Republic of Korea, where national surveys have been conducted every several years, high-resolution maps are expected with more surveys and the use of new mapping techniques (Qian et al., 2016a).

Second, the disability weight of *C. sinensis* infection has been estimated (Qian et al., 2011). Disability-adjusted life years (DALYs) are used to assess the health burden due to diseases, and consist of years of life lost (YLLs) and years lived with disability (YLDs) (King and Bertino, 2008). Thus, disability weight is a crucial parameter in calculation of YLDs, which reflects the average degree of loss of quality of life due to certain conditions. It ranges between 0 (complete health) and 1 (death). However, this parameter is not available for clonorchiasis. Additionally, although *C. sinensis* infection rarely causes death directly, it can lead to fatal cholangiocarcinoma (Bouvard et al., 2009). In a strategic diagram, Qian et al. argued that the burden of clonorchiasis could be calculated through the division of the sequelae of *C. sinensis* infection into two parts, namely fatal sequelae and non-fatal sequelae (Qian et al., 2011). The former includes early death and the period lived with cancer before death. Thus, the disability weight of early death is equal to the figure of one while that of living with cancer before death could be similar to those of other cancers. As to the non-fatal sequelae, the disability weight was captured through the multiplication of the probability and disability weight of single sequela caused by *C. sinensis* infection, then added together. The probability was captured by a field survey in a clonorchiasis endemic village in Guangdong province, while the disability weight of single sequela was found through literature searches. Then Monte Carlo simulation was applied. It was found that the overall disability weight is 0.075, and 0.101 in males and 0.050 in females. The disability weight is proportional to the geometric mean of eggs per gram of faeces (EPG), which is consistent to the fact that the morbidity of *C. sinensis* infection is positively correlated to the worm burden (Chen et al., 1994; Kim et al., 1982). The disability weight of *C. sinensis* infection is similar to that of hepatitis B, which also attacks the liver (Qian et al., 2013b).

Third, cholangiocarcinoma burden attributed to *C. sinensis* infection has also been calculated (Qian et al., 2012). Although the relationship between *C. sinensis* infection and cholangiocarcinoma has been observed long ago by scholars and verified in several cross-sectional and case-control studies, it is not until 2009 that *C. sinensis* was classified as a definite carcinogen to human beings (Bouvard et al., 2009). As to the cholangiocarcinoma burden attributed to *C. sinensis* infection, different approaches have been used (Furst et al., 2012; Plummer et al., 2016). In a meta-analysis by scholars from NIPD, it was found that the odds ratio between *C. sinensis* infection and cholangiocarcinoma reaches 4.47 (Qian et al., 2012). Then, based on the cholangiocarcinoma data in the Republic of Korea, it was estimated that the incidence of cholangiocarcinoma attributed to *C. sinensis* infection is 35/100000 in males and 25/100000 in females infected with *C. sinensis* (Qian et al., 2012). On the basis of this, it was estimated that in the early 2000s, nearly 5000 cholangiocarcinoma cases each year could be attributed to *C. sinensis* infection (Qian and Zhou, 2017).

5. Capturing social determinants

To allow a scientific design of interventions, the determinants of clonorchiasis should be first captured (Qian et al., 2016a). The national surveys demonstrate the spatial and temporal profiles, as well the distribution of clonorchiasis in populations (Technical Steering Panel for National Survey of Current Status of Major Human Parasitic Diseases, 2008; Xu et al., 2000; Zhou, 2018). While the transmission of clonorchiasis is caused by ingestion of *C. sinensis* metacercariae present in freshwater fish, infection may be acquired in various ways, such as the ingestion of raw or undercooked fish or through contamination of other foods (Fan et al., 2014). Usually, the direct ingestion of raw freshwater fish is believed to be the most important transmission route in China (Zheng, 2009). Thus, several field surveys have been carried out by scholars from the NIPD to further explore the epidemiology and practice of eating raw freshwater fish.

Hengxian county is located in the southeast part of China, and it is believed to be one of the most afflicted counties by clonorchiasis in China. Two earlier surveys embedded in the national surveys demonstrated the increasing prevalence of clonorchiasis in Hengxian county. To capture the update epidemiology, in 2011 another survey was organized by the NIPD (Qian et al., 2014a). 3437 villagers were enrolled in this survey from 24 villages. After comparison with two earlier surveys, it was demonstrated that the prevalence of clonorchiasis had increased from 19.7% in 1989 to 30.5% in 2002 and then 46.5% in 2011. What is a distinct comparison was the significant decrease of soil-transmitted helminthiases from 86.3% in 1989 to 25.1% in 2002 and 7.0 in 2011. Social and economic development promotes the great achievement in the control of soil-transmitted helminthiases in China, while the prevalence of clonorchiasis increases due to the development of pisciculture, which makes more people affordable to ingest raw freshwater fish.

In a cross-sectional survey in 16 counties in four major clonorchiasis endemic provinces, namely Guangdong, Guangxi, Heilongjiang and Jilin (Qian et al., 2019c), 16,230 participants were enrolled to explore the relationship between ingestion of raw freshwater fish and *C. sinensis* infection. There existed a linear relationship between the practice of eating of raw freshwater fish and *C. sinensis* infection by county level (coefficient of determination (R^2) = 0.44). The prevalence of clonorchiasis was 33.5% in those with raw-eating practice, while it was 2.6% in those without the practice. Thus, the relative risk of *C. sinensis* infection was 12.9. The attributable risk, attributable risk per cent, population attributable risk and population attributable risk per cent was 30.9%, 92.3%, 8.2% and 75.9%, respectively. This study clearly demonstrates the crucial role of ingestion of raw freshwater fish in the transmission of clonorchiasis in China. However, it also indicates the potential role of other transmission routes.

In another study, the knowledge and raw-eating practice in those with *C. sinensis* infection was analysed in different populations within one community in Guangdong province. First, it was demonstrated that a high proportion of persons ingested raw freshwater fish at restaurants (Qian et al., 2013a). It was also demonstrated that the proportion was higher in adult males (80.3%) than in adult females (54.9%) and children (25.0%). Thus, restaurants could not be neglected in the control of clonorchiasis in some endemic areas. Second, it was demonstrated that compared to children, adults had more knowledge on transmission and harm of clonorchiasis, but they also had more practice in ingesting raw freshwater fish. This disparity indicates that education in adults is likely to be less effective and that education should target children. Of course, adults sometimes have some misunderstanding on clonorchiasis, such as the killing effect of *C. sinensis* larvae in fish by alcohol, which still needs education to be rectified (Qian, 2014).

6. Applied research on diagnostics and drugs

6.1 Diagnostics

Both diagnosis and treatment of clonorchiasis caught the attention of the NIPD. Scholars from the NIPD have evaluated the effectiveness of different diagnostic methods. Li et al. conducted a study to compare the performance of serological diagnostics for clonorchiasis in China (Li et al., 2018). Four commercial tests were included. The performance was low for the one detecting IgM. The sensitivity was high for the other three detecting IgG and ranged from 81.51% (194/238) to 99.16% (236/238). It was also demonstrated that the sensitivity was higher in heavy infection compared to that in light infection. The specificity ranged from 73.42% (174/237) to 87.34% (207/237), and was lower in participants from clonorchiasis endemic areas as compared to those from non-endemic areas. Additionally, high cross reaction was demonstrated with paragonimosis. Overall, more research is expected to improve the sensitivity in light infection and the specificity for immunological tests.

Qian et al. have conducted two studies evaluating the performance of faecal examination techniques (Qian et al., 2013e, 2019e), of which one was nested in a drug evaluation trial (Qian et al., 2013e). In total 74 persons infected with *C. sinensis* were examined at baseline by faecal examination (the Kato-Katz method and the formalin-ether concentration technique (FECT)). Three weeks after treatment, the infection status was re-examined. 38 (51.4%) people were still infected as determined by a combination of the two methods. The number of positive cases was 23 (31.1%) with a single Kato-Katz smear, 29 (39.2%) with three Kato-Katz smears, 35 (47.3%) with six Kato-Katz smears from two samples, 13 (17.6%) with a single FECT examination and 17 (23.0%) with a duplicate FECT examination. In another natural setting, the diagnostic performance of Kato-Katz method in school children was evaluated, while the practice of ingestion of raw freshwater fish was also assessed (Qian et al., 2019e). With the "gold" standard of six smears from two faecal samples, 77 out of 397 students were detected with *C. sinensis* eggs. However, only 45 students were detected when a single smear was examined, which showed an underestimation of 41.6%, while the infection intensity in term of eggs per gram of faeces was overestimated. The practice of ingesting raw freshwater fish in false negative cases was comparable to those in the detected cases, while the raw-eating practice in both groups was significantly higher than that in the negative group, which indicates that the higher performance by multiple smears was due to improved diagnostic capacity rather than low specificity or contamination. Based on these two studies, it is argued that the underestimation of *C. sinensis* infection is difficult to be avoided in low endemicity situations when limited Kato-Katz smears are examined. Multiple smears from repeated samples should be used in such settings for the examination of those individuals ingesting raw freshwater fish, for the evaluation of drug efficacy or for the verification of elimination.

Xu et al. have compared the performance of the Kato-Katz method with a PCR technique in detecting *C. sinensis* eggs in human faeces (Xu et al., 2019). Out of 133 faecal samples, 103 (77.4%) were positive with the combination of Kato-Katz method and PCR technique. The positive rate was higher in PCR technique (70.7%) than Kato-Katz method (57.1%). Among 76 positive samples by Kato-Katz method, 67 (88.2%) were positive in PCR technique, while it was 47.4% (27/57) in the negative sample by Kato-Katz method. Although the difference was insignificant, the detection rate by PCR technique was a little higher in those with high infection intensity. The Kappa value of the consistency reached 0.73 between two methods. Thus, it is argued that in low-endemic areas, PCR technique is expected to improve the sensitivity.

The probability to introduce a rapid screening technique to identify people for treatment of clonorchiasis is important. By further analysing the data on *C. sinensis* infection and raw-freshwater fish-eating practice of villagers in 16 counties from four endemic provinces mentioned above (Qian et al., 2019c), Qian et al. evaluated the screening performance by practice of ingesting raw freshwater fish to potentially select those persons with *C. sinensis* infection (Qian et al., 2020). Overall, the sensitivity and false negative rate were 82.3% (1438/1747) and 17.7% (309/1747), respectively, while the specificity and false positive rate were 80.3% (11,626/14,483) and 19.7% (2857/14,483), respectively. Youden's index reached 0.6. The positive and negative predictive values were 33.5% (1438/4295) and 97.4% (11,626/11,935), respectively. Additionally, the sensitivity was related to the infection intensity: 71.1% in light intensity, 86.6% in moderate intensity and 86.0% in heavy intensity. Thus, the rapid screening for raw-freshwater fish-eating practice could be applied. However, it has also been noticed that other parameters should be explored to optimize the screening performance, e.g., treatment history and frequency and duration of the practice of eating raw freshwater fish.

6.2 Drugs

In the late 1980s, scholars from the NIPD initiated an efficacy trial of praziquantel against *C. sinensis* infection (Chen et al., 1982). In a lightly infected group of patients (EPG <1000) to whom praziquantel was administrated at a dose of 20 mg/kg trice in 1 day, the cure rate was 81.8%, the eggs reduction rate was 96.47% and the proportion of side effects was 50.0% (Table 1). In another group that received praziquantel 25 mg/kg trice in 1 day, the cure rate, egg reduction rate and proportion of side effects was 82.8%, 99.77%, 67.9%, respectively. The cure rate was 91.4% in light infection (EPG < 1000) and 65.4% in heavy infection (EPG ≥ 1000). In the heavily infected group (EPG ≥ 1000) treated with praziquantel 25 mg/kg trice in 1 day plus 15 mg/kg trice in another day, the cure rate, egg reduction rate and proportion of side effects was 94.4%, 99.995% and 80.6%, respectively. In the control group treated with hexachloropxylol (17.5 g divided into 10 times within 5 days), the cure rate, egg reduction rate and proportion of side effects was 79.5%, 99.81%, 87.5%, respectively. The cure rate was 80.8% in light infection (EPG < 1000) and 76.9% in heavy infection (EPG ≥ 1000). This trial demonstrated that praziquantel is efficacious and

Table 1 Efficacy and safety of praziquantel against *C. sinensis* infection in a clinical trial (Chen et al., 1982).

	Cure rate (%)[a]				
Regimens	**Light infection (EPG < 1000)**	**Heavy infection (EPG ≥ 1000)**	**Total**	**Eggs reduction rate(%)[a]**	**Side effects (%)**
PZQ (20 mg/kg, t.i.d)	81.8 (27/33)	–	81.8 (27/33)	96.47	50.0
PZQ (25 mg/kg, t.i.d)	91.4 (96/105)	65.4 (34/52)	82.8 (130/157)	99.77	67.9
PZQ (25 mg/kg, t.i.d + 15 mg/kg, t.i.d)	–	94.4 (34/36)	94.4 (34/36)	99.995	80.6
Hexachloropxylol (17.5 g divided into 10 times within 5 days)	80.8 (21/26)	76.9 (10/13)	79.5 (31/39)	99.81	87.5

[a]Evaluated 1 month after drug administration.

safe against *C. sinensis* infection. Both the efficacy and side effects are influenced by the infection intensity. A higher dose increases the efficacy but more side effects also occur. In the light of evidence also from other trials, praziquantel was gradually introduced as the treatment of clonorchiasis in China.

Tribendimidine is a new drug, developed in the 1970s by NIPD (Xiao et al., 2005, 2013). It was approved in 2004 by the Chinese Food and Drug Administration for treating soil-transmitted helminth infections in humans. Several experiments in vivo and in vitro also demonstrated the potential of tribendimidine against *C. sinensis* infection (Keiser et al., 2009; Xiao et al., 2008a,b, 2009a,b, 2011; Xue et al., 2009, 2010). Finally, in 2012, two clinical trials were conducted in China, which further demonstrated the potential of tribendimidine in the treatment of *C. sinensis* infection (Qian et al., 2013d; Xu et al., 2014).

In the first randomized open-label trial in Guangxi, the efficacy and safety of tribendimidine was assessed in two different arms, namely 400 mg once and 400 mg daily for 3 days compared to the standard dose of praziquantel (75 mg/kg, divided into three times in 1 day) (Qian et al., 2013d). In total 74 patients with *C. sinensis* were enrolled. The cure rate was 44% in the tribendimidine "once" group, 58% in the group that received tribendimidine for 3 days and 56% in the praziquantel group. The egg reduction rate was 97.6%, 98.8% and 98.8%, respectively. Especially, the cure rate was 100% in light infection (EPG < 1000) in all three groups. Additionally, the overall percentage of side effects was only 44% in the tribendimidine "once" group, 67% in the group that received tribendimidine for 3 days and 92% in the praziquantel group.

Another randomized clinical trial was implemented in Hunan province (Xu et al., 2014). A total of 156 patients were divided into four arms, 400 mg tribendimidine once, 200 mg tribendimidine twice a day, 75 mg/kg praziquantel divided into twice a day for 2 days, and 400 mg mebendazole single dose. The cure rate against *C. sinensis* was 50%, 33.3%, 56.8% and 0%, respectively, while the egg reduction rate was 98.3%, 97.1%, 98.0% and 71.2%, respectively. In a second part of the study, those uncured in the praziquantel group received the same regime of praziquantel, while those uncured in the tribendimidine groups (400 mg tribendimidine once and 200 mg tribendimidine twice per day) received 400 mg tribendimidine once. The cure rates were 75% and 78.1%, respectively, while the egg reduction rate was 75.8% and 74.2%, respectively. Additionally, it was also demonstrated that less side effect occurred in the tribendimidine groups as compared to the praziquantel group.

7. Exploration of an optimal control strategy and interventional measures

Between 2001 and 2004, an international cooperation between China and the Republic of Korea was implemented to tackle helminthiases. Heilongjiang province in northeastern China was selected for clonorchiasis (Choi et al., 2010). Seven villages were included, in which seven different chemotherapy strategies were tested. Among them, five were classified as heavy endemic areas in Zhaoyuan county with a prevalence of over 40%, while another two were moderately endemic areas with a prevalence between 20% and 30% (from Ningan county and Hailin county, respectively). After 3 years, the prevalence of *C. sinensis* infection decreased significantly in all seven groups, but to different degrees (Table 2). In all seven groups, 75 mg/kg praziquantel divided into three doses in 1 day was employed. The prevalence decreased by 72.9% from 69.5% in 2001 to 18.8% in 2004 in one heavily endemic village through mass treatment every 2 years. It decreased by 82.5% from 48.0% in 2001 to 8.4% in 2004 in one heavily endemic village through mass treatment every year. It decreased by 72.7% from 54.9% to 15.0% in one heavily endemic village

Table 2 Effectiveness of different chemotherapy strategies using praziquantel against *C. sinensis* infection in a community-based trial (Choi et al., 2010).

Endemic level	Prevalence in 2001 (%)	Prevalence in 2004 (%)	Declining rate (%)	Intervention measures
Heavily endemic villages	69.5	18.8	72.9	Mass treatment every 2 years
	48.0	8.4	82.5	Mass treatment every year
	54.9	15.0	72.7	Selective treatment every year
	73.2	12.3	83.2	Selective treatment and reservoir animal control every year
	59.5	7.5	87.4	Selective treatment every 6 months
Moderately endemic villages	24.8	1.9	92.3	Selective treatment every 2 years
	29.7	1.3	95.6	Selective treatment every year

through selective treatment every year. It decreased by 83.2% from 73.2% to 12.3% in one heavily endemic village through selective treatment and reservoir animal control every year. It decreased by 87.4% from 59.5% to 7.5% in one heavily endemic village through selective treatment every 6 months. In two moderate villages, the prevalence decreased by 92.3% from 24.8% to 1.9% through selective treatment every 2 years and by 95.6% from 29.7% to 1.3% through selective treatment every year. Thus, it was demonstrated that chemotherapy could effectively decrease the prevalence of *C. sinensis* infection. Particularly, high frequency of treatment (mass chemotherapy every year or selective treatment every 6 months in heavily endemic areas and selective treatment every year in moderately endemic areas) is indicated. Additional treatment of reservoir animals increased programme effectiveness.

Following increased awareness of the importance of clonorchiasis after the second national survey between 2001 and 2004, two control pilots were established for clonorchiasis between 2006 and 2009, and NIPD was in charge of the implementation of the projects (Ji et al., 2012; Xin-Hua et al., 2011). One pilot was established in Yangshan county, Guangdong province in southeastern China, and another one in Zhaoyuan county, Heilongjiang province in northeastern China. In Yangshan county, the prevalence of clonorchiasis was 14.0% (356/2541) in 2006 (Xin-Hua et al., 2011). A comprehensive intervention including education and selective chemotherapy (once every year with 75 mg/kg praziquantel divided into twice per day for 2 days or 64 mg/kg albendazole divided into eight times in 4 days) was used. The prevalence decreased to 6.9% (73/1062) in 2009. In Zhaoyuan county, the prevalence of clonorchiasis was 67.5% (1806/2677) in 2006 (Ji et al., 2012). Similarly, health education and chemotherapy (mass chemotherapy once every year with 75 mg/kg praziquantel divided into four times by 2 days) was employed. The persons who accepted treatment were 147,120, 190,991 and 146,644 in 2007, 2008 and 2009, respectively. The prevalence of clonorchiasis decreased to 7.1% (72/1011) in 2009.

The practice of eating raw freshwater fish is key in the transmission of clonorchiasis, and thus the interruption of this practice will determine the sustainability of control. Qian et al. explored a behavioural change strategy against clonorchiasis. To explore this strategy, behavioural change communication products were to be provided. Thus, a national screening was conducted (Qian et al., 2019d). The evaluation system was set up through experts' consultancy (Hui-Hui et al., 2018). A total of 96 groups of products on helminthiases were collected (Qian et al., 2019d).

However, most products targeted schistosomiasis and echinococcosis, and only five focused on clonorchiasis. Out of 34 products awarded, none targeted clonorchiasis. Based on the awarded products targeting schistosomiasis and echinococcosis, a clonorchiasis-specific product was designed and produced. Because of the difficulty to change adults' behaviour, a strategy targeting children was used. The product was a cartoon, which introduced the harm and prevention of clonorchiasis for children. Finally, a cluster trial was carried out (Qian et al., 2019b). In clonorchiasis endemic areas, one school was selected as intervention, and another one as control. Comprehensive education activities were implemented in interventional school, including the erection of a bulletin board, the presentation of the educational cartoon and distribution of educational brochures, as well as drawing and essay writing competitions. After 6 months, the effectiveness was evaluated. In interventional school (251 students), the percentage of students with knowledge on transmission route, early symptoms, complications and carcinogenicity of clonorchiasis increased from 15.1% to 92.4%, from 6.8% to 43.4%, from 4.8% to 94.8% and from 4.8% to 57.8% after 6 months. In the control group (153 students), the proportion increased from 9.2% to 56.9%, from 0.7% to 33.3%, from 2.6% to 42.5%, and from 2.6% to 20.9%. Although the percentage also increased in control school, the effect was lower compared to that in interventional school. Especially, the percentage of students determined not to eat raw freshwater fish in the future increased from 71.3% at baseline to 97.2% at evaluation survey in interventional school, but it didn't change significantly from 79.7% to 83.0% in control school.

8. Conclusions and perspectives

During the past 70 years, significant progresses have been achieved in the control of neglected tropical diseases in China owing to great social developments and control activities (Qian et al., 2019a). However, clonorchiasis is still highly prevalent in eastern China due to the deep-rooted habit of eating raw freshwater fish and the development of aquaculture. According to predictions of the national control programme, the prevalence of clonorchiasis should have decreased by 30% in major endemic areas in China in 2020, as compared to 2015 (National Health and Family Planning Commission of the People's Republic of China, 2016). In the near future, clonorchiasis is yet to be considered as still the most important food-borne parasitic disease in eastern China. It is therefore essential to take more actions to control clonorchiasis transmission in highly endemic areas. NIPD, the key organization in designing the strategic workplan to this purpose, will focus on several aspects to control clonorchiasis (Qian and Zhou, 2019). First, national surveillance will be strengthened. It is expected that a high resolution map, with details down to the county level will be available around 2020. Second, more research will be carried out to develop innovative diagnostic and treatment tools. Rapid screening tools both at the community and individual level will be operationalized. More research is also expected to validate the efficacy of tribendimidine against clonorchiasis. Third, detailed studies on the transmission mechanism will be explored to guide control. Especially, the determinants of *C. sinensis* transmission will be explored. Fourthly, control strategies and measurements will be further explored, including information, education and communication techniques, chemotherapeutic strategies, and integrated strategies. Additionally, NIPD will put more efforts on the cooperation with international institutions on other liver fluke infections, such as opisthorchiasis in southeastern Asia—Thailand, Laos, Cambodia, Vietnam and Myanmar (Qian and Zhou, 2019). As a result, NIPD is expected to become one of the leading institutions in the control of liver fluke infections globally.

References

Bouvard, V., Baan, R., Straif, K., Grosse, Y., Secretan, B., El Ghissassi, F., Benbrahim-Tallaa, L., Guha, N., Freeman, C., Galichet, L., Cogliano, V., WHO International Agency for Research on Cancer Monograph Working Group, 2009. A review of human carcinogens—part B: biological agents. Lancet Oncol. 10, 321–322.

Chen, M.G., Wan, Z.R., Hua, X.J., Wen, Y.Q., Wang, M.J., Zhu, P.J., Shi, Y.R., He, B.Z., Xu, M.Y., Zhong, J.X., 1982. Comparison of pyquiton (biltricied) with hexa-chloro-p-xylol in the treatment of *Clonorchis sinensis* infection. Zhongguo Yi Xue Ke Xue Yuan Xue Bao 4, 114–118. (in Chinese).

Chen, M.G., Lu, Y., Hua, X.J., Mott, K.E., 1994. Progress in assessment of morbidity due to *Clonorchis sinensis* infection: a review of recent literature. Trop. Dis. Bull. 91, R7–R65.

Chen, S., Chen, S., Wu, F., Chen, S., Ke, S., Chen, S., Pan, Z., 1997. Epidemiological survey on clonorchiasis sinensis in Yangxin County of Hubei Province of PR China. Southeast Asian J. Trop. Med. Public Health 28 (Suppl. 1), 51–53.

Chen, Y.D., Zhou, C.H., Xu, L.Q., 2012. Analysis of the results of two nationwide surveys on *Clonorchis sinensis* infection in China. Biomed. Environ. Sci. 25, 163–166.

Chen, J., Ding, W., Li, Z., Zhou, D.D., Yang, P., Wang, R.B., Zheng, B., Sheng, H.F., Guan, Y.Y., Xiao, N., Li, S.Z., Zhou, X.N., 2020. From parasitic disease control to global health: new orientation of the National Institute of Parasitic Diseases, China CDC. Acta Trop. 201, 105219.

Choi, D., Lim, J.H., Lee, K.T., Lee, J.K., Choi, S.H., Heo, J.S., Choi, D.W., Jang, K.T., Lee, N.Y., Kim, S., Hong, S.T., 2008. Gallstones and *Clonorchis sinensis* infection: a hospital-based case-control study in Korea. J. Gastroenterol. Hepatol. 23, e399–e404.

Choi, M.H., Park, S.K., Li, Z., Ji, Z., Yu, G., Feng, Z., Xu, L., Cho, S.Y., Rim, H.J., Lee, S.H., Hong, S.T., 2010. Effect of control strategies on prevalence, incidence and re-infection of clonorchiasis in endemic areas of China. PLoS Negl. Trop. Dis. 4, e601.
Cobbold, T.S., 1875. The new human fluke. Lancet 106, 423.
Doanh, P.N., Nawa, Y., 2016. *Clonorchis sinensis* and *Opisthorchis* spp. in Vietnam: current status and prospects. Trans. R. Soc. Trop. Med. Hyg. 110, 13–20.
Fan, S.Y., Shi, X.H., Niu, J.F., Lin, Z.P., Li, L.L., 2014. Investigation on *Clonorchis sinensis* infection and its risk factors in Futian District, Shenzhen City. Zhongguo Xue Xi Chong Bing Fang Zhi Za Zhi 26, 699–700.
Fang, Y.Y., Chen, Y.D., Li, X.M., Wu, J., Zhang, Q.M., Ruan, C.W., 2008. Current prevalence of *Clonorchis sinensis* infection in endemic areas of China. Zhongguo Ji Sheng Chong Xue Yu Ji Sheng Chong Bing Za Zhi 26 (99–103), 109. (in Chinese).
Faust, E.C., 1925. Some recent aspects of the epidemiology of *Clonorchis* infection in China. Chin Med J (Engl) 34, 287–296.
Furst, T., Keiser, J., Utzinger, J., 2012. Global burden of human food-borne trematodiasis: a systematic review and meta-analysis. Lancet Infect. Dis. 12, 210–221.
Hong, S.T., Choi, M.H., Kim, C.H., Chung, B.S., Ji, Z., 2003. The Kato-Katz method is reliable for diagnosis of *Clonorchis sinensis* infection. Diagn. Microbiol. Infect. Dis. 47, 345–347.
Hui-Hui, Z., Men-Bao, Q., Chang-Hai, Z., Ting-Jun, Z., Ying-Dan, C., 2018. Establishing evaluation system for health education products of parasitic diseases by Delphi method. Zhongguo Xue Xi Chong Bing Fang Zhi Za Zhi 30, 307–311. (in Chinese).
Jeong, Y.I., Shin, H.E., Lee, S.E., Cheun, H.I., Ju, J.W., Kim, J.Y., Park, M.Y., Cho, S.H., 2016. Prevalence of *Clonorchis sinensis* infection among residents along 5 major rivers in the Republic of Korea. Korean J. Parasitol. 54, 215–219.
Ji, Z., Ge, T., Yuan, S., Li, X.Z., Liu, G.H., Tang, L., 2012. Effect of community intervention model of clonorchiasis in Zhaoyuan County. Zhongguo Xue Xi Chong Bing Fang Zhi Za Zhi 24 (113), 115. (in Chinese).
Keiser, J., Xiao, S.H., Smith, T.A., Utzinger, J., 2009. Combination chemotherapy against *Clonorchis sinensis*: experiments with artemether, artesunate, OZ78, praziquantel, and tribendimidine in a rat model. Antimicrob. Agents Chemother. 53, 3770–3776.
Kim, M.S., Lee, J.S., Rim, H.J., 1982. Studies on the clinical aspects of clonorchiasis in Korea. Korea Univ. Med. J. 19, 107–121, (in Korean).
King, C.H., Bertino, A.M., 2008. Asymmetries of poverty: why global burden of disease valuations underestimate the burden of neglected tropical diseases. PLoS Negl. Trop. Dis. 2, e209.
Li, H.M., Zhu, H.H., Zhu, T.J., Zang, W., Chen, Y.D., 2012. Comparison on two methods for detecting metacercariae of *Clonorchis sinensis*. Guoji Yi Xue Ji Sheng Chong Bing Za Zhi 39, 152–154. (in Chinese).
Li, H.M., Qian, M.B., Yang, Y.C., Jiang, Z.H., Wei, K., Chen, J.X., Chen, J.H., Chen, Y.D., Zhou, X.N., 2018. Performance evaluation of existing immunoassays for *Clonorchis sinensis* infection in China. Parasit. Vectors 11, 35.
Lin, R.Q., Tang, J.D., Zhou, D.H., Song, H.Q., Huang, S.Y., Chen, J.X., Chen, M.X., Zhang, H., Zhu, X.Q., Zhou, X.N., 2011. Prevalence of *Clonorchis sinensis* infection in dogs and cats in subtropical southern China. Parasit. Vectors 4, 180.
Looss, A., 1907. On some parasites in the museum of the School of Tropical Medicine, Liverpool. Ann. Trop. Med. Parasitol. 1, 123–152.
McConnell, J.F.P., 1875. Remarks on the anatomy and pathological relations of a new species of liver-fluke. Lancet 106, 271–274.
National Health and Family Planning Commission of the People's Republic of China, 2016. National Control Plan on Echinococcosis and Other Important Parasitic Diseases in China (2016–2020). http://www.nhc.gov.cn/jkj/s5873/201702/dda5ffe3f50941a29fb0aba6233bb497.shtml. Accessed 10 August 2019. (in Chinese).

Plummer, M., de Martel, C., Vignat, J., Ferlay, J., Bray, F., Franceschi, S., 2016. Global burden of cancers attributable to infections in 2012: a synthetic analysis. Lancet Glob. Health 4, e609–e616.

Qian, M.B., 2014. Clonorchiasis control: starting from awareness. Infect. Dis. Poverty 3, 33.

Qian, M.B., Zhou, X.N., 2017. Global burden of cancers attributable to liver flukes. Lancet Glob. Health 5, e139.

Qian, M.B., Zhou, X.N., 2019. Human liver flukes in China and ASEAN: time to fight together. PLoS Negl. Trop. Dis. 13, e0007214.

Qian, M.B., Chen, Y.D., Fang, Y.Y., Xu, L.Q., Zhu, T.J., Tan, T., Zhou, C.H., Wang, G.F., Jia, T.W., Yang, G.J., Zhou, X.N., 2011. Disability weight of *Clonorchis sinensis* infection: captured from community study and model simulation. PLoS Negl. Trop. Dis. 5, e1377.

Qian, M.B., Chen, Y.D., Liang, S., Yang, G.J., Zhou, X.N., 2012. The global epidemiology of clonorchiasis and its relation with cholangiocarcinoma. Infect. Dis. Poverty 1, 4.

Qian, M.B., Chen, Y.D., Fang, Y.Y., Tan, T., Zhu, T.J., Zhou, C.H., Wang, G.F., Xu, L.Q., Zhou, X.N., 2013a. Epidemiological profile of *Clonorchis sinensis* infection in one community, Guangdong, People's Republic of China. Parasit. Vectors 6, 194.

Qian, M.B., Chen, Y.D., Yan, F., 2013b. Time to tackle clonorchiasis in China. Infect. Dis. Poverty 2, 4.

Qian, M.B., Chen, Y.D., Zhou, X.N., 2013c. Research priorities for the control and elimination of major helminthiases. Zhongguo Ji Sheng Chong Xue Yu Ji Sheng Chong Bing Za Zhi 31, 155–159. (in Chinese).

Qian, M.B., Yap, P., Yang, Y.C., Liang, H., Jiang, Z.H., Li, W., Tan, Y.G., Zhou, H., Utzinger, J., Zhou, X.N., Keiser, J., 2013d. Efficacy and safety of tribendimidine against *Clonorchis sinensis*. Clin. Infect. Dis. 56, e76–e82.

Qian, M.B., Yap, P., Yang, Y.C., Liang, H., Jiang, Z.H., Li, W., Utzinger, J., Zhou, X.N., Keiser, J., 2013e. Accuracy of the Kato-Katz method and formalin-ether concentration technique for the diagnosis of *Clonorchis sinensis*, and implication for assessing drug efficacy. Parasit. Vectors 6, 314.

Qian, M.B., Chen, Y.D., Yang, Y.C., Lu, M.F., Jiang, Z.H., Wei, K., Wei, S.L., Zhou, C.H., Xu, L.Q., Zhou, X.N., 2014a. Increasing prevalence and intensity of foodborne clonorchiasis, Hengxian County, China, 1989–2011. Emerg. Infect. Dis. 20, 1872–1875.

Qian, M.B., Chen, Y.D., Zhou, X.N., 2014b. From recognition to practice: the 140th anniversary of the discovery of *Clonorchis sinensis*. Zhongguo Ji Sheng Chong Xue Yu Ji Sheng Chong Bing Za Zhi 32, 247–252. (in Chinese).

Qian, M.B., Chen, Y.D., Zhou, X.N., 2016a. Gaps in clonorchiasis control in China. Zhongguo Ji Sheng Chong Xue Yu Ji Sheng Chong Bing Za Zhi 34, 373–376. (in Chinese).

Qian, M.B., Utzinger, J., Keiser, J., Zhou, X.N., 2016b. Clonorchiasis. Lancet 387, 800–810.

Qian, M.B., Chen, Y.D., Zhu, H.H., Zhu, T.J., Zhou, C.H., Zhou, X.N., 2018. Establishment and role of national clonorchiasis surveillance system in China. Zhonghua Liu Xing Bing Xue Za Zhi 39, 1496–1500. (in Chinese).

Qian, M.B., Chen, J., Bergquist, R., Li, Z.J., Li, S.Z., Xiao, N., Utzinger, J., Zhou, X.N., 2019a. Neglected tropical diseases in the People's Republic of China: progress towards elimination. Infect. Dis. Poverty 8, 86.

Qian, M.B., Jiang, Z.H., Gan, X.Q., Zhao, J.G., Li, W., Zheng, W.J., Lv, G.L., Zhu, T.J., Zhou, X.N., 2019b. Effect of health education on the awareness and control of clonorchiasis in primary school students. Zhongguo Ji Sheng Chong Xue Yu Ji Sheng Chong Bing Za Zhi 37, 202–206. (in Chinese).

Qian, M.B., Jiang, Z.H., Ge, T., Wang, X., Deng, Z.H., Zhou, C.H., Zhu, H.H., Chen, Y.D., Zhou, X.N., 2019c. Association of raw-freshwater fish-eating practice with the infection of *Clonorchis sinensis*. Zhongguo Ji Sheng Chong Xue Yu Ji Sheng Chong Bing Za Zhi 37, 296–301. (in Chinese).

Qian, M.B., Zhou, C.H., Zhu, H.H., Zhu, T.J., Huang, J.L., Chen, Y.D., Zhou, X.N., 2019d. Assessment of health education products aimed at controlling and preventing helminthiases in China. Infect. Dis. Poverty 8, 22.

Qian, M.B., Zhuang, S.F., Zhu, S.Q., Deng, X.M., Li, Z.X., Zhou, X.N., 2019e. Improving diagnostic performance of the Kato-Katz method for *Clonorchis sinensis* infection through multiple samples. Parasit. Vectors 12, 336.

Qian, M.B., Jiang, Z.H., Ge, T., Wang, X., Zhou, C.H., Zhu, H.H., Zhou, X.N., 2020. Rapid screening of *Clonorchis sinensis* infection: performance of a method based on the practice of eating raw freshwater fish. Acta Trop. 207, 105380.

Qiao, T., Ma, R.H., Luo, X.B., Luo, Z.L., Zheng, P.M., 2012. Cholecystolithiasis is associated with *Clonorchis sinensis* infection. PLoS One 7, e42471.

Solodovnik, D.A., Tatonova, Y.V., Burkovskaya, P.V., 2018. The geographical vector in distribution of genetic diversity for *Clonorchis sinensis*. Parasitol. Res. 117, 335–338.

Technical Steering Panel for National Survey of Current Status of Major Human Parasitic Diseases, 2008. Report on the National Survey of Current Status of Major Human Parasitic Diseases in China. People's Medical Publishing House, Beijing. (In Chinese).

Ting-Jun, Z., Chang-Hai, Z., Long-Qi, X., Bin, Z., Yan-Hong, X., Ying-Dan, C., 2018. Interpretation of detection of intestinal helminthes-the Kato-Katz method (WS/T 570-2017). Zhongguo Xue Xi Chong Bing Fang Zhi Za Zhi 30, 575–577. (in Chinese).

Wei, D.X., Yang, W.Y., Huang, S.Q., Lu, Y.F., Su, T.C., Ma, J.H., Hu, W.X., Xie, N.F., 1980. Parasitological studies on the ancient corpse of the Western Han Dynasty unearthed from tomb no. 168 on Phoenix Hill at Jiangling county. Wuhan Yi Xue Yuan Xue Bao 9 (1–6), 107. (in Chinese).

Whyte, G.D., 1908. Notes on the results of investigation of faeces in the Chao-Chow-Fu Prefectrre, Canton Province. Chin Med J (Engl) 22, 215–217.

World Health Organization, 1995. Control of Foodborne Trematode Infections, WHO Technical Report Series. vol. 849. Geneva.

World Health Organization, 2010. Working to Overcome the Global Impact of Neglected Tropical Diseases: First WHO Report on Neglected Tropical Diseases. First WHO Report on Neglected Tropical Diseases. https://www.who.int/neglected_diseases/resources/9789241564090/en/. Accessed 10 August 2019.

Xiao, S.H., Hui-Ming, W., Tanner, M., Utzinger, J., Chong, W., 2005. Tribendimidine: a promising, safe and broad-spectrum anthelmintic agent from China. Acta Trop. 94, 1–14.

Xiao, S.H., Jian, X., Tanner, M., Yong-Nian, Z., Keiser, J., Utzinger, J., Hui-Qiang, Q., 2008a. Artemether, artesunate, praziquantel and tribendimidine administered singly at different dosages against *Clonorchis sinensis*: a comparative in vivo study. Acta Trop. 106, 54–59.

Xiao, S.H., Xue, J., Tanner, M., Zhang, Y.N., Keiser, J., Utzinger, J., Qiang, H.Q., Liu, X.Y., 2008b. Effect of tribendimidine, artesunate, artemether and praziquantel, administered intragastrically at single, multiple or combined doses, to rats infected with *Clonorchis sinensis*. Zhongguo Ji Sheng Chong Xue Yu Ji Sheng Chong Bing Za Zhi 26, 321–326. (in Chinese).

Xiao, S.H., Keiser, J., Xue, J., Tanner, M., Morson, G., Utzinger, J., 2009a. Effect of single-dose oral artemether and tribendimidine on the tegument of adult *Clonorchis sinensis* in rats. Parasitol. Res. 104, 533–541.

Xiao, S.H., Xue, J., Xu, L.L., Zheng, Q., Qiang, H.Q., Zhang, Y.N., 2009b. The in vitro and in vivo effect of tribendimidine and its metabolites against *Clonorchis sinensis*. Parasitol. Res. 105, 1497–1507.

Xiao, S.H., Xue, J., Xu, L.L., Zhang, Y.N., Qiang, H.Q., 2011. Comparative effect of mebendazole, albendazole, tribendimidine, and praziquantel in treatment of rats infected with *Clonorchis sinensis*. Parasitol. Res. 108, 723–730.

Xiao, S.H., Utzinger, J., Tanner, M., Keiser, J., Xue, J., 2013. Advances with the Chinese anthelminthic drug tribendimidine in clinical trials and laboratory investigations. Acta Trop. 126, 115–126.

Xin-Hua, H., Zhi-Qi, L., Yue-Yi, F., Cai-Yun, R., Xue-Jun, Z., Di-Xuan, L., Shi-Li, L., 2011. Effect of comprehensive control in demonstration plot of clonorchiasis in Yangshan County, 2006–2009. Zhongguo Xue Xi Chong Bing Fang Zhi Za Zhi 23, 569–570. 574. (in Chinese).

Xu, L.Q., Yu, S.H., Xu, S.H., 2000. Distribution and Pathogenic Impact of Human Parasites in China. People's Medical Publishing House, Beijing. (In Chinese).

Xu, L.L., Jiang, B., Duan, J.H., Zhuang, S.F., Liu, Y.C., Zhu, S.Q., Zhang, L.P., Zhang, H.B., Xiao, S.H., Zhou, X.N., 2014. Efficacy and safety of praziquantel, tribendimidine and mebendazole in patients with co-infection of *Clonorchis sinensis* and other helminths. PLoS Neglected Tropical Diseases [electronic resource] 8, e3046.

Xu, M., Yin, J.H., Cao, S.K., Cao, J.P., Zhang, X.F., Shen, Y.J., 2019. Comparison of efficiency of Kato-Katz technique and PCR assay for detecting *Clonorchis sinensis* infection. Zhongguo Xue Xi Chong Bing Fang Zhi Za Zhi 31, 165–168. (in Chinese).

Xue, J., Xu, L.L., Qiang, H.Q., Zhang, Y.N., Xiao, S.H., 2009. Therapeutic effect of tribendimidine, artesunate and praziquantel administered to hamsters infected with *Clonorchis sinensis*. Zhongguo Ji Sheng Chong Xue Yu Ji Sheng Chong Bing Za Zhi 27, 215–218. (in Chinese).

Xue, J., Xu, L.L., Qiang, H.Q., Zhang, Y.N., Xiao, S.H., 2010. Treatment of rats infected with *Clonorchis sinensis* using clinical administration regimens of tribendimidine, praziquantel and artesunate. Zhongguo Ji Sheng Chong Xue Yu Ji Sheng Chong Bing Za Zhi 28, 166–171. (in Chinese).

Zheng, C.L., 2009. Research on Custom about Eating Sashimi. MSc Thesis, Guangxi University for Nationalities, Nanning. (in Chinese).

Zhou, X.N., 2018. Report on the National Survey of Important Human Parasitic Diseases in China (2015). People's Medical Publishing House, Beijing. (in Chinese).

Zhou, X.N., Chen, J.X., Chen, M.G., Bergquist, R., 2005. The National Institute of Parasitic Diseases, Chinese Center for Disease Control and Prevention: a new administrative structure for schistosomiasis control. Acta Trop. 96, 296–302.

Zhu, T.J., Chen, Y.D., Qian, M.B., Zhu, H.H., Huang, J.L., Zhou, C.H., Zhou, X.N., 2019. Surveillance of clonorchiasis in China in 2016. Acta Trop. 203, 105320.

CHAPTER EIGHT

Control of eosinopilic meningitis caused by *Angiostrongylus cantonensis* in China

Shan Lv[a,b,*], Yun-Hai Guo[a,b,c,d,e], Fu-Rong Wei[a,b,c,d,e], Yi Zhang[a,b], Ning Xiao[a,b], Xiao-Nong Zhou[a,b]

[a]National Institute of Parasitic Diseases, Chinese Center for Disease Control and Prevention, Shanghai, People's Republic of China
[b]Chinese Center for Tropical Diseases Research, Shanghai, People's Republic of China
[c]WHO Collaborating Centre for Tropical Diseases, Shanghai, People's Republic of China
[d]National Center for International Research on Tropical Diseases, Ministry of Science and Technology, Shanghai, People's Republic of China
[e]Key Laboratory of Parasite and Vector Biology, Ministry of Health, Shanghai, People's Republic of China
*Corresponding author: e-mail address: lvshan000@126.com

Contents

1. Introduction 270
2. Research on life cycle 272
   2.1 Larval morphology in snails 272
   2.2 Determinants on infection and larval development 273
   2.3 Molecular profile for infection in intermediate host 274
3. Detection and diagnosis 275
   3.1 Detection methods 275
   3.2 Diagnosis criteria 276
4. Epidemiology and surveillance of eosinopilic meningitis 277
   4.1 Epidemiology based on systematic review 277
   4.2 First national survey in 2006–2007 277
   4.3 Second national survey in 2016–2018 278
   4.4 Survey in Southeast Asia 278
   4.5 Surveillance system 279
5. Population genetics 280
   5.1 *Angiostrongylus cantonensis* 280
   5.2 *Pomacea* spp. 281
   5.3 *Achatina fulica* 282
6. Transmission modelling 282
7. Conclusions 283
References 285

Advances in Parasitology, Volume 110
ISSN 0065-308X
https://doi.org/10.1016/bs.apar.2020.02.002

Abstract

Rat lungworm *Angiostrongylus cantonensis* is the major infective agent of human eosinophilic meningitis (EM) in the world. The parasite was first noted in China in 1933. However, the public health importance was not realized until several EM outbreaks occurred recent years. Such disease is considered as emerging infectious disease in the People's Republic of China (P.R. China) since the major source of infection is invasive snail species, particularly *Pomacea* spp. National Institute of Parasitic Diseases (NIPD) initiated a systematic implementation research on this disease since 2003. Our researchers in NIPD developed the lung-microscopy for detecting *A. cantonensis* larvae in *Pomacea* snails and further accomplished the atlas of larval morphology by this method. We studied the determinants in infection, which helped the field collection of snails and improved the infection procedure in laboratory. Our researches promoted the promulgation of diagnosis criteria of angiostrongyliasis cantonensis by the Ministry of Health. We explored the molecular diversity of rat lungworm and its major snail host for development of source-tracing technique. The transmission modelling could provide the vulnerable area for surveillance. All the studies supported the surveillance system of EM caused by *A. cantonensis* in P.R. China. Such implementation research will provide a case study for control of emerging infectious diseases.

1. Introduction

The rat lungworm was first discovered from the pulmonary arteries of *Rattus rattus* and *R. norvegicus* in Guangzhou, China in 1933 (Chen, 1933) and named as *Pulmonema cantonensis* (Chen, 1935). The parasite was also reported as *Haemostrongylus ratti* in Taiwan in 1937 (Matsumoto, 1937; Yokogawa, 1937). Dougherty named rat lungworm as *Angiostrongylus cantonensis* (Dougherty, 1946). The first human case with rat lungworm infection was reported in Taiwan (Beaver and Rosen, 1964; Nomura and Lin, 1945). However, the importance of public health was not realized until 1960s (Alicata, 1962). Rat lungworm is now considered as the major cause of eosinophilic meningitis (EM) in tropic and subtropical regions (Wang et al., 2008). The main route of infection is ingesting terrestrial and freshwater mollusks containing the third-stage larvae of *A. cantonenis*. Infected lizard, frog, shrimps as well as contaminated vegetables could be source of infection (Fig. 1).

The first human case with discovery of larvae of *A. cantonensis* from cerebrospinal fluid in the mainland of the People's Republic of China (P.R. China) was reported in 1982 (He et al., 1984). Only few human cases of EM caused by *A. cantonensis* were reported before 1997. However, this disease occurred in many outbreaks in China due to the recent bio-invasion and geographical expansion of apple snail (*Pomacea* spp.) and African giant

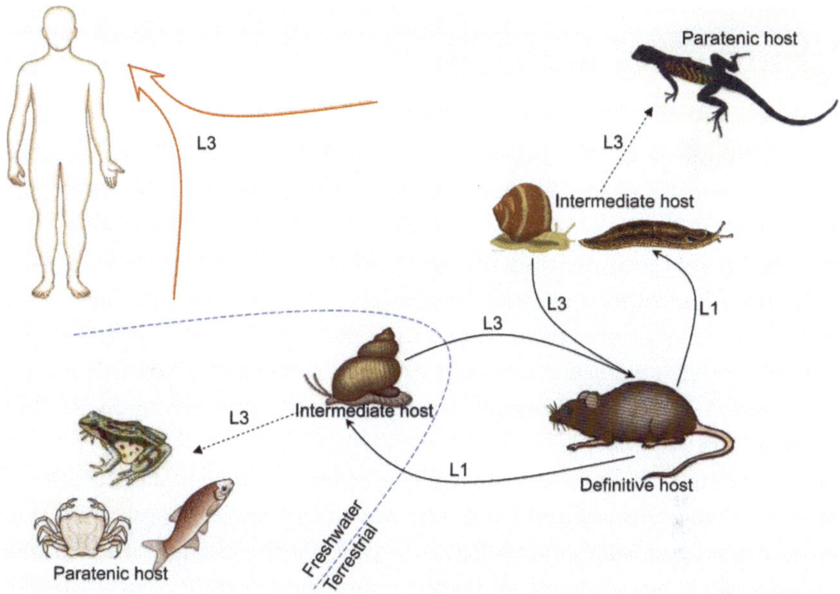

Fig. 1 Life cycle of *A. cantonensis*. L1: first-stage larva, L2: second-stage larva, L3: third-stage larva.

land snail (*Achatina fulica*) and hence considered as an emerging infectious disease (Lv et al., 2008, 2009b).

National Institute of Parasitic Diseases (NIPD), affiliated to Chinese Center for Disease Control and Prevention (China CDC), is the only national center for parasitic diseases control. Researchers from NIPD paid attention to EM caused by *A. cantonensis* as early as 1992 when the second human EM case in the mainland China was parasitologically confirmed in NIPD (Chang et al., 1997). In 1997 the first EM outbreak occurred in Wenzhou city located in the east coast area (Xue et al., 2000). The director of NIPD as the leader of expert committee set by Ministry of Health went to Wenzhou and urgently held a meeting to discuss the measures against the outbreak. Finally the source of infection was revealed and the invasive snail species, *Pomacea canaliculata*, was involved into that outbreak (Pan et al., 2002). The snail became a concern in public health for the first time.

NIPD conducted systematic research on *A. cantonensis* since 2003. Ministry of Science & Technology and Ministry of Health financially supported the research. In this chapter, we summarize the NIPD work pertaining to the research and control of EM caused by *A. cantonensis* in the mainland of P.R. China.

2. Research on life cycle

2.1 Larval morphology in snails

The morphology of *A. cantonensis* was described in many publications. However, most authors focused on the characteristics of adult and third-stage larvae (L3). Although the previous morphological description of L3 was helpful for confirmation of infection in mollusks, we indeed found some snails harbouring pre-third-stage larvae during field surveys, which implied an early infection. Therefore, we study the morphological evolution of *A. cantonensis* larvae in order to facilitate the identification of not only L3 but also second (L2) or first (L1) stage larvae in mollusks (Lv et al., 2009c).

We infected lab-breeding the freshwater snail *P. canaliculata* with the larvae of *A. cantonensis* isolated from fresh faeces of Sprague-Dawley (SD) rats and sacrificed two snails every 2 days to observe the change of larval morphology. We separated and opened the lung (left mantle cavity) of snails. The larval cysts on the lung wall were isolated and larvae were picked out from cysts. Morphology of the larvae was studied under a microscope. There were one and two complete sheaths surrounding L2 and L3, respectively. We found that the refractile granules filled the rear part of larval body at late first stage and second stage, but missed at early first stage and disappeared at third stage. We also observed the feature of larval movement at different stages in vitro. Early L1 obtained from fresh rat faeces actively moved with a coiled tail. When the early L1 invaded and migrate in snail, the coiled tail became straight. Late L1 was only able to shake the head mildly and the body became rigid. The L2 was almost static. L3 became active again and could be seen wriggling within sheaths. L3 normally broke through the sheaths after being isolated from tissue and was characterized by a mellifluous 'Q'-movement. The features mentioned above are helpful for identification of early infection and also facilitate the distinguishing them from free-living nematodes that are often observed to contaminate mollusc samples.

The atlas of larval morphology has been used in the first and second national survey as well as the surveillance in P.R. China. Meanwhile, the larval morphology is also used in some surveys in Brazil (Guerino et al., 2017; Maldonado et al., 2010; Thiengo et al., 2010), Colombia (Córdoba-R et al., 2017), Cuba (Vazquez and Sanchez, 2015), and USA (Qvarnstrom et al., 2013).

2.2 Determinants on infection and larval development

P. canaliculata could develop from a 1 mm egg to an 80 mm adult snail. We tested the susceptibility of *P. canaliculata* of different developmental stages to *A. cantonensis* in lab (Liu et al., 2005). We divided snails breeding at laboratory into four ranks (I–IV) according to the weight and exposed the snails to L1, respectively. Their mortality, infection rate, worm burden, and the size, development and distribution of larvae in snails were determined. The infection rate ranged from 76% to 100%. No significant difference was found among the groups. Snails at earlier developmental stage showed higher mortality. The heaviest worm burden and the largest number of snails harbouring more than 100 larvae were found in snails of rank III. In general the larvae in snails showed a synchronous development in the groups. The period before the L3 emergence and the peak time for L3 exhibited no significant difference among the four ranks. We concluded that the snails at different developmental stage showed similar susceptible to *A. cantonensis*. The snail age or weight is not the determinant of infection and development of *A. cantonensis*.

The developmental period of *A. cantonensis* larvae in snails could be considerably variable in different environment. *A. cantonensis* as well as the mollusc host are poikilothermal. Their developmental rate and thus population density relies on the environmental temperature. We determined the relationship between temperature and development of *A. cantonensis* larvae in *P. canaliculata* (Lv et al., 2006b). The threshold temperature for larval development is 15.04 °C, which means the *A. cantonensis* larvae cannot trigger development until the environmental temperature rise up to 15.04 °C. The thermal constant is 262.53 degree-days, which indicates that development from L1 to L3 needs an effective accumulated temperature of 262.53 degree-days. Therefore, increase in environmental temperature promotes the development and thus facilitates the transmission of *A. cantonensis*. Such parameters are useful in prediction of potential distribution range and the assessment of the impact of climate change on transmission of *A. cantonensis* (Kim et al., 2019; Lv et al., 2011).

Dormancy can facilitate the survival of *P. canaliculata* in adverse environment, such as low temperature and draught. We studied the influence of dormancy of snails on the development and infectivity of *A. cantonensis* (Liu et al., 2006). We found the period from L1 to L3 in dormant snails was shorter than that in the snails in aquarium under the same room temperature. We speculated that the lower water temperature probably decrease the growth of larvae. All L3 at various degree of activity recovered from snails in winter room conditions, including dormant snails and active snails in water, infected rats successfully. Although the weight of dormant snails decreased in winter, the survival rate was significantly higher than that of the snails in aquarium within 60 days. The findings from our study indicated that the freshwater snails *Pomacea* collected from field could be kept and brought to lab in dormant status.

Many mollusc species could serve as the intermediate host of *A. cantonensis* (Lv et al., 2008). However, few species play a role in transmission from snail to human. Three species of freshwater snails, i.e., *P. canaliculata*, *Cipangopaludina chinensis*, *Bellamya aeruginosa*, are commonly sold in food markets in P.R. China. We compared the compatibility of three snail species with *A. cantonensis* in lab (Lv et al., 2006a). We found that the mortality and infection rate were not related to exposure time and snail species. Worm burden of *P. canaliculata* was significantly higher than other two species. However, worm density (worm burden/snail weight) of *P. canaliculata* and *B. aeruginosa* was not significantly different, but higher than that of *C. chinensis*. In addition, the worm burden and worm density of *P. canaliculata* and *B. aeruginosa* depended on exposure time. All the three snail species could be infected by *A. cantonensis* in lab, but the compatibility of *P. canaliculata* is superior to the other two species. Since *C. chinensis* and *B. aeruginosa* are benthic, the possibility of contact with faeces of rats in field is less than *P. canaliculata* which living in shallow water or along bank of water body. Therefore, *P. canaliculata* is more important in transmission of *A. cantonensis*.

2.3 Molecular profile for infection in intermediate host

Pomacea spp. is one of few freshwater snails that play a role in transmission of rat lungworm. We study the change of gene expression post-infection with *A. cantonensis* to explore the potential molecular interaction between parasite and host (Yue, 2019). We collected snail 1 day, 10 day, 20 day post-infection. Meanwhile we collected control snail at the same intervals. We extract total RNA and then reverse-transcribed into cDNA. The cDNA were sequenced. A total of 983 genes were identified due to differential expression. Only 13 genes differentially expressed at all three stages. The verified genes were mostly involved in cell growth and metabolism, signal transduction, immune response and stress response, which indicated a complex molecular interface between host and parasite involved in infection. The down-regulated genes were more than the up-regulated.

G lysozyme and α-tubulin were demonstrated the key molecules in infection. We found the two genes of *Pomacea* spp. were down-regulated after exposure to *A. cantonensis*. We further study the expression profile in different organs or tissues by time. We found that G lysozyme gene was highly expressed in liver of both non-infected female and male snails, which significantly higher than that in kidneys, lung, digestive gut, and foot muscles (Yue et al., 2019b). However, α-tubulin gene showed the differential expression in different organs or tissues. The α-tubulin gene was highly expressed in liver of male and female snails. In addition, this gene was also highly expressed in lung and digestive gut of female snails.

In infected *Pomacea* snails, G lysozyme gene was significantly down-regulated in liver, kidneys, digestive gut, lung, and foot muscles at early and middle stage of infection, but the expression become normal in liver and kidneys at later stage (20 day post-infection). Digestive gut, lung, and foot muscles were sites of invasion by *A. cantonensis* larvae. Most larvae dwell in these organs. As an immune-related gene, the expression of G lysozyme gene was probably influenced by circulating antigens released from larval development. The larvae could develop to third-stage larvae surrounded by two sheathes approximately 3 weeks after infection. The third-stage larvae will maintain diapause for a long time in snail. Therefore, the decreasing circulating antigens in liver and kidneys cause increasing expression of G lysozyme gene.

The α-tubulin gene was normally down-regulated in all organs with an exception of food muscles at the middle and later stage (10 day and 20 day post-infection) (Yue et al., 2019a). Interestingly, such gene expression did not show different before and after exposure. Alpha-tubulin is a protein associated with cell growth. Our result indicated infection with *A. cantonensis* will influence the growth and development of snails.

3. Detection and diagnosis

3.1 Detection methods

Artificial digestion is the main method for detection of *A. cantonenis* larvae in mollusks. However, this method is time-costing and shows less effectiveness at low worm burden. PCR is a newly developed and sensitive method. However, sampling from a large-sized snail, such as *P. canaliculata*, lead to an uncertainty in detection result, since only part of tissue was submitted to DNA extraction. In addition, PCR is not accessible in some settings. The large-sized snails, *Pomacea* spp. and *A. fulica*, are the major source of

A. cantonensis infection. The special anatomical structure of their lung makes it possible to directly observe the larval cysts under a dissecting microscope (Lv et al., 2009c). To make sure the cyst is due to *A. cantonensis* larvae, we picked the cyst under the dissecting microscope and isolate the larvae. We identify the species of larvae according to the morphology and moving features mentioned above. We call this method as lung-microscopy. Compared to the conventional artificial digestion, lung-microscopy significantly save the time and showed the similar effectiveness (Liu et al., 2007). We also established the relationship between worm burden in lung tissue and that in other tissues. The result showed larvae burden in lung was positively correlated with that in other tissues. Therefore, we recommended this method to be applied in large-scale survey. The method is not only a routine method in surveillance of snail food for *A. cantonensis* but also was recommended in other countries (Córdoba-R et al., 2017; Giraldo et al., 2019; Solorzano Alava et al., 2014).

3.2 Diagnosis criteria

The outbreak of angiostrongyliasis occurred in Beijing in 2016 was the biggest one up to date, which involved 160 individuals. The outbreak implied that infections can occur in non-endemic area through the transportation of snail food. Since EM caused by *A. cantonensis* is an emerging infectious disease, the awareness and diagnosis capacity of physicians is limited. Therefore, a diagnosis guide is necessary for the clinics to look for the evidences to support the reliable diagnosis. Shortly after the outbreak in Beijing the Ministry of Health held a consultant conference. Diagnosis criteria is one of the most important topics. In fact, an empirical diagnosis criteria had been implemented during the Beijing outbreak (Yin et al., 2006). In the early 2008 another outbreak of EM caused by *A. cantonensis* occurred in Dali city, Yunnan province (Lv et al., 2009a). The authors conduct a survey on the patients and collected the epidemiological and clinic data of inpatients and outpatients. The patients mainly distributed in two hospitals. However, the diagnosis criteria were not identical. Since the discovery of parasite from human body is very difficult, a diagnosis should standardized to guide the physicians to find the most reliable evidence to support their diagnosis. NIPD took the task from the Ministry of Health to organize experts to draft the diagnosis criteria for EM due to *A. cantonensis*. We categorized diagnosis into suspected, clinically diagnosed, pathologically confirmed cases. The diagnosis criteria was eventually issued by Ministry of

Health in 2010 after several rounds of revisions (Zhang and Lv, 2010). We also propagated the criteria in national training course. It became the only official standard for diagnosis of EM due to *A. cantonensis* in China.

4. Epidemiology and surveillance of eosinopilic meningitis

4.1 Epidemiology based on systematic review

The first human case of angiostrongyliasis in the mainland of P.R. China was reported in 1984. Only three more cases were recorded before 1997 when the first outbreak occurred in Wenzhou city, Zhejiang province located in east coast. Some sporadic human cases were reported in the following years. The importance of public health was not realized until the large outbreak occurred in Beijing in 2006. There were 160 cases involved in that outbreak, which indicate EM caused by *A. cantonensis* could occur in a non-endemic area by transportation of snail food. In order to understand the epidemiology of *A. cantonensis* and EM in China, we performed a systematic review based on literature in 2006 (Lv et al., 2008). A total of 334 cases including four dead cases were identified and 86.5% cases were involved in 7 outbreaks. They distributed in 9 out of 31 provinces. The major source of infection was *P. canaliculata* and *A. fulica*. The rat lungworm was endemic in six provinces. We revealed the distribution of EM due to *A. cantonensis* in P.R. China first time. Such findings directly guided our first national survey on *A. cantonensis* with the focus on *P. canaliculata* and *A. fulica* and facilitated the determination of survey range.

4.2 First national survey in 2006–2007

Although some areas were confirmed endemic with *A. cantonensis* in south China (Lv et al., 2008), the current status remained unknown due to long time span and small scale surveys. In order to reveal the endemic range of *A. cantonensis*, the Ministry of Health decided to conduct a national survey soon after the Beijing outbreak. NIPD took this task and carried out the first national survey between 2006 and 2007 (Lv et al., 2009b). We determined the survey region based on potential distribution of *P. cantonensis* according to a degree-day model (Zhou et al., 2003). The sampling sites were selected by a method of geographical grid sampling. We collected terrestrial and freshwater mollusks as well as rats. All the specimens were brought back to lab for detection. In addition, we also sampled and investigated the snails in markets, restaurants and snail farms. Finally, 164 counties in 19 provinces were involved in this survey. We identified 59 counties in 7 provinces were endemic of *A. cantonensis*. The major intermediate hosts were *Pomacea* spp. and *Achatina fulica* in China. They were widely distributed in south China and the infection rate was 6.8% and 13.4%, respectively. The survey for the first time revealed the present distribution of *A. cantonensis* and two invasive snail species. The findings implied that *Pomacea* spp. and *Achatina fulica* as the vector played an important role in a local endemic infectious diseases, which highlighted the health importance of invasive species. The survey also provided the basic information for establishment of surveillance.

4.3 Second national survey in 2016–2018

Biological invasion and the impact are recognized in China. The Ministry of Science and Technology set a series of survey on invasive species, including vectors of pathogen. NIPD lead a team including seven institutes to conduct a national survey on *Pomacea* spp. (unpublished data). Based on the results of first national survey and the transmission modelling (Lv et al., 2011) we chose 49 counties. The survey indicated that the range of *Pomacea* has extended to the basin of Yangtze River. The snails were commonly found in Hunan, Hubei, Jiangxi, Zhejiang as well as Shanghai. We draw the map of *Pomacea* snail based on our two surveys and other published data (Fig. 2). Since the snail is an important intermediate host of *A. cantonensis*, the further survey on the infection should be push forward.

4.4 Survey in Southeast Asia

EM caused by rat lungworm is endemic in Southeast Asia, but little is known about its distribution in Laos, Cambodia and Vietnam. We performed a multicountry survey for *A. cantonensis* in these countries and estimated the prevalence in snails along the Mekong River and the east coast of Vietnam (Lv et al., 2018). We visited 34 villages and found *Pomacea* spp. in most of villages. A total of 1291 *Pomacea* snails were detected for larvae of rat lungworm. The overall prevalence of *A. cantonensis* was 2.5% with a highest prevalence of 16.2%. Our morphological study and DNA sequencing indicated that only *A. cantonensis* was endemic in the study area. The wide accessibility of *Pomacea* snails as well as infection with *A. cantonensis* indicates that this snail species could be used as an index for risk of EM outbreaks.

Freshwater snails are commonly food items in Southeast Asia and the Far East. According to our surveys in P.R. China and Southeast Asia *Pomacea*

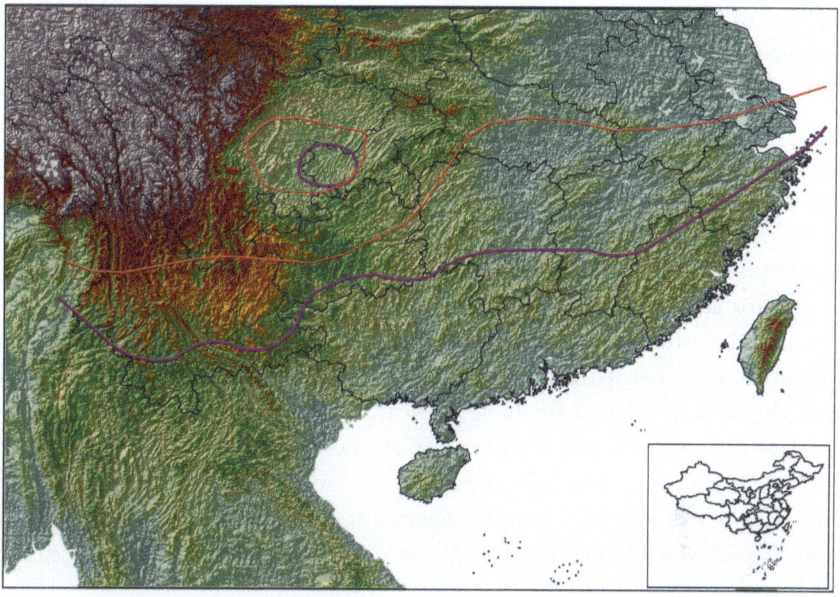

Fig. 2 Geographical range of *Pomacea* spp. in P.R. China. Purple line denotes the distribution of *Pomacea* snails by first national survey 2006–2007; red line denotes the distribution by second national survey and other public data source.

snails could be an indicator for surveillance for epidemicity of *A. cantonensis*. First, *Pomacea* snails have been widely spread and saw in food markets in Southeast Asia and the Far East (Joshi and Sebastian, 2006). Second, *Pomacea* snails are much easier to find than *Pila* and other freshwater snails since they lay highly visible bright pink eggs above the water surface on emergent hard surfaces. Third, the time for collecting *Pomacea* snails is not dependent on weather and light that influence the activity of land snails and slugs. Finally, a simple detection technique for *Angiostrongylus* in *Pomacea* has been developed (Liu et al., 2007; Lv et al., 2009c).

4.5 Surveillance system

After outbreaks in Beijing and Dali, the surveillance system of EM caused by *A. cantonensis* was established and implemented in Beijing, Yunnan, Fujian, Guangdong and Shanghai. Two parallel surveillances are done in the trial city. Physicians in the pilot hospitals identify the EM due to *A. cantonensis* according the diagnosis criteria issued by Ministry of Health and report cases through an internet-based reporting system to NIPD. On the other hand, the staff in local CDCs routinely collected and examined the snail samples (i.e. *Pomacea* spp., *Achatina fulica*, *Bellamya* spp. and *Cipangopaludina* spp.) from food markets and presented the data to NIPD. The hospitals and CDCs regularly exchange the information pertaining to the patients and surveillance results on snails. In the last 10 years, we identified 265 patients in pilot hospitals and also found an outbreak through this surveillance system. No parasitologically confirmed patient was reported. Infected snails were commonly found in markets in Fuzhou and Xiamen, Fujian province and Dali, Yunnan province. The results showed seasonality. In addition, we conducted training for the staff from pilot hospitals and CDCs every year and promoted the capacity of response to the outbreaks. However, we realized the efficacy of surveillance became weak since the EM due to *A. cantonensis* was rare in China after all. For instance, we found infected snails in markets in some cities, but recently few human cases were reported. The occurrence of EM due to *A. cantonensis* is dependent on the preparedness of snail food to some extent. We are considering to extent the disease to the surveillance of hypereosinophilic syndrome.

5. Population genetics

5.1 *Angiostrongylus cantonensis*

A. cantonensis and *A. costaricensis* are two known species of *Angiostrongylus* that can cause illness in human beings. We characterized their mitochondrial (mt) genomes (Lv et al., 2012). Their mt genomes were 13,497 and 13,585 bp, respectively, and represented the smallest in the class of Chromadorea so far. Each mt genome encodes 12 proteins, 22 transfer RNAs, and 2 ribosomal RNAs. All genes are located on the same strand. Nucleotide difference of the two mt genomes is 18.4%. To our knowledge we first characterized the mt genome of genus *Angiostrongylus*. The mt genomes of other *Angiostrongylus* species or stains were determined shortly (Aghazadeh et al., 2015; Cervena et al., 2019; Gasser et al., 2012; Yong et al., 2015, 2016). The information provided by complete mt genomes is useful to screen the biomarker for population genetic study.

P.R. China is the presumptive home range of the rat lungworm (Prociv et al., 2000). However, the genetic diversity was little known. We conducted a study of the genetic variation of *A. cantonensis* in P.R. China to deepen the current knowledge pertaining to its origin and global spread from a molecular perspective (Lv et al., 2017). The specimens used in this study were collected from 32 collection sites during first national survey 2006–2007 as well as following complementary investigations. We characterized the complete cox1 and nad1 genes of 130 specimens and obtained 357 nuclear DNA sequences containing two complete ITS (ITS1 and ITS2) and 5.8S rRNA of the same 130 samples. Only *A. cantonensis* was identified in our study. There were two major groups of *A. cantonensis* in P.R. China. The first group showed higher diversity and could be further divided into six distinct clades, while lower diversity was observed in the other group. The most common genotypes were Group Ia (47.7%) and Group II (22.3%).

We also compared the different gene markers and sequencing policies. Genotyping showed a striking difference between mitochondrial DNA and ITS. Since intragenomic variation of ITS sequences was found in a large proportion of the samples, we thought ITS was not ideal marker for population genetic study of *A. cantonensis*. In addition, we observed a considerable proportion of singlets in nad1 gene. It was probably caused by the sequencing strategy, i.e., clone sequencing. Therefore, clone sequencing cannot be recommended.

5.2 *Pomacea* spp.

Pomacea spp. is the most important source of *A. cantonensis* infection in China. Many EM outbreaks were attributed to the snail (Odermatt et al., 2010). In addition, the snail is also notorious for crop damage and ecological degradation (Joshi and Sebastian, 2006). We determine the genetic diversity of invasive snails (*Pomacea* spp.) to reveal their distribution pattern in P.R. China (Lv et al., 2013). We collected *Pomacea* snails in the first national survey on *A. cantonensis* (Lv et al., 2009b). A total of 544 snails from 54 sites were used for our study. We successfully obtained 521 cox1 sequences and identified 24 unique haplotypes, of which 6 haplotypes were commonly found in P.R. China. *P. canaliculata, P. insularum* and two cryptic groups were observed. The distribution of the 24 haplotypes across P.R. China shows a mosaic pattern, which indicated that multiple and secondary introductions occurred in P.R. China.

We also compared our haplotypes to the information available in GenBank. Globally, only 6 of 112 haplotypes of *P. canaliculata, P. insularum, P. dolioides, P. lineata* and *P. paludosa* are shared between introduced and native (South

America) snail populations. We categorized the 112 haplotypes into 16 clusters, 5 of which occur in mainland P.R. China. Three of the five clusters could be traced back to South America. The remaining two clusters were unique to P.R. China. Therefore, there is still a gap in knowledge of population genetic of *Pomacea* in South America.

5.3 *Achatina fulica*

A. fulica is a large-sized land snail species. The emergence in the mainland of P.R. China was noted in 1931 in a university campus located in south-east coast (Jarrett, 1931). The first national survey of *A. cantonensis* showed that the snail infested in Fujian, Guangdong, Guangxi, Hainan, Yunnan and Guizhou provinces (Lv et al., 2009b). According to Fontanilla (Fontanilla et al., 2014) the genetic variation of *A. fulica* outside East Africa and Indian Ocean islands was low. Indeed, the invasion history of *A. fulica* is well known (Mead, 1961). A bottleneck effect could be an explanation of low genetic diversity in Southeast Asia. We hence did not research the population genetics of *A. fulica* in P.R. China. Instead, we characterized the whole genome of *A. fulica* (Guo et al., 2019). The genome size of *A. fulica* was estimated 2.12 Gb with a high repeat content up to 71%. A total of 101.6 Gb genomic long-read data were generated and finally a genome of 1.85 Gb was assembled. We predicted 23,726 protein-coding genes in the whole genome, of which 96.3% genes were functionally annotated. To our knowledge, this was the first complete characterized genome of terrestrial mollusks to date. The chromosome sequence will provide a valuable genetic resource for research on environmental adaptation of the invasive snail as well as chromosome-level of evolution within mollusks. Since *A. fulica* played an important role of global spread of rat lungworm at early stage, global invasion history of *A. fulica* based on whole genome should be helpful in clarification of global transmission of *A. cantonensis*.

6. Transmission modelling

Climate change is a global concern and is imposing negative biotic impact. Vector- and snail-borne tropical diseases are being influenced by climate change. Rat lungworm is transmitted by poikilothermal mollusks. Only few freshwater snail species play an important role in transmission of *A. cantonensis*, which is partially explained by the low likelihood of contact between snails in the water body and infected rat faeces. *Pomacea* spp. is an exception due to their ecological characteristics. The snail was introduced into China in 1981 and has become the key intermediate host for *A. cantonensis*. Thus far, the snail has been recorded in 13 provinces of southern China (Lv et al., 2009b). We developed a biological model and assessed potential impacts of climate change on the distribution of *P. canaliculata* and hence the transmission of *A. cantonensis* (Lv et al., 2011). The transmission of *A. cantonensis* is dependent on both definitive and intermediate hosts. Therefore, the interaction between parasite and hosts as well as the dependence of hosts on environment were considered in our model. We identified the mean January temperature and snail generation intensity (generation number) as the key factors determining *P. canaliculata* distribution, while the generation intensity in the intermediate host, the longevity of *A. cantonensis*-infected rats and the dormant period of *P. canaliculata* are the major factors for *A. cantonensis* transmission. Our models predicted a remarked increase for the 'spread' regions of *Pomacea* snails and potential endemic area of *A. cantonensis* by 2030s. We also identified the vulnerable areas for *A. cantonensis* transmission, which help us to set the surveillance sites. Our modelling provided a case study pertaining the impact of climate change on parasite transmission (Cizauskas et al., 2017; Okamura and Feist, 2011).

7. Conclusions

Since 2003 we conducted a systematic implementation research on *A. cantonensis* and its hosts, covering field surveys, laboratory studies, modelling and surveillance. All the studies were performed for the establishment of surveillance system and disease control. Additionally, our researches supported five master and one PhD students and provided a postdoc position in last 15 years. A total of 45 peer-review articles were published. More than 1000 staff from local CDCs and hospitals was trained. Our research promoted the elevation of control capacity. The findings from our research were also instructive for other researches by other Chinese and abroad institutes. Such implementation research will provide a case study for control of emerging infectious diseases (Fig. 3).

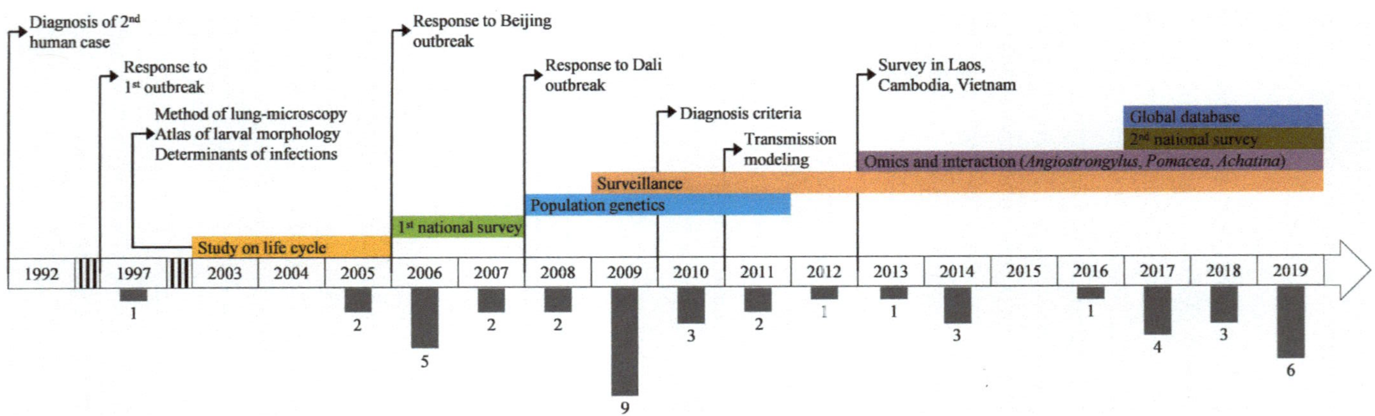

Fig. 3 Main outputs pertaining to *Angiostrongylus cantonensis* since 1992. Grey bar denote the number of publication by NIPD.

References

Aghazadeh, M., Traub, R.J., Mohandas, N., Aland, K.V., Reid, S.A., McCarthy, J.S., Jones, M.K., 2015. The mitochondrial genome of *Angiostrongylus mackerrasae* as a basis for molecular, epidemiological and population genetic studies. Parasit. Vectors 8 (1), 473.

Alicata, J.E., 1962. *Angiostrongylus cantonensis* (Nematoda: Metastrongylidae) as a causative agent of eosinophilic meningoencephalitis of man in Hawaii and Tahiti. Can. J. Zool. 40 (1), 5–8.

Beaver, P.C., Rosen, L., 1964. Memorandum on the first report of *Angiostrongylus* in man, by Nomura and Lin, 1945. Am. J. Trop. Med. Hyg. 13 (4), 589–590.

Cervena, B., Modry, D., Feckova, B., Hrazdilova, K., Foronda, P., Alonso, A.M., Lee, R., Walker, J., Niebuhr, C.N., Malik, R., Slapeta, J., 2019. Low diversity of *Angiostrongylus cantonensis* complete mitochondrial DNA sequences from Australia, Hawaii, French Polynesia and the Canary Islands revealed using whole genome next-generation sequencing. Parasit. Vectors 12 (1), 241.

Chang, Z.S., Jiang, Z.X., Ni, X.J., Zhang, Y.N., Tang, M.G., Zhao, Y.M., 1997. Coexistence of a fifth-stage larva and a developing female of *Angiostrongylus cantonensis* in the cerebrospinal fluid of a patient. Chin. J. Parasitol. Parasit. Dis. 15 (5), 31–33.

Chen, H.T., 1933. A preliminary report on a survey of animal parasites of Canton, China, rats. Lingnan Sci. J. 12 (1), 65–74.

Chen, H.T., 1935. Un nouveau nematode Pulmonaire, *Pulmonema cantonensis*, n. g., n. sp. des rats de Canton. Ann. Parasitol. 13 (4), 312–317.

Cizauskas, C.A., Carlson, C.J., Burgio, K.R., Clements, C.F., Dougherty, E.R., Harris, N.C., Phillips, A.J., 2017. Parasite vulnerability to climate change: an evidence-based functional trait approach. R. Soc. Open Sci. 4 (1), 160535.

Córdoba-R, D., Patiño-Montoya, A., Giraldo, A., 2017. Prevalence of Strongylida nematodes associated with African snail, *Achatina fulica*, in Valle del Cauca, Colombia. Rev. MVZ Córdoba 22 (3), 6276–6286.

Dougherty, E.C., 1946. The genus *Aelurostrongylus* cameron, 1927 (Nematoda: Metastrongylidae), and its relatives; with descriptions of *Parafilaroides*, gen. nov., and *Angiostrongylus gubernaculatus*, sp. nov. Proc. Helminthol. Soc. Wash. 13 (1), 16–25.

Fontanilla, I.K., Sta Maria, I.M., Garcia, J.R., Ghate, H., Naggs, F., Wade, C.M., 2014. Restricted genetic variation in populations of *Achatina* (*Lissachatina*) *fulica* outside of East Africa and the Indian Ocean Islands points to the Indian Ocean Islands as the earliest known common source. PLoS One 9 (9), e105151.

Gasser, R.B., Jabbar, A., Mohandas, N., Schnyder, M., Deplazes, P., Littlewood, D.T., Jex, A.R., 2012. Mitochondrial genome of *Angiostrongylus vasorum*: comparison with congeners and implications for studying the population genetics and epidemiology of this parasite. Infect. Genet. Evol. 12 (8), 1884–1891.

Giraldo, A., Garzón, C., Castillo, A., Córdoba-Rojas, D.F., 2019. Confirmation of the presence of *Angiostrongylus cantonensis* in lung tissue of the African giant snail (*Lissachatina fulica*) in Colombia. Infect. 23 (2), 129–132.

Guerino, L.R., Pecora, I.L., Miranda, M.S., Aguiar-Silva, C., Carvalho, O.D., Caldeira, R.L., Silva, R.J., 2017. Prevalence and distribution of *Angiostrongylus cantonensis* (Nematoda, Angiostrongylidae) in *Achatina fulica* (Mollusca, Gastropoda) in Baixada Santista, Sao Paulo, Brazil. Rev. Soc. Bras. Med. Trop. 50 (1), 92–98.

Guo, Y.H., Zhang, Y., Liu, Q., Huang, Y., Guangyao Mao, G.Y., Yue, Z.Y., Abe, E.M., Li, J., Wu, Z.D., Li, S.Z., Zhou, X.N., Hu, W., Xiao, N., 2019. A chromosomal-level genome assembly for the giant African snail *Achatina fulica*. Giga Sci. 8 (10), 1–8.

He, J.Z., Zhu, S.H., Yang, S.Q., Yu, B.W., Chen, Y.S., Hu, X.X., Wang, X.B., Wang, L., 1984. First discovery and evidence of *Angiotrongylus cantonensis* in the cerebrospinal fluid from the case of the population of the mainland of China. J. Guangzhou Med. Coll. 12 (3), 1–4.

Jarrett, V.H.C., 1931. The spread of the snail *Achatina fulica* to South China. Hong Kong Nat. 2 (4), 262–264.

Joshi, R.C., Sebastian, L.S., 2006. Global Advances in Ecology and Management of Golden Apple Snails. Philippine Rice Research Institute, Nueva Ecija.

Kim, J.R., Wong, T.M., Curry, P.A., Yeung, N.W., Hayes, K.A., Cowie, R.H., 2019. Modelling the distribution in Hawaii of *Angiostrongylus cantonensis* (rat lungworm) in its gastropod hosts. Parasitology 146 (1), 42–49.

Liu, H.X., Zhang, Y., Zhou, X.N., Lv, S., Zhu, D., Lin, J.X., Li, L.S., Li, Y.S., Yin, W.G., 2005. Studies on susceptibility of *Pomacea canaliculata* of different developmental stages to infection with *Angiostrongylus cantonensis*. Chin. J. Parasitol. Parasit. Dis. 23 (5), 262–265.

Liu, H.X., Zhang, Y., Zhou, X.N., Lv, S., Zhu, D., Lin, J.X., LI, L.S., Li, Y.S., 2006. Studies on the growth-development and infectivity of *Angiostrongylus cantonensis* in dormant *Pomacea canaliculata*. Chin. J. Parasitol. Parasit. Dis. 24 (4), 269–272.

Liu, H.X., Zhang, Y., Lv, S., Zhu, D., Wang, X.H., Hu, L., Zhou, X.N., 2007. A comparative study of three methods in detecting *Angiostrongylus cantonensis* larvae in lung tissue of *Pomacea canaliculata*. Chin. J. Parasitol. Parasit. Dis. 25 (1), 53–56.

Lv, S., Zhang, Y., Wang, X.H., Liu, H.X., Zhu, D., Yin, W.G., Zhou, X.N., 2006a. Experimental study on compatibility of three species of freshwater snails with *Angiostrongylus cantonensis*. Chin. J. Parasitol. Parasit. Dis. 24 (4), 277–280.

Lv, S., Zhou, X.N., Zhang, Y., Liu, H.X., Zhu, D., Yin, W.G., Steinmann, P., Wang, X.H., Jia, T.W., 2006b. The effect of temperature on the development of *Angiostrongylus cantonensis* (Chen 1935) in *Pomacea canaliculata* (Lamarck 1822). Parasitol. Res. 99 (5), 583–587.

Lv, S., Zhang, Y., Steinmann, P., Zhou, X.N., 2008. Emerging angiostrongyliasis in Mainland China. Emerg. Infect. Dis. 14 (1), 161–164.

Lv, S., Zhang, Y., Chen, S.R., Wang, L.B., Fang, W., Chen, F., Jiang, J.Y., Li, Y.L., Du, Z.W., Zhou, X.N., 2009a. Human angiostrongyliasis outbreak in Dali, China. PLoS Negl. Trop. Dis. 3 (9), e520.

Lv, S., Zhang, Y., Liu, H.X., Hu, L., Yang, K., Steinmann, P., Chen, Z., Wang, L.Y., Utzinger, J., Zhou, X.N., 2009b. Invasive snails and an emerging infectious disease: results from the first national survey on *Angiostrongylus cantonensis* in China. PLoS Negl. Trop. Dis. 3 (2), e368.

Lv, S., Zhang, Y., Liu, H.X., Zhang, C.W., Steinmann, P., Zhou, X.N., Utzinger, J., 2009c. *Angiostrongylus cantonensis*: morphological and behavioral investigation within the freshwater snail *Pomacea canaliculata*. Parasitol. Res. 104 (6), 1351–1359.

Lv, S., Zhang, Y.I., Steinmann, P., Yang, G.J., Yang, K., Zhou, X.N., Utzinger, J., 2011. The emergence of angiostrongyliasis in the People's Republic of China: the interplay between invasive snails, climate change and transmission dynamics. Freshwater Biol. 56 (4), 717–734.

Lv, S., Zhang, Y., Zhang, L., Liu, Q., Liu, H.X., Hu, L., Wei, F.R., Steinmann, P., Graeff-Teixeira, C., Zhou, X.N., Utzinger, J., 2012. The complete mitochondrial genome of the rodent intra-arterial nematodes *Angiostrongylus cantonensis* and *Angiostrongylus costaricensis*. Parasitol. Res. 111 (1), 115–123.

Lv, S., Zhang, Y., Liu, H.X., Hu, L., Liu, Q., Wei, F.R., Guo, Y.H., Steinmann, P., Hu, W., Zhou, X.N., 2013. Phylogenetic evidence for multiple and secondary introductions of invasive snails: *Pomacea* species in the People's Republic of China. Divers. Distrib. 19 (2), 147–156.

Lv, S., Zhang, Y., Steinmann, P., Utzinger, J., Zhou, X.N., 2017. The genetic variation of *Angiostrongylus cantonensis* in the People's Republic of China. Infect. Dis. Poverty 6 (1), 125.

Lv, S., Guo, Y.H., Nguyen, H.M., Sinuon, M., Sayasone, S., Lo, N.C., Zhou, X.N., Andrews, J.R., 2018. Invasive *Pomacea* snails as important intermediate hosts of *Angiostrongylus cantonensis* in Laos, Cambodia and Vietnam: implications for outbreaks of eosinophilic meningitis. Acta Trop. 183, 32–35.

Maldonado, A.J., Simoes, R.O., Oliveira, A.P., Motta, E.M., Fernandez, M.A., Pereira, Z.M., Monteiro, S.S., Torres, E.J., Thiengo, S.C., 2010. First report of *Angiostrongylus cantonensis* (Nematoda: Metastrongylidae) in *Achatina fulica* (Mollusca: Gastropoda) from Southeast and South Brazil. Mem. Inst. Oswaldo Cruz 105 (7), 938–941.

Matsumoto, T., 1937. On a nematode found in the lungs, especially the pulmonary artery of the wild rat. J. Formos. Med. Assoc. 36 (12), 32–35.

Mead, A.R., 1961. The Giant African Snail: A Problem in Economic Malacology. University of Chicago Press.

Nomura, S., Lin, P.H., 1945. First case report of human infection with *Haemostrongylus ratti* Yokogawa. Taiwan No Ikai 3, 589–592.

Odermatt, P., Lv, S., Sayasone, S., 2010. Less common parasitic infections in Southeast Asia that can produce outbreaks. Adv. Parasitol. 72, 409–435.

Okamura, B., Feist, S.W., 2011. Emerging diseases in freshwater systems. Freshwater Biol. 56 (4), 627–637.

Pan, C.W., Liang, S.H., Huang, H.C., Lin, H.B., Liu, Q.Z., Zheng, X.Y., Qin, Q., Jin, Y.G., Yi, W.P., Lin, B.C., Zhou, Z.L., Wang, K.H., Lin, D.X., 2002. Study on *Angiostrongyliasis cantonensis* in Wenzhou. J. Wenzhou Med. Coll. 32 (6), 341–343.

Prociv, P., Spratt, D.M., Carlisle, M.S., 2000. Neuro-angiostrongyliasis: unresolved issues. Int. J. Parasitol. 30 (12 − 13), 1295–1303.

Qvarnstrom, Y., Bishop, H.S., da Silva, A.J., 2013. Detection of rat lungworm in intermediate, definitive, and paratenic hosts obtained from environmental sources. Hawaii J. Med. Public Health 72 (6 Suppl. 2), 63–69.

Solorzano Alava, L.F., Martini Robles, L., Hernandez Alvarez, H., Sarracent Perez, J., Muzzio Aroca, J., Rojas Rivero, L., 2014. *Angiostrongylus cantonensis*: an emerging parasite in Ecuador. Rev. Cubana Med. Trop. 66 (1), 20–33.

Thiengo, S.C., Maldonado, A., Mota, E.M., Torres, E.J., Caldeira, R., Carvalho, O.S., Oliveira, A.P., Simoes, R.O., Fernandez, M.A., Lanfredi, R.M., 2010. The giant African snail *Achatina fulica* as natural intermediate host of *Angiostrongylus cantonensis* in Pernambuco, Northeast Brazil. Acta Trop. 115 (3), 194–199.

Vazquez, A.A., Sanchez, J., 2015. First record of the invasive land snail *Achatina (Lissachatina) fulica* (Bowdich, 1822) (Gastropoda: Achatinidae), vector of *Angiostrongylus cantonensis* (Nematoda: Angiostrongylidae), in Havana, Cuba. Molluscan Res. 35 (2), 139–142.

Wang, Q.P., Lai, D.H., Zhu, X.Q., Chen, X.G., Lun, Z.R., 2008. Human angiostrongyliasis. Lancet Infect. Dis. 8 (10), 621–630.

Xue, D.Y., Ruan, Y.Z., Lin, B.C., Zheng, R.Y., Fang, J.Q., Zhao, Q.X., Li, F.M., Pan, C.W., 2000. Epidemiological investigation on an outbreak of angiostrongyliasis cantonensis in Wenzhou. Chin. J. Parasitol. Parasit. Dis. 18 (3), 176–178.

Yin, C.H., Gan, S.B., Liu, J., Han, X.Q., Feng, M.L., Ji, A.P., Guo, Z.Z., Wang, F., Li, S.G., Wu, Z.Y., 2006. Guide for diagnosis and treatment of angiostrongyliasis cantonenis. Chin. J. Intern. Med. 45 (12), 1051–1052.

Yokogawa, S., 1937. A new species of nematode found in the lungs of rats *Haemostrongylus ratti* sp. nov. Trans. Nat. Hist. Soc. Formosa 27 (170), 247–250.

Yong, H.S., Song, S.L., Eamsobhana, P., Goh, S.Y., Lim, P.E., Chow, W.L., Chan, K.G., Abrahams-Sandi, E., 2015. Mitochondrial genome supports sibling species of *Angiostrongylus costaricensis* (Nematoda: Angiostrongylidae). PLoS One 10 (7), e0134581.

Yong, H.S., Song, S.L., Eamsobhana, P., Lim, P.E., 2016. Complete mitochondrial genome of *Angiostrongylus malaysiensis* lungworm and molecular phylogeny of Metastrongyloid nematodes. Acta Trop. 161, 33–40.

Yue, Z.Y., 2019. Identification of Differentially Expressed Genes Before and After Pomacea canaliculata Infection With Angiostrongylus cantontensis. Chinese Center for Diseases Control and Prevention, Beijing.

Yue, Z.Y., Zhang, Y., Guo, Y.H., Qin, Z.Q., Huang, Y., Zhang, W., 2019a. Quantitative analysis of gene expression in *Pomacea canaliculata* infected with *Angiostrongylus cantonensis* and α-tubulin gene expression in various tissues. Chin. J. Schisto. Control 31 (4), 404–409.

Yue, Z.Y., Zhang, Y., Guo, Y.H., Qin, Z.Q., Huang, Y., Zhang, W., Mao, G.Y., 2019b. Effect of *Angiostrongylus cantonensis* infection on expression of G-type lysozyme in *Pomacea canaliculata*. Chin. J. Parasitol. Parasit. Dis. 37 (3), 326–331.

Zhang, Y., Lv, S., 2010. Interpretation to diagnosis criteria for angiostrongyliasis cantonensis. China Health. Stand. Manag. 1 (4), 64–65.

Zhou, W.C., Wu, Y.F., Yang, J.Q., 2003. Viability of the ampullaria snail in China. Fujian J. Agri. Sci. 18 (1), 25–28.

CHAPTER NINE

Control of taeniasis and cysticercosis in China

Men-Bao Qian[a,b,c,d,e], Ning Xiao[a,b,c,d,e], Shi-Zhu Li[a,b,c,d,e], Bernadette Abela-Ridder[f], Hélène Carabin[g], Anna Sophie Fahrion[f], Dirk Engels[h], Xiao-Nong Zhou[a,b,c,d,e,*]

[a]National Institute of Parasitic Diseases, Chinese Center for Disease Control and Prevention, Shanghai, People's Republic of China
[b]Chinese Center for Tropical Diseases Research, Shanghai, People's Republic of China
[c]Key Laboratory of Parasite and Vector Biology, Ministry of Health, Shanghai, People's Republic of China
[d]National Center for International Research on Tropical Diseases, Ministry of Science and Technology, Shanghai, People's Republic of China
[e]WHO Collaborating Centre for Tropical Diseases, Shanghai, People's Republic of China
[f]Department of Control of Neglected Tropical Diseases, World Health Organization, Geneva, Switzerland
[g]Department of Biostatistics and Epidemiology, College of Public Health, University of Oklahoma Health Sciences Center, Oklahoma City, OK, United States
[h]Uniting to Combat NTDs Secretariat, Geneva, Switzerland
*Corresponding author: e-mail address: zhouxn1@chinacdc.cn

Contents

1. Introduction 290
2. Historical records 291
3. Epidemiological map 292
   3.1 Taeniasis 292
   3.2 Cysticercosis 294
4. Socio-ecological determinants 296
5. Clinical manifestations 298
   5.1 Taeniasis 298
   5.2 Cysticercosis 298
6. Diagnosis 299
   6.1 Taeniasis 299
   6.2 Cysticercosis 301
7. Treatment 303
   7.1 Taeniasis 303
   7.2 Cysticercosis 304
8. Control 305
9. Gap analysis 307
10. Conclusions and future perspectives 308
References 309

Advances in Parasitology, Volume 110
ISSN 0065-308X
https://doi.org/10.1016/bs.apar.2020.04.005

Abstract

National surveys suggest that the prevalence of taeniasis has considerably decreased in China, while reported cases indicated *T. solium* cysticercosis was historically highly endemic in northeastern, central and southwestern China. The high prevalence of taeniasis and cysticercosis there was driven by socio-ecological determinants. Cysticercosis may occur in the central nervous system, spinal cord, subcutaneous muscle, eyes, heart and oral cavity. Neurocysticercosis, the clinically most important type, causes epilepsy, increased intracranial pressure and neuropsychiatric symptoms. New molecular diagnostic techniques have been introduced for high sensitivity and discrimination of *Taenia* species. Immunological methods remain useful in the diagnosis of cysticercosis, especially neurocysticercosis. The introduction of imaging techniques including computed tomography and magnetic resonance imaging has significantly improved the diagnosis of neurocysticercosis. Recently, a combination of pumpkin seeds and areca nut has been explored against taeniasis, while praziquantel and albendazole are administrated simultaneously against cysticercosis, with promising efficacy and low side-effects. The widespread adoption of deworming protocols and techniques for inspection, management and treatment of pigs as well as improved sewage management has contributed to the significant decrease of taeniasis and cysticercosis in northern China. The positive results of these techniques should now be extended to highly endemic areas in western China to achieve the national elimination target for taeniasis and cysticercosis. Elimination of taeniasis and cysticercosis in China will not only benefit public health within China but also set an important example for less developed countries.

1. Introduction

Human taeniasis is caused by three *Taenia* spp., namely *T. solium*, *T. saginata* and *T. asiatica* (Ito et al., 2003). The life cycle of all three species involves humans as definitive hosts, who harbour the adult stage of the hermaphrodite worms in their intestines, and either pigs (*T. solium* and *T. asiatica*) or cattle (*T. saginata*) as the intermediate hosts. The intermediate hosts harbour the larval stages of the worm in the form of cysts which establish themselves in several highly oxygenated tissues, such muscles, a condition referred to as cysticercosis. The intermediate hosts become infected by consuming infected human faeces or ingesting food contaminated with infected human faeces. Humans become infected with *Taenia* spp. after consuming undercooked pork meat (*T. solium*), pork viscera (*T. asiatica*) or beef meat (*T. saginata*). Taeniasis caused by these three species has been reported to be generally asymptomatic or associated with very mild symptoms. In the case of *T. solium* infection, humans can serve as accidental intermediate hosts when consuming food contaminated by faeces of humans with taeniasis or through self-infection. Human cysticercosis occurs when the larvae of *T. solium* migrate to different body tissues. When the larvae migrate to the central nervous system, the cysts establish and cause neurocysticercosis (NCC) (Garcia et al., 2003). NCC is believed to be the most common helminthic infection of the brain and is considered a public health issue in areas endemic for *T. solium* due to its severe sequelae such as epilepsy and severe chronic headaches.

Taeniasis and *T. solium* cysticercosis are believed to be endemic in China and a threat to public health. The goal of this work is to review Chinese data and literature on the epidemiology, risk factors, clinical manifestations, diagnosis and treatment of, and mass control activities for taeniasis and cysticercosis. This review will also serve to identify gaps in knowledge regarding the epidemiology and control of taeniasis and cysticercosis while also identify successes in control which may inspire other developing countries afflicted by taeniasis and cysticercosis.

2. Historical records

Cases of taeniasis have been documented since the early 1900s in China. Dr. C.K. Roys reported that, during 1899–1909, dozens of cases of human *T. solium* and *T. saginata* infections occurred in Qingdao, Shandong province in eastern China, but it is likely that all of these cases were of European origin (Roys, 1910). In 1910, Dr. Henry S. Houghton reported a local *T. saginata* infection from Wuhu General Hospital, Anhui province in central China (Houghton, 1910). During the same period, correspondence issued to members of the China Medical Missionary Association reported several *T. solium* infections in some areas of China according to the Research Committee of China Medical Journal (Research Committee of China Medical Journal, 1910). It is however believed that taeniasis has been present in China for at least 2000 years, as demonstrated by the discovery of *Taenia* spp. eggs in the intestine of an excavated corpse of the Western Han Dynasty in Jiangling county, Hubei province, China in 1975 (Wei et al., 1980). Although differentiation for *Taenia* species was unsuccessful, traces of beef and piglets' skeletons were found buried with the dead.

The first documented case of human cysticercosis occurred in Weihaiwei (now Weihai), Shandong province, China in 1922. Cysts affected various parts of the patient's body (Barnes, 1922). In 1924, Dr. Ralph G. Mills reported cysticercosis of the right side of a patient's chest in Beijing, northern China (Mills, 1924). The case was recorded in 1924 but the symptoms may have appeared up to 12 years before the case was diagnosed.

3. Epidemiological map

3.1 Taeniasis

Three national surveys on parasitic diseases have been conducted in China (Qian et al., 2019). In the first national survey (1988–1992), 2449 out of 1,477,742 sampled individuals in 30 provinces were found to be infected with *Taenia* spp. in 28 provinces (Fig. 1) (Xu et al., 2000). These data led to an estimation of about 1.3 million total *Taenia* spp. infections in the country (Hotez et al., 1997). In the second national survey (2001–2004), 983 from 12 provinces out of 356,629 individuals samples from 31 provinces were found to be infected with *Taenia* spp. (Fig. 2) (Technical Steering Panel for National Survey of Current Status of Major Human Parasitic Diseases, 2008). These data suggested that a total population of 0.55 million was infected (Technical Steering Panel for National Survey of Current Status of Major Human Parasitic Diseases, 2008). In the third national survey between 2014 and 2015, out of 484,210 participants from 31 provinces, 1752 ones were detected with *Taenia* spp. from 12 provinces (Fig. 3). After adjustment by the population structure in total population,

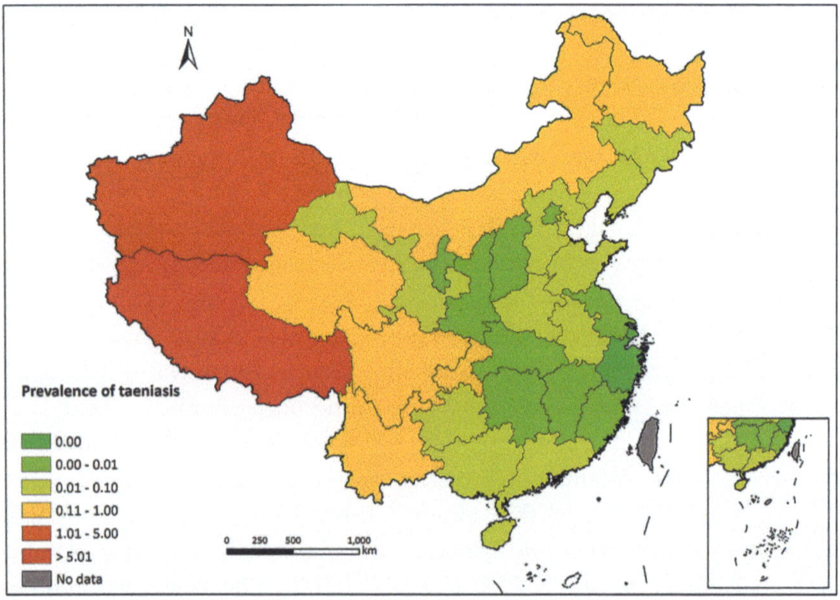

Fig. 1 The prevalence of taeniasis by provinces in China between 1988 and 1992.

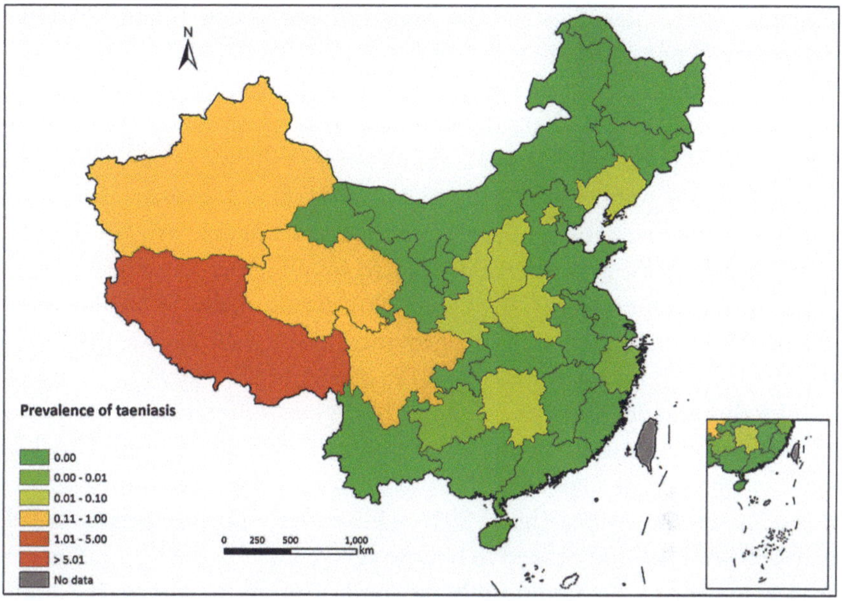

Fig. 2 The prevalence of taeniasis by provinces in China between 2001 and 2004.

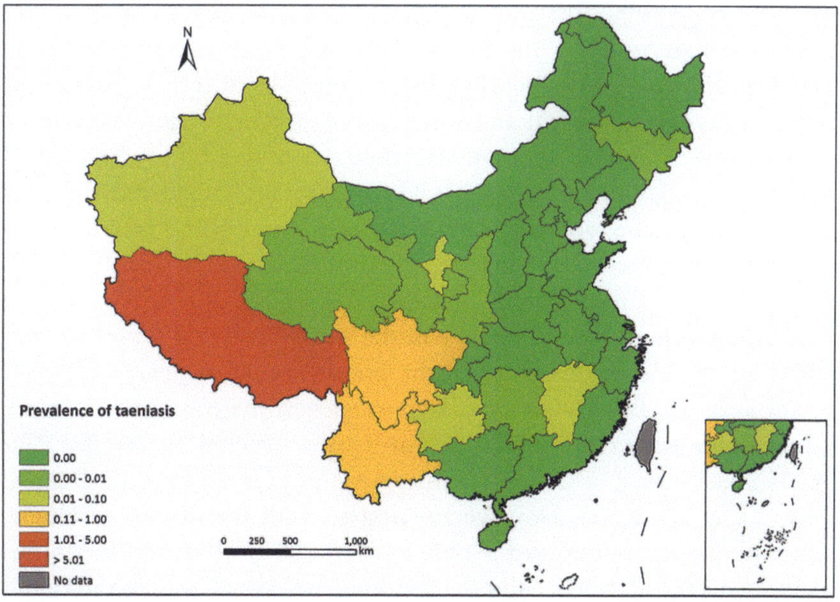

Fig. 3 The prevalence of taeniasis by provinces in China between 2014 and 2015.

a weighted prevalence of 0.06% was estimated, which indicated that 366 thousand persons were infected (Zhou, 2018). In the first survey, species identification techniques were carried out for some patients: 162 cases were identified as *T. solium* distributed across 10 provinces, notably in northeastern China, while 1399 cases were identified as *T. saginata* distributed across four provinces, mostly in western China (Xu et al., 2000). On the basis of the first national survey and the literature reports, *T. solium* and *T. saginata* infections were considered to be endemic in at least 20 and 18 provinces, respectively (Xu et al., 2000). Additionally, *T. asiatica* infection has also been reported in five provinces (Eom et al., 2002; Fan, 1991; Mo et al., 2005; Nkouawa et al., 2012; Zhang et al., 1999).

The three national surveys found no obvious difference in the prevalence of taeniasis among males and females (Technical Steering Panel for National Survey of Current Status of Major Human Parasitic Diseases, 2008; Xu et al., 2000; Zhou, 2018). However, the prevalence usually increased with age and peaked in middle aged people in the two earlier surveys, while it was not significantly different in the third survey (Technical Steering Panel for National Survey of Current Status of Major Human Parasitic Diseases, 2008; Xu et al., 2000; Zhou, 2018).

3.2 Cysticercosis

A retrospective survey in the first national survey of 16,508 cysticercosis cases from 671 counties of 29 provinces demonstrated that most cases occurred in northeastern, central and southwestern regions (Fig. 4) (Xu et al., 1999). The second national survey included a serological study for cysticercosis: of 96,008 individuals sampled in 31 provinces, 553 individuals in 25 provinces tested positive for cysticercosis (prevalence proportion of 0.58%; Fig. 5) (Technical Steering Panel for National Survey of Current Status of Major Human Parasitic Diseases, 2008).

Although the serological survey in the second national survey demonstrated no significant difference in the prevalence of cysticercosis by sex, the proportion of collected cysticercosis cases in the first national survey in males was more than double that in females (Technical Steering Panel for National Survey of Current Status of Major Human Parasitic Diseases, 2008; Xu et al., 1999). However, consistent with the highest serological positivity rate in middle aged people in the serological survey, the highest proportion of collected cases in the retrospective survey was also in middle aged people (Technical Steering Panel for National Survey of Current Status of Major Human Parasitic Diseases, 2008; Xu et al., 1999).

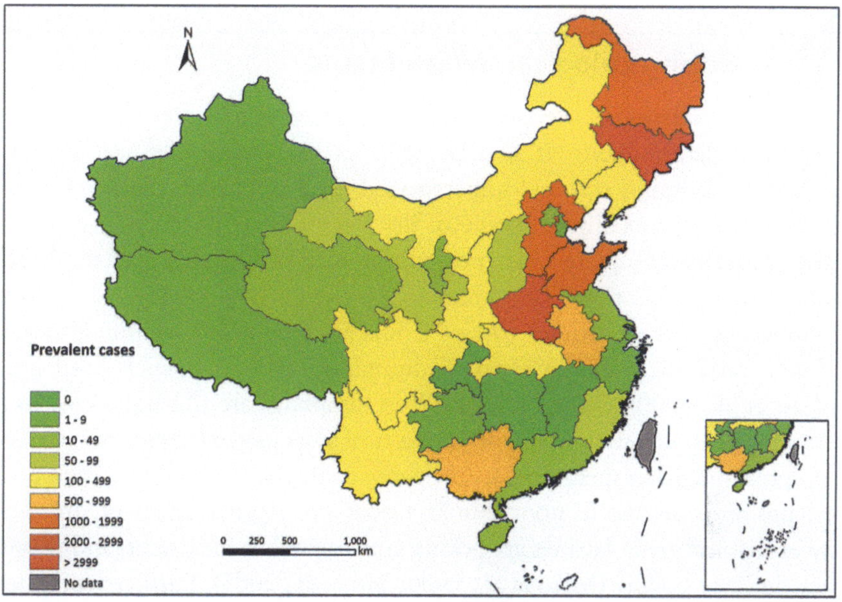

Fig. 4 The distribution of cysticercosis cases by provinces in China in a retrospective survey between 1988 and 1992*.

* The onset or diagnosis time of the cases was not specified.

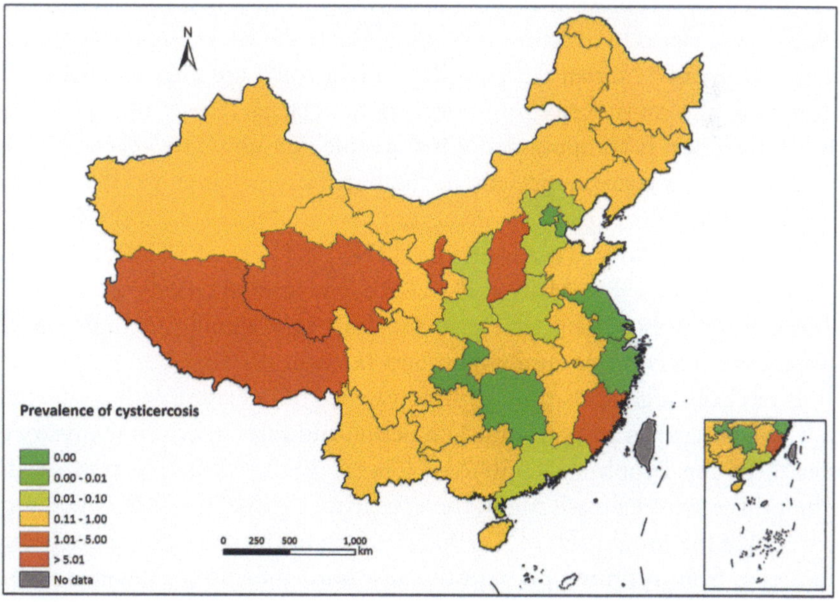

Fig. 5 The positive rate of cysticercosis by provinces in China in the serological survey between 2001 and 2004.

4. Socio-ecological determinants

The endemicity of taeniasis and cysticercosis has complex socio-ecological determinants. It is suspected that the practice of and attitudes towards consuming raw and undercooked meat is an important risk factor for taeniasis. For example, people of the *Bai* ethnicity in Yunnan province often consume completely raw pork, known as *Shengpi* (Yang et al., 2007). While people of the *Han* ethnicity rarely eat completely raw pork, populations in the northern part of China are at risk of consuming undercooked pork meat in the form of dumplings and quick-stick fried pork (Miao et al., 2000; Wang et al., 1999). Dumplings are not boiled for long enough and quick-stick fried pork consists of large pieces of meat not cooked at high enough temperatures to properly cook the meat. Pig cysticercosis remains an issue due to poor human faeces and pig management. In some lower income areas, latrines are poorly constructed or inexistent, and people may defecate openly (Miao et al., 2000; Mo et al., 2005). Further worsening of the situation and cycle of disease is due to the fact that some pigpens are connected with human toilets (Miao et al., 2000). Lastly, the absence of rigorous meat inspection practices in some rural areas leads to contaminated pork meat being sold for human consumption (Liu et al., 1996).

Human cysticercosis is also of concern due to poorly constructed toilets which cause faeces to overflow and contaminate the environment, especially water. Untreated human faeces, called "night soil", are used as fertilizer in local areas, also causing environmental contamination (Pan et al., 1995). The risk of infection is compounded when people consume raw vegetables and drink untreated water, especially in the hot summer (Ji et al., 1997; Miao et al., 2000). Poor personal hygiene can cause subsequent morbidity from cysticercosis in those with taeniasis (Table 1) (Ji et al., 1997; Li et al., 1998; Ma et al., 1988; Miao et al., 2000; Zhang et al., 1990; Zhao et al., 1999). Poor personal hygiene can also lead to a significant high risk of subsequent infection of family members (Li et al., 1998).

It has been argued that the high prevalence of infection with cysticercus in *T. solium* carriers arises from self-infection and auto-infection (Ministry of Health of the People's Republic of China, 2012a). Self-infection occurs when those with taeniasis ingest the eggs from themselves. It has been suggested that the proglottids of *T. solium* may enter the stomach reversely and then eggs from ruptured proglottids cause auto-infection. However, some

Table 1 Co-infection of taeniasis and cysticercosis from field surveys in China.

Location (References)	Time	Number examined	Number with both taeniasis and cysticercosis	Number with taeniasis (percentage with cysticercosis)	Number with cysticercosis (percentage with taeniasis)
Huaiyang county, Henan province (Ma et al., 1988)	1986–87	26,655	33	208 (15.9%)	52 (63.5%)
Kaifeng county, Henan province (Zhang et al., 1990)	1988	2677	11	27 (40.7%)	16 (68.8%)
Haerbin city, Heilongjiang province (Ji et al., 1997)	1993–94	10,525	10	56 (17.9%)	18 (55.6%)
Xiayi county, Henan province (Li et al., 1998)	1994–96	122,310	24	55 (43.6%)	147 (16.3%)
Licheng district, Jinan city, Shandong province (Zhao et al., 1999)	1996–98	238,643	28	75 (37.3%)	73 (38.4%)
Five counties (Yangxin, Qihe, Gaotang, Linqing and Dongying) in Shandong province (Miao et al., 2000)	1999	31,124	1	10 (10.0%)	16 (6.3%)

researchers have cast doubt on the latter assertion, because individuals harbouring a high number of encysted subcutaneous infections had no significantly increased trend of taeniasis (Wei et al., 1986). Additionally, although highly occasional, vertical transmission from mother to foetus has also been documented (Liu, 1987; Zhang, 1998).

5. Clinical manifestations

5.1 Taeniasis

The symptoms of taeniasis arising from all three species are usually not severe. The most typical manifestation is expulsion of proglottids, which is useful in diagnosis (Lang et al., 2004; Ma et al., 1988; Mo et al., 2005). Other nonspecific symptoms include pruritus ani, debilitation, dizziness, headache, drowsiness, anorexia, nausea, abdominal discomfort, diarrhoea and weight loss. *T. solium* occasionally parasitizes the thighs and the thyroid ectopically (Liu, 2013; Zhu and Gao, 1989). An enclosed mass of parasitized tissues is presented and the worms do not usually exceed 20 cm in length.

5.2 Cysticercosis

Cysticercosis is classified on the basis of the tissues affected. It may include but is not limited to NCC, subcutaneous muscle cysticercosis, ocular cysticercosis, heart cysticercosis and oral cavity cysticercosis (Xu et al., 1998). NCC, the most important infection in terms of incidence and morbidity, could account for 65% of total hospital-based cysticercosis cases, and may occur concurrently with other types (Xu et al., 1998). NCC may involve the cerebral parenchyma, cerebral ventricles, meninges or a mixture of these locations (Ma, 2004).

Epilepsy and increased intracranial pressure are the most commonly reported manifestations of NCC (Fang et al., 2009; Ge et al., 1993; Liu et al., 2008). The type of epilepsy is related to the location of cysticercus in the brain, while the frequency of seizure is related to how active the cysticercus is (Xie et al., 1995). In the cerebral cortex, especially precentral gyrus, generalized tonic clonic seizures and simple partial seizures are seen, while complex partial seizures are seen in deep temporal lobe involvement; diverse seizure types may occur when several lesions located in various parts of the brain are present. The frequency of seizure is highest in the degrading stage, reaching as many as one episode per month (Xie et al., 1995). High intracranial pressure is most common when cerebral ventricles are involved (Guo et al., 2001) and is associated positively with the occurrence of seizure (Ma et al., 1998); it also causes severe headache (Guo et al., 2001). Neuropsychiatric symptoms are often reported among people with NCC (Ma et al., 1996; Shen and Li, 2013). Neurasthaenia and anxiety depression are higher in schizophrenia patients with NCC than among those without

NCC (Shen and Li, 2013). Other manifestations (paraesthesia, disturbance of consciousness, dysaudia, aphasia, haemiplegia, hypomnesia and dementia) may also be present in lesser frequencies (Ge et al., 1993; Liu et al., 2008), so do cerebral infarctions (Hu et al., 1994). Spinal cysticercosis is infrequent and usually occurs concurrently with cerebral cysticercosis (Huang et al., 2013). Compression on nerves leads to numbness of the bilateral lower extremities, paraesthesia, changes in muscle strength and dysuria (Huang et al., 2013).

Subcutaneous nodules, the typical sign present in subcutaneous muscle cysticercosis, is most frequently distributed in the trunk, and then the limbs and the neck (Xu et al., 1998). Usually, no obvious symptoms are present when only subcutaneous muscles are involved, but severe infections (multiple nodules) may cause muscular pseudohypertrophy (Guo et al., 1996). Hypermyotrophy is more obvious in the gastrocnemius muscle, and fever and myosalgia are present in the early stage.

Ocular cysticercosis may occur in any part of the eyes, but intravitreal and subretinal cysticercosis are most frequent (Xu et al., 1998). It can lead to the loss of sight and even acroisa from vitreous opacity, detachment of retina, uveitis, endophthalmitis and cataract (Yang et al., 1997). NCC without direct involvement of the eyes may also impact sight (Cui et al., 2000).

Heart cysticercosis is rare (Xu et al., 1998). The order of frequency of cysticercus location in the heart is interventricular septum, left ventricular wall, chordae tendineae and bicuspid valve (Kong et al., 1995). Symptoms are usually mild including cardiopalmus, chest distress, discomfort in praecordium and fear.

Despite being rare, cysticercosis may also occur in the oral and maxillofacial region; the tongue and cheeks are the most frequent locations (Yu, 1988). Clinical presentation is an enclosed mass with a clear boundary, while symptoms may include pain, itching and swelling.

6. Diagnosis

6.1 Taeniasis

The Kato-Katz method and the perianal swab technique were applied in national surveys of parasitic diseases in China (Technical Steering Panel for National Survey of Current Status of Major Human Parasitic Diseases, 2008; Yu, 1991; Zhou, 2018). Although neither techniques specifically targeted *Taenia* spp. (the Kato-Katz method was used to detect many intestinal helminths, especially soil-transmitted helminths, and the perianal swab technique was used to detect *Enterobius vermicularis*), their application has allowed the mapping of taeniasis epidemiology in China. The direct smear technique is also usually applied because of its simplicity (Han et al., 1981; Yang, 1955). The sensitivity of diagnostic techniques varies with the different *Taenia* species. The perianal swab technique is more sensitive than the direct smear technique for *T. saginata* infection (Yang, 1955), but the reverse is true (the direct smear technique being more sensitive than the perianal swab technique) for *T. solium* infection (Han et al., 1981). However, the major limitation of all these methods is their inability to differentiate *Taenia* species.

When screening for cases in large field surveys, a history of expelling proglottids is the most convenient screening tool, and it can be employed to detect large numbers of cases in a short time (Wang et al., 1987; Wu, 1981). This strategy is usually more sensitive in *T. saginata* infection because the proglottids are expelled automatically without defecation (Yang, 1955). In suspected cases, administration of cesticidal drugs, called "tentative deworming", is also useful because detection of worms after drugs administration could make definite diagnosis (Ito et al., 2013; Ministry of Health of the People's Republic of China, 2012b). This technique can differentiate species, especially when the scolex of intact worm is obtained.

Following the development of molecular methods, many new techniques have been introduced for the diagnosis of taeniasis and species discrimination. When parasite materials become available, multiplex polymerase chain reaction (PCR) techniques based on the cytochrome C oxidase subunit 1 gene (cox1) can be used for the simultaneous detection of any of the three species (Li et al., 2012, 2013). The coproPCR technique, which uses faecal samples for analysis, can also be used for detection of infection and species discrimination (Li et al., 2012, 2013). These PCR tests are expensive and complicated and are thus applied only in the laboratory setting. Recently, the loop-mediated isothermal amplification method (LAMP) was introduced for species discrimination, enabling rapid species identification in field surveys (Nkouawa et al., 2012).

Immunological tests, such as the indirect haemagglutination test (IHA), have also been developed (Qin et al., 1985). Furthermore, enteroscopy is helpful in clinical settings (Meng et al., 2007).

To promote the management of taeniasis in China, the former Ministry of Health issued a diagnostic standard in 2012 (Ministry of Health of the People's Republic of China, 2012b). Three levels of evidence are introduced (epidemiology, clinical manifestations and aetiology) and are employed to draw two categories of diagnostic criteria (suspected cases and definitive cases). The recommended techniques for etiological diagnosis are faecal examination (for proglottids and eggs), perianal swabs (for eggs) and tentative deworming (for proglottids and adult worms).

6.2 Cysticercosis

Immunodiagnostic techniques have been extensively explored for the diagnosis of cysticercosis in China; these include intradermal test, latex agglutination test, indirect charcoal agglutination test, IHA, enzyme-linked immunosorbent assay (ELISA), dot-immunobinding assay, indirect fluorescent antibody test, enzyme-linked immunoelectro transfer blot and colloid gold immunochromatographic test (Liu, 1988; Liu et al., 2002; Qiu et al., 1998; Shi et al., 1989; Sun et al., 1985; Wu et al., 1994; Yin et al., 1990); IHA and ELISA are mostly applied, especially to support the diagnosis of NCC (Lian et al., 2003; Yang et al., 2008b). In NCC, both sera and cerebrospinal fluid can be applied in IHA and ELISA (Lian et al., 2003; Yang et al., 2008b). The sensitivity of IHA is similar to that of ELISA, and analysis of sera is more sensitive than that of cerebrospinal fluid in both IHA and ELISA tests (Lian et al., 2003; Yang et al., 2008b). The sensitivity of these techniques is positively related to the intensity of infection, namely the number of cysts (Lian et al., 2003). Furthermore, IHA and the classical ELISA detecting total antibody are not suitable to apply in evaluating treatment efficacy due to the persistence of positive results (Yang et al., 2008b). Thus, detection of special IgG4 antibody or cysticercus circulating antigen can be employed (Yang et al., 2008b). The positivity proportion and reaction intensity are higher in those patients in the active stage than those in the inactive stage when circulating antigen is detected (Zhang et al., 1992). Generally, immunological tests are auxiliary because of their insensitivity in some low-intensity infections as well as the false-positive results to other parasitic diseases, especially echinococcosis.

The introduction of imaging techniques including computed tomography (CT) and magnetic resonance imaging (MRI) in the early and late 1980s in China, respectively, has significantly improved the diagnosis of NCC (Ma et al., 1989; Sui et al., 1990). These two techniques are useful not only in definitive diagnosis but also in quantifying, positioning and staging of cysticercosis (Ma et al., 1989; Sui et al., 1990; Wei et al., 1999; Yan et al., 2004; Zhu et al., 1994). Usually, based on CT and/or MRI, four stages are defined (active stage, degeneration stage, inactive (or calcification) stage, and mixed stage which includes two or three stages simultaneously) (Sui et al., 1990; Zhu et al., 1994). Overall, MRI is preferable (to CT) for diagnosis in active and degeneration stages, while CT is preferable (to MRI) in the calcification stage (Sui et al., 1990). The staging by CT and MRI guides the appropriate choice of treatment (Sui et al., 1990; Zhu et al., 1994).

Electroencephalogram is also helpful in diagnosis of NCC (Wang et al., 2000). Abnormality of electroencephalogram is related to the epilepsy and the characteristics include diffuse or focal irregular slow waves in the frontal, central and anterior temporal lobes (Wang et al., 2004a). The abnormality disappears or improves after treatment (Wang et al., 2004b). Thus, electroencephalogram can be applied with imaging techniques in diagnosis of NCC (Wang et al., 2004a,b).

For suspected NCC cases, "tentative treatment" has been applied (Jia and Ou-yang, 2009; Ministry of Health of the People's Republic of China, 2012a). A high dose of praziquantel is administered and a definite diagnosis is then judged according to efficacy after treatment. Many final ascertained cases with cysticercosis appear with such adverse reactions as headache, dizziness, acroanesthesia, local convulsion, urticaria, pyrexia during the "tentative treatment", which are important indicators for diagnosis. However, ethical questions should be paid attention to.

Additionally, ultrasound can be used in the diagnosis of cysticercosis in the heart, eyes and subcutaneous muscle (Jiang et al., 1999; Li et al., 2011; Shi et al., 2000).

Cysticercosis is complex and involves diverse locations, clinical appearances and types, which could lead to misdiagnosis especially in NCC (Ge et al., 1993; Yang et al., 2008a). Some reports estimate that over 20% of cysticercosis cases have been misdiagnosed before reaching a definitive diagnosis of cysticercosis (Ge et al., 1993). Cysticercosis is usually misdiagnosed as epilepsy unrelated to cysticercosis, viral encephalitis, tuberculous meningitis, perencephaly, cerebrovascular malformation and brain tumour (Ge et al., 1993). Furthermore, many other diseases especially brain tumour can be misdiagnosed as NCC (Yang et al., 2008a).

In 1985 an unofficial diagnostic standard for cysticercosis was first proposed by scholars during the first national seminar on cysticercosis in China (Ma et al., 1989), and then revised several times (Ma, 2004). In 2012 an official diagnostic standard was issued by the former Ministry of Health (Ministry of Health of the People's Republic of China, 2012a). It includes six types of evidence (epidemiology, clinical manifestations, aetiology, immunology, imaging and tentative treatment) to enable classification of cases into three groups (suspected, clinical and definitive).

7. Treatment

7.1 Taeniasis

In the 1970s, Chinese researchers found that extract from the plant *Agrimonia pilosa Ledeb.* was effective against *Taenia* spp. infection. The effective cesticidal component is agrimophol (No authors, 1974). The commercial product (*Shataoling*) has been widely used for treatment of *T. solium* infection in combination with cathartic medicines such as dihydroxyphthalophenone and magnesium sulphate in the field (Wang et al., 1987; Wu, 1981). The cure rate varies from 55% to 95% (No authors, 1974; Wang et al., 1987; Wu, 1981). Another brand (*Qutaojiaonang*) is made also from the plant extract to which cathartic Chinese medicines are added to facilitate administration (Wen et al., 1993). However, it was reported that out of 10,459 cases treated with *Shataoling*, one lethal case occurred due to allergic shock (Wu, 1981). Thus, *Shataoling* should be used with caution.

More than 1000 years ago the Chinese discovered the efficacy of areca nut against taeniasis. In the 1950s, Chinese scientists found that pumpkin seeds acted synergistically with the areca nut extract against taeniasis (Feng, 1956). Recently, it has been demonstrated that a combination of pumpkin seeds and areca nut extract has higher efficacy and lesser time for deworming than any single component (Li et al., 2012). Sequential administration of pumpkin seeds, areca nut extract and cathartic compounds (e.g. magnesium sulphate) usually resulted in a cure rate of 90–100% in those infected with *T. saginata* (Feng, 1956; Gan et al., 1992; Li et al., 1996). However, varying efficacy has been reported with *T. solium* infection, with high efficacy in some cases (Zhang et al., 1985) and in others only 30–40% (Wang and Fang, 2008; Zhang, 2003).

Praziquantel is also administered but mostly in clinical setting (Wang and Fang, 2008; Zhang, 2003). A single dosage (30 mg/kg) followed by a cathartic compound (manicol or natrii sulfas) within 30 min can have a cure rate of over 90%. Furthermore, niclosamide (Zhang et al., 1985), the combination of niclosamide and bitin (Xue and Han, 1981), and paromomycin (Feng et al., 1994) are also effective and are administered. Tribendimidine, a new drug developed in China, has also been documented to be effective against taeniasis in field tests, and may be of great value for use in settings with extensive intestinal polyparasitism, especially soil-transmitted helminthiases (Steinmann et al., 2008).

7.2 Cysticercosis

The high incidence and morbidity of NCC warrant its priority for research on treatment by Chinese scientists. Before the advent of praziquantel, many traditional Chinese medicines were introduced to treat cysticercosis, most of which were either inefficacious or controversial (Liu and Ge, 1978). It has been reported that *Ganwusan* (a combination of lacca sinica exsiccata, pasta ulmi and cinnabar) shows promising efficacy (Ge et al., 1989; Liu and Ge, 1978; Lu et al., 2003). The pharmacodynamic action is mild and side-effects are few. Thus, hospitalization in treatment is not compulsory. However, the major challenge is the need for long term treatment, usually over 1 year.

In the late 1970s, praziquantel was introduced to treat cysticercosis in China (Pan et al., 1981). Initially, efficacy was low because praziquantel was administrated as a single course of treatment and at a low dose (e.g. a total dose of 120–240 mg/kg) (Li et al., 1986; Pan et al., 1981). On the one hand, animal experiment studies showed that the development of the larvae of *T. solium* are not synchronous (Ma et al., 1991). On the other hand, pharmacokinetics studies found that the concentration of praziquantel is lower in cerebrospinal fluid than in blood plasma (Zhu et al., 1993). Thus, a high total dose was gradually applied, of up to 360 mg/kg per course, with administration of two or more courses (Huo et al., 2004; Wu et al., 2013). However, given the intense anthelminthic action of praziquantel, especially in high doses, side-effects are common and severe. These usually appear during the first course of medication (Huo et al., 2004). Major side-effects include the rapid increase of intracranial pressure, and induction or aggravation of seizures (Huo et al., 2004; Meng et al., 1986; Wu et al., 2013). In cases of heavy infection with diffuse location of cysts in the brain, death may result from the severe cysticidal reaction (Wang, 2009). Thus, hospitalization is compulsory when praziquantel is administrated against cysticercosis.

Praziquantel is also administered to treat schistosomiasis and clonorchiasis, both of which are endemic in China (Qian et al., 2016; Zhou et al., 2007). Due to the overlapping distribution with cysticercosis, treatment and mass chemotherapy for schistosomiasis and clonorchiasis in endemic areas could cause serious reactions in those with NCC, and caution must be exercised in these cases (Bi and Wu, 1996; Shu et al., 1992).

In the early 1980s, albendazole was introduced to treat cysticercosis in China (Liu et al., 1987). As with praziquantel, doses have increased from a low dose (e.g. 10–20 mg/kg per day for 10 days and repetition for three courses) to a high dose (e.g. 60 mg/kg per day for 10 days and repetition for three to five courses) (Liu et al., 1987; Wu et al., 2013). Fewer side-effects occur with albendazole than with praziquantel, but hospitalization is needed with both medicines (Gao et al., 2010). Similarly, side-effects usually occur during the first course of medication (Liu et al., 1987).

In later years, albendazole was combined with praziquantel for treatment of cysticercosis with consideration of the differences in cysticidal mechanisms as well as their sequential side-effects (Li et al., 1995). During the past 20 years, this combination has been increasingly accepted by Chinese scientists (Wu et al., 2013). The following combination strategy is usually applied (Jia et al., 2005; Xu et al., 2002): albendazole and praziquantel are sequentially applied in the first course (albendazole: 20 mg/kg per day for 12 days; praziquantel: 30 mg/kg per day for 12 days); only praziquantel is used in second course (30 or 50 mg/kg per day for 12 days); and only praziquantel is used in the third course (50 mg/kg per day for 12 days). There is an interval of about 2–3 months between the two courses.

Surgery is compulsory in severe cases to decrease intracranial pressure and resolve hydrocephalus (Bu et al., 2009; Yuan et al., 2005). However, antiparasitic therapy cannot be substituted. Given the advances in minimally invasive techniques, minimally invasive neurosurgery can be successfully applied in cases of mild NCC (i.e. only a few cysts), especially in the case of a single cyst (Zhang et al., 2001). Because of the risk of loss of sight from antiparasitic reactions, surgery is the only choice in cysticercosis of the eyes (Cui et al., 2000; Yang et al., 1997).

8. Control

In response to awareness of the harm of pig cysticercosis to the economy and of human cysticercosis to public health, an "Office for Controlling and Eliminating Taeniasis and Cysticercosis" was established in many endemic provinces in northern China during the 1970s–90s involving the departments of commerce, agriculture and health (Liu et al., 1996; Ma et al., 1992). Four major interventions were applied: deworming for taeniasis, inspection of pork, management of human faeces and pigs, and treatment of pigs (Liu et al., 1996; Ma et al., 1992).

Large-scale active screening in villages detected and treated those with taeniasis. The detection method was to request the history of expelling proglottids, especially during the previous 6 months, while *Shataoling* was widely applied for treatment (Wang et al., 1987; Wu, 1981). For example, during 1979–83 in Hunan province of central China over 140,000 people were found to have a history of expelling proglottids, nearly 100,000 cases of whom accepted treatment (Wang et al., 1987). The expelled worms were collected and managed collectively. It was then found that the incidences of human taeniasis, human cysticercosis and porcine cysticercosis decreased consistently (Wang et al., 1987). Additionally, by tracing the whereabouts of pork infected with larvae of *T. solium*, those eating the pork could be included for treatment (Liu et al., 1996; Ma et al., 1992).

The selling of infected pork is an important component in the transmission of taeniasis. Thus, pork inspection is included among the four arms of the national strategy against taeniasis and cysticercosis. Pork cannot be sold without inspection and those carcasses infected with cysticercosis should be destroyed.

Blocking contact between human faeces and pigs reduces transmission and promotes disease control. To do so, pigs are penned and not allowed to roam freely, and new closed toilets were built. Before applying fertilizer, human faeces are processed through non-hazardous treatment such as thermophilic fermentation.

To avert economic losses, pigs are examined during rearing and treated if infected with larvae of *T. solium*. Many immunological techniques were developed to detect the infection (Gu and Shen, 1998; Liu et al., 1996); albendazole and praziquantel were adopted to treat infected pigs (Han et al., 1983; Liu et al., 1996). Usually, the drugs have no residual effects and the resulting pork is equally nutritious to that without previous infection or treatment (Chen et al., 1991) However, pigs with severe infection are unsuitable for treatment because it may kill the animal and reduce nutritional value (Liu et al., 1996).

Lack of knowledge is an important risk factor of taeniasis and cysticercosis (Cao et al., 1997). Thus, health education is effective in community interventions (Li et al., 1998, 2000). Health education not only promotes knowledge of the disease but also enhances capacity to distinguish infected pork, leading to positive changes in the management of human faeces and rearing of pigs.

In later years the prevalence of taeniasis and cysticercosis decreased significantly and the Office for Controlling and Eliminating Taeniasis and Cysticercosis has been scaled back gradually in many areas. Although no widespread screening and treatment for taeniasis is yet available, proper management of human faeces and pigs and inspection of pigs are occurring owing to social and economic development. More hygienic toilets are being built and human faeces are no longer an important fertilizer thanks to the production of chemical fertilizers (Department of Comprehensive Statistics of National Bureau of Statistics, 2009). Pigs are usually raised collectively and inspection is more standardized, especially in developed areas.

9. Gap analysis

Despite the achievements in epidemiology, diagnosis and treatment, and control, many gaps still hamper the final elimination of taeniasis and cysticercosis in China. Firstly, the national epidemiological map is inaccurate and the disease burden is unclear. Three national surveys were implemented to include detection of, but not focus on taeniasis (Hotez et al., 1997; Technical Steering Panel for National Survey of Current Status of Major Human Parasitic Diseases, 2008; Xu et al., 2000; Zhou, 2018). On the one hand, national sampling mainly targeted on soil-transmitted helminthiases and thus its representativeness was inadequate for taeniasis. On the other hand, the faecal examinations applied were not very sensitive for *Taenia* spp. However, the decreasing trend of disease incidence provides evidence of social and economic development as well as massive control activities. Although serological investigation was carried out for cysticercosis in the second survey, a national figure for the total infected population was not estimated in the survey report, due to the inaccuracy of serological detection (Technical Steering Panel for National Survey of Current Status of Major Human Parasitic Diseases, 2008). In other literature, a national estimation of 3 million individuals infected was usually mentioned (Lu et al., 2002; Xu, 2002). However, these were only simple estimations and detailed descriptions could not be provided. Another estimation of 7 million infections with cysticercosis was also referred to, based on the rude calculation of national serological data (Yang et al., 2014). According to the Global Burden of Disease Study 2013, the disability-adjusted life years (DALYs) attributed to cysticercosis were 435,100 in 1990 and 341,200 in 2013 worldwide, and the corresponding data in China were 88,900 and 47,000, respectively (GBD DALYs and Hale Collaborators et al., 2015). However, these figures may be severely underestimated, because WHO Foodborne Disease Burden Epidemiology Reference Group estimated that DALYs due to cysticercosis reached 2,780,000 globally in 2010 (Torgerson et al., 2015). Obviously, the unavailability of accurate epidemiological maps in China, especially for cysticercosis, hampers the formulation of further intervention strategies.

Second, research remains inadequate. The cross-reaction of immunological diagnosis with other parasitic diseases especially echinococcosis—another important parasitic disease in China—weakens its large-scale application in fields. No national treatment standard is available for taeniasis and cysticercosis. Although many advances have been made in treatment, gaps still exist. Nowadays, the traditional Chinese medicines *Shataoling* and *Qutaojiaonang* are no longer produced because of the limited demand for treatment of taeniasis in the field. The new method combining pumpkin seeds and areca nut extract should be standardized for the treatment of taeniasis, which is also expected to be commercialized. Although the material is easy to obtain, the quality of effective components should be guaranteed during the process. Although the combination of albendazole and praziquantel is gradually being accepted for the treatment of NCC, large standardized, randomized and controlled clinical studies are still needed, which will promote the wide application of this treatment strategy.

Third, no special intervention programme is being carried out. Following the achievements of control in hyperendemic areas during the 1970s–90s, the Office for Controlling and Eliminating Taeniasis and Cysticercosis was gradually dismantled. Although social and economic development has promoted improvements in environment, hygiene, and pig management and inspection, taeniasis and cysticercosis have not been eliminated and remain endemic in some remote areas, e.g., Yunnan and Sichuan provinces in western China (Li et al., 2019; Zang et al., 2019a, b). Their control has not been included in the updated national control programme for parasitic diseases, although some control pilots are being implemented (Qian et al., 2019). Although active surveillance for human beings has just been established for taeniasis, the coverage is low (Qian et al., 2019). Especially, taeniasis and cysticercosis are excluded from the National Infectious Diseases Reporting System through passive surveillance, which includes many parasitic diseases (amebiasis, schistosomiasis, malaria, visceral leishmaniasis, echinococcosis and filariasis), hampering the acknowledgement of the importance of and subsequent interventions against taeniasis and cysticercosis (Qian et al., 2019).

10. Conclusions and future perspectives

Both in China and globally, cysticercosis has been listed among the eradicable diseases (Ma et al., 1992; Maurice, 2014). However, taeniasis and cysticercosis remain endemic in some poor areas of China. To achieve the final elimination target, the gaps mentioned above should be filled. Taeniasis and cysticercosis should be included in the National Infectious Diseases Reporting System not only to benefit the establishment of a national epidemiological map but also to promote the clinical management of cases. Similar case management of malaria in China should be used as a reference. For taeniasis, the source of transmission should be traced and those contracting the disease should be examined along with their family members. Cases of cysticercosis should be examined for taeniasis and their family members also included for surveillance. The Office for Controlling and Eliminating Taeniasis and Cysticercosis is expected to be re-established in some hyperendemic areas in order to integrate the efforts of health, agriculture and commerce. New sensitive diagnosis techniques are expected to be developed and national treatment standards should be issued.

WHO has selected China among the first countries in a network to further control and finally eliminate cysticercosis (Maurice, 2014). Thus, under this framework, the National Institute of Parasitic Diseases, Chinese Center for Disease Control and Prevention is exploring the strategies for the control and elimination of taeniasis and cysticercosis in China. The pilots include one elimination pilot in Henan province, where taeniasis and cysticercosis were highly endemic historically but quite low nowadays, through information, education, and communication and establishment of surveillance system. Another two control pilots are established in Yunnan and Sichuan provinces, respectively, where taeniasis and cysticercosis are still highly endemic. Information, education, and communication, detection and treatment of cases, improvement of sanitation and provision of water are integrated. The control and eventual elimination of taeniasis and cysticercosis in China will promote the health of local people. Successful research and control of taeniasis and cysticercosis in China during the past 40 years have boosted our confidence and provided important experiences. These experiences should be summarized and generalized to benefit those endemic areas in China, as well as other endemic countries.

References

Barnes, J.E., 1922. Chinese patient with *Cystisercus cellulosae*. Chin Med J (Engl) 36, 546.

Bi, S.Y., Wu, H.X., 1996. A case report on severe reaction of neurocysticercosis due to the massive chemotherapy with praziquantel. Zhongguo Xue Xi Chong Bing Fang Zhi Za Zhi 8, 256 (In Chinese).

Bu, X.Y., Liu, M., Zhang, J.G., Chai, C., Zhou, W., Li, Y., Guo, S.C., 2009. Surgical treatment of brain cysticercosis patients with hydrocephalus. Henan Yi Xue Yan Jiu 18, 193–195 (In Chinese).

Cao, W., van der Ploeg, C.P., Xu, J., Gao, C., Ge, L., Habbema, J.D., 1997. Risk factors for human cysticercosis morbidity: a population-based case-control study. Epidemiol. Infect. 119, 231–235.

Chen, X.P., Chen, Y., Han, Y., Hu, Y.H., Gu, W.J., 1991. Nutrition values in albendazole cured *Cysticercus cellulosae* pig body. Jilin Nong Ye Da Xue Xue Bao 13, 60–64, 96 (In Chinese).

Cui, G.Y., Zhang, E.D., Ding, X.Q., Wang, H., Hu, J.Q., 2000. Ocular complications of brain cysticercosis and its treatment. Yan Ke Xin Jin Zhan 20, 50–51 (In Chinese).

Department of Comprehensive Statistics of National Bureau of Statistics, 2009. China Compendium of Statistics 1949–2008. China Statistics Press, Beijing (In Chinese).

Eom, K.S., Jeon, H.K., Kong, Y., Hwang, U.W., Yang, Y., Li, X., Xu, L., Feng, Z., Pawlowski, Z.S., Rim, H.J., 2002. Identification of Taenia asiatica in China: molecular, morphological, and epidemiological analysis of a Luzhai isolate. J. Parasitol. 88, 758–764.

Fan, P.C., 1991. Asian Taenia saginata: species or strain? Southeast Asian J. Trop. Med. Public Health 22 (Suppl), 245–250.

Fang, W., Chen, F., Wang, Y.Y., 2009. Epidemiological features of in-patients with cysticercosis from 1999 to 2006 in Dali. Zhongguo Bing Yuan Sheng Wu Xue Za Zhi 4, 298–300 (In Chinese).

Feng, L.Z., 1956. Treatment of human taeniasis with pumpkin seeds combined with areca nut extract. Zhonghua Yi Xue Za Zhi 42, 138–147 (In Chinese).

Feng, M.L., Zhao, G.H., Xu, C.B., Tian, J., Huang, Y.P., Li, W., Xu, Y.Y., 1994. The efficacy of paromomycin in the treatment of taeniasis: a report of 415 cases. Beijing Yi Xue 16, 321–322 (In Chinese).

Gan, Y.C., Lai, Y.Y., Qin, Q.S., Huang, H.Q., Chen, S.X., 1992. Efficacy of pumpkin seeds and areca nut extract against *Taenia saginata* infection. Zhongguo Ji Sheng Chong Xue Yu Ji Sheng Chong Bing Za Zhi 10, 238 (In Chinese).

Gao, P., Cui, A.H., Li, J.D., 2010. Analysis of therapeutic efficacy and immunodetection in 60 patients with cerebral cysticercosis. Zhongguo Bing Yuan Sheng Wu Xue Za Zhi 5, 696–698 (In Chinese).

Garcia, H.H., Gonzalez, A.E., Evans, C.A., Gilman, R.H., Cysticercosis Working Group in Peru, 2003. Taenia solium cysticercosis. Lancet 362, 547–556.

GBD DALYs and Hale Collaborators, Murray, C.J., Barber, R.M., Foreman, K.J., Abbasoglu Ozgoren, A., Abd-Allah, F., Abera, S.F., Aboyans, V., Abraham, J.P., Abubakar, I., Abu-Raddad, L.J., Abu-Rmeileh, N.M., Achoki, T., Ackerman, I.N., Ademi, Z., Adou, A.K., Adsuar, J.C., Afshin, A., Agardh, E.E., Alam, S.S., Alasfoor, D., Albittar, M.I., Alegretti, M.A., Alemu, Z.A., Alfonso-Cristancho, R., Alhabib, S., Ali, R., Alla, F., Allebeck, P., Almazroa, M.A., Alsharif, U., Alvarez, E., Alvis-Guzman, N., Amare, A.T., Ameh, E.A., Amini, H., Ammar, W., Anderson, H.R., Anderson, B.O., Antonio, C.A., Anwari, P., Arnlov, J., Arsic Arsenijevic, V.S., Artaman, A., Asghar, R.J., Assadi, R., Atkins, L.S., Avila, M.A., Awuah, B., Bachman, V.F., Badawi, A., Bahit, M.C., Balakrishnan, K., Banerjee, A., Barker-Collo, S.L., Barquera, S., Barregard, L., Barrero, L.H., Basu, A., Basu, S., Basulaiman, M.O., Beardsley, J., Bedi, N., Beghi, E., Bekele, T., Bell, M.L., Benjet, C., Bennett, D.A., Bensenor, I.M., Benzian, H., Bernabe, E., Bertozzi-Villa, A., Beyene, T.J., Bhala, N., Bhalla, A., Bhutta, Z.A., Bienhoff, K., Bikbov, B., Biryukov, S., Blore, J.D., Blosser, C.D., Blyth, F.M., Bohensky, M.A., Bolliger, I.W., Bora Basara, B., Bornstein, N.M., Bose, D., Boufous, S., Bourne, R.R., Boyers, L.N., Brainin, M., Brayne, C.E., Brazinova, A., Breitborde, N.J., Brenner, H., Briggs, A.D., Brooks, P.M., Brown, J.C., Brugha, T.S., Buchbinder, R., et al., 2015. Global, regional, and national disability-adjusted life years (DALYs) for 306 diseases and injuries and healthy life expectancy (HALE) for 188 countries, 1990-2013: quantifying the epidemiological transition. Lancet 386, 2145–2191.

Ge, L.Y., Wang, Z.H., Cao, Y.C., Sun, Q.L., Yang, S.F., 1989. Clinical studies on treatment of cysticercosis with Gan-Wu pulvis—the Chinese herbal medicine. Zhongguo Ji Sheng Chong Bing Fang Zhi Za Zhi 2 (Suppl), 61–63 (In Chinese).

Ge, L.Y., Zhang, J.J., Wang, C.Y., Xu, H.X., Kong, Q.A., Cao, Y.C., Yang, S.F., Kong, Q.Y., Jia, F.J., Yang, Y.J., 1993. Analysis of 3956 out-patients with taeniasis and cysticercosis. Zhongguo Ji Sheng Chong Bing Fang Zhi Za Zhi 6, 23–26 (In Chinese).

Gu, Y.F., Shen, Y.L., 1998. Advances in immunological diagnosis of cysticercosis in pigs. Zhongguo Shou Yi Ji Sheng Chong Bing 6, 41–44 (In Chinese).

Guo, J.X., Zhou, F., Chen, S.L., 1996. Clinical analysis on 24 cases with muscular pseudohypertrophy caused by cysticercosis. Zhongguo Ji Sheng Chong Bing Fang Zhi Za Zhi 9, 154 (In Chinese).

Guo, D.M., Xie, S.P., Liu, A.H., 2001. Clinical analysis of the relation factors of cysticercosis with headache. Shoudu Yi Ke Da Xue Xue Bao 22, 349–351 (In Chinese).

Han, F., Ding, S.T., Shi, Z.M., Sheng, Z.P., Guo, N.G., 1981. Comparison on detection of *Taenia solium* infection by direct smear technique and perianal swab technique. Hebei Yi Yao 3, 8–9 (In Chinese).

Han, Y., Zhu, M.L., Liu, D.H., Liu, S.Z., Chen, Y.X., Li, P.Y., 1983. Preliminary report on efficacy of albendazole and praziquantel against cysticercosis in pigs. Jilin Nong Ye Da Xue Xue Bao 5, 44–48 (In Chinese).

Hotez, P.J., Zheng, F., Long-qi, X., Ming-gang, C., Shu-hua, X., Shu-xian, L., Blair, D., McManus, D.P., Davis, G.M., 1997. Emerging and reemerging helminthiases and the public health of China. Emerg. Infect. Dis. 3, 303–310.

Houghton, H.S., 1910. Report from the Wuhu general hospital. Chin Med J (Engl) 24, 86–90.

Hu, W.M., Liang, Q.C., Hu, W.X., 1994. Cysticercotic cerebral infraction. Haerbin Yi Ke Da Xue Xue Bao 28, 225–227 (In Chinese).

Huang, L., Yang, J., Wang, G.H., 2013. Report of three spinal cysticercosises and review of literatures. Lin Chuang Shen Jing Wai Ke Za Zhi 10, 257–259 (In Chinese).

Huo, H.Y., Dong, Y.H., Cheng, W.L., Yin, L.X., Zhang, Z.H., 2004. Cysticercidal reaction of praziquantel and the strategy against it in treatment of cysticercosis. Zhongguo Yao Ye 13, 27–28 (In Chinese).

Ito, A., Nakao, M., Wandra, T., 2003. Human Taeniasis and cysticercosis in Asia. Lancet 362, 1918–1920.

Ito, A., Li, T., Chen, X., Long, C., Yanagida, T., Nakao, M., Sako, Y., Okamoto, M., Wu, Y., Raoul, F., Giraudoux, P., Craig, P.S., 2013. Mini review on chemotherapy of taeniasis and cysticercosis due to *Taenia solium* in Asia, and a case report with 20 tapeworms in China. Trop. Biomed. 30, 164–173.

Ji, K.B., Tong, S.F., Meng, Y.L., Liu, Y.L., Li, M.Y., 1997. Sero-epidemiological survey of human cysticercosis in Haerbin. Zhongguo Ren Shou Gong Huan Bing Za Zhi 13, 77–78 (In Chinese).

Jia, F.J., Ou-yang, A.M., 2009. Diagnostic and therapeutic effects of praziquantel on suspected cerebral cysticercosis. Zhongguo Xue Xi Chong Bing Fang Zhi Za Zhi 21, 52–55.

Jia, F.J., Ou-yang, A.M., Wu, X.Y., Hu, Y.X., Zhang, Z.H., Wei, Y.B., 2005. Changes of CT images and clinical significance in cerebral cysticercosis treated by PQT and ALB with dose increasing method. Di Fang Bing Tong Bao 20, 7–9 (In Chinese).

Jiang, S.K., Wang, Y., Kong, Q.A., Ge, L.Y., 1999. Study on ultrasonography in diagnosing cardiac cysticercosis. Yi Xue Ying Xiang Xue Za Zhi 9, 95–96 (In Chinese).

Kong, Q.A., Ge, L.Y., Fu, Z.Y., Wang, C.Y., Ruan, S.Z., 1995. Study on mechanism of electrocardiogram changes of patients with heart cysticercosis. Zhongguo Ji Sheng Chong Bing Fang Zhi Za Zhi 8, 275–277 (In Chinese).

Lang, S.Y., Mou, R., Zhu, W.J., Zhang, K., Ling, H.Y., Bao, H.E., 2004. Epidemiological investigation of *Taenia saginata* infection in Congjiang area of Guizhou province. Re Dai Yi Xue Za Zhi 4, 738–739 (In Chinese).

Li, Y.B., Hu, D.F., Dong, Y., Li, F.X., Ji, Y.L., Zhang, J.Y., 1986. Treatment of neurocysticercosis with praziquantel. Hebei Yi Xue Yuan Xue Bao 7, 67–69 (In Chinese).

Li, Q.S., Ge, L.Y., Xu, H.X., Yang, Y.J., Zhang, J.J., Wang, C.Y., Kong, Q.A., Kong, Q.Y., Yu, Z.H., Jia, F.J., Hou, L.Q., Sun, G.P., Ma, Q.R., Li, H.J., 1995. Observation on treatment of 415 neurocysticercosis cases with praziquantel and albendazole administered on increasing-dose regime. Zhongguo Ji Sheng Chong Bing Fang Zhi Za Zhi 8, 124–126 (In Chinese).

Li, H.B., Che, Y., Gan, S., 1996. Efficacy of pumpkin seeds and areca nut extract against *Taenia saginata* infection. Shi Yong Ji Sheng Chong Bing Za Zhi 4, 144 (In Chinese).

Li, Y., Zhao, X.D., Mu, Q.X., Wang, H., Yin, G.S., Zheng, H.M., Zhou, X.H., 1998. Studies on the efficacy of health education for control of taeniasis and cysticercosis. Zhongguo Ji Sheng Chong Bing Fang Zhi Za Zhi 11, 29–31 (In Chinese).

Li, W., Kou, J.X., Wang, J.X., Liu, X., Zhao, Z.P., Gao, C.L., Li, G.P., 2000. The role of health education in the control of taeniasis and cysticercosis. Zhongguo Chu Ji Wei Sheng Bao Jian 14, 59–60 (In Chinese).

Li, X.D., Cui, Y., Zhang, L., Huan, Y.C., 2011. Diagnostic value of color doppler ultrasound on cutaneous muscle cysticercosis. Lin Chuang Chao Sheng Yi Xue Za Zhi 13, 271 (In Chinese).

Li, T., Ito, A., Chen, X., Long, C., Okamoto, M., Raoul, F., Giraudoux, P., Yanagida, T., Nakao, M., Sako, Y., Xiao, N., Craig, P.S., 2012. Usefulness of pumpkin seeds combined with areca nut extract in community-based treatment of human taeniasis in northwest Sichuan province, China. Acta Trop. 124, 152–157.

Li, T., Chen, X., Yanagida, T., Wang, H., Long, C., Sako, Y., Okamoto, M., Wu, Y., Giraudoux, P., Raoul, F., Nkouawa, A., Nakao, M., Craig, P.S., Ito, A., 2013. Detection of human taeniases in Tibetan endemic areas, China. Parasitology 140, 1602–1607.

Li, T., Chen, X., Wang, H., Openshaw, J.J., Zhong, B., Felt, S.A., Ito, A., Luby, S.P., 2019. High prevalence of taeniasis and Taenia solium cysticercosis in children in western Sichuan, China. Acta Trop. 199, 105133.

Lian, C., Liu, C., Zhao, X.H., She, J.X., Zheng, X.C., Li, Z.Y., 2003. Study on diagnosis of cerebral cysticercosis by antibody detection and imaging techniques. Zhongguo Ji Sheng Chong Xue Yu Ji Sheng Chong Bing Za Zhi 21, 207–209 (In Chinese).

Liu, L., 1987. A cysticercosis case in newborne. Zhongguo Ji Sheng Chong Xue Yu Ji Sheng Chong Bing Za Zhi 5, 199 (In Chinese).

Liu, G.J., 1988. Diagnosis on human cysticercosis by indirect charcoal agglutination test. Zhongguo Ren Shou Gong Huan Bing Za Zhi 4, 57 (In Chinese).

Liu, X., 2013. An ectopical case of *Taenia solium*. Zhongguo Ji Sheng Chong Xue Yu Ji Sheng Chong Bing Za Zhi 31, 302 (In Chinese).

Liu, H.A., Ge, L.Y., 1978. Efficacy of traditional Chinese medicine Ganwusan against cysticercosis: a report of 66 cases. Shandong Yi Yao 18, 22–23 (In Chinese).

Liu, D.H., Liu, S.Z., Zhang, Y.J., Li, S.S., Lang, Z.J., Wang, Y.C., Li, F.W., Ge, J.X., Zhao, J.C., 1987. Clinical observation on treatment of 300 cases of cysticercosis with albendazole. Zhongguo Ji Sheng Chong Xue Yu Ji Sheng Chong Bing Za Zhi 5, 141–143 (In Chinese).

Liu, D.H., Zhao, M.H., Geng, J.M., 1996. The epidemiology and research survey of swine cysticercosis in China. Jilin Nong Ye Da Xue Xue Bao 18, 7–12 (In Chinese).

Liu, Y.B., Dai, W., Liu, B., Yin, K.X., Xu, H.X., Li, G.L., Ge, L.Y., 2002. Application of colloid gold immunochromatographic test in diagnosis of neurocysticercosis. Zhongguo Re Dai Yi Xue 2, 144–146 (In Chinese).

Liu, M., Chen, J.F., He, Z.Y., Li, S.H., 2008. Clinical characteristic of neurocysticercosis. Zhong Feng Yu Shen Jing Ji Bing Za Zhi 25, 478–481 (In Chinese).
Lu, G., Han, M., Meng, Q.B., 2002. The prevalence of cysticercosis and the current situation of its treatment. Neimenggu Min Zu Da Xue Xue Bao (Zi Ran Ke Xue Ban) 17, 79–81 (In Chinese).
Lu, G.Y., Ge, L.Y., Xin, F.M., 2003. Long-term observation on the therapeutic effects of Ganwusan on cysticercosis patients. Zhongguo Re Dai Yi Xue 3, 354–355 (In Chinese).
Ma, Y.X., 2004. The creation and development of three criteria on cerebral cysticercosis. Zhongguo Ji Sheng Chong Xue Yu Ji Sheng Chong Bing Za Zhi 22, 371–373 (In Chinese).
Ma, Y.X., Han, C.R., He, Z.J., Xue, X.L., Li, G.L., Hao, J.L., Liu, D.Y., Li, S.Z., Wang, J.J., Wu, S.R., Zhang, F.Y., 1988. Research on the control and prevention of taeniasis and cysticercosis. Wei Sheng Fang Yi 17, 1–6 (In Chinese).
Ma, Y.X., Xue, X.L., Yu, Q.L., 1989. Suggestions for the diagnosis, clinical types and treatment judgment of cerebral cysticercosis. Zhongguo Ji Sheng Chong Xue Yu Ji Sheng Chong Bing Za Zhi 7, 134–135 (In Chinese).
Ma, Y.X., Liu, J.H., Wang, Y.Z., Li, G.L., Liu, W.B., Wei, G.R., Qiao, S.J., Jiang, H.G., Chen, S., Ma, X.A., Zhu, C.H., 1991. Experimental study on the development of *Cysticercus cellulosae*. Henan Yu Fang Yi Xue Za Zhi 2, 657–660 (In Chinese).
Ma, Y.X., Xu, C.B., Shao, L., Zhu, Y.G., Xu, J.T., Kang, Q.D., 1992. Achievements, status and countermeasures on control of taeniasis and cysticercosis during past 40 years in China. Zhongguo Gong Gong Wei Sheng 8, 244–246 (In Chinese).
Ma, E.X., Chen, J.Q., Wang, X.J., Li, Q.S., 1996. Neuropsychiatric disorders caused by neurocysticercosis. Zhongguo Shen Jing Jing Shen Ji Bing Za Zhi 22, 294–296 (In Chinese).
Ma, Y., Jiang, H., Cao, D.B., 1998. Nurse on the secondary epilepsy due to cerebral cysticercosis with intracranial hypertension. Hu Li Xue Za Zhi 13, 22–23 (In Chinese).
Maurice, J., 2014. Of pigs and people–WHO prepares to battle cysticercosis. Lancet 384, 571–572.
Meng, J.M., Xie, S.P., Zhao, C.Y., Guo, L.Y., Liu, Y.P., Zhang, J., Zhu, Y.X., Cheng, N.P., Li, C.Y., 1986. The treatment of brain cysticercosis with intracranial hypertension. Beijing Di Er Yi Xue Yuan Xue Bao 7, 128–131, 170 (In Chinese).
Meng, Q.D., Xiao, S.J., Zhu, Y.J., Xu, S.F., Hu, Q., 2007. 120 cases with taeniasis diagnosed by enteroscopy. Zhonghua Xiao Hua Nei Jing Za Zhi 24, 393 (In Chinese).
Miao, F., Zhao, Z.P., Liu, X., Wan, G.Q., Li, G.P., Fu, Z.Y., Fu, B., Li, W., Zhao, C.L., Deng, X.L., 2000. Epidemiological survey on taeniasis and cysticercosis in major epidemic areas of Shandong province. Zhongguo Xue Xi Chong Bing Fang Zhi Za Zhi 12, 217–220 (In Chinese).
Mills, R.G., 1924. Notes on the occurrence of *Taenia solium* and *Cysticercus cellulosae* in North China. Chin Med J (Engl) 38, 465–471.
Ministry of Health of the People's Republic of China, 2012a. Diagnosis of Cysticercosis. WS 381–2012. (In Chinese).
Ministry of Health of the People's Republic of China, 2012b. Diagnosis of Taeniasis. WS 379–2012. (In Chinese).
Mo, X.Z., Bao, H.E., Rong, J.Q., Chen, L.H., Wang, H.Y., Wu, G.P., Li, B., Xie, H.S., Wang, H., He, F.G., 2005. Infection status and epidemiological investigation of *Taenia asiatica* in Duyun. Qiannan Min Zu Yi Zhuan Xue Bao 18, 159–160, 163 (In Chinese).
Nkouawa, A., Sako, Y., Li, T., Chen, X., Nakao, M., Yanagida, T., Okamoto, M., Giraudoux, P., Raoul, F., Nakaya, K., Xiao, N., Qiu, J., Qiu, D., Craig, P.S., Ito, A., 2012. A loop-mediated isothermal amplification method for a differential identification of Taenia tapeworms from human: application to a field survey. Parasitol. Int. 61, 723–725.

No authors, 1974. Clinical observation on the treatment of taeniasis with herba agrimoniae. Zhonghua Yi Xue Za Zhi 54, 344–345 (In Chinese).

Pan, X.Z., Sun, T., Weng, X.H., Xu, Z.Y., Dai, Z.Y., Huang, M.Y., 1981. Praziquantel treatment of cysticercosis. Shanghai Di Yi Yi Xue Yuan Xue Bao 8, 127–132 (In Chinese).

Pan, S.C., Xu, G.H., Wu, Y.Z., Li, J.H., Yan, W.A., Wang, G.X., Sun, F.Y., 1995. A background survey and future strategies of latrines and nightsoil treatment in rural in China. Wei Sheng Yan Jiu 24 (Suppl), 1–10 (In Chinese).

Qian, M.B., Utzinger, J., Keiser, J., Zhou, X.N., 2016. Clonorchiasis. Lancet 387, 800–810.

Qian, M.B., Chen, J., Bergquist, R., Li, Z.J., Li, S.Z., Xiao, N., Utzinger, J., Zhou, X.N., 2019. Neglected tropical diseases in the People's Republic of China: progress towards elimination. Infect. Dis. Poverty 8, 86.

Qin, Q.X., Li, S.C., Wang, H.Y., Luo, L.H., Huang, D.J., 1985. Preliminary report on the diagnosis of *Taenia solium* infection by indirect hemagglutination test. Ji Sheng Chong Xue Yu Ji Sheng Chong Bing Za Zhi 3, 226 (In Chinese).

Qiu, L.S., Zhang, Y.H., Xue, H.C., Li, H., Zhou, H.J., Chen, S.X., Fu, X.L., Jiang, B.Q., Shao, S.C., Yang, X.C., 1998. Observations on several cysticercus antigens and its use in EITB for diagnosis of cysticercosis. Zhongguo Xue Xi Chong Bing Fang Zhi Za Zhi 10, 29–32 (In Chinese).

Research Committee of China Medical Journal, 1910. Reports of worm infections arranged according to the provinces of China. Chin Med J (Engl) 24, 93–98.

Roys, C.K., 1910. Extract from a letter from Dr. C. K. Roys, Weihsien, with statistics from Tsingtau. Chin Med J (Engl) 24, 93.

Shen, L.C., Li, Z.Q., 2013. The control study of clinical neuropsychiatric symptoms of schizophrenia patients with cerebral cysticercosis. Zhongguo Bing Yuan Sheng Wu Xue Za Zhi 8, 753–755 (In Chinese).

Shi, F.M., Ge, L.Y., Yang, S.H., Xu, W.H., 1989. Study on the immunodiagnosis of neurocysticercosis. Zhongguo Ji Sheng Chong Bing Fang Zhi Za Zhi 2 (Suppl), 65–66 (In Chinese).

Shi, D.P., Li, S.Y., Guo, Y., Guo, X.R., 2000. A diagnostic analysis of imaging in ocular cysticercosis. Zhonghua Yan Ke Za Zhi 36, 56–58 (In Chinese).

Shu, J., Wen, J.S., Li, Y.H., 1992. A case report on intracranial hypertension in treatment of clonorchiasis with praziquantel due to the concurrence of neurocysticercosis. Zhongguo Ji Sheng Chong Xue Yu Ji Sheng Chong Bing Za Zhi 10, 289 (In Chinese).

Steinmann, P., Zhou, X.N., Du, Z.W., Jiang, J.Y., Xiao, S.H., Wu, Z.X., Zhou, H., Utzinger, J., 2008. Tribendimidine and albendazole for treating soil-transmitted helminths, Strongyloides stercoralis and Taenia spp.: open-label randomized trial. PLoS Negl. Trop. Dis. 2, e322.

Sui, B.S., Sun, J., Chen, Y.B., Shi, M.S., Jin, X.K., 1990. Application of MRI in staging of neurocysticercosis and its clinical value. Zhonghua Yi Xue Za Zhi 70, 502–505 (In Chinese).

Sun, H.B., Tang, S.Y., Li, X.B., Shao, S.X., Zhao, B.Z., Xue, X.L., Qiao, G.X., 1985. Study on the diagnosis of cysticercosis by latex agglutination test. Ji Sheng Chong Xue Yu Ji Sheng Chong Bing Za Zhi 3, 221–223 (In Chinese).

Technical Steering Panel for National Survey of Current Status of Major Human Parasitic Diseases, 2008. Report on the National Survey of Current Status of Major Human Parasitic Diseases in China. People's Medical Publishing House, Beijing (In Chinese).

Torgerson, P.R., Devleesschauwer, B., Praet, N., Speybroeck, N., Willingham, A.L., Kasuga, F., Rokni, M.B., Zhou, X.N., Fevre, E.M., Sripa, B., Gargouri, N., Furst, T., Budke, C.M., Carabin, H., Kirk, M.D., Angulo, F.J., Havelaar, A., de Silva, N., 2015. World Health Organization estimates of the global and regional disease burden of 11 foodborne parasitic diseases, 2010: a data synthesis. PLoS Med. 12, e1001920.

Wang, R.F., 2009. A case of brain herniation and death following treatment to eliminate cerebral cysticercosis. Zhongguo Bing Yuan Sheng Wu Xue Za Zhi 4, insert 4 (In Chinese).

Wang, A.H., Fang, Z.R., 2008. Analysis on the efficacy of combination of traditional Chinese medicine and western medicine against *Taenia solium* infection. Sichuan Zhong Yi 26, 69–70 (In Chinese).

Wang, Z.Z., Yang, Q.Y., Shan, X.G., Xue, R.Y., 1987. Control on human *Taenia solium* infection in Henan province. Zhongguo Ren Shou Gong Huan Bing Za Zhi 3, 33–35 (In Chinese).

Wang, Z.Q., Ma, Y.X., Cui, J., Ji, X.X., Yan, X.X., Zhu, W., Xu, B.L., Zhang, Y.L., 1999. Sero-epidemiological survey of human cysticercosis in Weishi county, Henan province. Zhongguo Ji Sheng Chong Bing Fang Zhi Za Zhi 12, 236–237 (In Chinese).

Wang, Z.L., Ge, L.Y., Yang, S.F., Li, G.L., 2000. Observation on EEGs in patients with cysticercosis. Lin Chuang Nao Dian Xue Za Zhi 9, 94–95 (In Chinese).

Wang, Z.L., Dai, W., Li, H., Yang, S.F., Yue, J.E., Mao, D.H., 2004a. Comparative study on electroencephalogram (EEG) and imaging in cerebra cysticercosis patients. Zhongguo Re Dai Yi Xue 4, 175–177 (In Chinese).

Wang, Z.L., Hu, Y.X., Mao, D.H., Li, G.L., Zhao, X.L., 2004b. Observation of development of EEG and CT in patients with cerebral cysticercosis before and after treatment. Zhongguo Ji Ceng Yi Yao 11, 790–791 (In Chinese).

Wei, D.X., Yang, W.Y., Huang, S.Q., Lu, Y.F., Su, T.C., Ma, J.H., Hu, W.X., Xie, N.F., 1980. Parasitological studies on the ancient corpse of the Western Han dynasty unearthed from tomb no. 168 on Phoenix Hill at Jiangling county. Wuhan Yi Xue Yuan Xue Bao 9, 1–6, 107 (In Chinese).

Wei, G.Z., Li, C.J., Xiao, Z.X., Meng, J.M., Ding, M.C., 1986. Cysticercosis of central nervous system (a clinical study of 1400 cases). Beijing Di Er Yi Xue Yuan Xue Bao 7, 135–140, 171 (In Chinese).

Wei, D.P., Zhao, X.C., Zhang, B.X., Han, L.P., Jiang, X.P., 1999. Diagnostic value of cerebral CT scanning in cerebral cysticercosis. Zhongguo Lin Chuang Yi Xue Ying Xiang Za Zhi 10, 117–119 (In Chinese).

Wen, G.Z., Yu, Y.X., Tan, S.Q., 1993. Preparation and therapeutic effect of tenifugal capsule. Jiamusi Yi Xue Yuan Xue Bao 16, 38–39 (In Chinese).

Wu, Z.D., 1981. Preliminary analysis on the efficacy of Shataoling against *Taenia solium* infection. Hebei Yi Yao 3, 17–18 (In Chinese).

Wu, F., Jin, X.X., Mao, F.R., Wang, Z.Q., Cui, J., 1994. Detection of antibody in 198 cases with cysticercosis. Henan Yi Xue Yan Jiu 3, 73–75 (In Chinese).

Wu, W., Jia, F., Wang, W., Huang, Y., Huang, Y., 2013. Antiparasitic treatment of cerebral cysticercosis: lessons and experiences from China. Parasitol. Res. 112, 2879–2890.

Xie, S.P., Pang, S.Q., Wang, Y.J., Wei, G.Z., 1995. A retrospective analysis of epilepsy and related factors in the patients with cerebral cysticercosis. Zhonghua Shen Jing Jing Shen Ke Za Zhi 28, 370–372 (In Chinese).

Xu, L.Q., 2002. The harm of important parasitic diseases in western areas of China and rethink on the control activities. Zhongguo Ji Sheng Chong Bing Fang Zhi Za Zhi 15, 1–3 (In Chinese).

Xu, L.Q., Zhou, C.H., Jiang, Z.X., Zhang, X.Q., 1998. Clinical characteristics of cysticercosis in China. Xian Dai Kang Fu 2, 67–68 (In Chinese).

Xu, L.Q., Jiang, Z.X., Zhou, C.H., Zhang, X.Q., Ren, Z.X., Chang, J., Xu, J.T., Li, Y.B., Zhang, L.L., Yu, S.H., Zhang, G.Y., Ma, C.J., 1999. Distribution of cysticercosis in China. Zhongguo Ji Sheng Chong Bing Fang Zhi Za Zhi 12, 30–32 (In Chinese).

Xu, L.Q., Yu, S.H., Xu, S.H., 2000. Distribution and Pathogenic Impact of Human Parasites in China. People's Medical Publishing House, Beijing (In Chinese).

Xu, H.X., Yin, K.X., Xu, J., Liu, Y.B., 2002. Study of measures on reducing the sequelae in patients with neurocysticercosis. Di Fang Bing Tong Bao 17, 69–72 (In Chinese).

Xue, J.D., Han, Z.X., 1981. Efficacy of combination of niclosamide and bitin against *Taenia solium* infection. Shan'xi Xin Yi Yao 10, 26–28 (In Chinese).
Yan, J.J., Liu, J., Geng, Z.W., Tong, X.Y., Bai, Y.H., Cui, Z.T., 2004. Imaging analysis of cerebral cysticercosis on MRI. Heilongjiang Yi Xue 28, 420–421 (In Chinese).
Yang, W.Y., 1955. Preliminary report on the detection methods for *Taenia saginata* infection. Zhonghua Yi Xue Za Zhi 41, 147–149 (In Chinese).
Yang, J.X., Zhang, C., Jiang, X., Zhang, X.F., 1997. The clinical analysis of 52 cases of ocular cysticercosis. Yan Wai Ke Zhi Ye Yan Bing Za Zhi 19, 324–326 (In Chinese).
Yang, Y.M., Li, Z., Shi, W.X., 2007. Epidemiological investigation on the prevalence of Taeina around Erhai Lake of Dali, China. Zhongguo Bing Yuan Sheng Wu Xue Za Zhi 2, 109, 114 (In Chinese).
Yang, Y.J., Sun, G.P., Kong, Q.A., Kong, Q.Y., Fu, T.X., Ma, Q.R., Yin, L.X., 2008a. Misdiagnostic analysis of 177 cases of cerebral cysticercosis. Zhongguo Bing Yuan Sheng Wu Xue Za Zhi 3, 770–772 (In Chinese).
Yang, Y.J., Wu, X.Y., Yang, S.F., Jia, F.J., Shi, F.M., Wei, Y.B., 2008b. Value of four immunological techniques in diagnosis and treatment of cerebral cysticercosis. Zhongguo Re Dai Yi Xue 8, 1088–1090 (In Chinese).
Yang, G.J., Liu, L., Zhu, H.R., Griffiths, S.M., Tanner, M., Bergquist, R., Utzinger, J., Zhou, X.N., 2014. China's sustained drive to eliminate neglected tropical diseases. Lancet Infect. Dis. 14, 881–892.
Yin, S., Yang, S.H., Shi, F.M., Ge, L.Y., 1990. Application of dot-ELISA in immunodiagnosis of cerebral cysticercosis. Zhongguo Ji Sheng Chong Bing Fang Zhi Za Zhi 3, 31–32 (In Chinese).
Yu, S.M., 1988. Cysticercosis in the oral and maxillofacial region. Shi Yong Kou Qiang Yi Xue Za Zhi 4, 93–95 (In Chinese).
Yu, S.H., 1991. Illustrations on the protocol for national distribution survey on human parasitic diseases. Ji Sheng Chong Xue Yu Ji Sheng Chong Bing Za Zhi Suppl, 3–5 (In Chinese).
Yuan, Z., Ren, H.J., Wu, X.L., Pan, Y.W., Ding, Y.Z., Zhang, J.S., Qiu, M.D., 2005. Necessity of surgical intervention in the treatment of severe neurocysticercosis. Lanzhou Da Xue Xue Bao (Yi Xue Ban) 31, 30–32 (In Chinese).
Zang, X.Z., Li, H.Z., Qian, M.B., Chen, Y.D., Zhou, C.H., Liu, H.K., Liu, Y.H., Li, S.Z., 2019a. Extensive disseminated cysticercosis: a case report in Yunnan province, China. BMC Infect. Dis. 19, 535.
Zang, X.Z., Li, H.Z., Liu, H.K., Liu, Y.H., Zhou, C.H., Chen, Y.D., Qian, M.B., Li, S.Z., 2019b. Epidemiological analysis of new cases of cysticercosis in Dali prefecture, Yunnan Province from 2014 to 2017. Zhongguo Xue Xi Chong Bing Fang Zhi Za Zhi 31, 143–147 (in Chinese).
Zhang, S.F., 1998. Ten cysticercosis cases in infants. Zhongguo Ji Sheng Chong Xue Yu Ji Sheng Chong Bing Za Zhi 16, 171 (In Chinese).
Zhang, S.F., 2003. Report on 60 cases with *Taenia solium* infection treated with praziquantel. Zhongguo Ji Sheng Chong Xue Yu Ji Sheng Chong Bing Za Zhi 21, 196 (In Chinese).
Zhang, W.Z., Lu, Y.L., Zhang, Q.Y., Xiao, A.X., Wang, C.J., Tian, Q.Y., Yin, R.H., Liang, Y.P., Xin, Z.M., Zhao, B., 1985. Treatment of 150 cases with taeniasis. Ji Sheng Chong Xue Yu Ji Sheng Chong Bing Za Zhi 3, 271 (In Chinese).
Zhang, D.S., Shi, R.Z., Ma, Z.K., Ma, J.S., Pu, X.J., Cao, Y.X., Song, X.X., Si, J.C., Mao, J.F., Hu, W.Y., 1990. Epidemiological survey on human cysticercosis and taeniasis in Kaifeng county, Henan province. Henan Yu Fang Yi Xue Za Zhi 1, 298–299, 282 (In Chinese).
Zhang, H.H., Wang, C.Y., Ge, L.Y., Yin, S., Liu, Y.B., 1992. Studies on preparation of kits for diagnosis of neurocysticercosis by detection of CAg with McAb. Zhongguo Ji Sheng Chong Bing Fang Zhi Za Zhi 5, 178–180 (In Chinese).

Zhang, L., Tao, H., Zhang, B., Wang, H., Wang, Y., Li, Z., Yang, J., Yang, B., Li, Y., Pang, Y., Zhang, H., Li, Y., Wu, Y., 1999. First discovery of *Taenia saginata asiatica* infection in Yunnan province. Zhongguo Ji Sheng Chong Xue Yu Ji Sheng Chong Bing Za Zhi 17, 95–96 (In Chinese).

Zhang, J.N., Zhang, X., Cao, W.D., Fei, Z., Fu, L.A., Liu, W.P., Wang, Z.X., Guo, Y., Qu, Y., Zhen, H.N., 2001. Minimally invasive neurosurgery for neurocysticercosis. Di Si Jun Yi Da Xue Xue Bao 22, 2191–2194 (In Chinese).

Zhao, Z.P., Liu, X., Wan, G.Q., Gao, C.L., Li, G.P., Li, W., Deng, X.L., Li, D.J., Chen, X.X., Xu, F.Q., Yang, B.J., 1999. Studies on community control strategies for taeniasis/cysticercosis in Shandong province. Zhongguo Ji Sheng Chong Bing Fang Zhi Za Zhi 12, 111–114 (In Chinese).

Zhou, X.N., 2018. Report on the National Survey of Important Human Parasitic Diseases in China (2015). People's Medical Publishing House, Beijing (in Chinese).

Zhou, X.N., Guo, J.G., Wu, X.H., Jiang, Q.W., Zheng, J., Dang, H., Wang, X.H., Xu, J., Zhu, H.Q., Wu, G.L., Li, Y.S., Xu, X.J., Chen, H.G., Wang, T.P., Zhu, Y.C., Qiu, D.C., Dong, X.Q., Zhao, G.M., Zhang, S.J., Zhao, N.Q., Xia, G., Wang, L.Y., Zhang, S.Q., Lin, D.D., Chen, M.G., Hao, Y., 2007. Epidemiology of schistosomiasis in the People's Republic of China, 2004. Emerg. Infect. Dis. 13, 1470–1476.

Zhu, K.H., Gao, J.G., 1989. Two ectopical *Taenia solium* cases. Zhongguo Ji Sheng Chong Xue Yu Ji Sheng Chong Bing Za Zhi 7, 251 (In Chinese).

Zhu, Y.X., Jia, J.Y., Zhao, S.J., Chen, Y.L., Tang, Z.M., 1993. Experimental and clinical study of cerebral cysticercosis treated with praziquantel. Zhongguo Ji Sheng Chong Bing Fang Zhi Za Zhi 6, 20–22 (In Chinese).

Zhu, T.M., Li, S.F., Pan, G.F., 1994. CT staging of cerebral cysticercosis and its clinical significance. Zhonghua Shen Jing Jing Shen Ke Za Zhi 27, 150–152 (In Chinese).

CHAPTER TEN

Epidemiological big data and analytical tools applied in the control programmes on parasitic diseases in China: NIPD's sustained contributions in 70 years

Shang Xia[a,b,c,d,e], Jin-Xin Zheng[a,b,c,d,e], Xin-Yi Wang[a,b,c,d,e], Jing-Bo Xue[a,b,c,d,e], Jian-Hong Hu[a,b,c,d,e], Xue-Qiang Zhang[a,b,c,d,e], Xiao-Nong Zhou[a,b,c,d,e], Shi-Zhu Li[a,*]

[a]National Institute of Parasitic Diseases, Chinese Center for Disease Control and Prevention, Shanghai, People's Republic of China
[b]Chinese Center for Tropical Diseases Research, Shanghai, People's Republic of China
[c]Key Laboratory of Parasite and Vector Biology, Ministry of Health, Shanghai, People's Republic of China
[d]National Health Commission of China, Shanghai, People's Republic of China
[e]WHO Collaborating Centre for Tropical Diseases, Shanghai, People's Republic of China
*Corresponding author: e-mail address: lisz@chinacdc.cn

Contents

1. Introduction 320
2. Empirical analysis 322
   2.1 Statistical methods for analysing the causative factors and disease incidences 323
   2.2 Descriptive methods for characterizing the temporal and spatial disease distribution 325
   2.3 Experimental methods for test epidemiological hypotheses 328
3. Theoretical and mathematical analysis 333
   3.1 Theoretical modelling for analysing the dynamics of disease transmission 333
   3.2 Mathematical analysis for estimated the effectiveness of diseases intervention measures 334
4. Computational analysis 335
   4.1 Computational modelling for analysing the dynamics of disease transmission 335
   4.2 Data mining and machine learning 337
   4.3 Systems thinking for disease control 339

Advances in Parasitology, Volume 110
ISSN 0065-308X
https://doi.org/10.1016/bs.apar.2020.04.009

5. Remaining challenges and future opportunities 339
5.1 Factors influencing diseases are rooted in multiple sources and at multiple scales 342
5.2 Complexity in epidemiological system of parasitic diseases needs integrated more factors in the modelling 342
5.3 Systems thinking plays an emerging important role for the prediction and control of parasitic diseases 343
6. Conclusion 344
References 344

Abstract

The analysis of epidemiological data has played an important role for the academic research carried out by the National Institute of Parasitic Diseases, China CDC, since its foundation in 1950s. Those researches, e.g., the temporal-spatial patterns of disease transmission and the identification of risk factors, have contributed significantly to the national parasitic disease control and elimination programmes in China. With the development and application of epidemiological data analysis in the last decade, all research results improve our understanding of parasitic diseases epidemiology and related health issues through the application platform of epidemiological big data and analytical tools. In particular, implementation research on analytical predictions on disease outbreak or epidemic risks have provided references to the scientific guidance on effective preventions and interventions in the parasitic disease elimination in China, such as fliariasis, malaria and schistosomiasis. This review has reflected the function of data accumulation and application of temporospatial tools in parasitic diseases control, and the ways of the NIPD's sustained contributions to the disease control programmes in China.

1. Introduction

Epidemiology serves as the cornerstone of public health. According to the definition of World Health Organization (WHO), epidemiology is "the study of the distribution and determinants of health-related states or events (including disease), and the application of this study to the control of diseases and other health problems". Since the main tasks for the National Institute of Parasitic Diseases at China CDC (NIPD) is to organize or coordinate national parasitic diseases control efforts with more works on surveillance, scientific research on parasitic diseases, emergency response to outbreaks of parasitic and tropical diseases, services on detection and consultation on parasite infections, and international cooperation, etc. Therefore, the epidemiological investigation and field surveillance are two major components for the NIPD's routine work, with supported by big databases and analytical tools. In this regards, epidemiological studies carried out in NIPD mainly focus on the transmission patterns of parasitic and tropical diseases, e.g., the temporal-spatial patterns of parasitic disease transmission and the identification of risk factors, etc.

The major epidemiological researches in NIPD are driven by the problem-solving approach, particularly on development of new control strategy, application of modern tools in surveillance and response to outbreaks, and providing the updated references as baseline data to the decision making for formulation of 5-year's working plan or standards in the control programmes. All of these epidemiological studies are only enabled by the confluence of data from various sources, including (i) surveillance data, e.g., records of when, where, and who has been infected, (ii) biological data of parasites and etiological data of disease vectors, e.g., mosquitos biting rate in malaria, data on variables of parasites' lifecycle, (iii) environmental data, e.g., temperature, precipitation, vegetation, land usage, (iv) socio-economic data, e.g., income level or poverty level, and (v) population's behavioural data, e.g., host population mobile and migrant activities. The variety of epidemiological data will break through the exiting scientific research paradigms, enable a new generation of scientific research. For instance, the goal of understanding the impacts of climate change on transmission of schistosomiasis transmission is a driving research for problem-solving, then the questions is what kinds of datasets we need to accumulated, what kind of analytical tools we could use for the study, and how to integrate the datasets and analytical tools to explain the what extents of climate changes' impacts on the schistosomiasis transmission or spreading in the future. So that all datasets, such as plant physiology, soil science, meteorology, oceanography, hydrology, and fluvial geomorphology, are necessary for the analysis, and the tools of theoretical and mathematical analysis, and computational analysis are essential for the study, as well as bridging the diverse length and time scales involved will require a collection of cooperating models. Thus, the purpose of this article is to review the application of various analytical tools in the control programmes on parasitic diseases in China, in particular, the NIPD's sustained contributions in her 70 years' development, through application of epidemiological methodologies on empirical analysis, theoretical and mathematical analysis, and computational analysis.

The search strategy for literature review for this investigation was done through the NCBI PubMed (http://www.ncbi.nlm.nih.gov/pubmed/) and CNKI (http://www.cnki.net/, one of the biggest biomedical literature databases in China), by searching the related English and Chinese literature for application of epidemiological analysis in parasitic disease studies of China.

Based on the science of epidemiology that is of interdisciplinary nature, e.g., disease aetiology, vector biology, surveillance and monitoring, meteorological and geographical environments, ecological evaluation, molecular and genetic analysis, and pharmaceutical studies on treatment effects including clinical trials, our review has focus on the development of epidemiological approaches, which are typically on three types of research paradigms, as illustrated in Fig. 1 and 2, namely empirical analysis, theoretical and mathematical analysis, and computational analysis.

2. Empirical analysis

In the earlier stage of epidemiological investigations, epidemiologists from NIPD were mainly relying on expertise from multiple scientific disciplines (Lv et al., 2011; Xia et al., 2017). For example, biologists can help provide a better understanding of parasitic disease processes and the attributes of parasites to the disease development. Empirical observations or experiences may be analysed quantitatively or qualitatively. By taking

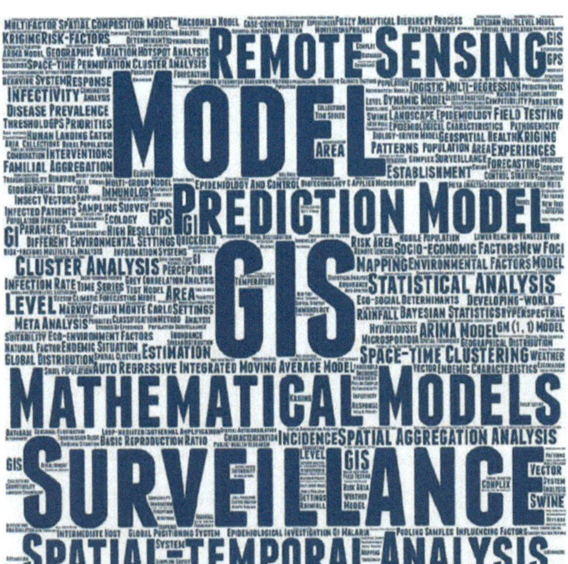

Fig. 1 The cloud mapping from the frequency of key words collected by our literature review.

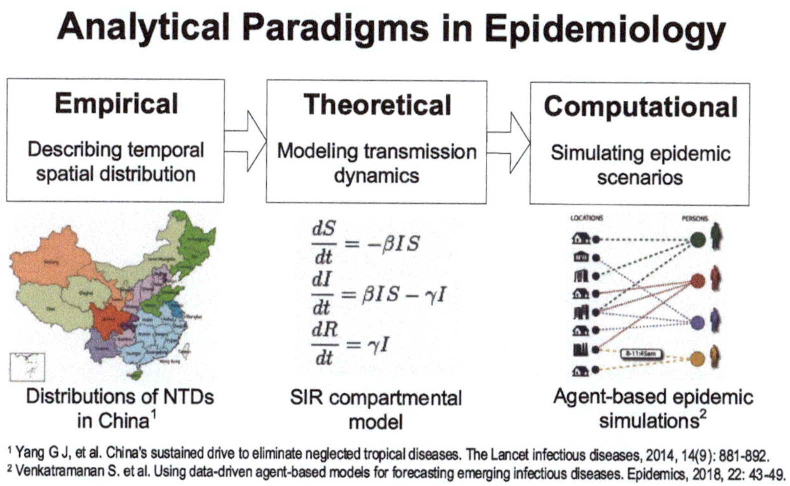

Fig. 2 Three types of research paradigms in epidemiological data analysis, namely empirical analysis, theoretical and mathematical analysis, and computational analysis.

different disciplines into consideration, the investigations on empirical epidemiology carried out in NIPD can be divided into three catalogues, such as (1) analytical epidemiology, in which statistical analysis of epidemiological data in an attempt to establish relationships between causative factors and incidences of a disease; (2) descriptive epidemiology, in which information about the distribution of a disease, and some of its mathematical descriptions; (3) experimental epidemiology, in which prospective population experiments are designed to test epidemiological hypotheses, and usually attempt to relate the postulated causes to the observed effects. Trials of new antiviral drugs and vaccines are an example.

2.1 Statistical methods for analysing the causative factors and disease incidences

In the history, scientists from NIPD did a lot of epidemiological studies to link the multi-impact factors with the disease prevalence. For instance, the impacts of the Three Gorges Dam on the schistosomiasis transmission supported by the ecological studies, the results were turned in formulation of a policy to take the surveillance in the Three Gorges region of Chongqian city. Up to now, the surveillance measures are still carried out in the Three Gorges region to monitor the potential risks of schistosomiasis transmission there. In addition, more tools of ecology and geographic information systems have introduced to the schistosomiasis epidemiological studies, which improve the prediction capacity of NIPD in the risk assessment of schistosomiasis transmission, and now those modern tools have been used for routine application for the field evaluation of the schistosomiasis elimination programmeme every year.

In the field of malaria studies, the impact of climate changes on malaria transmission were explored in Yunnan province by the joint-team from NIPD and Yunnan Provincial Institute on Parasitic Diseases, the logistic regression models employed with climate data and malaria incidence data were developed. The major results of these studies indicated a positive correlation between warming climate or El Nino events and malaria incidence in Yunnan province, therefore, those impact of climate changes can used for the risks prediction to help mitigation strategy formulation. At the same time, a risk prediction model was established by employed the environmental data, malaria incidence data, and mosquito density data in Yunnan province. In addition, the impact of earthquake and social economic status on malaria transmission were evaluated in Yunnan and Hainan provinces by statistic models, which help local government to take consideration in the decision making (Feng et al., 2015b); More recent study aiming to understand the dynamics of malaria vectors in implementing malaria control strategies was undertaken in the Three Gorges Reservoir areas. So a total of six villages were selected from different sections in the Three Gorges Reservoir for exploring the relationship between the climatic factors and its malaria vector density from 1997 to 2007 using the auto-regressive linear model regression method. The result indicated that both temperature and precipitation were better modelled as quadratic rather than linearly related to the density of Anopheles sinensis (Wang et al., 2014).

For other parasitic diseases, several studies have been performed to correlate the social economic and environmental factors with the prevalence of echinococcosis in western and northwestern China, either through the cross-sectional surveys or case-control studies. Most of those results showed that the prevalence was highly correlated to the ecosystems and social behaviours (Xue et al., 2018; Yang et al., 2015a). Thus more interventions to reduce transmissibility through the staydogs, improve sanitation, and promote health education have put forward to control echinococcosis both in human and animals. An epidemiological analysis of the cases of visceral leishmaniasis in the non-endemic province of China, conducted a case investigation in the form of telephone and letter, and made a descriptive analysis of the obtained data in the region, time and population, gender, age and occupation distribution; Another epidemiological study to understand impact factors of Blastocystis hominis infection among specific population, human immunodeficiency virus (HIV), carried out in Yunnan province by logistic regression model, results illustrated that among 3.7% (12/324) positives in SSU-RNA gene of Blastocystis hominis, the quality of drinking water, livestock raised at home, and immunocompromised status were the major factors related to the blastocystosis (Teng et al., 2018).

2.2 Descriptive methods for characterizing the temporal and spatial disease distribution

More researches from NIPD were focus on establishment of temporal spatial database for various parasitic diseases together with local environmental data, and then applied the GIS tools to integrate the epidemiological features to presented the transmission patterns with both space and time dimensions.

In schistosomiasis studies, scientists from NIPD have identified the cluster patterns and transmission risk areas where schistosomiasis heavily transmitted, by using temporal-spatial models. First, the scale of the study ranged from the national, provincial, county and village levels. Based on the local settings, both human and animal transmission patterns have been explored and compared, it was found the animal transmission more focalized along the lakes or rivers, particularly in the marshland and lake ecozoons. Second, another study on correlation between location of patients and distribution of *Oncomelania* snails found the spatial autocorrelation relationship strongly existed between these two variables, indicating people's infection was highly related to the environmental factors by exposure to the infested water in the lake region. Third, more researches on the detection of snail habitats by using remote sensing technology, i.e., Landcet, QuickBird images (Huang et al., 2007; Yang and Guo, 2006). It also found both *Oncomelania* snails or infected snails presented in cluster distribution in the marshland where snail habitats appeared as under water in summer and being wetland in winter (Gao et al., 2013b). These characters of the snail habitats are able to detect or layout from analysed remote sensing images after performing the temporal-spatial modelling in image analysis. Fourth, the landscape genetics on *Oncomelania* snail was also undertaken to illustrated the correlation of spatial and genetic distances of *Oncomelania hupensis* in mainland China (Guan et al., 2016; Li et al., 2009) using microsatellite loci and mitochondrial DNA techniques.

In malaria investigations, several studies have been done to provide a baseline for the ongoing national malaria elimination programmeme in China by illustrating the correlation between malaria incidences or prevalence and other factors. First, a study on a 6-year longitudinal surveillance on density of malaria vectors was carried out in the 62 national malaria surveillance sites, to reveal the spatio-temporal distribution of malaria vectors in the national malaria surveillance sites from 2005 to 2010. The spatial and temporal analyses of the four primary vectors, including *Anopheles sinensis*, *An. lesteri*, *An. dirus* and *An. minimus* distribution were conducted by the methods of kernel k-means and the cluster distribution of the most widely distribution vector of *An. sinensis* was identified. It was revealed that *An. sinensis* was the most widely distributed, accounting for 96.25% of all collections, and its distribution was divided into three different clusters with a significant increase of density observed in the second cluster which located mostly in the central parts of China (Huang et al., 2015). Second, one report on spatial aggregation of malaria cases and specific vectors in the areas along the Huang-Huai River of central China where malaria had outbreak during 2006–2008, the data on geographical, meteorological and vectorial factors were analysed to determine the key factors related to malaria re-emergence in the areas (Zhou et al., 2012). It was found the spatial distribution between malaria cases and water-body, the changing of meteorological factors, and increasing vectorial capacity and basic reproductive rate of *An. sinensis* leaded to malaria re-emergency. Third, to identify the effective strategy to control the malaria re-emergency, a study was done to control the sources of malaria infection, based on the identification of risk areas and population. The geographic information of malaria cases and their surrounding water bodies were collected from Suixi, Guoyang, Guzhen, Yingshang, Fengyang and Yongqiao County in Anhui province, Yongcheng and Tongbai County in Henan province. Data on environment and malaria cases were mapped and analysed in 113 villages of the eight counties to identify the spatial correlation between malaria cases and water bodies. A total of 357 malaria cases and their geographic data as well as surrounding water bodies were collected and analysed. It was found that 74% of malaria cases were located within the extent of 60 m proximity to the water bodies, and distribution of water bodies is an important factor influencing the occurrence and distribution of malaria cases in the *An. sinensis* distributed areas. Therefore, the intervention for malaria case management needs to mainly target on the population living within 60 m around water bodies (Yan et al., 2014). Fourth, studies in mosquito vectors, a spatial-temporal variation and primary ecological drivers of *An. sinensis* human biting rates in malaria epidemic-prone regions of China was performed, and the temporal variation in biting rates was much larger than the spatial variation, with 19.3% of temporal variation attributable to differences in minimum temperature and enhanced vegetation index and 16.9% of spatial variance due to distance to the nearest river and the paddy index. The results indicated that substantial spatial-temporal variation in *An. sinensis* human biting rates exists in malaria epidemic-prone regions of China, with minimum temperature and enhanced vegetation index accounting for the greatest proportion of temporal variation and distance to nearest river and paddy index accounting for the greatest proportion of spatial variation among observed ecological drivers (Huang et al., 2011; Ren et al., 2015).

In epidemiological studies on echinococcosis, two major areas on methodologies of temper-spatial distribution of human echinococcosis as well as well cluster patterns of the endemic areas were explored, which can provide the precision information on high risk areas where interventions need to be strengthened both in human and animals. For instance, the spatial scan analysis and spatial autocorrelation methods were used to identify the spatial clusters on the detection rate of cystic echinococcosis and alveolar echinococcosis patients who were newly detected based on the screening data of echinococcosis among population in 18 counties of Ganzi Tibetan Autonomous Prefecture, Sichuan province of China, in 2014. It was found that both the primary cluster of cystic and alveolar echinococcosis was majorly located in the northwest part of the Ganzi Tibetan Autonomous Prefecture, but the former one more concentrated in the northeastern Dege county and northeastern Shiqu county, while later one more focalized in the northeastern Dege county, northern Ganzi county, and northeastern Shiqu (Zhao et al., 2015). Then, the prediction of echinococcosis prevalence was also tried by use of spatial interpolation methods in Qinghai province, and the Kriging interpolation got the best estimation of echinococcosis prevalence which arranged from 2.04% to 5.28% in two counties (Fang et al., 2014).

In the other parasitic diseases, more studies were to identify the cluster distribution of infections. For example, one study on food-borne helminthic zoonosis, angiostrongyliasis, caused by infections with the third-stage larvae of *Angiostrongylus cantonensis*, was performed in three villages from Nanao Island, Guangdong Province to understand the spatial distribution of hosts of the parasites as a reference to the control strategy formulation. The density and natural infection of *Pomacea canaliculata*, and the snail intermediate host of *A. cantonensis*, and various rat species, the reservoir host of *A. cantonensis*, were surveyed from December 2015 to September 2016, and the survey data was analysed by ArcGIS, scan statistics, ordinary least squares (OLS) and geographically weighted regression (GWR) models. Results indicated that there was a risk of angiostrongyliasis in this region of China, with 6.1% (72/1190) larval infection rate in snails and 29.1% (32/110) infection rate in the rate definitive hosts. The distribution of positive *P. canaliculata* and rats exhibited spatial correlation, and the GWR model had advantage over the OLS model in the spatial analysis of hosts of *A. cantonensis* (Hu et al., 2017).

Another study was on the spatial analysis of visceral leishmaniasis in the oases of the plains of Kashi Prefecture, Xinjiang Uygur Autonomous Region, results showed that the spatial distribution of visceral leishmaniasis-infected families by Poisson model spatial aggregation of the Scan statistical analysis was clustered, with three clustered zones, including Zone A located in a centre point of 76.153447°E, 39.528477°N within its 1.11 mile radius, where the cumulative life-incidence of leishmaniasis was 1.95 times as high as that in surrounding areas ($P<0.05$); Zone B located in a centre point of 76.111968°E, 39.531895°N within its 0.54 mile radius, where the cumulative life-incidence of leishmaniasis was 1.82 times as high as that in surrounding areas ($P<0.01$); and Zone C located in a centre point of 76.195427°E, 39.563835°N within its 0.68 mile radius, where the cumulative life-incidence of leishmaniasis was 1.31 times as high as that in surrounding areas ($P<0.05$) (Wang et al., 2016).

2.3 Experimental methods for test epidemiological hypotheses

Some prospective studies on transmission risk assessment in high endemic areas of parasitic diseases, including schistosomiasis, malaria, echinococcosis and leishmaniasis, have been carried in China in order to verify current transmission risks and identify the potential new foci of those diseases driven by environmental changes or social features, which may provide assistance for the local decisions on control or elimination strategy (Table 1).

In schistosomiasis, active surveillance in selected risk areas of three provinces, namely Jiangsu, Anhui and Shandong, near the norther border of schistosomiasis endemic areas, was carried out from 2008 to 2010, including screening of potential infections in human and snail detection. Sentinel surveillance showed eight, nine and five confirmed cases of acute schistosomiasis in mobile populations (fishermen, migrant workers) in 2008–2010, respectively. All these cases were detected in Chaohu County, which must therefore be deemed an area at risk. In sentinel sites of Anhui, for example, the local number of acute cases account for 43% (19/44) in 2008, 33%

Table 1 Characteristics of studies using experimental analysis in the control programmes on parasitic diseases by NIPD.

Disease	Study aim	Study scales	Data set	Methodology: Method	Methodology: Algorithm/Model/Technique/Tools	References
Malaria	To study the impact of earthquake and social economic status on malaria transmission	Province (Yunnan & Hainan)	Earthquake data, social economic status data and malaria data	Analytical study	Statistic models	Feng et al. (2015a)
Malaria	To explore the relationship between the climatic factors and its malaria vector density	Village (the Three Gorges Reservoir areas)	Climate data and malaria vector density data	Analytical study	Auto-regressive linear model	Wang et al. (2014)
Echinococcosis	To study the correlation between the social economic & environmental factors and the prevalence of echinococcosis	Country	Social economic data, environmental factor data and echinococcosis incidence data	Descriptive study	Cross-sectional surveys & case-control studies	Xue et al. (2018) and Yang et al. (2015a)
Visceral leishmaniasis	To study visceral leishmaniasis cases	Country	Leishmaniasis cases incidence data	Descriptive analysis	Case investigation	Chen et al. (2011)
Blastocystis hominis disease	To study the impact factors of Blastocystis hominis infection among specific population	Province (Yunnan)	Gene data and impact factors data	Analytical study	Logistic regression models	Teng et al. (2018)
Schistosomiasis	To explore a new method of Oncomelania snails distribution and density prediction	County and Village	Oncomelania snails distribution GIS database	Descriptive method	QuickBird images	Huang et al. (2007)

Continued

Table 1 Characteristics of studies using experimental analysis in the control programmes on parasitic diseases by NIPD.—cont'd

Disease	Study aim	Study scales	Data set	Methodology: Method	Methodology: Algorithm/Model/Technique/Tools	References
Schistosomiasis	To study on correlation between location of patients and distribution of Oncomelania snails	Province	Schistosomias epidemiological data, GIS database	Descriptive method	Arc View3.x,S-PLUS and Spatial Statistics software	
Schistosomiasis	To detection of Oncomelania snails habitats	County and Village	Environmental data, malaria incidence data, and mosquito density data	Descriptive method	Temporal-spatial model	Yang ang Guo (2006)
Schistosomiasis	To illustrated the correlation of spatial and genetic distances of Oncomelania hupensis	National	Spatial distribution data, gene data	Descriptive method	Microsatellite loci and mitochondrial DNA techinquies	Guan et al. (2016) and Li et al. (2009)
Malaria	To survey the density of malaria vectors	National	Malaria vectors temporal distribution data	Descriptive method	Kernel k-means and the cluster distribution	Huang et al. (2015)
Malaria	To determine the key factors related to malaria re-emergence in the areas	Area research	Geographical, meteorological and vectorial factors	Descriptive method	Spatial analysis tools	Zhou et al. (2012)
Malaria	To identify the effective strategy to control the malaria re-emergency	County	Malaria cases spatial aggregation data	Descriptive method	Statistic analysis	Yan et al. (2014)
Malaria	To determine the primary ecological drivers of malaria vector human biting rates in malaria epidemic-prone regions of China	Area research	62 Malaria surveillance sentinel sites data	Descriptive method	Linear mixed effects model	Ren et al. (2015) and Huang et al. (2015)

Echinococcosis	To the spatial clusters on the detection rate of echinococcosis	Area research	Screening spatial data	Descriptive method	Spatial scan analysis and spatial autocorrelation methods	Zhao et al. (2015)
Echinococcosis	To estimate the echinococcosis prevalence	Province	Echinococcosis prevalence data	Descriptive method	Spatial interpolation method	Fang et al. (2014)
Schistosomiasis	To carry out active surveillance	Province	Potential infections in human and snail detection Sentinel surveillance data	Experimental method	Sample survey	Qian et al. (2013)
Malaria	To study on the epidemic factors of malaria	County	Malaria epidemiological investigation data	Experimental method	ArcGIS, excel	Feng et al. (2015a)
Echinococcosis	To determine the distribution of the intermediate hosts of Echinococcus multilocularis	County	Density and antigen positive rate of Intermediate hosts	Experimental method	Systematic sampling method, remote sensing (RS), regression analysis	Xue et al. (2018)
Echinococcosis	To study epidemiological characteristics and risk factors of human hydatidosis	Area research	The prevalence of cystic echinococcosis in slaughtered sheep	Experimental method	Logistic regression analysis	Yang et al. (2015a)
Leishmaniasis	To predict the risks and high risk areas for leishmaniasis transmission	National	Data in the region, time and population, gender, age and occupation distribution	Experimental method	Case investigation	Chen et al. (2011)

(25/75) in 2009 and 40% (17/42) in 2010, compare to the total number in the country. While the proportion of newly detected snail breeding areas in Anhui accounted for 42% (5.03/11.98) in 2008, 95% (8.39/8.79) in 2009 and 79% (8.52/10.81) in 2010, compared a total of 31.58 km^2 snail breeding sites that were newly detected nationwide through the year 2008–2010 (Qian et al., 2013).

In echinococcosis, to determine the distribution of the intermediate hosts of *Echinococcus multilocularis* in Shiqu County, Sichuan, China. A systematic sampling method was used to investigate the density of burrows of rodents and lagomorphs at 97 pasture sites in winter and summer pastureland and remote sensing (RS) technology was used to correlate their densities to the distribution of these animals in different landscape types. The regression relationship between the densities of *Ochotona curzoniae* and *Microtus fuscus* and their burrow densities indicated that burrow densities could reflect the animal densities and was greater in summer pastureland than that in winter pastureland. The main distribution areas of the intermediate hosts were in bareland and half-bareland. Xue et al. used a stratified cluster sampling method to select 364 villages in 70 counties (district) from August to October 2016. In each village, 200 residents were randomly selected for B-ultrasound examination, accompanied by a serological test for those suspected. Results showed that the highest risk for the transmission was the family with antigen positive rate in dog feaces, and human ages (Xue et al., 2018); Yang et al. analyse epidemiological characteristics and risk factors of human hydatidosis in Habahe, Xinjiang uygur autonomous region, and provide evidence for local prevention and control of human hydatidosis. Results demonstrated that the prevalence of cystic echinococcosis in slaughtered sheep may be used as an indicator to assess and monitor the transmission status during and after control programme providing information for betterment of performance to sustain control strength (Yang et al., 2015a).

In malaria, Feng et al. analysed the epidemic factors of malaria after earthquake in the recent 3 years in Cangyuan County of Yunnan Province, results indicated the western Cangyuan County had higher malaria transmission risk which related to the natural environment, social factors (Feng et al., 2015a).

In leishmaniasis, Chen et al. made an epidemiological analysis of the cases of visceral leishmaniasis in the non-endemic province of China, conducted a case investigation in the form of telephone and letter, and made a descriptive analysis of the obtained data in the region, time and population, gender, age and occupation distribution. Results showed that among 13 non-endemic provinces, the top three provinces had higher risks for the transmission were Chongqing, Hunan and Guangxi. Where those infections were occurred mainly in Gangshu, Sichuan and Xingjiang (Chen et al., 2011).

3. Theoretical and mathematical analysis

When applied mathematics introduced into the research on epidemiology of parasitic diseases, the use of mathematical models to explain and reproduce the empirical observations in the national control or elimination programmes on parasitic diseases that will improve the precision effectiveness in the implementation of control strategies or diseases surveillance. Mathematical models can project how parasites infections progress and show the likely outcomes of an epidemic for parasitic diseases, and thus help inform public health decisions on interventions in the national programmes, technically supported by NIPD (Chen et al., 2018; Xia et al., 2017; Zhu et al., 2015). Two approaches developed for specific purposes, such as modelling for the dynamics of diseases transmission, and the intervention evaluation, were performed in NIPD up to now.

3.1 Theoretical modelling for analysing the dynamics of disease transmission

Several studies were carried out to apply autoregressive integrated moving average (ARIMA) model to predict the monthly reported malaria cases in China, for example, results showed that ARIMA is a good model for the short-term prediction, which can provide a reference for prevention and control of malaria (Hou et al., 2017); ARIMA model was established based on the month malaria incidences from 1996 to 2005 in Huaiyuan County of Anhui Province. Parameters of the model were estimated through maximum likelihood method; the structure was determined according to criteria of residual uncorrelation and concision, and the order of model was confirmed through Akaike Information Criterion (AIC) and Schwarz Bayesian Criterion (BSC). The constructed model was used to predict the month incidence in 2006 and the result was compared with the actual incidence (Zhou et al., 2007). While other models also used for the imported malaria prediction, such as Xia builds a GM (1, 1) model for forecasting of imported falciparum malaria cases in provinces free of indigenous falciparum malaria (Xia, 2009).

In another field of active surveillance on malaria guided by the modelling results under scenario of global warming, which have been integrated with surveillance system in the national malaria elimination programmes (Wang et al., 2014). For instance, Tian et al. applied the prediction modelling to understand the impact of global warming on malaria transmission, and Yang et al. established the prediction model by using integrated biology-driven and statistical models to understand the risk areas distribution which located in the border of China, under scenario of global warming.

In helminthiasis, biology driven two-host model and grey model were used in helminthiasis transmission prediction (Xu et al., 2015). For instance, barbour's two-host model was applied to predict and evaluate the effect of different control measures by computer simulation. Pilot samples in two villages of Shanghai suburb in 1950s, and three villages in Jiangxi provinces in 2000s were applied for the analysis (Gao et al., 2013a, 2017), all results indicated the basic reproduction (R0) can been changed due to intervention. In addition, construction of grey model for prediction of *Clonorchis sinensis* infection (Zang et al., 2015), and establishment of the mathematical model by the turning point of filariasis transmission were also established, as well as application of transmission dynamic model for echinococcosis and ascariasis.

3.2 Mathematical analysis for estimated the effectiveness of diseases intervention measures

The effectiveness of intervention on interrupting parasitic diseases transmission is also can be estimated or simulated by mathematical analysis (Wu, 2005). In schistosomiasis transmission was monitored after anti-flooding policy implemented in lake region after 1989 when the big floods occurred along whole Yangtze river. In comparison of predicting effect of schistosomiasis prevalence, five statistical models was established in the areas of "breaking dikes or opening sluice for water store" in Dongting Lake, it was found this policy had high impact to the snail habitats re-established (Sai et al., 2006). A drug-based intervention was studied on interrupt the transmission from buffaloes to human Schistosoma japonicum infection around Poyang Lake, which showing the priorities of controlling buffalo transmission in lake region (Guo et al., 2006). Optimizing molluscicide treatment strategies in different control stages of schistosomiasis in China (Yang et al., 2012b).

After the national malaria elimination programmeme launched in 2010, the surveillance is the major intervention to reach the goal of malaria free in

China (Yang et al., 2012a). Therefore, how to understand and to establish the surveillance-response systems with focus on the mobile populations as the basis of the elimination efforts (Chen et al., 2018; Yin et al., 2013; Zhou et al., 2015). One of big surveillance studies was carried out in Tengchong County of Yunnan province, where highest number of malaria cases were observed in the beginning of the malaria elimination programmeme launched (Li et al., 2016). Not only the surveillance of the "1-3-7" approach at the China-Myanmar Border were adopted (Feng et al., 2016; Wang et al., 2017), but also management of imported cases were strengthened (Wang et al., 2015). All these countermeasures produced great outcomes which help Tengchong County to reach the free malaria status in 2015 (Li et al., 2016) (Table 2).

4. Computational analysis

Along with the development of computer science, artificial intelligence, data mining, and geographic science, epidemiology becomes a research domain that applies multidisciplinary techniques to better understand the processes of disease spread, the patterns of disease infection and transmission, as well as the effectiveness of disease control and interventions from public health authorities. In computational epidemiology, the primary focus is the application of computational concepts and resources to disease related issues, while, providing analytical tools to public health authorities and epidemiological scientists. With the tools of modelling, simulation, and visualization, epidemiologists will expand their capabilities in predicting and analysing disease infection and transmission in a given population. Furthermore, the public health authorities can more effectively conduct scenario analysis and, therefore, facilitate their strategic planning of on allocating public health resources.

4.1 Computational modelling for analysing the dynamics of disease transmission

The computational modelling approaches have been used to understand the dynamics of parasitic diseases transmission, such as endemic types of echinococcosis transmission, infection or prevalence rates of various parasitic infections, severity of the disease and mortality (Zhang et al., 2018).

In schistosomiasis epidemiology, Li et al. used the endemic data of schistosomiasis in Xinjian District which were collected by taking the village as a unit from 2009 to 2014 to establish an endemic database, and the

Table 2 Characteristics of studies using theoretical and mathematical analysis in the control programmes on parasitic diseases by NIPD.

Disease	Study aim	Study scales	Data set	Methodology		References
				Method	Algorithm/Model/Technique/Tools	
Malaria	To predict the monthly reported malaria cases	National	Time series of monthly reported cases of malaria	Theoretical method	Autoregressive integrated moving average model (ARIMA)	Hou et al. (2017)
Malaria	To predict the month incidence	County	Month malaria incidences	Theoretical method	Maximum likelihood method	Zhou et al. (2007)
Malaria	To forecast imported falciparum malaria cases	National	Imported falciparum malaria incidence	Theoretical method	GM (1, 1) model	Xia (2009)
Malaria	To understand the impact of global warming on malaria transmission	National, area research	Meteorological data and malaria incidence	Theoretical method	Prediction model, integrated biology-driven and statistical models	Wang et al. (2014)
Clonorchis sinensis	To predict the Clonochis sinensis infection	Province	. Epidemiological survey data	Theoretical method	Grey model	Zang et al. (2015)
Schistosomiasis	To compare predicting effect of schistosomiasis prevalence	Area research	Prevalence of schistosomiasis	Mathematical method	Moving average method, exponential smoothing method, autoregression method, ARIMA model and grey prediction method	Sai et al. (2006)
Schistosomiasis	To estimate the effectiveness of drug intervention	Area research	Surveillance data	Mathematical method	Predicted modelling	Guo et al. (2006)
Schistosomiasis	To estimate molluscicide treatment strategies	National	Surveillance data	Mathematical method	Predicted modelling	Yang et al. (2012b)
Malaria	To survey the malaria cases	County	Surveillance data	Mathematical method	Sample survey	Yang et al. (2012a), Zhou et al. (2015), Yin et al. (2013), Chen et al. (2018), Li et al. (2016), Feng et al. (2016) and Wang et al. (2017)

SaTScan software was applied to analyse the spatiotemporal aggregation areas of Schistosoma japonicum infection in people, Oncomelania hupensis snails and cattle (Li et al., 2018; Xu et al., 2015); Application of climatic forecasting models to rank the various risk areas or patterns for schistosomiasis transmission in Chin Transmission risks of schistosomiasis japonica were catalogued in lake or mountainous regions, by extraction from back-propagation artificial neural network and logistic regression models, and by Barbour's two-host schistosomiasis model with seasonal fluctuations (Gao et al., 2017; Xu et al., 2013; Yang et al., 2009, 2015). Xue et al. were collected and analysed the schistosomiasis epidemiological data in Jiangling County from 2009 to 2013 together with the related geographical information. The tempo-spatial distribution patterns were analysed by the spatial autocorrelation analysis and spatial clustering analysis (Xue et al., 2016).

In malaria transmission, models were set up to predict the monthly distribution of malaria cases or classify the distribution zones based on the transmission risks. For instance, Zhou et al. collected the data of malaria incidence of 156 counties or cities along the Yellow River and Huai River in 2005 to establish the geographical information system data base by Arcgis 9.0 software. Mapping the malaria probability distribution based on the GIS data base by the spatial local interpolation method in the extension function. The predictive incidence probability map and semivariance function was produced by unbiased criterion. Cross-validation technique was used to evaluate the fitness of the distribution maps by mapping the error distribution map (Zhou et al., 2007); Another study by using multifactor spatial composite model to predict transmission tendency of malaria at national level (Yang et al., 2002). Many field investigations were carried out in Hainan provinces where malaria was heavily epidemic in early 1980s, so the transmission features were computerized for understand the epidemiological features, impacts of the social development projects.

4.2 Data mining and machine learning

With the complicated data derived from the large-scaled epidemiological studies, the machine learning approach starts to use for computerizing the problems solving, for instance, what is the optimized intervention for the control or elimination of a parasitic disease in certain settings, and how long or how much fund needed. For a given task, these tasks all seek to learn from data, based on several or huge number of parameters, either by supervised or unsurprised learners. In schistosomiasis, we have applied these methods in identification of transmission risk patters in schistosomiasis transmission in China by back-propagation artificial neural network model (Xu et al., 2013) This neural network consists of two layers, one is the input layer and the other one is output layer. The former one is the input values supplied by the user in terms of the age of a person, the sex of a person and the region where the person lives. The system automatically generates weights for the input. A transformation function is applied on the weighted sum of the input to produce an output. The later one gives a prediction of if the person stands a high or low risk of being infected. So that we can understand the risk of infections in the study areas and population.

In addition to the application of machine learning in epidemiological studies, the large amount of genetic sequence information generated has provided massive data banks that lift the ability of researchers to examine and process this information with the aid of computational methods. For instance, computational methods can be used for identification of the pathogen origins and co-evaluation with their hosts. An integrated immunoproteomics and bioinformatics approach for the analysis of Schistosoma japonicum tegument proteins has been studied in NIPD (Chen et al., 2014). In this study, an integrated immunoproteomics and bioinformatics approach was used to profile the tegument of the human blood fluke Schistosoma japonicum, with an assistance of machine learning. It was found that 30 highly immunoreactive tegument proteins and 10 antigens with an AUC value greater than 0.90 were identified for the first time. In particularly, we found 17 of tegument immunoproteomes having putative interaction networks with other proteins of *S. japonicum*. The results will provide clues of potential target molecules for vaccine development and biomarkers for diagnostics of schistosomiasis.

In malaria, the impact of heterogeneous factors on geographic variations of *Plasmodium vivax* incidences is systematically investigate in Tengchong, Yunnan province, by data mining of geographic variations of *P. vivax* for active surveillance (Shi et al., 2015). The transmission model is built upon relevant demographic, environmental, and biophysical factors to describe the local infections of *P. vivax*. While the hidden time-dependent process is assessed by several socioeconomic factors to account for the imported cases of *P. vivax*. To quantitatively assess the impact of heterogeneous factors on geographic variations of *P. vivax* infections, a Markov chain Monte Carlo (MCMC) simulation method is developed to estimate the model parameters by fitting the space-time model to the reported spatial-temporal disease incidences. Although this study focuses mainly on the transmission of *P. vivax*, the proposed space-time model is general and can readily be extended to investigate geographic variations of other diseases. Practically, such a computational model will offer new insights into active surveillance and strategic planning for disease surveillance and control.

4.3 Systems thinking for disease control

Systems thinking is a method for analysing and understanding the dynamic behaviours of complex systems in virtue of computer simulation technology, and has incomparable advantages in the study of complex nonlinear systems, but less in the fields of medicine and public health. Therefore, NIPD together with Yunnan Provincial Institute for Endemic Diseases, explored an innovative system model for schistosomiasis elimination in Erhai Lake in combination with ecological approaches based on the characteristics of schistosomiasis transmission in Eryuan County. The system modelling method was used to evaluate and predict the effectiveness of interventions, based on the outcomes of the multifarious interventions applied in the study. Results found that ecological approaches to control schistosomiasis in Eryuan County consist of three major components: (i) implementing precise interventions to stop schistosomiasis transmission by means of controlling the source of infection, blocking the biological transmission chains and cutting off the route of disease transmission; (ii) employing ecological approaches to improve the co-effectiveness of environmental protection and schistosomiasis prevention in the study area; and (iii) strengthening the professional skills of personnel involving in the schistosomiasis control programme. Therefore, ecological approaches implemented in schistosomiasis endemic areas of the Eryuan region are able to improve the co-effectiveness of environmental protection and schistosomiasis control, providing a new avenue for eliminating schistosomiasis thanks to the application of precise interventions (Table 3).

5. Remaining challenges and future opportunities

It has been nearly 70 years since the beginning of applying epidemiological data analysis techniques for the control and prevention of parasitic diseases in China. These researches carried out by the NIPD have made great developments and contributions. The achievements in using epidemiological data analysis in parasitic diseases' control and prevention are obvious, but novel developments in methodology and filed practices are still limited. Here some key problems and challenges are highlighted as follows:

Table 3 Characteristics of studies using computational analysis in the control programmes by NIPD.

				Methodology		
Disease	**Study aim**	**Study scales**	**Data set**	**method**	**Algorithm/Model/ Technique/Tools**	**References**
Schistosomiasis	To analyse the spatiotemporal aggregation areas of Schistosoma japonicum infection in people	County (Xinjian)	The endemic data of schistosomiasis	Computational modelling	Spatial clustering model	Li et al. (2018)
Schistosomiasis	To rank the various risk areas or patterns for schistosomiasis transmission	Country	Climate data	Computational modelling	Forecasting models	Xu et al. (2013), Yang et al. (2009, 2015b), Gao et al. (2017)
Schistosomiasis	To catalogue transmission risks of schistosomiasis japonica	In lake or mountainous regions	Climate data and he endemic data of schistosomiasis	Computational modelling	Back-propagation artificial neural network and logistic regression models, and Barbour's two-host schistosomiasis model with seasonal fluctuations	Xu et al. (2013), Yang et al. (2015b), Gao et al. (2013a)
Schistosomiasis	To analyse the tempo-spatial distribution patterns of schistosomiasis	County (Jiangling)	The endemic data of schistosomiasis	Computational modelling	Spatial clustering model	Xue et al. (2016)

Malaria	To map the malaria probability distribution	The Yellow River and Huai River	The data of malaria incidence of 156 counties or cities	Computational modelling	The spatial local interpolation method	Zhou et al. (2007)
Malaria	To predict transmission tendency of malaria	Country	The endemic data of malaria	Computational modelling	Multifactor spatial composite model	Yang et al. (2002)
Schistosomiasis	To identify the transmission risk patters in schistosomiasis transmission	Country	The endemic data of schistosomiasis	Data mining and machine learning	Back-propagation artificial neural network model	Xu et al. (2013)
Schistosomiasis	To analysis the Schistosoma japonicum tegument proteins	–	The large amount of genetic sequence information	Data mining and machine learning	An integrated immunoproteomics and bioinformatics approach ， the application of machine learning	Chen et al. (2014)
Malaria	To quantitatively assess the impact of heterogeneous factors on geographic variations of *P. vivax* infections	County (Tengchong)	Demographic, environmental, and biophysical factors	Data mining and machine learning	Markov chain Monte Carlo (MCMC) simulation method	Shi et al. (2015)

5.1 Factors influencing diseases are rooted in multiple sources and at multiple scales

The driven forces root in two emerging observations, in terms of multiple factors, and the coupled interactions among them. At the micro level, e.g., molecules and pathogens, biologists carry out fundamental research to explore and explain the mechanism of disease infection and transmission, as well as develop the antiviral drugs and vaccines. As for middle level, e.g., host individuals and vector species, epidemiologists conduct front filed interviews and collect samples for analyses. Ecologists analyse and study the relationships among diseases, vectors and the host population, by exploring their mutual interactions as well as the correlations with their environments. As for a higher middle level, e.g., population or communities, health authorities are in charge of the implementation of disease control and interventions by evaluating the effectiveness of different intervention strategies, as well as balancing the costs and benefits associated. At the macro level, e.g., nations and continents, the global climate and the landscape usage are changing, and epidemiologists must quantify significant linkages between atmospheric, oceanic, and terrestrial processes to properly study the influence on the emergence and varnish of a certain disease.

5.2 Complexity in epidemiological system of parasitic diseases needs integrated more factors in the modelling

Parasitic diseases, intermedia vector species, as well as their relied environments are both complicatedly interacted and act with complexity. Due to such a complexity, the emerging epidemiological issues requires large synthesis studies being attempted by collaborations beyond the traditional domain-based studies. To address these challenges in epidemiological studies, researchers first need to realize that impact factors are represented at multiple levels, including the micro level biochemical, etiological, genetic, and the macro level environmental, socio-economical, and behavioural, as well as at multiple scales at temporal and spatial aspects, such as from the seasons, years to decades, and from the villages, nations to continental regions. Second, these factors interrelate with each other and the influences are also represented at multiple levels and scales. An epidemiological system consists of all organizations, people and actions whose primary intent is to promote, restore or maintain health.

5.3 Systems thinking plays an emerging important role for the prediction and control of parasitic diseases

In many parts, the problem in the existing research paradigms of epidemiological data analysis, namely, lack of a comprehensive systems perspective. The spread of infectious diseases and its controls are, by and large, studied in separate disciplines. The policies and decisions on disease preventions and interventions, i.e., reducing morbidity and mortality, are developed within each research domain, which encumbering efforts from the public health authorities for an efficiently allocation of scarce resources. Diseases are not stand-alone studies with clear boundaries among multiple displaces. Disease parasites, human population, as well as the applied interventions are coupled with each other with the complexities in their causes and interactions at multiple levels and scales. The interrelationships in terms of their environmental, biological, socio-economical as well as behavioural conditions, which defies a simple representation of epidemiological systems. They call for novel ways of thinking to improve our ability to predict and control infectious diseases in host population. As illustrated in Fig. 3, a holistic systems framework is needed to capture diseases, host population, and interventions, and thus their intricate relationships in-to a unified platform. Such frameworks are developed using complex systems the problem in the existing research paradigms of epidemiological studies, namely, lack of a comprehensive systems perspective.

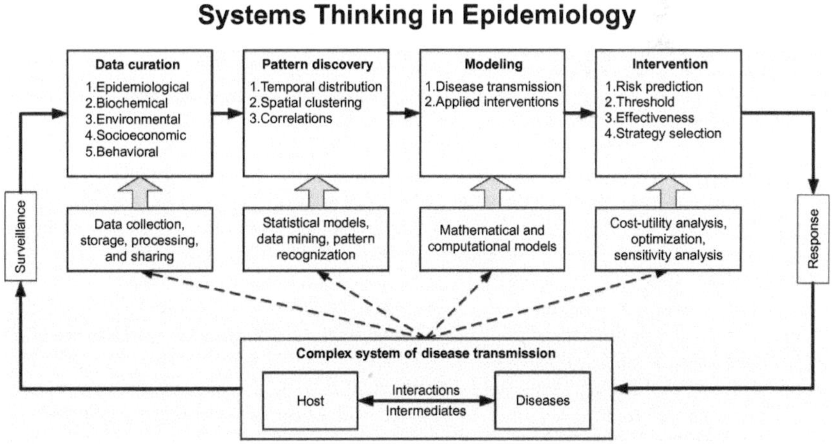

Fig. 3 Systems thinking for the control and prevention of parasitic diseases.

6. Conclusion

Since 1950s, the 70s years' continuous efforts by the NIPD on the analysis of epidemiological data have contributed significantly for the control and elimination of parasitic diseases in China. These mentioned paradigms and methods of epidemiological data analysis have significantly accelerated the in explorations in the prediction and control studies of parasitic diseases. While, prompt adaptations in epidemiological studies are also required as a response to the rapid changes in the understanding of disease related ecological environments and human related social systems. We have provided a complete review on epidemiological data analysis in parasitic diseases control and prevention by the NIPD's 70 years efforts and have identified the possible challenges and opportunities, hoping to shed lights on future research directions.

References

Chen, H.T., Wang, J.Y., Zhu, X.H., 2011. Epidemiological analysis of cases with visceral leishmaniasis reported from non-epidemic provinces in China. Int. J. Med. Parasit. Dis. 38, 177–179.

Chen, J.H., Zhang, T., Ju, C., Xu, B., Lu, Y., Mo, X.J., Chen, S.B., Fan, Y.T., Hu, W., Zhou, X.N., 2014. An integrated immunoproteomics and bioinformatics approach for the analysis of Schistosoma japonicum tegument proteins. J. Proteomics 98, 289–299.

Chen, T.M., Zhang, S.S., Feng, J., Xia, Z.G., Luo, C.H., Zeng, X.C., Guo, X.R., Lin, Z.R., Zhou, H.N., Zhou, S.S., 2018. Mobile population dynamics and malaria vulnerability: a modelling study in the China-Myanmar border region of Yunnan Province, China. Infect. Dis. Poverty 7.

Fang, Q., Wu, W.P., Wang, L.Y., Zeng, X.M., 2014. Use of spatial interpolation to predict the prevalence of hydatidosis. J. Parasit. Biol. 203–206 (in Chinese).

Feng, J., Xia, Z.G., Zhou, S.S., Xia, S., Wang, Q., Xiao, N., Zhou, X.N., 2015a. Preliminary evaluation of potential transmission risk of malaria after earthquake in Cangyuan County, Yunnan Province. Chin. J. Parasitol. Parasit. Dis. 33, 301–305.

Feng, J., Xiao, H., Xia, Z., Zhang, L., Xiao, N., 2015b. Analysis of malaria epidemiological characteristics in the People's Republic of China, 2004–2013. Am. J. Trop. Med. Hyg. 93 (2), 293–299.

Feng, J., Liu, J., Feng, X., Zhang, L., Xiao, H., Xia, Z., 2016. Towards malaria elimination: monitoring and evaluation of the "1-3-7" approach at the China-Myanmar border. Am. J. Trop. Med. Hyg. 95, 806–810.

Gao, S.J., He, Y.Y., Liu, Y.J., Yang, G.J., Zhou, X.N., 2013a. Field transmission intensity of Schistosoma japonicum measured by basic reproduction ratio from modified Barbour's model. Parasit. Vectors 6, 141.

Gao, F.H., Zhang, S.Q., He, J.C., Li, T.T., Wang, T.P., Zhang, G.H., Li, S.Z., 2013b. Spatial-temporal analysis on the distribution of infected snail at the village level in Anhui province, 2006–2012. Chin. J. Epidemiol. 34, 1101–1104 (in Chinese).

Gao, S.J., Cao, H.H., He, Y.Y., Liu, Y.J., Zhang, X.Y., Yang, G.J., Zhou, X.N., 2017. The basic reproductive ratio of Barbour's two-host schistosomiasis model with seasonal fluctuations. Parasit. Vectors 10, 42.

Guan, W., Li, S.Z., Abe, E.M., Webster, B.L., Rollinson, D., Zhou, X.N., 2016. The genetic diversity and geographical separation study of Oncomelania hupensis populations in mainland China using microsatellite loci. Parasit. Vectors 9, 28.

Guo, J.G., Li, Y.S., Gray, D., Ning, A., Hu, G.H., Chen, H.G., Davis, G.N., Sleigh, A.C., Feng, Z., McManus, D.P., Williams, G.M., 2006. A drug-based intervention study on the importance of buffaloes for human Schistosoma japonicum infection around Poyang Lake, People's Republic of China. Am. J. Trop. Med. Hyg. 74, 335–341.

Hou, H.Y., Song, H.Q., Zhang, S.X., Ai, L., Lu, Y., Cai, Y.C., Li, S.Z., Teng, X.J., Yang, C.L., Hu, W., Chen, J.X., 2017. Application of ARIMA model to predict number of malaria cases in China. Chin. J. Schistosom. Control 29, 436–440. 458 (in Chinese).

Hu, Q.A., Zhang, Y., Guo, Y.H., Lv, S., Xia, S., Liu, H.X., 2017. Investigation of Pomacea canaliculata and rodent densities and their infection status with Angiostrongylus cantonensis in Nan'ao island of Guangdong Province. Chin. J. Parasitol. Parasit. Dis. 35, 239–245.

Huang, Q.N., Tang, L.L., Jiang, X.G., Chen, Z., Zhou, X.N., 2007. Retrieving eco-environment factors relevant to Oncomelania snail distribution based on QuickBird image. Chin. J. Parasitol. Parasit. Dis. 304–309 (in Chinese).

Huang, F., Zhou, S.S., Zhang, S.S., Zhang, H.W., Li, W.D., 2011. Meteorological factors–based Spatio-temporal mapping and predicting malaria in Central China. Am. J. Trop. Med. Hyg. 85, 560–567.

Huang, J.X., Xia, Z.G., Zhou, S.S., Pu, X.J., Hu, M.G., Huang, D.C., Ren, Z.P., Zhang, S.S., Yang, M.N., Wang, D.Q., 2015. Spatio-temporal analysis of malaria vectors in national malaria surveillance sites in China. Parasit. Vectors 8, 146.

Li, S.Z., Wang, Y.X., Yang, K., Liu, Q., Wang, Q., Zhang, Y., Wu, X.H., Guo, J.G., Bergquist, R., Zhou, X.N., 2009. Landscape genetics: the correlation of spatial and genetic distances of Oncomelania hupensis, the intermediate host snail of Schistosoma japonicum in mainland China. Geospat. Health 3, 221–231.

Li, S.G., Yin, S.Q., Wang, J.Z., Li, X.S., Feng, J., 2016. Shifting from control to elimination: analysis of malaria epidemiological characteristics in Tengchong County around China-Myanmar border, 2005-2014. Malar. J. 15, 1–7.

Li, Y.L., Xue, J.B., Hu, W.C., Mao, Y.H., Zhang, Z.S., Zhou, L., Zhang, P., Yu, Z.K., Xu, J., 2018. Endemic features and space-time clustering of schistosomiasis in Xinjian District, Nanchang City from 2009 to 2014. Chin. J. Schistosom. Control 30, 260–263.

Lv, S., Zhang, Y., Steinmann, P., Yang, G.J., Yang, K., Zhou, X.N., Utzinger, J., 2011. The emergence of angiostrongyliasis in the People's Republic of China: the interplay between invasive snails, climate change and transmission dynamics. Freshw. Biol. 56, 717–734.

Qian, Y.J., Li, S.Z., Xu, J., Yang, K., Huang, Y.X., Cao, Z.G., Feng, M., Dang, H., Zhang, L.J., Zhang, L., Wang, Q., Bergquist, R., Xiao-Nong, Z., 2013. Potential schistosomiasis foci in China: a prospective study for schistosomiasis surveillance and response. Acta Trop. 141, 342–348.

Ren, Z.P., Wang, D.Q., Hwang, J., Bennett, A., Sturrock, H.J.W., Ma, A., Huang, J.X., Xia, Z.G., Feng, X.Y., Wang, J.F, 2015. Spatial-temporal variation and primary ecological drivers of Anopheles sinensis human biting rates in malaria epidemic-prone regions of China. PLoS One 10, e0116932.

Sai, X.Y., Xing, Q.J., Meng, D.R., Jia, Y.R., Cai, K.P., Li, Y.S., 2006. Comparison of predicting effect of schistosomiasis prevalence by 5 statistical models in the areas of "breaking dikes or opening sluice for water store" in Dongting Lake. J. Fourth Mil. Med. Univ. 1603–1605 (in Chinese).

Shi, B.Y., Tan, Q., Zhou, X.N., Liu, J.M., 2015. Mining geographic variations of plasmodium vivax for active surveillance: a case study in China. Malar. J. 14, 216.

Teng, X.J., Chu, Y.H., Zhai, C.C., Yu, Y.F., Cai, Y.C., Chen, S.H., Ai, L., Tian, L.G., Chen, J.X., 2018. The epidemiological characteristics and influencing factors for Blastocystis hominis infection among human immunodeficiency virus seropositive individuals in Tengchong of Yunnan Province. Chin. J. Parasitol. Parasit. Dis. 36, 129–134 (in Chinese).

Wang, D.Q., Gu, Z.C., Zheng, X., Guo, Y., Tang, L.H., 2014. Application of autoregressive linear model in understanding the effect of climate on malaria vectors dynamics in the three gorges reservoir. Biomed. Environ. Sci. 27, 811–814.

Wang, D.Q., Li, S.G., Cheng, Z.B., Xiao, N., Cotter, C., Hwang, J., Li, X.S., Yin, S.Q., Wang, J.Z., Bai, L., Zheng, Z., Wang, S.B., 2015. Transmission risk from imported plasmodium vivax malaria in the China-Myanmar border region. Emerg. Infect. Dis. 21, 1861.

Wang, L.Y., Wu, W.P., Fu, Q., Guan, Y.Y., Han, S., Niu, Y.L., Tong, S.X., Osman, I., Zhang, S., Kaisar, K., 2016. Spatial analysis of visceral leishmaniasis in the oases of the plains of Kashi prefecture, Xinjiang Uygur autonomous region, China. Parasit. Vectors 9, 1–7.

Wang, D.Q., Cotter, C., Sun, X.D., Bennett, A., Gosling, R.D., Xiao, N., 2017. Adapting the local response for malaria elimination through evaluation of the 1-3-7 system performance in the China-Myanmar border region. Malar. J. 16.

Wu, K.C., 2005. Math ematical model in prediction and evaluation of the effects on control measures for schistosomiasis. Chin. J. Parasitol. Parasit. Dis. 23, 408–414.

Xia, Z.G., 2009. Forecasting of imported falciparum malaria cases in provinces free of indigenous falciparum malaria using GM(1,1)model. Pract. Prev. Med. 16, 1744–1746.

Xia, S., Zhou, X.-N., Liu, J., 2017. Systems thinking in combating infectious diseases. Infect. Dis. Poverty 6 (5), 57–63.

Xu, J.F., Xu, J., Li, S.Z., Jia, T.W., Huang, X.B., Zhang, H.M., Chen, M., Yang, G.J., Gao, S.J., Wang, Q.Y., Zhou, X.N., 2013. Transmission risks of schistosomiasis japonica: extraction from back-propagation artificial neural network and logistic regression model. PLoS Negl. Trop. Dis. 7, e2123.

Xu, J.F., Lv, S., Wang, Q.Y., Qian, M.B., Liu, Q., Bergquist, R., Zhou, X.N., 2015. Schistosomiasis japonica: modelling as a tool to explore transmission patterns. Acta Trop. 141, 213.

Xue, J.B., Xia, S., Zhang, X., Hu, H.H., Wang, Q., Zhou, Y.B., Li, S.Z., 2016. Pattern analysis of tempo-spatial distribution of schistosomiasis in marsh-land epidemic areas in stage of transmission control. Chin. J. Schistosom. Control 28, 624–629 (in Chinese).

Xue, C.Z., Wu, W.P., Han, S., Zheng, C.J., Wang, Y., Wang, L.Y., Yu, Q., Zhang, S.S., Zhou, X.N., Xiao, N., 2018. Prevalence and influencing factors of hydatid disease among children in Tibet Autonomous Region. Chin. J. Parasitol. Parasit. Dis. 36, 20–25 (in Chinese).

Yan, R.Z., Zhou, S.S., Xia, Z.G., Zhang, S.S., Xiao, H.H., 2014. Spatial-temporal characteristics of malaria transmission in China. J. Parasit. Biol. 198–202.

Yang, G.J., Zhou, X.N., Malone, J.B., Mccarroll, J.C., Liu, J.X., 2002. Application of multifactor spatial composite model to predict transmission tendency of malaria at national level. Chin. J. Parasitol. Parasit. Dis. 20, 145–147 (in Chinese).

Yang, D.P., Guo, J.G., 2006. The application of remote sensing in rapid identification of the snails' breeding areas. Int. J. Med. Parasit. Dis. 33 (05), 268–272 (in Chinese).

Yang, G.J., Sun, L.P., Hong, Q.B., Yang, K., Deng, Y., Li, S.Z., Lv, S., 2009. Application and perspective of climatic forecasting models related to schistosomiasis transmission. Chin. J. Schistosom. Control 21, 432–436.

Yang, G.J., Sun, L.P., Hong, Q.B., Zhu, H.R., Yang, K., Gao, Q., Zhou, X.N., 2012a. Optimizing molluscicide treatment strategies in different control stages of schistosomiasis in the People's Republic of China. Parasit. Vectors 5, 260.

Yang, G.J., Tanner, M., Utzinger, J., Malone, J.B., Bergquist, R., Chan, E.Y.Y., Gao, Q., Zhou, X.N., 2012b. Malaria surveillance-response strategies in different transmission zones of the People's Republic of China: preparing for climate change. Malar. J. 11, 426.

Yang, S.J., Li, M., Tian, T., Zhao, J.S., Wu, W.P., 2015a. Epidemiological characteristics of human hydatidosis and risk factors in Habahe County, Xinjiang. Dis. Surveillance 30, 485–488 (in Chinese).

Yang, K., Zhou, X.N., Jia, T.W., Yang, G.J., Wu, X.H., Shi, X.W., Li, H.J., Steinmann, P., Utzinger, J., Bergquist, R., 2015b. Eco-social determinants of Schistosoma japonicum infection supported by multi-level modelling in Eryuan county, People's Republic of China. Acta Trop. 141, 391–398.

Yin, J.H., Yang, M.N., Zhou, S.S., Wang, Y., Feng, J., Xia, Z.G., 2013. Changing malaria transmission and implications in China towards National Malaria Elimination Programmeme between 2010 and 2012. PLoS One 8, e74228.

Zang, W., Lu, G.H., Zhu, T.J., Chen, Y.D., 2015. Construction of grey model for prediction of Cllaborating sinensis infection. Int. J. Med. Parasit. Dis. 42, 170–172.

Zhang, M.Y., Wang, L.Y., Guan, Y.Y., Wu, W.P., 2018. Meta-analysis on severity of hydatidosis. Chin. J. Parasitol. Parasit. Dis. 36, 156–160.

Zhao, Y., Wu, W.P., Li, W., Xu, K.J., Wang, L.Y., Wang, Y., 2015. Spatial clustering analysis of echinococcosis in Ganzi Tibetan autonomous prefecture, Sichuan Province. Int. J. Med. Parasit. Dis. 42, 164–169.

Zhou, S.S., Huang, F., Shen, Y.Z., 2007. Application of ARIMA model on prediction of malaria incidence. J. Pathogen Biol. 2, 284–286 (in Chinese).

Zhou, S.S., Zhang, S.S., Wang, J.J., Zheng, X., Huang, F., Li, W.D., Xu, X., Zhang, H.W., 2012. Spatial correlation between malaria cases and water-bodies in Anopheles sinensis dominated areas of Huang-Huai plain, China. Parasit. Vectors 5, 106.

Zhou, S.S., Zhang, S.S., Li, Z., Rietveld, A.E.C., Ramsay, A.R., Zachariah, R., Bissell, K., Bergh, R.V.D., Xia, Z.G., Zhou, X.N., 2015. China's 1-3-7 surveillance and response strategy for malaria elimination: is case reporting, investigation and foci response happening according to plan? Infect. Dis. Poverty 4, 1–9.

Zhu, H.R., Liu, L., Zhou, X.N., Yang, G.J., 2015. Ecological model to predict potential habitats of Oncomelania hupensis, the intermediate host of Schistosoma japonicum in the mountainous regions, China. PLoS Negl. Trop. Dis. 9, e0004028.

CHAPTER ELEVEN

Construction and application of surveillance and response systems for parasitic diseases in China, led by NIPD-CTDR

Yu-Wan Hao, Qiang Wang*, Chun-Li Cao, Tian Tian, Ze-Lin Zhu, Jing Xu, Shuisen Zhou, Weiping Wu, Yingdan Chen, Yi Zhang, Jia-Xu Chen, Shi-Zhu Li*, Ning Xiao, Xiao-Nong Zhou

National Institute of Parasitic Diseases, Chinese Center for Disease Control and Prevention; National Center for International Research on Tropical Diseases, Ministry of Science and Technology; Key Laboratory of Parasite and Vector Biology, Ministry of Health; WHO Collaborating Centre for Tropical Diseases, Shanghai, People's Republic of China

*Corresponding authors: e-mail address: wangqiang@nipd.chinacdc.cn; lisz@chinacdc.cn

Contents

1. Background 350
2. History of the parasitic disease surveillance system in China 352
   2.1 Surveillance site monitoring 352
   2.2 Thematic surveillance 353
   2.3 Risk surveillance 356
   2.4 Case report based on active information reporting system 357
3. Surveillance work organized by NIPD and its role 359
   3.1 Identifying public health problems 359
   3.2 Grasping the distribution and trends of parasitic diseases 361
   3.3 Discovering outbreaks and epidemic anomalies 362
   3.4 Evaluating the effects of interventions 362
   3.5 Identifying risk factors in different populations and regions 363
4. Emergency response mechanisms for parasitic diseases 363
5. Role of the emergency response system in managing parasitic diseases 364
   5.1 Case study I: Schistosomiasis outbreaks 365
   5.2 Case study II: Malaria outbreaks 365
   5.3 Case study III: Food-borne parasitic disease outbreaks 366
   5.4 Case study IV: Filariasis outbreaks 366
   5.5 Case study V: Imported Africa human trypanosomiasis 367
6. Conclusion 367
Acknowledgements 368
Competing interest 368
Authors' contributions 368
References 368

Advances in Parasitology, Volume 110
ISSN 0065-308X
https://doi.org/10.1016/bs.apar.2020.04.001

Abstract

Parasitic diseases have been widely epidemic in China with a long history. Great endeavours made in past 70 years led to significant decrease in morbidity and mortablity caused by several major parasitic diseases, while challenges existed to eliminate parasitic diseases. Surveillance-response system has play a crucial role in identifying public health problems, ascertaining the distribution and epidemic dynamics, discovering outbreaks and epidemic anomalies, evaluating the effects of on-site intervention activities and identifying risk factors. In this article, we reviewed the progress of the surveillance system for parasitic diseases, analysed the role of NIPD in the construction and application of surveillance-response system of parasitic diseases through elaborating the surveillance activities and typical surveillance-response events led by NIPD. Suggestion and comments for improve the surveillance-response system were put forward for further control or elimination of parasitic diseases.

1. Background

Surveillance refers to the continuous and systematic collection of data and information, followed by analysis and interpretation of information related to public health plans and assessments, as well as the timely use of information for disease prevention and control (Groseclose and Buckeridge, 2017). The purposes of parasitic disease surveillance, specifically, is to monitor the occurrence and development patterns of diseases, provide prompt feedback and analysis of effective information, as well as to respond to and initiate rapid emergency responses (German, 2000; Yang et al., 2012). Additionally, an early warning consists of an advanced prediction that an epidemic may reach a dangerous level or an epidemic event is going to be occurred (Ameme et al., 2016). Such warnings are made with respect to a series of epidemic indicators or a possible epidemic event, based on long-term surveillance data and other basic information. It also involves the timely forecasting of development trends regarding an epidemic at a spatiotemporal level and minimizing the harm caused by the spread or occurrence of the epidemic. Alternatively, a response refers to the actions taken by the local government or agency based on the level of early warning, involving the felicitous organization of resources to carry out the emergency response plan, prevent the spread of the epidemic or reduce the extent of harm caused, to ensure the health safety of the population (Perry et al., 2007; Ren et al., 2016). Therefore, surveillance, early warnings, and responses must be strategically combined, well-coordinated, and enable a collaborative division of labour to improve the sensitivity of

Fig. 1 Relationships between surveillance, early warnings, and responses.

the surveillance system, thereby achieving the expected outcome of controlling transmission or spread of parasitic diseases (Fig. 1) (Tambo et al., 2014).

The transmission of parasitic diseases is influenced by many factors, such as the ecological environment, distribution of intermediate host, as well as natural factors including temperature, humidity, rainfall, and social factors including economic development, population mobilization, entertainment or production behaviours of the population (Cheng et al., 2016). Therefore, surveillance has always been a key aspect for parasitic diseases control in China. Since the establishment of our institute, most notably since it was renamed the National Institute of Parasitic Diseases, Chinese Center for Disease Control and Prevention (Zhou et al., 2005), we have gradually established and continuously improved a surveillance-response system for parasitic diseases based on the control stages of key parasitic diseases. This system has played a crucial role in understanding the epidemic dynamics, transmission patterns, and influencing factors of parasitic diseases in China, thereby facilitating the early detection of epidemics and guiding their management; while promoting on-site prevention and control activities (Feng et al., 2016b). Herein, we review the work carried out by our institute on constructing a national key parasitic disease surveillance network and developing emergency responses to epidemic events.

2. History of the parasitic disease surveillance system in China

According to the aims and surveillance contents, infectious disease surveillance in China primarily underwent three different stages, namely, disease surveillance, epidemiological surveillance, and public health surveillance (Zhou et al., 2013). Parasitic disease surveillance has also developed continuously alongside advances in surveillance technologies and changes in endemicity, which has been accompanied by expansion in terms of the scope and contents of surveillance. Monitoring surveillance sites, thematic surveillance, risk surveillance, and case reporting all served as key elements in the parasitic disease surveillance system and played unique roles over time.

2.1 Surveillance site monitoring

As schistosomiasis was given high priority by central government since the founding of the People's Republic of China, from 1950 to 1970, disease surveillance was adopted as the primary approach to carry out snail surveys focused on the occurrence and outcome (recovery or death) of schistosomiasis. This provided a reliable basis guiding for treating patients and controlling schistosomiasis outbreaks. Since 1990, the Bureau of Disease Control of the former Ministry of Health commissioned the Shanghai Medical University to conduct schistosomiasis surveillance in 14 villages of eight provinces (municipalities), which were enlarged to 21 epidemic surveillance sites from 2000 to 2004. In addition to epidemic indicators, the surveillance content also included epidemiological factors, which provided a scientific basis for understanding the epidemic changes of schistosomiasis in China and evaluating the effectiveness of different interventions (Yuan et al., 2002).

In 2002, the National Institute of Parasitic Diseases (NIPD), Chinese Academy of Preventive Medicine, renamed the National Institute of Parasitic Diseases, Chinese Center for Disease Control and Prevention, accompanying a reform from a research institute to a national institute responsible for providing technical support and guidance for national parasitic disease control and elimination through organizing national surveillance, data collection and analysis, response to parasitic disease outbreaks, etc. Construction of the infectious disease and vector surveillance network was initiated in 2005, including surveillance on schistosomiasis, malaria, soil- and food-borne parasitic diseases, and other related diseases (Chen et al., 2020). In response to changes in the strategies used for preventing and controlling each disease, our institute has organized and updated the surveillance programs in a timely manner, with adjustment or expansion of content and scope of surveillance (Xu et al., 2015).

For schistosomiasis, the national surveillance scheme evolved three versions, first developed in 2005 and upgraded in 2011 followed by revision in 2014. The surveillance scope was expanded from 80 surveillance sites in 10 provinces to the current 100% coverage rate of all schistosomiasis-endemic counties plus four counties (cities, districts) in the Three Gorges Reservoir Region which were regarded as having potential risk of schistosomiasis transmission (Table 1) (Zhu et al., 2013). For malaria, it was switched to the National Malaria Elimination Surveillance Program from sentinel surveillance since 2015 (Feng et al., 2016a). The National Soil-Borne Nematodiasis Surveillance Program was piloted in 2006, which was further revised in 2011 and 2016, respectively. In addition, our institute has established the National Lymphatic Filariasis Post-elimination Surveillance Program, the Pilot Program for Angiostrongyliasis Symptom Surveillance, and the Early Warning Dissemination Program (2008 version) as well as the National Echinococcosis Surveillance Program (2016 version).

During the process of updating and improving the surveillance programs and implementing on-site organization, our institute actively fulfilled the roles of technical support, data management, and quality control. This not only enabled the continuous enrichment and improvement of the National Key Parasitic Disease Surveillance Network, but also provided a scientific basis for grasping the epidemic dynamics of parasitic diseases in China, understanding their epidemiological patterns and influencing factors, while also formulating and adjusting prevention and control measures.

2.2 Thematic surveillance

During various stages of prevention and control, our institute led and organized several national thematic surveys. Three national epidemiological-sampling surveys for schistosomiasis were conducted in 1989, 1995, and 2004, before, during and after the implementation of the World Bank Loan project for schistosomiasis control in P.R. China, respectively (Table 2) (Wu et al., 2007; Xu et al., 2016). The data obtained from these surveys presented the endemic status and explored the transmission patterns of schistosomiasis nationwide at that time, thus provided reference for developing the national control programs and modifying the control strategy or intervention activities.

Table 1 Upgrades and adjustments to the schistosomiasis surveillance program.

No.	Version (issue no.)	Surveillance objectives	Surveillance scope	Main content
1	2005 pilot version (MOH-DDC [2005]74)	1. To understand the epidemiological dynamics and influencing factors of schistosomiasis, and predict its epidemiological trends 2. To provide a scientific basis for formulating prevention and control measures, and outcome evaluation	A total of 80 surveillance sites in 10 provinces, autonomous regions, and municipalities	Routine surveillance, sudden outbreak surveillance, and sentinel surveillance (including survey on infection situation of residents, livestock and snails using serological test and microscopy techniques, as well as epidemiological factors)
2	2011 revised edition (CCDC [2011]336)	1. To understand the epidemiological dynamics of schistosomiasis and the relevant factors of schistosomiasis epidemiology, and to predict its epidemiological trends 2. To evaluate the effects of schistosomiasis prevention and control 3. To provide a scientific basis for formulating schistosomiasis prevention and control measures	A total of 81 surveillance sites in 12 provinces, autonomous regions, and municipalities nationwide	Routine surveillance, sudden outbreak surveillance, and sentinel surveillance (including survey on infection situation of residents, livestock and snails using serological test and microscopy techniques, as well as epidemiological factors)

3	2014 version (CCDC [2014]420)	1. To understand the epidemiological dynamics of schistosomiasis and changes in its epidemiological factors, and to investigate its epidemiological trends 2. To discover the epidemic status and potential transmission risk of schistosomiasis, and to adopt timely interventions 3. To scientifically evaluate the effects of schistosomiasis prevention and control, and to provide a scientific basis for formulating and improving prevention and control measures for schistosomiasis	National schistosomiasis surveillance sites were set up in all schistosomiasis-endemic counties nationwide and some counties in the Three Gorges Reservoir Region. Three types of regions were classified: transmission-uninterrupted counties, transmission-interrupted counties, the and Three Gorges Reservoir Region	Routine surveillance (including case reports, case studies, early warning of acute schistosomiasis, sudden outbreak reports), sentinel surveillance (including survey on infection situation of residents, livestock and snails using serological test and microscopy techniques, as well as epidemiological factors), and risk surveillance (including infected snail surveillance and wild faeces surveillance using molecular biological methods)

Table 2 Parasitological data of the three national sampling survey of schistosomiasis in China.

Year of national sampling survey	No. villages sampled	Estimated no. people infected	Prevalence in human (%)	Prevalence in bovine (%)
1989	353	1,638,103	10.2	13.3
1995	364	865,084	4.9	9.1
2004	239	726,112	5.1	5.7

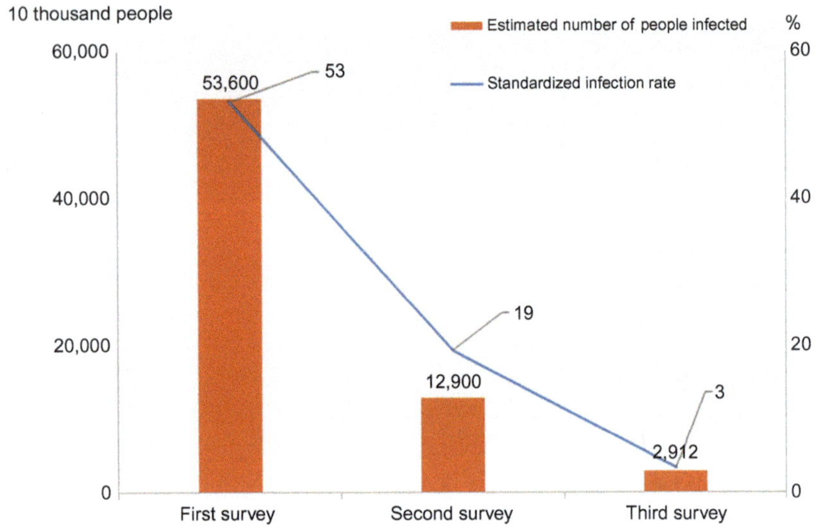

Fig. 2 Three rounds national surveys on key human parasitic diseases vs the prevalence of soil-borne nematode infections.

Moreover, three national surveys on the current status of key human parasitic diseases were carried out in 1988, 2002, and 2016 (Fig. 2) (Zhu et al., 2015), and a national survey on angiostrongyliasis epidemic foci was conducted in 2006 (Lv et al., 2009). These national cross-sectional surveys provided objective scientific data supporting comprehensive evaluation of the prevalence of parasitic diseases at different stages and progress in their prevention and control, while also providing basic information for formulating prevention and control plans at the various stages.

2.3 Risk surveillance

At the turn of the 21st century, ecological changes in the environment caused by global warming, large-scale water conservation projects, increased population mobility, economic and trade development together with other social factors increased the transmission risks of parasitic diseases. Accordingly, our institute has organized and conducted several thematic risk surveillance programs, primarily focused on schistosomiasis. To evaluate the impact of climate change and large-scale water conservation projects on schistosomiasis transmission, especially for former non-endemic regions, risk surveillance was performed on snails' spread and infection status of schistosomiasis in mobile populations in 11 counties (cities) of 5 provinces (municipalities), near or along the Three Gorges Dam and the South–North Water Diversion Project, from 2008 to 2011 (Fig. 3). The surveillance showed that although no snail was spread to non-endemic areas, *Oncomelania hupensis* could survive in Caohu and Jining city in non-endemic areas and several stool positives were found in mobile populations in surveillance sites. Surveillance should be strengthened and continued in such regions (Dang et al., 2014).

To improve the sensitivity of schistosomiasis surveillance and the early warning system, in 2010–16, the sentinel-mouse method was used for monitoring and identifying the transmission risk in major water bodies susceptible to schistosomiasis in seven provinces (Hunan, Hubei, Jiangxi, Anhui, Jiangsu, Yunnan, and Sichuan). The annual results indicated that the potential risk still exist in the lake and marshland regions although the prevalence of schistosomiasis in these regions has decreased to below the transmission control threshold (Zheng et al., 2013).

For each type of risk surveillance, the content and scope were configured and adjusted according to the actual needs of the prevention and control site. This not only helped us understand the potential transmission risks and its distribution, but also further enriched and improved on-site surveillance systems.

2.4 Case report based on active information reporting system

In 2004, an epidemic-network, direct reporting system based on infectious disease case reports was established and officially launched nationwide. This system greatly improved the coverage range, timeliness, sensitivity, and accuracy of reports on epidemic events, essentially achieving the overall goal of complete epidemic surveillance reporting (Yang et al., 2011). Thereafter, the Chinese Disease Prevention and Control Information System (with nationwide coverage) was gradually improved, with the Infectious Disease and Public Health Emergency Surveillance Report Information

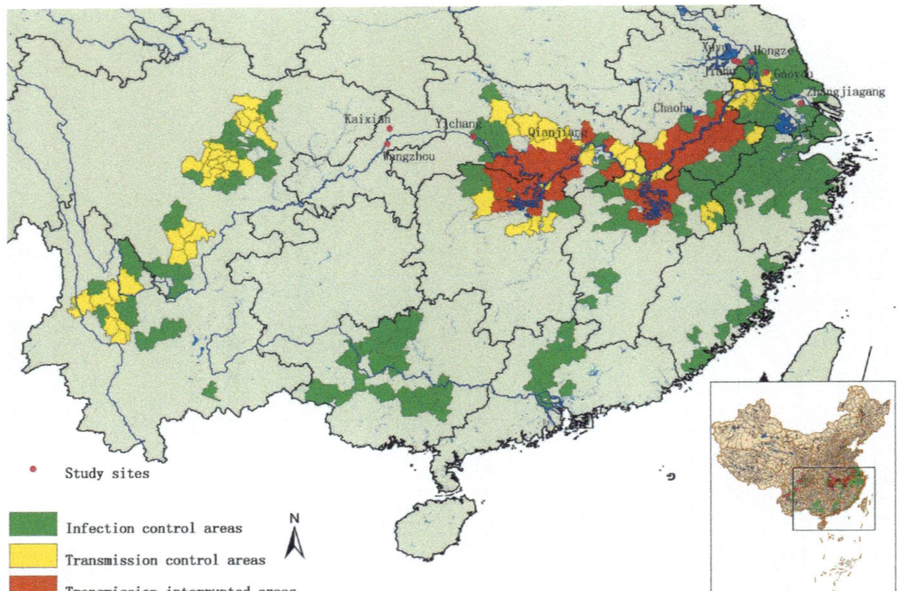

Fig. 3 Distribution of surveillance sites for potential schistosomiasis-endemic regions.

System as its core (Lai et al., 2014). This created a precedent for the implementation of "case, real-time, online" reporting in the field of public health in China, which fundamentally differed from the approach of the previous reporting system based on monthly reports and data aggregation by county (Li et al., 2012).

To meet the needs of key parasitic disease prevention and control strategies, The Information-management System for Parasitic Disease Control was officially launched in 2010, which was spearheaded by the Chinese Center for Disease Control, designed by NIPD, and assisted by Sinosoft Co., Ltd. (Li et al., 2010). The online system was an advance over paper-based reports in terms of data collection processes, reporting, querying statistics, analysis, and other aspects. Alongside the in-depth development of the surveillance work, this system was upgraded in 2012 and 2016. In May 2017, it was incorporated into the Chinese disease prevention and control information system platform allowing for implementation of unified system user authentication and authorization management, which not only enhanced the system security, but also improved the timeliness of epidemic reporting for key parasitic diseases, thereby playing important roles in the standardization of programmatic epidemic management and surveillance.

3. Surveillance work organized by NIPD and its role

Since 2002, NIPD has become the technical center for the national programs of parasitic diseases control, undertaking the technical guidance, organizing disease monitoring and surveillance, assessing the effects of national programs or strategies for parasitic disease control, as well as the relevant emergency responses to outbreaks of parasitic diseases and public health emergencies. The surveillance network which established by NIPD plays a crucial role in identifying public health problems, ascertaining the distribution and epidemic dynamics of major parasitic diseases nationwide, discovering outbreaks and epidemic anomalies, evaluating the effects of on-site intervention activities and identifying risk factors as well as at-risk populations and regions. Collectively, these activities effectively promote national programs of key parasitic diseases control. In this section, case studies will be presented to introduce the different roles served by the surveillance work organized by NIPD (Fig. 4).

3.1 Identifying public health problems

In June 2006, a food-borne outbreak of angiostrongyliasis occurred in Beijing due to eating raw snails infected with *Pomacea canaliculata*

Fig. 4 Major roles of public health surveillance.

(Wang et al., 2010). NIPD cooperated with the relevant agencies to undertake epidemiological surveys. A total of 160 clinically diagnosed cases were discovered, and an outbreak of food-borne angiostrongyliasis was confirmed. However, at that time, limited information was available regarding the prevalence of angiostrongyliasis and the distribution of epidemic foci in China. In light of this, after the epidemic was effectively managed, NIPD, with the support of the former Ministry of Health, organized a systematic national survey pertaining to the epidemic foci of angiostrongyliasis in 2007 and 2008 consecutively (Fig. 5) (Lv et al., 2009) to investigate the distribution and diffusion of *A. cantonensis* as well as its hosts. By clarifying the epidemic foci of this disease, the distribution of intermediate and definitive hosts and their infection status we gained further insights into the public health risks posed by angiostrongyliasis. In 2009, a pilot program for symptom surveillance and early warnings of angiostrongyliasis transmission was launched, which drove the surveillance, prevention, and control of angiostrongyliasis in various regions, thus controlling or decreasing the public health risk caused by angiostrongyliasis.

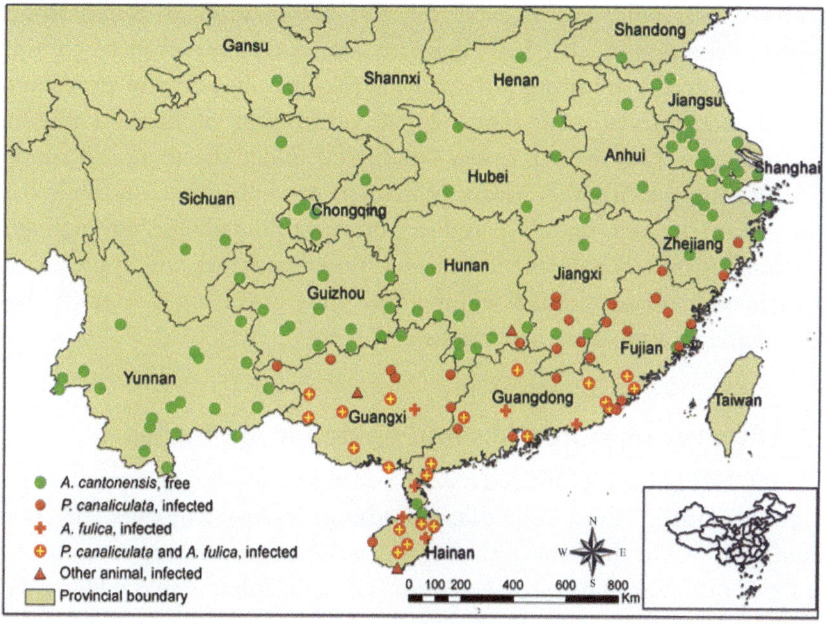

Fig. 5 Survey results of angiostrongyliasis epidemic foci.

3.2 Grasping the distribution and trends of parasitic diseases

Following the large-scale prevention and control strategy for filariasis in China, surveillance became an important component for the control and elimination of filariasis (De-Jian et al., 2013). In the 1970s, surveillance began the next year since the endemic province reached the threshold of basic elimination of filariasis, primarily focusing on the discovery and elimination of residual sources of infection to consolidate the results of prevention and control. In the 1980s, after clarifying the indications for interrupting the transmission of filariasis, longitudinal surveillance was performed to observe dynamic changes of microfilaremia situation in human populations, and to confirm that microfilaremia and natural larval infections could no longer be found in human populations and mosquito vectors (Wang et al., 1998). By the 1990s, horizontal surveillance, longitudinal surveillance, and surveillance of floating populations were conducted to understand the changes of microfilaremia rates, the transmission trends of filariasis in human populations, as well as the status of microfilaria infections in floating populations and their impact on the elimination of filariasis. These studies provided evidence for the elimination of filariasis within a region (Fang and Zhang, 2019). In 2001, NIPD participated in the preparation of the

Regional Surveillance Program for the Elimination of Filariasis and reviewed the regions that had eliminated filariasis in an effort to continue identifying and monitoring potentially weak links in terms of prevention, control, and surveillance. Since 2004, verification of filariasis elimination based on case reporting was conducted under the technical support of NIPD. Reviewing the history of filariasis surveillance can provide basic information for understanding its distribution in China while also highlighting changes in human infection rates at different stages and consolidating the achievements of filariasis elimination, thus providing a scientific basis for sustained elimination of filariasis in China.

3.3 Discovering outbreaks and epidemic anomalies

In 2005, NIPD experts noticed a sharp rise in the number of malaria cases in Anhui Province. The number of reported cases increased by over 70%, compared with that in the same period of the previous year, and appeared to be a localized outbreak. The cases were primarily concentrated in the Huanghuai River Basin, specifically in the counties (districts) of Guoyang, Mengcheng, Huaiyuan, and Guzhen, where *Anopheles sinensis* was the main vector of *Plasmodium* (Zhang et al., 2008). After reporting to the former Bureau of Disease Control of the Ministry of Health, NIPD provided technical support to local Diseases Control Center for spring interregnum treatment and health education. The Anhui Spring Interregnum Radical Treatment Program was formulated with the support of NIPD, followed by on-site supervision of treatment by NIPD experts. After implementation of effective intervention over two epidemic seasons, the reported number of malaria cases in Anhui Province decreased significantly and malaria was deemed to be effectively controlled (Pan et al., 2012).

3.4 Evaluating the effects of interventions

In 2006, the former Ministry of Health established Demonstration Pilots of Comprehensive Control for Parasitic Diseases. To evaluate the effects of integrated interventions in these demonstration areas and explore the sustainable development mechanisms for parasitic diseases control, NIPD actively conducted longitudinal observations in 12 demonstration pilots from 2011 to 2013. This enabled the continuous implementation of comprehensive prevention and control measures, as well as effective development of surveillance programs for monitoring infections in the human population, which facilitated evaluation of prevention and control efforts in the demonstration zones while also providing reference for formulating the national plan for prevention and control of major parasitic diseases (Ying-Dan et al., 2019).

3.5 Identifying risk factors in different populations and regions

For the timely detection of high-risk regions during the schistosomiasis transmission season, in the first half year of 2009, NIPD collaborated with the Jiangsu and Jiangxi Provincial Institutes for Schistosomiasis Control to conducting risk analysis based on snail surveillance data collected throughout the spring in combination with epidemic data from previous years. Early warnings were put forward and indicated that the Jiangsu Yangtze River and Poyang Lake regions of Jiangxi Province were still with high risk of schistosomiasis transmission through spatio-temporal analysis (Li et al., 2014).

Surveillance serves as the foundation for guiding effective performance of parasitic disease emergency responses, as well as an effective means for evaluating the effects of emergency management. Natural disasters may alter the natural, social, and biological factors, leading to reoccurrence or spread of parasitic diseases, thus, NIPD organized risk assessments for major parasitic diseases, such as schistosomiasis, malaria, echinococcosis, and kala-azar in disaster areas based on the accumulated surveillance data when disasters occurred, such as the Wenchuan earthquake occurred in 2008, the Gansu Zhouqu mudslide in 2010 and the Sichuan Jiuzhaigou earthquake happened in 2017, etc. Assessment showed that high transmission risk occurred as a result of the disasters, as well as persistence of infection sources in endemic areas especially those vector- or food-borne parasitic diseases (Ren et al., 2016; Tambo et al., 2014). NIPD also coordinated with local disease control and prevention agencies to carry out symptom surveillance, health education, mosquito and lacewing control, and the promotion of self-protection in disaster areas. These measures prevented the occurrence or outbreaks of major parasitic diseases following the disasters. In addition, during a series of major events such as the Beijing Olympics held in 2008, the Shanghai World Expo in 2010, and the China International Import Expo, emergency preparedness was actively conducted to ensure the success and health safety of these events.

4. Emergency response mechanisms for parasitic diseases

Following the severe acute respiratory syndrome (SARS) outbreak in 2003, stemming from high concerns over safeguarding human health and safety, the Central Committee of the Chinese Communist Party and the

State Council assigned high priority status to the construction and development of emergency response systems. Once these systems were established by the state, the establishment and development of health-related emergency response systems followed. Substantial progress was made in the construction of "one plan and three systems" (i.e. contingency plan, legal system, and institutional system and mechanisms), while the emergency response teams and their capabilities were also enhanced.

In the field of parasitic disease prevention and control, the contingency plan for the emergency response to major epidemics of schistosomiasis (Pilot) and the contingency plan for the emergency response to malaria outbreaks (Pilot) were drafted by NIPD. These were the first two emergency response-contingency plans issued by the former Ministry of Health after the SARS outbreak. After assessing their validity in pilot studies over 2 years, the contingency plan for the emergency response to schistosomiasis outbreaks and the contingency plan for the emergency response to malaria outbreaks were officially published in 2006, which provided guidance for increasing the rapidity and timeliness of response in various regions, while also providing guidance for proper preparation of epidemic emergency responses.

In combination with the needs of the health emergency support work performed during the 2008 Olympics, NIPD established an emergency reserve mechanism. Materials including diagnostic reagents, medicine and other related emergency supplies for parasitic disease control and prevention were stored and managed under the guide of China CDC. The emergency reserves are updated and renewed each year, thus providing material security for emergency responses to various outbreaks. In 2012, NIPD formulated and issued a contingency plan for the emergency response to sudden outbreaks of parasitic diseases. Accordingly, a leading group, an expert technical guidance group, and an emergency response team for health-related emergency response work were set up in NIPD and served as the inception of the emergency response-contingency plan, mechanism, and teams. The emergency response-contingency plan follows the principles of unified command, tiered responsibility, rapid response, reliance on science, and management based on the law, thus providing a mechanistic assurance for the emergency response to parasitic disease outbreaks.

5. Role of the emergency response system in managing parasitic diseases

In the 70 years since our institute was established, the parasitic disease emergency response system has undergone gradual improvements and has played important roles in the effective prevention and timely management of emergency responses to various, sudden epidemic outbreaks.

5.1 Case study I: Schistosomiasis outbreaks

Between August 4th and 10th, 2003, 10 acute schistosomiasis cases were reported in Shitai County, Anhui Province. In response, NIPD organized relevant experts in a timely manner to perform an epidemiological survey, and execute rapid prevention and control measures in collaboration with experts from the Anhui Provincial Institute for Schistosomiasis Control. This case represented the first emergency response management of an epidemic outbreak that our institute had participated in after its restructuring, leading to prevent the rebound and spread of schistosomiasis effectively.

In July 2005, five acute schistosomiasis cases were reported from Wa'er Village of Xide County, Sichuan Province, through the National Schistosomiasis Direct Reporting Network System. NIPD dispatched an expert team to conduct surveys on-site and supervise the prevention and control. By implementing a series of effective prevention and control measures, the acute schistosomiasis outbreak was rapidly under control (Li et al., 2009).

5.2 Case study II: Malaria outbreaks

In July 2007, a malaria outbreak occurred in Congjiang County, Guizhou Province. NIPD organized experts to perform an on-site survey, investigate the local malaria prevention and control status, and supervise the local implementation of comprehensive measures including case findings, treatment of patients and their family members, mosquitoes control, *Anopheles* density monitoring, and health education. The implementation of these measures effectively controlled the epidemic, and prevented its spread, while also targeted opinions and suggestions for strengthening the local malaria prevention and control measures, were provided (Sheng et al., 2007).

In July 2010, Neijiang City of Sichuan Province reported an adverse event related to the medication in an prophylactic malaria group through the National Public Health Incident Report Management Information System (Xuelan, 2019). NIPD rapidly organized experts to supervise the investigation and management in Neijiang City. The cause of the incident was ascertained, while guidance and suggestions were offered regarding the exposed problems in terms of formulating the corresponding technical plans and emergency response-contingency plan, strengthening capacity of grassroots programs through training, and improving the quality of group medications.

5.3 Case study III: Food-borne parasitic disease outbreaks

In February 2009, surveillance data from the National Public Health Emergency Report Management Information System indicated that a serious unexplained poisoning incident had occurred in Lanping County, Nujiang, Yunnan Province. As this incident was suspected to be caused by *Trichinella spiralis* infection, NIPD organized experts to carry out epidemiological surveys, pathogen examination, and serological testing, along with the local Disease Control Center, to identify its cause and verify the diagnoses. The feature and scope of the epidemic was accurately determined leading to the effective control of the trichinellosis outbreak within a short time.

On February 1, 2012, an epidemic of a suspected food-borne parasitic disease was reported in Binchuan County of the Dali Bai Autonomous Prefecture, Yunnan Province. Conventional anthelmintics such as praziquantel and albendazole were ineffective, and the cause of disease was unclear. NIPD dispatched experts to the field on February 2nd and 12th to conduct active case exploration, surveys on hosts and vectors, and diagnosis through pathogen, immunological, and molecular biological testing. A specific drug against fascioliasis (triclabendazole), reserved at NIPD, was used to perform interventional treatment. We confirmed that this epidemic was caused by *Fasciola gigantica*, and determined that Binchuan County of Dali Bai was a natural epidemic focus of *F. gigantica*. The NIPD experts supervised the treatment of cases and provided suggestions for further prevention and control measures (Chen et al., 2013).

5.4 Case study IV: Filariasis outbreaks

In August 2007, the Fuchuan County in Guangxi Province reported a case of filariasis through the Infectious Disease Epidemic Surveillance Network System. Subsequently, NIPD organized experts to conduct field investigation on this filariasis event and found that Ganshang and Yinshan Natural Villages in Fuchuan County, Guangxi Province were residual epidemic foci of Bancroft's filariasis (Xue-Ming et al., 2008). The experts suggested that post-elimination surveillance of filariasis should be strengthened and a national review of residual epidemic foci of filariasis should be conducted to ensure its elimination nationwide. Until now, no new case was detected through decades of surveillance.

5.5 Case study V: Imported Africa human trypanosomiasis

In August 2017, Fujian Provincial Center for Disease Control reported an imported case of African trypanosomiasis (Chen et al., 2019). Subsequently, NIPD led a team from the World Health Organization headquarters and the Fujian Center for Disease Control to review the case and conduct additional testing. Based on analysis of the patient's clinical symptoms, epidemiological history, and laboratory test results, the patient was regarded as the first imported case of *Trypanosoma brucei* in China. During the same month, Huashan Hospital (affiliated with Fudan University) reported another imported case of African trypanosomiasis. Experts from NIPD conducted investigations and timely testing allowing for the determination that the patient was the second imported case infected with *Trypanosoma brucei*. NIPD experts provided professional guidance for case treatment and follow-up, while also actively seeking the support from the World Health Organization to establish a drug reserve at NIPD for potential imported cases in future (Liu et al., 2018).

During the management of these various epidemic outbreaks, experts of NIPD fulfilled the role of being the "final voice" as a national team in case investigations and follow-up, field survey, on-site vector control, health education, etc. In addition, NIPD ensured the timely and effective management of epidemic outbreaks, while also safeguarded the security of public health through several measures, including the on-site emergency response management of various epidemic outbreaks, improving the material, techniques, and expert reserves of NIPD, adjusting and improving the emergency response commands and organization as well as coordination mechanisms, continuously improving emergency response-contingency plans, establishing multi-departmental cooperation mechanisms, and improving the level of emergency response.

6. Conclusion

With 70 years of relentless efforts, NIPD contributed greatly to parasitic disease control in China, with substantial developments and improvements in surveillance and response capabilities. However, due to the vast territory of China and the wide range and various parasitic species, certain neglected parasitic diseases remain endemic in remote and impoverished settings. Further, the parasitic diseases which have been controlled continue to face risks of transmission or rebound. With socioeconomic development and increase in tourism, trade exchanges have become more frequent, which implies that there is also a continuous rise in the threat of imported epidemics (Zhou et al., 2018). In addition, owing to dietary and lifestyle habits, outbreaks of certain food-borne parasitic diseases may also occur locally in some areas. Therefore, NIPD should strengthen the capability of surveillance and response to parasitic diseases, while continuously improving the sensitivity and rapidity of surveillance-response systems that are compatible with the progress of disease prevention and control systems (Chun-Li et al., 2017), all of which continue to offer important practical significance for prevention and control of parasitic diseases in the future.

Acknowledgements

This study was supported by the National Special Science and Technology Project for Major Infectious Diseases of China (Grant No. 2016ZX10004222-004), the fourth round of Three-Year Public Health Action Plan (GWIV-29).

Competing interest

The authors declare that they do not have competing interests.

Authors' contributions

Y.-W. Hao, Q. Wang and S.-Z. Li conceived the study; Y.-W. Hao, Q. Wang, J. Xu and S.-Z. Li wrote and revised the manuscript; C.-L. Cao, T. Tian, Z.-L. Zhu, S.-S. Zhou, W.-P. Wu, Y.-D. Chen, Y. Zhang, J.-X. Chen, N. Xiao and X.-N. Zhou revised the manuscript and gave approval of the version to be published. All the authors read and approved the final version of the manuscript.

References

Ameme, D.K., Nyarko, K.M., Kenu, E., Afari, E.A., 2016. Strengthening surveillance and response to public health emergencies in the West African sub-region: the role of Ghana FELTP. Pan Afr. Med. J. 25 (Suppl. 1), 1.

Chen, J.X., Chen, M.X., Ai, L., Xu, X.N., Jiao, J.M., Zhu, T.J., Su, H.Y., Zang, W., Luo, J.J., Guo, Y.H., Lv, S., Zhou, X.N., 2013. An outbreak of human fascioliasis gigantica in Southwest China. PLoS One 8 (8), e71520.

Chen, N., Jin, K., Xu, J., Zhang, J., Weng, Y., 2019. Human African trypanosomiasis caused by Trypanosoma brucei gambiense: the first case report in China. Int. J. Infect. Dis. 79, 34–36.

Chen, J., Ding, W., Li, Z., Zhou, D.D., Yang, P., Wang, R.B., Zheng, B., Sheng, H.F., Guan, Y.Y., Xiao, N., Li, S.Z., Zhou, X.N., 2020. From parasitic disease control to global health: new orientation of the National Institute of Parasitic Diseases, China CDC. Acta Trop. 201, 105219.

Cheng, G., Li, D., Zhuang, D., Wang, Y., 2016. The influence of natural factors on the spatio-temporal distribution of Oncomelania hupensis. Acta Trop. 164, 194–207.

Chun-Li, C., Le-Ping, S., Qing-Biao, H., Bian-Li, X.U., Xiao-Nong, Z., 2017. Establishment of response system to emergency parasitic disease affairs in China. Chin. J. Schistosomiasis Control 29 (4), 397–401.

Dang, H., Xu, J., Li, S.Z., Cao, Z.G., Huang, Y.X., Wu, C.G., Tu, Z.W., Zhou, X.N., 2014. Monitoring the transmission of Schistosoma japonicum in potential risk regions of China, 2008–2012. Int. J. Environ. Res. Public Health 11 (2), 2278–2287.

De-Jian, S., Xu-Li, D., Ji-Hui, D., 2013. The history of the elimination of lymphatic filariasis in China. Infect. Dis. Poverty 2 (1), 30.

Fang, Y., Zhang, Y., 2019. Lessons from lymphatic filariasis elimination and the challenges of post-elimination surveillance in China. Infect. Dis. Poverty 8 (1), 66.

Feng, J., Liu, J., Feng, X., Zhang, L., Xiao, H., Xia, Z., 2016a. Towards malaria elimination: monitoring and evaluation of the "1-3-7" approach at the China-Myanmar border. Am. J. Trop. Med. Hyg. 95 (4), 806–810.

Feng, Y., Liu, L., Xia, S., Xu, J.F., Bergquist, R., Yang, G.J., 2016b. Reaching the surveillance-response stage of schistosomiasis control in the People's Republic of China: a modelling approach. Adv. Parasitol. 92, 165–196.

German, R.R., 2000. Sensitivity and predictive value positive measurements for public health surveillance systems. Epidemiology 11 (6), 720–727.

Groseclose, S.L., Buckeridge, D.L., 2017. Public health surveillance systems: recent advances in their use and evaluation. Annu. Rev. Public Health 38, 57–79.

Lai, S., Liao, Y., Zhang, H., Li, X., Ren, X., Li, F., Yu, J., Wang, L., Yu, H., Lan, Y., Li, Z., Wang, J., Yang, W., 2014. Comparing the performance of temporal model and temporal-spatial model for outbreak detection in China infectious diseases automated-alert and response system, 2011–2013, China. Zhonghua Yu Fang Yi Xue Za Zhi 48 (4), 259–264.

Li, S.Z., Luz, A., Wang, X.H., Xu, L.L., Wang, Q., Qian, Y.J., Wu, X.H., Guo, J.G., Xia, G., Wang, L.Y., Zhou, X.N., 2009. Schistosomiasis in China: acute infections during 2005–2008. Chin Med J (Engl) 122 (9), 1009–1014.

Li, H.Z., Qiang, W., Zheng, C.J., Qian, Z., Qian, Y.J., Li, S.Z., Wu, X.H., Dong, Y.F., 2010. Development and application of information system for parasitic diseases control and prevention. Chin. J. Schistosomiasis Control 22 (04), 301–303.

Li, Z., Yin, W., Clements, A., Williams, G., Lai, S., Zhou, H., Zhao, D., Guo, Y., Zhang, Y., Wang, J., Hu, W., Yang, W., 2012. Spatiotemporal analysis of indigenous and imported dengue fever cases in Guangdong province, China. BMC Infect. Dis. 12, 132.

Li, S.Z., Zheng, H., Abe, E.M., Yang, K., Bergquist, R., Qian, Y.J., Zhang, L.J., Xu, Z.M., Xu, J., Guo, J.G., Xiao, N., Zhou, X.N., 2014. Reduction patterns of acute schistosomiasis in the People's Republic of China. PLoS Negl. Trop. Dis. 8 (5), e2849.

Liu, Q., Chen, X.L., Chen, M.X., Xie, H.G., Liu, Q., Chen, Z.Y., Lin, Y.Y., Zheng, H., Chen, J.X., Zhang, Y., Zhou, X.N., 2018. Trypanosoma brucei rhodesiense infection in a Chinese traveler returning from the Serengeti National Park in Tanzania. Infect. Dis. Poverty 7 (1), 50.

Lv, S., Zhang, Y., Liu, H.X., Hu, L., Yang, K., Steinmann, P., Chen, Z., Wang, L.Y., Utzinger, J., Zhou, X.N., 2009. Invasive snails and an emerging infectious disease: results from the first national survey on Angiostrongylus cantonensis in China. PLoS Negl. Trop. Dis. 3 (2), e368.

Pan, J.Y., Zhou, S.S., Zheng, X., Huang, F., Wang, D.Q., Shen, Y.Z., Su, Y.P., Zhou, G.C., Liu, F., Jiang, J.J., 2012. Vector capacity of Anopheles sinensis in malaria outbreak areas of central China. Parasit. Vectors 5, 136.

Perry, H.N., McDonnell, S.M., Alemu, W., Nsubuga, P., Chungong, S., Otten Jr., M.W., Lusamba-dikassa, P.S., Thacker, S.B., 2007. Planning an integrated disease surveillance and response system: a matrix of skills and activities. BMC Med. 5, 24.

Ren, Z., Wang, D., Ma, A., Hwang, J., Bennett, A., Sturrock, H.J., Fan, J., Zhang, W., Yang, D., Feng, X., Xia, Z., Zhou, X.N., Wang, J., 2016. Predicting malaria vector distribution under climate change scenarios in China: challenges for malaria elimination. Sci. Rep. 6, 20604.

Sheng, H.F., Zheng, X., Shi, W.Q., Xu, J.J., Jiang, W.K., Wang, D.Q., Tang, L.H., 2007. Factors affecting malaria outbreak in Congjiang county of Guizhou Province. Zhongguo Ji Sheng Chong Xue Yu Ji Sheng Chong Bing Za Zhi 25 (3), 225–229.

Tambo, E., Ugwu, E.C., Ngogang, J.Y., 2014. Need of surveillance response systems to combat Ebola outbreaks and other emerging infectious diseases in African countries. Infect. Dis. Poverty 3, 29.

Wang, P., Wang, F., Sun, Y., Wang, J., Kou, J., Huang, Y., Kang, X., Wan, G., Liang, Y., Wang, Q., 1998. Study on the transmission threshold value of bancroftian filariasis. Chin Med J (Engl) 111 (3), 261–264.

Wang, J., Qi, H., Diao, Z., Zheng, X., Li, X., Ma, S., Ji, A., Yin, C., 2010. An outbreak of angiostrongyliasis cantonensis in Beijing. J. Parasitol. 96 (2), 377–381.

Wu, X.H., Wang, X.H., Utzinger, J., Yang, K., Kristensen, T.K., Berquist, R., Zhao, G.M., Dang, H., Zhou, X.N., 2007. Spatio-temporal correlation between human and bovine schistosomiasis in China: insight from three national sampling surveys. Geospat. Health 2 (1), 75–84.

Xu, J., Xu, J.F., Li, S.Z., Zhang, L.J., Wang, Q., Zhu, H.H., Zhou, X.N., 2015. Integrated control programmes for schistosomiasis and other helminth infections in P.R. China. Acta Trop. 141 (Pt. B), 332–341.

Xu, J., Yu, Q., Tchuente, L.A., Bergquist, R., Sacko, M., Utzinger, J., Lin, D.D., Yang, K., Zhang, L.J., Wang, Q., Li, S.Z., Guo, J.G., Zhou, X.N., 2016. Enhancing collaboration between China and African countries for schistosomiasis control. Lancet Infect. Dis. 16 (3), 376–383.

Xuelan, Z.J.Y., 2019. Analysis on the characteristics of emergency public health events in Neijiang City from 2004 to 2018. Parasitoses Infect. Dis. 17 (03), 152–156.

Xue-Ming, L.I., Yang, Y.C., Huang, K.L., Hong-Bo, X.U., Xie, Z.Y., Mai, F.Z., Shang, S.M., Ai-Rong, S.U., Shu-Lin, L.I., Fang-Qi, O.U., 2008. Discovery of a residual focus of bancroftian filariasis after declaration of its transmission interruption in Guangxi. Chin. J. Parasitol. Parasit. Dis. 26 (6), 2–408.

Yang, W., Li, Z., Lan, Y., Wang, J., Ma, J., Jin, L., Sun, Q., Lv, W., Lai, S., Liao, Y., Hu, W., 2011. A nationwide web-based automated system for outbreak early detection and rapid response in China. Western Pac. Surveill. Response J. 2 (1), 10–15.

Yang, K., Sun, L.P., Huang, Y.X., Yang, G.J., Wu, F., Hang, D.R., Li, W., Zhang, J.F., Liang, Y.S., Zhou, X.N., 2012. A real-time platform for monitoring schistosomiasis transmission supported by Google Earth and a web-based geographical information system. Geospat. Health 6 (2), 195–203.

Ying-Dan, C., Hui-Hui, Z., Ji-Lei, H., Ting-Jun, Z., Xiao-Nong, Z., 2019. Status and working principals of soil-transmitted nematodiasis during new period in China. Chin. J. Schistosomiasis Control 31 (1), 23–25.

Yuan, H., Jiang, Q., Zhao, G., He, N., 2002. Achievements of schistosomiasis control in China. Mem. Inst. Oswaldo Cruz 97 (Suppl. 1), 187–189.

Zhang, W., Wang, L., Fang, L., Ma, J., Xu, Y., Jiang, J., Hui, F., Wang, J., Liang, S., Yang, H., Cao, W., 2008. Spatial analysis of malaria in Anhui province, China. Malar. J. 7, 206.

Zheng, H., Li, S.Z., Cao, C.L., Zhang, L.J., Sun, L.P., Yang, K., Tu, Z.W., Li, Y.Y., Yang, W.P., Gu, X.N., Wu, Z.S., Feng, X.G., Zhu, R., Xu, J., Xiao, N., Zhou, X.N., 2013. Surveillance and response for Schistosomiasis japonica based on sentinel mice examination for cercariae-infested water in risk region, 2012. Zhongguo Ji Sheng Chong Xue Yu Ji Sheng Chong Bing Za Zhi 31 (6), 428–432.

Zhou, X.N., Chen, J.X., Chen, M.G., Bergquist, R., 2005. The National Institute of Parasitic Diseases, Chinese Center for Disease Control and Prevention: a new administrative structure for schistosomiasis control. Acta Trop. 96 (2–3), 296–302.

Zhou, X.N., Bergquist, R., Tanner, M., 2013. Elimination of tropical disease through surveillance and response. Infect. Dis. Poverty 2 (1), 1.

Zhou, X.N., Qian, M.B., Priotto, G., Franco, J.R., Guo, J.G., 2018. Tackling imported tropical diseases in China. Emerg. Microbes Infect. 7 (1), 12.

Zhu, R., Qin, Z.Q., Feng, T., Dang, H., Zhang, L.J., Xu, J., 2013. Assessment of effect and quality control for parasitological tests in national schistosomiasis surveillance sites. Zhongguo Xue Xi Chong Bing Fang Zhi Za Zhi 25 (1), 11–15.

Zhu, H.H., Zhou, C.H., Chen, Y.D., Zang, W., Xiao, N., Zhou, X.N., 2015. SWOT analysis of the national survey on current status of major human parasitic diseases in China. Zhongguo Ji Sheng Chong Xue Yu Ji Sheng Chong Bing Za Zhi 33 (5), 377–381.

CHAPTER TWELVE

Establishment and application of the National Parasitic Resource Center (NPRC) in China

Shao-Hong Chen[a,c,d,e], Hai-Mo Shen[a,c,d,e], Yan Lu[a,c,d,e], Lin Ai[a,c,d,e], Jia-Xu Chen[a,c,d,e], Xue-Nian Xu[a,c,d,e], Peng Song[a,c,d,e], Yu-Chun Cai[a,c,d,e], Xiao-Nong Zhou[a,b,c,d,e,*]

[a]National Institute of Parasitic Diseases, Chinese Center for Disease Control and Prevention, Shanghai, People's Republic of China
[b]School of Global Health, Chinese Center for Tropical Diseases Research, Shanghai Jiao Tong University School of Medicine, Shanghai, People's Republic of China
[c]WHO Collaborating Centre for Tropical Diseases, Shanghai, People's Republic of China
[d]National Center for International Research on Tropical Diseases, Ministry of Science and Technology, Shanghai, People's Republic of China
[e]Key Laboratory of Parasite and Vector Biology, Ministry of Health, Shanghai, People's Republic of China
*Corresponding author: e-mail address: zhouxn1@chinacdc.cn

Contents

1. Background 374
2. History for the NPRC development 376
   2.1 Phase I: Initial platform 376
   2.2 Phase II: Operational platform 377
   2.3 Phase III: Sharing services platform 377
   2.4 Phase IV: Resource centre 377
3. Main function of the NPRC 379
   3.1 Formulation of resource management system and standards 379
   3.2 Resource collection and preservation 380
   3.3 Resource sharing and services 382
   3.4 Management of resources facilities 385
   3.5 Capacity building 385
4. Achievements of the NPRC 386
   4.1 Development of resource management systems and standards 386
   4.2 Resource collection and preservation 386
   4.3 Resource sharing 391
   4.4 Resource services 392
   4.5 Capacity building 395
5. Looking forward 395
6. Conclusion 396
Acknowledgements 397

Advances in Parasitology, Volume 110
ISSN 0065-308X
https://doi.org/10.1016/bs.apar.2020.04.006

Competing interests 397
Authors' contributions 397
References 397

Abstract

The National Parasitic Resource Center (NPRC) was created in 2004. It is a first-level platform under the Basic Condition Platform Center of the Ministry of Science and Technology of China. The resource centre involves 21 depository institutions in 15 regions of the country, including human parasite and vector depository, animal parasite depository, plant nematode characteristic specimen library, medical insect characteristic specimen library, trematode model specimen library, parasite-vector/snail model specimen library, etc. After nearly 15 years of operation, the resource centre has been built into a physical library with a database of 11 phyla, 23 classes, 1115 species and 117,814 pieces of parasitic germplasm resources, and three live collection bases of parasitic germplasm resources. A variety of new parasite-related immunological and molecular biological detection and identification technologies produced by the resource centre are widely used in the fields of public health responses, risk assessments on food safety, and animal or plant quarantine. The NPRC is the largest and top level resource centre on parasitology in China, and it is a leading technology platform for collecting and identifying parasitic resources.

1. Background

The National Parasite Resource Center (NPRC) was created in 2004 supported by the Ministry of Science and Technology (MOST) of China (Zhou et al., 2006). It was originally started with the research project on "*Standardization of Parasite Germplasm Resources Management, Integration and Sharing Pilot Project*" funded by the MOST of China, and became the "*National Parasite Germplasm Resource Sharing Service Platform*" in 2017, it renamed as "National Parasite Resource Center (NPRC)" in 2019 as one of 30 national resource centres in the country, located in the National Institute of Parasitic Diseases at China CDC (NIPD) based in Shanghai, China. After 15 years of construction and operation, this resource library has gained three major achievements, including (i) for the first time, a national level, most complete and comprehensive parasite germplasm resource bank has been established, covering eight types of physical libraries and databases, e.g. protozoa, trematodes, tapeworms, nematodes, arthropods, molluscs, crustaceans, and important other parasites, etc.; (ii) a total of 28 new technologies for immunological and molecular biological detection tests and identification tools for various parasites have been developed through this germplasm resource centre, and these tests or tools have been widely used in the fields of public health response, risk assessments of food safety, and animal or plant quarantine, which improved the accuracy of parasite species detection and identification, thus the NPRC has become a leading technological platform in parasite detection and identification in the country; (iii) a platform for sharing the parasite and tropical disease germplasm resources has been created on website: http://www.tdrc.org.cn, this web-based platform can share the resource information for dissemination and provide services in domestic and international scientific researches, teaching and training activities, science popularization, and medical treatment, etc. The above three achievements have also become the main features of the resource centre (Shen et al., 2007; Zhou, 2008, 2009).

With the rapid development of world economic integration and the implementation of the "Belt and Road" initiative (Chen et al., 2019a; Qian et al., 2017), the global population movements and animal trading has caused more pathogens and vectors to spread widely, thus pathogens and their vectors have become important resources in prevention and control of infectious diseases which requires the joint efforts of countries around the world (Bahl et al., 2013; Dugan, 2012; Galsworthy et al., 2011; Saegerman et al., 2018; Si et al., 2009). For example, the outbreak of Ebola virus disease in West Africa in 2014 (Carias et al., 2018; Li et al., 2018a), a new species of schistosome newly discovered in France which has infected to humans in 2015 (Horak et al., 2015; Jouet et al., 2015), and the first report of imported Africa human trypanosomiasis cases reported in China in 2016 (Chen et al., 2019b; Liu et al., 2018; Wang et al., 2018), all those response activities to those emerging and re-emerging diseases need the cooperation of global experts to timely and accurately isolate and identify pathogens or vectors in control the spread of disease risk. At the same time, animal and plant trade will cause more diseases to spread among animals and plants around the world, causing huge economic losses (Bai et al., 2018; Jesus et al., 2018; Torgerson et al., 2018; Yang et al., 2015). Therefore, preventing the spread of parasitic pathogens has become one of the main contents for the global health security, thus the primary task of the NPRC is to identify the pathogen or vector host in a timely and accurate manner during the response activities in combating the disease outbreak (Brown et al., 2018; Campeau et al., 2018; Li et al., 2018a). Therefore, we aim to build a world-class parasitic germplasm resource centre which can not only ensure the stability and availability in supporting to the national researches on parasitology, but also serve in supporting to the "Belt and Road" initiative in terms of involving more activities on the global health security (Stothard et al., 2018). For this purpose, this article summarizes the achievements of the NPRC in the past 15 years, and proposes future development directions based on the analysis on future needs and deficiencies in sharing our resources.

2. History for the NPRC development

The development of the NPRC is divided into following four stages, such as initial platform, operational platform, sharing services platform, and resource centre.

2.1 Phase I: Initial platform

From 2004 to 2008, the "*Standardization pilot project on management, integration and sharing of important parasite germplasm resources*" was funded by the MOST of China for the first time in 2004, followed by the second project of "*Standardization of consolidation, integration, and sharing of important parasite species resources*" was continued for a period of 3 years granted in 2005. The entire project has accomplished the tasks on construction of a physical library and database of 11 phyla, 23 classes, 1115 species with 117,814 pieces of parasitic germplasm resources, and establishment of three live collection bases of parasitic germplasm resources (Chen et al., 2013; Shen et al., 2007; Zhou et al., 2006). The resource platform on parasitic germplasm involves 21 depository institutions in 15 regions of the country, including human parasite and vector depository, animal parasite depository, plant nematode characteristic specimen library, medical insect characteristic specimen library, trematode model specimen library, parasite-vector/snail model specimen library, etc. The resource platform consisting of 16 institutions from the country was performed as a way of distributing resources in different institutions and centralized the information in the NPRC of the NIPD, and safely storing the resources in duplicate by matching different species for a medium- and long-term designing. During this stage, a total of 11 new species of brook crab and 1 new species of nematode were discovered, the National Parasite Species Resource Network was established which formed as the national parasite germplasm collection centre. All those activities in the resource platform on parasitic germplasm by use of modern technologies such as information sharing and internet technology have established a national parasite resource sharing system, thereby paved a path to success for the long-term development of the resource platform.

2.2 Phase II: Operational platform

From 2009 to 2013, the recourse platform on parasitic germplasm have being kept in the NIPD by the way of operation with self-raised funds. Three exhibition halls or museums under the resource platform have been open to the public (Zhou et al., 2015). Among them, the Museum of Parasites and Vectors at the NIPD has become one of science education bases in Huangpu District of Shanghai Municipality, and has received 37,000 visits from home and abroad. The resource centre provided 16,720 resources to various research institutions and research projects for sharing, becoming an important source to share living resources for major national research projects, and training a group of young scientists in the identification and detection of parasite germplasm resources.

2.3 Phase III: Sharing services platform

In 2014, after the resource platform on parasitic germplasm was evaluated by the MOST of China, the resource platform on parasitic germplasm was included as one of components in the National Scientific Data Sharing Platform for Population and Health (National Scientific Data, 2018), under the MOST of China. Thus the resource platform on parasitic germplasm started the sharing services to use parasitic germplasm in the fields of public health responses, risk assessments on food safety, and animal or plant quarantine, etc., with the funds from MOST through the National Population and Health Science Data Sharing Service Platform (Chen et al., 2020). This sharing services platform has been maintained until 2016.

2.4 Phase IV: Resource centre

In 2017, after the resource platform on parasitic germplasm taken the second evaluation by the MOST of China, the platform was formally upgraded into the first-level platform as one of resource centres under the administration by the Basic Facility Platform Center of the MOST, namely "National Parasite Germplasm Resource Bank" with sustainable funds supported by the MOST of China. The newly official name was awarded as the "National Parasitic Resource Center (NPRC)" by the MOST of China in 2019, as one of 30 National resource centres in the country. Up to date, the NPRC composes of 16 sub-platforms involving the operation to collect more parasitic resources, maintain the classification and identification of various parasites and vectors, and share the resources for different purposes requested by research institutes and universities (Fig. 1). As of the end of

2018, the NPRC (http://www.tdrc.org.cn) holds a physical library and database with resources consisting of 11 phyla, 23 classes, 1115 species and 117,814 pieces of parasitic specimens. The NPRC has become the largest and top level parasite species bank in China (Chen et al., 2020). The physical library is an important facility in providing resources for hundreds of scientific research projects and research institutions at home and abroad. Those services, including scientific researches, training and teaching, science popularization, government decision-making, major research projects, emergency response events, corporate innovation, serving people's livelihood and information dissemination, has become an essential task of the NPRC, as a leading technological platform for collecting and identifying parasitic resources.

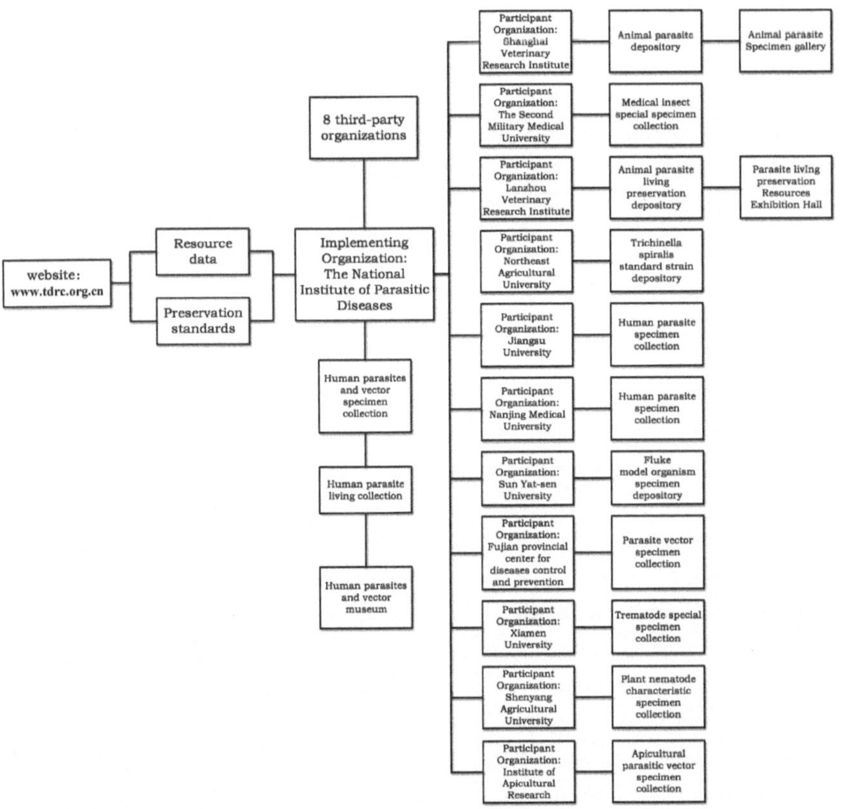

Fig. 1 The framework of the physical resource library in the National Parasitic Resource Center in China.

3. Main function of the NPRC

The main function of the NPRC is to formulate resource management system and standards, collect and preserve the parasitic resources, provide resource sharing and services, manage the resources facilities, and strengthen capacity building for the professionals in the field of parasitic resource identification and preservation (Dougherty et al., 2016).

3.1 Formulation of resource management system and standards

3.1.1 Establishment of a resource management system

The resource database has established in a complete set of operation management systems, which includes nine sub-systems, such as expert group review system, financial management system, work quality supervision system, joining system, information and data network reporting system, resource application registration system, platform user registration system, resource utilization filing and registration system, the rules and regulations of the parasites and vectors museum and the depository, in order to ensure the normal operation of the resource centre. At present, the resource centre implements the system of responsibility of the main person in charge, regularly communicates the progress of the work, timely discovers problems and proposes solutions. At the same time, an expert group is set up to be responsible for the review of various standards and the quality assessment of the implementation projects twice a year. During the implementation of the resource centre quality control is carried out according to the "Work Quality Scoring Standard", which is used as an indicator for allocating work funds for the next stage, and an assessment dynamic management mechanism is implemented. Each participating institutions under the resource network has a resource working group responsible for daily maintenance. The participating institutions cooperate and complement each other, and exchange information from time to time.

3.1.2 Formulate the description protocols, preservation procedures and inventory of parasite species resources

3.1.2.1 Principles in developing the protocol documents

Prioritized work for developing the platform documents is to develop the description protocols and standards for the existing database accompanied with the current needs, such as resource research and resources used for teaching. At the same time, taking into account the needs of resource platform's development, priority is given to specimen collection based on current settings and international information extracted from the published descriptor table of foreign parasite germplasm resources (Kumar et al., 2017).

3.1.2.2 Description of the protocols or norms

Several documents to describe the specimen during collection and management have been formulated, such as the "General Description Table of Animal Germplasm Resources", "Description Table of Common Characteristics of Parasite Germplasm Resources", and the individual description table of various types of parasite germplasm resources one by one, in accordance with the descriptions and standards in the "Animal Germplasm Resource Common Description Specification", "Parasite Germplasm Resource Common Description Specification", and various description on specifications of the parasite germplasm.

3.1.2.3 Formulation of relevant standards

A total of 29 preservation specifications have been developed: including "Specifications for Parasite Germplasm Resource Description", "Technical Regulations for the Conservation of Chicken Coccidia", "Technical Regulations for the Conservation of Piroplasma", "Technical Regulations for Conservation of *Brugia malayi*", "Technical Regulations for Conservation of *Trichinella spiralis*", etc. The standards and specifications basically cover the main parasite species in China, including all kinds of parasite species resources, so as to achieve the goal of comprehensive and refined management in resource collection and preservation quality control, resource sharing and collation standardized management.

3.2 Resource collection and preservation

3.2.1 Resource screening

The purpose of resource screening is to ensure the scientifically, systematically and operationally establishing the information database, to standardize the procedures in collection, preservation and biosafety of resources, and to promote the efficient sharing and sustainable use of resources collected. The resource screening is mainly aimed at parasites and related vectors that infected in humans and animals, including those specimens of living parasites, slide specimens, liquid immersion specimens, genetic resources and biological materials, which can be included in the scope of parasite species resources in this resource centre. Highly pathogenic pathogens are not included in the resource centre. The living parasites collected or operated by the resource collection institutions must have a biosafety level laboratory required by the state for the corresponding parasite operation. The representativeness of resource selection have to give full play to the advantages of each institution, and collect resources according to the collaborative division of labour to ensure the representativeness and integrity of the resource database.

3.2.2 Physical library construction

Systematic research on the species and distribution of global parasite resources has been carried out, and surveys of global parasite resources have been completed through literature searching and other means. At the same time, a survey of parasite resources in China has been completed, and a gap analysis was performed with reference to the existing types and quantities of the resource centre (Poulin, 2014).

We also established a directory that covers the world's important parasitic pathogens and vectors, and planned to undertake specimen collection based on gaps existed (Spencer and Zuk, 2016). On the basis of maintaining the existing sorting scope, we have expanded the scope and varieties of parasite specimen collection and sorting, and strengthened the cooperation and promoted exchanges between the resource bank and parasitoid-related research institutions around the world. Supplementary collection of new, rare, invasive parasites or vector species worldwide are encouraged through field collection, specimen exchange, etc. The long-term work is to make local resources covering more than 50% of the total species of parasites and vectors.

3.2.3 Information database construction

3.2.3.1 Digitization of numbers

Each depository institution establishes a "Parasite Classification List of Parasite Germplasm Resources" according to the "Specifications of Common Description of Natural Science and Technology Resources (Trial)" and the "Specifications of Data Reported by Resource Institutions (Trial)", and renumbers different parasite resources uniformly. Perfect records, captured image data are saved as. JPG files (pixels > 300 dpi).

3.2.3.2 Digitization of resources

The digitization of resources includes following steps, such as inputting data, establishing and assessing the database, upload the database into system and feedback, etc. These steps provide the bases for construction of the database, by the way of cross-regional integration of important parasite species resources covering medical parasites, veterinary parasites, medical arthropods and medical shellfish in China, in order to achieve complementary advantages and resource sharing.

3.2.3.3 Construction of the database

The integrated data has been used to establish eight databases in a unified format on the shared platform. The main databases include parasitic protozoa, parasitic flukes, parasitic nematodes, parasitic maggots, other important parasites, medical arthropods, medical molluscs, and medical crustaceans. The database was established according to the relationship of various parasites and the relationship between attributes in order to ensure the integrity, redundancy, security of the data, and to ensure all types of databases to be stored, retrieved and analysed in the form of both pictures and text (Nunn and Altizer, 2005).

The primary task of the resource platform is to provide services for scientific researches (Stephens et al., 2017). For parasites infected in humans, animals and plants as well as relevant vectors, priority of the specimen collection is to collect those endangered species or species that are extremely rare in the region (Campeau et al., 2018; Mordecai et al., 2016; Stringer and Linklater, 2014). We have not only vigorously strengthened the collection and preservation of rare, emerging or re-emerging parasites, but also established animal models, provided sample to be used in the research projects, such as projects on pathogenic mechanisms, diagnostics development and drug screening of rare parasites (Kumar et al., 2017; Li, 1989). It is ensured that the data in the resource library are shared in a unified manner through the resource sharing platform, and it is also ensured that the established resource depository, specimen exhibition hall, living deposit centre can be shared across institutions or universities through specific mechanism of sharing data and specimen.

3.3 Resource sharing and services

3.3.1 Establish a set of parasite species resource service sharing system

With the increasing needs to share the resources, a parasite species resource alliance system was established that covers 21 national depository institutions across the three major fields of animals, humans and plants. The distribution of depository institutions across the country is shown in Fig. 2.

Fig. 2 The distribution of deposits across the country (1. The Northeast Agriculture University; 2. Shenyang Agriculture University; 3. Institute of Apiculture Research, Chinese Academy of Agricultural Sciences (CAAS); 4. Lanzhou Veterinary Research Institute, CAAS; 5. Nanjing Medical University; 6. Jiangsu University; 7. The Second Military Medical University; 8. National Institute of Parasitic Diseases, China CDC; 9. Shanghai Veterinary Research Institute, CAAS; 10. Fujian Provincial Center for Disease Control and Prevention; 11: Xiamen University; 12. Sun Yat-sen University).

3.3.2 Resource services

In terms of resource services, we strive to add more applications to the old resources collected early, use new DNA identification technology, obtain bioinformatics data with clear background information, and provide new applications for resources. For newly collected resources, we obtain multiple types of information from the same resource through data mining according to the needs of subject development and subsequent applications, so that the potential information of the resources becomes possible for multiple applications to improve resource utilization. We are carrying out a large number of biological determinations on important samples, that comprehensively covered parasitic protozoa, trematodes, nematodes, cestodes, arthropods, molluscs, crustaceans and other invasive parasitic species. At the same time, we are further providing intensive services to expand resource utility in the research projects on development of diagnostic and therapeutic related products, and providing a more comprehensive database of sequencing or mutation sites in identification and detection of the invasive species and in tracing the rare or important parasites.

3.3.3 Resource sharing

In terms of sharing, we have improved the working mechanism of the parasite resource bank, and carried out resource collection, integration specifications, germplasm resource sharing service system and formulation of corresponding national standards. In the future, the resource bank will continue to improve various technical standards and specifications, and strive to basically cover the main parasite species in China, covering the entire operation process of the species resources from quality control to standardization.

Our key services on resource sharing are as follows:

- *Special services for the public through the internet and new media:* We undertake the identification of human parasites and set up the network of diagnostic testing services in major clinical hospitals across the country, which help clinical hospitals in precision diagnosis of parasitic diseases. We use new media to provide public-oriented knowledge dissemination, and also provide test results query services.
- *Participation in the emergency response and disease epidemic control services.* We participate in the national public health emergency responses, disseminate various information, and provide identification and detection services through the internet to ensure public health safety (Stringer and Linklater, 2014).
- *Implementation of research cooperation services through sharing of data and physical resources.* We open a service port for major scientific research projects, and open a download information window related to standards to achieve a bridge connection between germplasm resources and research users, thereby further promoting physical resource services for major scientific research projects (Bell et al., 2018).

- *Providing professional training and education services through on-site training and new media methods.* We provide a teaching service for continuing education training courses, through providing parasite teaching resources. We continue to interact with various media to provide health education videos on parasitic diseases.

3.4 Management of resources facilities

We have established high-quality preservation venues for the resource centre. The resource depository is distributed in 15 provinces and cities across the country. There are 21 professional depository institutions participating in the resource network. All participating institutions have their own sites and maintain independent depository. Each depository institution constructs a physical repository of parasite resources in accordance with unified standards, the largest of which is the human parasite and vector depository of the NIPD. The exhibition hall and three supporting specimen libraries cover a total area of more than 1000 square meters, and its exhibition hall is regularly open to the public. In terms of databases, we built a germplasm resource service platform website, and based on this, we have produced a number of supporting software or systems such as resource standardization management software, project management software, parasitic disease and parasite species picture recognition software, parasitic disease outpatient information software, parasitic disease health education and outpatient result query public numbers. At present, our portal website and system management software have been transformed into BS mode, and due to the requirements of offsite backup, we have used the cloud servers of other participating institutions to undertake. In addition to the main management system, 21 participating institutions have their own specimen library management and query systems, using the platform's unified resource number as the unique identification code to exchange data between the systems. The system is managed and operated by the information centre or network centre of each unit, and the maintenance costs are shared by the parasite germplasm resource service platform and the host institution.

3.5 Capacity building

The NPRC is implemented by the director-responsibility system under the leadership of the board of the NPRC facilitated by the NIPD. The director of the NIPD serves as the general manager of the resource base platform, and the chief of the main deposit base institution is a member. There are currently nine expert groups and a total of 171 staff members working in the NPRC. Among them, the head of the resource bank and the key senior participants are 30 people. This team involves multidisciplinary fields such as medical parasitology, veterinary parasitology, pathogen biology, medical malacology/vector biology, molecular biology, veterinary science, etc., and is responsible for reviewing and assessing the system specifications of the resource centre.

4. Achievements of the NPRC

4.1 Development of resource management systems and standards

Based on the biological characteristics and classification system of parasites and vectors, the "Description protocol of the parasite species from human, animal and plant" was issued firstly, followed by "Collection protocol of China's parasite species resources", e.g. *Taenia rapae* collection protocol, "Technical regulations for parasite preservation (Volume 1)" and "Directory of China's livestock and poultry parasite", etc. (Zhou, 2008, 2009) All of these documents laid the foundation for the taxonomic research of parasites. In the process of developing the inventory of parasitic resources, 11 new species of vectors, 1 nematode and 1 rare parasite have been discovered after we established the technical procedures for preservation, labelling and integration of parasitic resources. For example, the rare parasite of *Armillifer agkistrodontis* which causes visceral pentastomiasis has also been investigated, the life history of *Armillifer agkistrodontis* was revealed for the first time in the world with animal models established which can provide scientific evidence for the pathogenic mechanism and drug screening of this rare parasite (Chen et al., 2010).

4.2 Resource collection and preservation

A total of 21 parasite species resources institutions, 3 physical exhibition halls and 3 parasite live collection centres from 15 provinces and cities have been established nationwide. A total of 1200 species/125,000 pieces of parasite resources in 11 phyla were recorded (Table 1), including human parasites and vector depository, animal parasite depository, plant nematode characteristic specimen library, medical insects characteristic specimen library, trematode pattern specimen library, and parasite vector snail pattern specimen library. Resources were stored in a manner of network with all of

Table 1 Table of co-constructed institutions of the National Parasite Resource Bank.

No.	Area	Depository	Resource preservation repository type	No. of data exchanged	Reason for inclusion	Number of participants (person)
1	Shanghai	Institute of Parasitic Disease Control and Prevention, Chinese Center for Disease Control and Prevention (undertaking unit)	Human parasite and vector library	315/29 , 059 pieces	The only national parasite research institute	37
2		Shanghai Institute of Veterinary Medicine, Chinese Academy of Agricultural Sciences	Animal parasite depository	754 species/ 30751 pieces	Covers the largest variety of animal parasite banks, participating in the construction of parasite platforms for 15 years	18
3		Chinese People's Liberation Army Navy Medical University	Medical insect specimen collection	69 species/ 5290 pieces	The museum has a large collection of parasites and vectors that are characteristic of medical insects, and has participated in the construction of the platform for 15 years.	2
4		Fudan University	Human parasite gene bank	27 species/ 645 pieces	Specializing in genomics of parasites, revealing the whole genome of schistosomiasis for the first time in the world.	6
5		Shanghai Customs Animal and Plant and Food Inspection and Quarantine Technology Center	Invasion vectors and parasite-specific specimen libraries	80 species/ 2071 pieces	As the nation's largest port, a large number of invading vectors and parasite specimens have been collected to ensure national security.	5

Continued

Table 1 Table of co-constructed institutions of the National Parasite Resource Bank.—cont'd

No.	Area	Depository	Resource preservation repository type	No. of data exchanged	Reason for inclusion	Number of participants (person)
6	Gansu	Lanzhou Veterinary Research Institute, Chinese Academy of Agricultural Sciences	Animal parasite living depository	114 species/ 9805 pieces	The leading institution of national veterinary parasite research has the most complete collection of animal parasite living deposits and has participated in the construction of the platform for 15 years	22
7	Heilongjiang	Northeast Agricultural University	Trichinella spiralis standard collection bank	158 species/ 6399 pieces	The construction of the platform was last for 15 years, and has obtained many patents in the research of Trichinella spiralis living preservation technology and veterinary drugs	10
8	Jiangsu	Jiangsu University	Human parasite collection bank	71 species/ 4758 pieces	The construction of the platform was last for 15 years, and has a full set of human parasite teaching specimens	3
9		Nanjing Medical University	Human parasite collection bank	46 species/ 3645 pieces	The construction of the platform was last for 10 years, and has a full set of human parasite teaching specimens	3

10	Guangdong	Sun Yat-sen University	Trematode specimen collection bank	72 species/ 3771 pieces	Participated in the construction of the platform for 15 years, and is a large base for parasite research and preservation in southern China	13
11		Shenzhen Center for Disease Control and Prevention	Invasive parasite vector and pathogen depository	4 species/ 781 pieces	It has a variety of invasive parasite vectors and pathogens in China	5
12	Fujian	Fujian Provincial Center for Disease Control and Prevention	Trematode and vector specimen collection bank	80 species/ 4597 pieces	Participated in the construction of the platform for 15 years, and has a base of human parasitic flukes in the southern part of China	5
13		Xiamen University	Trematode special specimen collection	112 species/ 3906 pieces	Participated in the construction of the platform for 15 years, and is the animal parasite base and storage in southern China	6
14	Jilin	Shenyang Agricultural University	Plant nematode specimen collection	138 species/ 5037 pieces	Participated in the construction of the platform for 15 years and is the only depository institution that studies plant parasites	7
15	Beijing	Chinese Academy of Agricultural Sciences	Bee parasite vector collection bank	4 species/ 2699 pieces	Participated in the construction of the platform for 10 years, and is a preservation research institution featuring research on insect parasites	2

Continued

Table 1 Table of co-constructed institutions of the National Parasite Resource Bank.—cont'd

No.	Area	Depository	Resource preservation repository type	No. of data exchanged	Reason for inclusion	Number of participants (person)
16	Jiangxi	College of Medicine, Nanchang University	Parasite vector crustacean depository	67 species/ 1582 pieces	Participated in the construction of the platform for 15 years, and is a depository institution that specializes in studying the transmission of parasites	5
17	Sichuan	Sichuan Academy of Animal Science	Animal Parasite Specimen Bank of Southwest China	115 species/ 2449 pieces	Participated in the construction of the platform for 5 years and is an animal parasite preservation institution in southwest China	6
18	Henan	Henan Agricultural University	Zoonotic parasite bank	21 species/ 3890 pieces	It is a parasite preservation institution that is specialized in the research of parasite protozoa	3
19	Qinghai	Qinghai Institute of Animal Husbandry and Veterinary Medicine	Animal parasite depository	89 species/ 2520 pieces	Participated in the construction of the platform for 5 years and has a large number of animal parasite preservation institutions	6
20	Yunnan	Yunnan Provincial Institute of Parasitic Diseases	Invasive parasite vector and pathogen depository	1 specie/642 pieces	With a large number of invasive parasite vectors and pathogens, it is the base for research and preservation of human parasites in southwest China	3
21	Guangxi	Guangxi Provincial Center for Disease Control and Prevention	Invasive parasite vector and pathogen depository	6 species/ 703 pieces	It has a large number of invasive parasite vectors and pathogens	4

information centralized in the NIPD, and stored securely in duplicate locations. The amount of resources have accounted for 45% of the total in the country.

Meanwhile, eight major databases of the resource database, including parasite species resource database, a parasitic fluke species resource database, a parasitic nematode species resource database, a parasitic tapeworm species resource database, other important parasite resource databases, medical arthropod germplasm resource databases, medical molluscs and medical crustacean germplasm resource database, have been developed which become the largest and top database for parasite species collection in China.

4.3 Resource sharing

The shared services are the main function of the NPRC through sharing port platform (http://www.tdrc.org.cn), which mainly used in three major areas, such as public health responses, risk assessment of food safety, and animal breeding or quarantine. The research work on resource sharing and services have produced a number of patents and technical standards.

In the field of public health response, several technical methods developed through the NPRC have been used for the rapid clinical diagnosis as well as epidemiological response. For instance, the parasite egg recognition system based on multi-features of eggs' morphology can be applied to diagnostic services through remote network, which will provide diagnostic references for clinical hospitals and disease control centres. Established detection methods for *Clonorchis sinensis* and *Angiostrongylus cantonensis* have provided rapid and accurate diagnosis for epidemiological screening and control of *A. cantonensis* and *C. sinensis* infections (Chen et al., 2012, 2016a; Cheng et al., 2018; Chu et al., 2013; Li et al., 2018b; Qian et al., 2019). A protozoa staining method has been established to improve the long-term preservation of protozoa specimens. In addition, several devices have been developed for parasite identification, e.g. an improved Bellman's device, a digestive tract nematode culture device, a venom aspirator and stool sampling, and a parasite concentrator device, etc. A total of nine detection and identification patents, including detection methods for *C. sinensis* and *Toxoplasma gondii*, have been invented (Chen et al., 2016b; Cheng et al., 2018; Liu et al., 2016). All of those innovative activities have not only helped the collection of parasites and provided a convenient means for the accumulating more parasites in the resource bank, but also laid a foundation for the control of parasitic diseases (Ying-Dan et al., 2017).

In the field of risk assessment of food safety, six parasite species detection and identification technologies have been developed, of which patent authorizations have been awarded, such as "Universal real-time fluorescent PCR detection method for *Trichinella spiralis*", "High-density fermentation process of recombinant *Pichia yeast*", "Recombinant gene, vaccine for porcine cysticercosis and preparation method", "Indirect ELISA detection kit for porcine cysticercosis disease and preparation method", "A test methods for *Trichinella spiralis* and *Cysticercus cellulosae* in food", "Primers and probes and kits for detecting bradyzoite of *Toxoplasma gondii*" (Wei et al., 2011). Three industry standards have been formulated for the detection and identification of aquatic products, namely: "Identification methods for *Clonorchis sinensis* in fish" (SN/T 2975-2011), "Quarantine technology of metacercaria from trematode in crustacean aquatic products" (SN/T 3504-2013) and"Technical specifications for *Gnathostoma quarantine* in aquatic products" (SN/T 3497-2013). All aforementioned patented techniques and standard products have improved the risk assessment of food safety.

In the field of animal breeding and quarantine, about four tests and identification patents have invented, including "Kit and method for differential diagnosis of *Theileria annulata*", "Diagnostic kit and application recombinant antigen of *Babesia jerkii*", "Test method and kit for the detection of lyme disease pathogens in ticks", "A method for biosafety evaluation of genetically modified animals in a hard tick model", all above kits have provided fast methods for field diagnosis, prevention and control of animal diseases, in particular, to block the spread of zoonotic parasitic diseases (Gao et al., 2007).

4.4 Resource services

Since 2008, the NPRC has made important contributions to the scientific research, teaching, information dissemination and diagnostic services, through its networks.

4.4.1 Scientific research services

The physical library provided physical resources for 100 scientific research projects from 52 scientific research institutions in domestic and abroad. Up to date, there were 16,720 resources in 7 categories for physical sharing to support various research projects, including 2 for the major national infectious disease projects, 16 for the national supporting-plan research projects, 2 for the national 863 research projects, 2 for the national 973 research projects, 6 for the international projects, 24 for the research projects supported by the National Natural Science Foundation of China, 48 for other R&D projects. The resource sharing ensured smooth performance of research projects. A total of 288 publications, including 178 and 110 articles were published in Chinese and English, respectively. Twelve monographs have been published at the same time.

4.4.2 Education and training services

The resource bank provided teaching services for 36 training courses on parasitic disease. The participants of those training course were from 20 countries, hosted by 5 institutions and universities. A total of 12,494 person-times participants were trained with a provision of 2298 parasitic specimen in teaching.

4.4.3 Information dissemination

The three exhibition halls under the NPRC were open to the public since their establishment. A total of five health education films have been published, including "*Beware of Prevention from Paragonimiasis*", "*Beware of Prevention from Liver Fluke Infections*", "*Beware of Prevention from Angiostrongyliasis cantonensis Infections*", "*Beware of Prevention from Pentastomiasis*" and "*Trichinelliasis on the Tip of the Tongue*". Great impacts of the series films on food-borne parasitic infections have gained in health education and information disseminations. For example, "*Beware of Prevention from Paragonimiasis*" won the 2013 Medical Science Award of the Chinese Medical Association. We also provided those health education materials to assist media, such as CCTV10, CCTV7, Hunan Satellite TV, Heilongjiang Satellite TV and China Rural Magazine, etc., to disseminate the health information by shooting videos for promotion of prevention measures to parasitic infections, and by conducting a series of lectures (Table 2).

4.4.4 Consultation and testing services

The NPRC has provided social-oriented services specialized in parasitic infections. For instance, it has provided consultation and testing services to clinical or medical institutions for 54,558 person-times, animal quarantine for 8791 animal tests, and plant protection for 642 times, respectively. The platform has created a parasite egg identification system applied to diagnosis through remote network, providing technical inspection, egg identification, and species identification thematic service entrances to help clinicians with differential diagnosis.

Table 2 Services on information dissemination from the NPRC, 2008–2018.

Year	The number of services on information dissemination	Number of people served for dissemination
2008	1	1000
2008	1	1200
2010	2	1600
2011	4	2200
2012	5	More than 200 million
2013	5	More than 200 million
2014	4	More than 200 million
2015	7	More than 60 million
2016	12	More than 200 million
2017	13	More than 200 million
2018	13	More than 200 million

4.4.5 Database services

The NPRC has provided services supported by the important parasitic and biological information database, human pathogen gene and biomarker database, Chinese *Anopheles* species graphic retrieval system, parasite monitoring data sharing and analysis platform, and parasite disease literature abstract database services. Among them, the important parasitic and biological information database, the first comprehensive parasitic biological information database in China, has help to analysis genome data covering 22 parasites, 24 transcriptome databases, 7 proteome databases, such as *Schistosoma japonicum*, *Schistosoma mansoni*, *Brugia malayi*, *Leishmania spp.*, *Cryptosporidium spp.*, *T. gondii*, *Trichinella*, *Giardia lamblia*, *Plasmodium*, and *Trypanosomes*. Meanwhile, the system integrated related analysis software and analysis modules, including membrane protein prediction, signal peptide protein prediction, secreted protein prediction, epitope prediction, multiple sequence alignment, homology alignment, domain analysis, and phylogenetic analysis (Hu et al., 2003). In addition, parasitology literature database provided articles, abstracts and other information related to tropical diseases, forming a unique parasitology literature summary database. The database also provided the Tropical Disease Academic Briefings with 128 hot-topics analysis issues.

4.5 Capacity building

Through the establishment of the network on parasite resource collection and shared platform in China, the NPRC has trained a huge number of professional teams engaged in parasitology research in multiple fields.

In last 15 years, the NPRC has developed a well-balanced talent team structure. There were a total of 171 staff members working in the network of the NPRC, of which 30 at the top level are taking the leading role in each field of parasitic resources, or the main backbone of the NPRC. The specialized staff of the NPRC network are working in eight positions, including parasite germplasm resource collection staff, collation and integration staff, separation and identification staff, data mining analysts, shared service staff and technical support staff, in which contained 40 people at senior level (24%), 74 at medium level (43%), and 57 with junior titles. In addition, a consultation group consisting of nine famous experts who familiar with the parasitic resources work provided various consultancy and academic evaluation for the NPRC. During the operation period, a total of 288 research papers and 12 monographs were published, and 1575 postgraduates and 630 doctoral students were trained and graduated. All these achievements have a great impact to the capacity building in the fields of tropical medicine, public health, inspection and quarantine, food safety, and animal husbandry and veterinary medicine.

5. Looking forward

In the future, the NPRC will further improve the means in collection and integration of parasitic resources, sharing services, and management mechanisms based on recommendation from the consultation group and a series of standard specifications (Chang, 2006; Ying-Dan et al., 2017). More works will focus on supplementary collection of new, rare, and invasive biological species resources (Chen et al., 2010; Liu and Zhou, 2015; Molina et al., 2019; Scare et al., 2018; Song et al., 2018). It is expected that the NPRC will be one of advanced resource centres in the world by following actions: (i) transitioning from national level to the global level in resources collection and preservation, scientific research services, and intelligence of resource sharing; (ii) building a national top platform or resource bank by integrating functions on research, dissemination, specimen exhibition, training, networking and information sharing to improve parasites;

(iii) promoting the R&D activities in modern biology and relevant fields (Cable et al., 2017).

The detail activities to be a world well-known NPRC will be arranged in the following seven actions, including (i) improving the working mechanism of the parasite germplasm resource bank; (ii) expanding the scope and variety of parasitic species collection by cooperation and exchanges; (iii) further standardization of the parasitic specimen library by formulation of more standards or norms; (iv) building a genetic material library for human, animal, and plant parasitic resources; (v) construction of gene tag information database for parasitic resources by sequencing more important parasites and vector organisms; (vi) building a parasitic resource data warehouse; (vii) improving the public impacts on using information resources and services of the NPRC.

In order to keep the same space with the international development on parasite collection and resource preservation, we need to learn the rich experience from the world well-known institutions on parasite resource preservation, such as the British Museum of Natural History of UK (True, 1885), the Smithsonian Institute of USA (Rubinoff and Leigh, 1990; Smithsonian Institute, 1920), and the U.S. National Parasite Collection Center (Hoberg and Phillips, 2014), etc. We also need to keep cooperation with other international emerging institutions in the field of parasite collection, such as Japan, South Korea, the Philippines, etc. Therefore, a gap analysis between the NPRC and other international parasitic resource centres is warranted to perform in order to better design the further directions for the NPRC development (Hai-Juan et al., 2019).

6. Conclusion

The parasitic resource bank is an important and basic facility on science and technology. The NPRC has been built as the most comprehensive and largest parasite resource bank in China led by the NIPD based in Shanghai, supported by the Ministry of Science and Technology of China since 2004. In spite of the NPRC has well developed last 15 year to break through the zero-blank situation in China's systematic parasite resource bank, but compared with the internationally advanced parasite resource bank, huge gaps are still existed in terms of scale and number of collections, the level of resource services, and research capacity in resource utilization. Therefore, there is still a long way to go for further collection and utilization of the parasitic resources. It believes that in the near future, the NPRC will definitely become the leading parasitic resource centre in the country and the well-known one in the world.

Acknowledgements

This study was supported by Ministry of Science and Technology (2019-194-30), the National Special Science and Technology Project for Major Infectious Diseases of China (Grant No. 2016ZX10004222-004), and The fourth round of Three-Year Public Health Action Plan (GWIV-29) of Shanghai Municipality.

Competing interests

The authors declare that they do not have competing interests.

Authors' contributions

S.-H.C., H.-M.S., Y.L., L.A. and X.-N.Z. conceived the study; S.-H.C., H.-M.S. and X.-N.Z. wrote and revised the manuscript; S.-H.C., H.-M.S., Y.L., L.A., J.-X.C., X.-N.X., P.S., Y.-C.C. and X.-N.Z. revised the manuscript and gave approval of the version to be published. All the authors read and approved the final version of the manuscript.

References

Bahl, J., Krauss, S., Kuhnert, D., Fourment, M., Raven, G., Pryor, S.P., Niles, L.J., Danner, A., Walker, D., Mendenhall, I.H., Su, Y.C., Dugan, V.G., Halpin, R.A., Stockwell, T.B., Webby, R.J., Wentworth, D.E., Drummond, A.J., Smith, G.J., Webster, R.G., 2013. Influenza a virus migration and persistence in North American wild birds. PLoS Pathog. 9 (8), e1003570.

Bai, Z., Lee, M.R.F., Ma, L., Ledgard, S., Oenema, O., Velthof, G.L., Ma, W., Guo, M., Zhao, Z., Wei, S., Li, S., Liu, X., Havlik, P., Luo, J., Hu, C., Zhang, F., 2018. Global environmental costs of China's thirst for milk. Glob. Chang. Biol. 24 (5), 2198–2211.

Bell, K.C., Carlson, C.J., Phillips, A.J., 2018. Parasite collections: overlooked resources for integrative research and conservation. Trends Parasitol. 34 (8), 637–639.

Brown, A.A., Penrith, M.L., Fasina, F.O., Beltran-Alcrudo, D., 2018. The African swine fever epidemic in West Africa, 1996-2002. Transbound. Emerg. Dis. 65 (1), 64–76.

Cable, J., Barber, I., Boag, B., Ellison, A.R., Morgan, E.R., Murray, K., Pascoe, E.L., Sait, S.M., Wilson, A.J., Booth, M., 2017. Global change, parasite transmission and disease control: lessons from ecology. Philos. Trans. R. Soc. Lond. B Biol. Sci. 372 (1719), 20160088.

Campeau, L., Degroote, S., Ridde, V., Carabali, M., Zinszer, K., 2018. Containment measures for emerging and re-emerging vector-borne and other infectious diseases of poverty in urban settings: a scoping review. Infect. Dis. Poverty 7 (1), 95.

Carias, C., Adhikari, B.B., Ravat, F., Meltzer, M.I., Marston, B.J., 2018. Resources needed for US CDC's support to the response to post-epidemic clusters of Ebola in West Africa, 2016. Infect. Dis. Poverty 7 (1), 113.

Chang, Z.S., 2006. Collection of parasite samples and preservation of specimens. Chin. J. Parasitol. Parasit. Dis. 24 (S1), 76–81 (in Chinese).

Chen, S.H., Liu, Q., Zhang, Y.N., Chen, J.X., Li, H., Chen, Y., Steinmann, P., Zhou, X.N., 2010. Multi-host model-based identification of Armillifer agkistrodontis (Pentastomida), a new zoonotic parasite from China. PLoS Negl. Trop. Dis. 4 (4), e647.

Chen, J.X., Chen, M.X., Ai, L., Chen, J.H., Chen, S.H., Zhang, Y.N., Cai, Y.C., Zhu, X.Q., Zhou, X.N., 2012. A protein microarray for the rapid screening of patients suspected of infection with various food-borne helminthiases. PLoS Negl. Trop. Dis. 6 (11), e1899.

Chen, S.H., Hu, W., Shen, H.M., Zhang, Y., Chen, J.X., Li, H., Zhang, Y.N., Zhou, K., Tong, X.M., Zhou, X.N., 2013. Construction and look forward of the national parasitic germplasm resources in China. Chin. J. Zoonoses 29 (5), 427–432 (in Chinese).

Chen, M.X., Chen, J.X., Chen, S.H., Huang, D.N., Ai, L., Zhang, R.L., 2016a. Development of lateral flow immunoassay for antigen detection in human Angiostrongylus cantonensis infection. Korean J. Parasitol. 54 (3), 375–380.

Chen, M.X., Ai, L., Chen, J.H., Feng, X.Y., Chen, S.H., Cai, Y.C., Lu, Y., Zhou, X.N., Chen, J.X., Hu, W., 2016b. DNA microarray detection of 18 important human blood protozoan species. PLoS Negl. Trop. Dis. 10 (12), e0005160.

Chen, J., Bergquist, R., Zhou, X.N., Xue, J.B., Qian, M.B., 2019a. Combating infectious disease epidemics through China's Belt and Road Initiative. PLoS Negl. Trop. Dis. 13 (4), e0007107.

Chen, N., Jin, K., Xu, J., Zhang, J., Weng, Y., 2019b. Human African trypanosomiasis caused by Trypanosoma brucei gambiense: the first case report in China. Int. J. Infect. Dis. 79, 34–36.

Chen, J., Ding, W., Li, Z., Zhou, D.D., Yang, P., Wang, R.B., Zheng, B., Sheng, H.F., Guan, Y.Y., Xiao, N., Li, S.Z., Zhou, X.N., 2020. From parasitic disease control to global health: new orientation of the National Institute of Parasitic Diseases, China CDC. Acta Trop. 201, 105219.

Cheng, N., Xu, X.N., Zhou, Y., Dong, Y.T., Bao, Y.F., Xu, B., Hu, W., Feng, Z., 2018. Cs1, a Clonorchis sinensis-derived serodiagnostic antigen containing tandem repeats and a signal peptide. PLoS Negl. Trop. Dis. 12 (8), e0006683.

Chu, Y.H., Cai, Y.C., Ai, L., Lu, Y., Zhang, J., Chen, J.X., 2013. Effect comparison between two ELISA kits in IgG antibody detection of Echinococcus granulosus. Zhongguo Xue Xi Chong Bing Fang Zhi Za Zhi 25 (3), 284–286.

Dougherty, E.R., Carlson, C.J., Bueno, V.M., Burgio, K.R., Cizauskas, C.A., Clements, C.F., Seidel, D.P., Harris, N.C., 2016. Paradigms for parasite conservation. Conserv. Biol. 30 (4), 724–733.

Dugan, V.G., 2012. A robust tool highlights the influence of bird migration on influenza A virus evolution. Mol. Ecol. 21 (24), 5905–5907.

Galsworthy, S.J., ten Bosch, Q.A., Hoye, B.J., Heesterbeek, J.A., Klaassen, M., Klinkenberg, D., 2011. Effects of infection-induced migration delays on the epidemiology of avian influenza in wild mallard populations. PLoS One 6 (10), e26118.

Gao, J., Luo, J., Fan, R., Guan, G., Ren, Q., Ma, M., Sugimoto, C., Bai, Q., Yin, H., 2007. Molecular characterization of a myosin alkali light chain-like protein, a "concealed" antigen from the hard tick Haemaphysalis qinghaiensis. Vet. Parasitol. 147 (1–2), 140–149.

Hai-Juan, D., Li-Yong, W., Xiao-Nong, Z., 2019. Current status and new challenges of three important imported parasitic diseases. Zhongguo Xue Xi Chong Bing Fang Zhi Za Zhi 31 (4), 353–355.

Hoberg, E.P., Phillips, A.J., 2014. Transfer of the U.S. national parasite collection. Comp. Parasitol. 81 (2), 300–301.

Horak, P., Mikes, L., Lichtenbergova, L., Skala, V., Soldanova, M., Brant, S.V., 2015. Avian schistosomes and outbreaks of cercarial dermatitis. Clin. Microbiol. Rev. 28 (1), 165–190.

Hu, W., Yan, Q., Shen, D.K., Liu, F., Zhu, Z.D., Song, H.D., Xu, X.R., Wang, Z.J., Rong, Y.P., Zeng, L.C., Wu, J., Zhang, X., Wang, J.J., Xu, X.N., Wang, S.Y., Fu, G., Zhang, X.L., Wang, Z.Q., Brindley, P.J., McManus, D.P., Xue, C.L., Feng, Z., Chen, Z., Han, Z.G., 2003. Evolutionary and biomedical implications of a Schistosoma japonicum complementary DNA resource. Nat. Genet. 35 (2), 139–147.

Jesus, V., Martins, D., Branco, T., Valerio, N., Neves, M., Faustino, M.A.F., Reis, L., Barreal, E., Gallego, P.P., Almeida, A., 2018. An insight into the photodynamic approach versus copper formulations in the control of Pseudomonas syringae pv. actinidiae in kiwi plants. Photochem. Photobiol. Sci. 17 (2), 180–191.

Jouet, D., Kolarova, L., Patrelle, C., Ferte, H., Skirnisson, K., 2015. Trichobilharzia anseri n. sp. (Schistosomatidae: Digenea), a new visceral species of avian schistosomes isolated from greylag goose (*Anser anser* L.) in Iceland and France. Infect. Genet. Evol. 34, 298–306.

Kumar, N., Das, B., Solanki, J.B., Jadav, M.M., Menaka, R., 2017. Plastination of macroparasites: an eco-friendly method of long-term preservation. Vet. World 10 (11), 1394–1400.

Li, J., 1989. Establishment of animal models for the periodic type of *Brugia malayi* from different endemic areas of China. J. Dalian Med. Univ. 11 (3), 45 (in Chinese).

Li, Y., Wang, H., Jin, X.R., Li, X., Pender, M., Song, C.P., Tang, S.L., Cao, J., Wu, H., Wang, Y.G., 2018a. Experiences and challenges in the health protection of medical teams in the Chinese Ebola treatment center, Liberia: a qualitative study. Infect. Dis. Poverty 7 (1), 92.

Li, H.M., Qian, M.B., Yang, Y.C., Jiang, Z.H., Wei, K., Chen, J.X., Chen, J.H., Chen, Y.D., Zhou, X.N., 2018b. Performance evaluation of existing immunoassays for *Clonorchis sinensis* infection in China. Parasit. Vectors 11 (1), 35.

Liu, Q., Zhou, X.N., 2015. Preventing the transmission of American trypanosomiasis and its spread into non-endemic countries. Infect. Dis. Poverty 4, 60.

Liu, Z., Yin, L., Li, Y., Yuan, F., Zhang, X., Ma, J., Liu, H., Wang, Y., Zheng, K., Cao, J., 2016. Intranasal immunization with recombinant Toxoplasma gondii actin depolymerizing factor confers protective efficacy against toxoplasmosis in mice. BMC Immunol. 17 (1), 37.

Liu, Q., Chen, X.L., Chen, M.X., Xie, H.G., Liu, Q., Chen, Z.Y., Lin, Y.Y., Zheng, H., Chen, J.X., Zhang, Y., Zhou, X.N., 2018. Trypanosoma brucei rhodesiense infection in a Chinese traveler returning from the Serengeti National Park in Tanzania. Infect. Dis. Poverty 7 (1), 50.

Molina, G.A., Palacios Gordon, M.F., Jerez Ortiz, J.R., Chong Menendez, R., Constante Ruiz, J.E., Padilla Paredes, H.I., Guadalupe Rodriguez, R.A., 2019. Bowel obstruction due to the migration of the deflated intragastric balloon, a rare and potentially lethal complication. J. Surg. Case Rep. 2019 (4), rjz091.

Mordecai, E.A., Gross, K., Mitchell, C.E., 2016. Within-host niche differences and fitness trade-offs promote coexistence of plant viruses. Am. Nat. 187 (1), E13–E26.

National Scientific Data, 2018. National scientific data sharing platform for population and health. Chin. Med. Ethics 31 (5), 681 (in Chinese).

Nunn, C.L., Altizer, S.M., 2005. The global mammal parasite database: an online resource for infectious disease records in wild primates. Evol. Anthropol. Issues News Rev. 14 (1), 1–2.

Poulin, R., 2014. Parasite biodiversity revisited: frontiers and constraints. Int. J. Parasitol. 44 (9), 581–589.

Qian, M.B., Abela-Ridder, B., Wu, W.P., Zhou, X.N., 2017. Combating echinococcosis in China: strengthening the research and development. Infect. Dis. Poverty 6 (1), 161.

Qian, M.B., Zhuang, S.F., Zhu, S.Q., Deng, X.M., Li, Z.X., Zhou, X.N., 2019. Improving diagnostic performance of the Kato-Katz method for *Clonorchis sinensis* infection through multiple samples. Parasit. Vectors 12 (1), 336.

Rubinoff, I., Leigh Jr., E.G., 1990. Dealing with diversity: the Smithsonian tropical research Institute and tropical biology. Trends Ecol. Evol. 5 (4), 115–118.

Saegerman, C., Bertagnoli, S., Meyer, G., Ganiere, J.P., Caufour, P., De Clercq, K., Jacquiet, P., Fournie, G., Hautefeuille, C., Etore, F., Casal, J., 2018. Risk of introduction of lumpy skin disease in France by the import of vectors in animal trucks. PLoS One 13 (6), e0198506.

Scare, J.A., Steuer, A.E., Gravatte, H.S., Kalman, C., Ramires, L., Dias de Castro, L.L., Norris, J.K., Miller, F., Camargo, F., Lawyer, A., De Pedro, P., Jolly, B., Nielsen, M.K., 2018. Management practices associated with strongylid parasite prevalence on horse farms in rural counties of Kentucky. Vet. Parasitol. Reg. Stud. Rep. 14, 25–31.

Shen, H.M., Hu, W., Chen, S.H., Zhen, Q., Zhou, X.N., Shen, H.M., Hu, W., Chen, S.H., Zhen, Q., Zhou, X.N., 2007. The development of parasite resourse information command system using J2EE technique. Chin. J. Vet. Parasitol. 15 (1), 8–12 (in Chinese).

Si, Y., Skidmore, A.K., Wang, T., de Boer, W.F., Debba, P., Toxopeus, A.G., Li, L., Prins, H.H., 2009. Spatio-temporal dynamics of global H5N1 outbreaks match bird migration patterns. Geospat. Health 4 (1), 65–78.

Smithsonian Institute, 1920. International exchange service. Bull. Med. Libr. Assoc. 9 (4), 46.

Song, L.G., Zeng, X.D., Li, Y.X., Zhang, B.B., Wu, X.Y., Yuan, D.J., He, A., Wu, Z.D., 2018. Imported parasitic diseases in mainland China: current status and perspectives for better control and prevention. Infect. Dis. Poverty 7 (1), 78.

Spencer, H.G., Zuk, M., 2016. For Host's sake: the pluses of parasite preservation. Trends Ecol. Evol. 31 (5), 341–343.

Stephens, P.R., Pappalardo, P., Huang, S., Byers, J.E., Farrell, M.J., Gehman, A., Ghai, R.R., Haas, S.E., Han, B., Park, A.W., Schmidt, J.P., Altizer, S., Ezenwa, V.O., Nunn, C.L., 2017. Global mammal parasite database version 2.0. Ecology 98 (5), 1476.

Stothard, J.R., Littlewood, D.T.J., Gasser, R.B., Webster, B.L., 2018. Advancing the multidisciplinarity of parasitology within the British Society for Parasitology: studies of host-parasite evolution in an ever-changing world. Parasitology 145 (13), 1641–1646.

Stringer, A.P., Linklater, W., 2014. Everything in moderation: principles of parasite control for wildlife conservation. Bioscience 64 (10), 932–937.

Torgerson, P.R., Ruegg, S., Devleesschauwer, B., Abela-Ridder, B., Havelaar, A.H., Shaw, A.P.M., Rushton, J., Speybroeck, N., 2018. zDALY: an adjusted indicator to estimate the burden of zoonotic diseases. One Health 5, 40–45.

True, F.W., 1885. The British museum of natural history. Science 6 (132), 127–130.

Wang, X., Ruan, Q., Xu, B., Gu, J., Qian, Y., Chen, M., Liu, Q., Lu, Q., Zhang, W., 2018. Human African trypanosomiasis in emigrant returning to China from Gabon, 2017. Emerg. Infect. Dis. 24 (2), 400–404.

Wei, J., Gu, Y., Yang, J., Yang, Y., Wang, S., Cui, S., Zhu, X., 2011. Identification and characterization of protective epitope of Trichinella spiralis paramyosin. Vaccine 29 (17), 3162–3168.

Yang, G.J., Utzinger, J., Zhou, X.N., 2015. Interplay between environment, agriculture and infectious diseases of poverty: case studies in China. Acta Trop. 141 (Pt. B), 399–406.

Ying-Dan, C., Ting-Jun, Z., Long-Qi, X., Bin, Z., Yan-Hong, X., Chang-Hai, Z., 2017. Interpretation of diagnostic criteria for clonorchiasis. Zhongguo Xue Xi Chong Bing Fang Zhi Za Zhi 29 (5), 538–540.

Zhou, X.N., 2008. Standards for Description of the Parasitic Germplasm Resources. Shanghai Science and Technology Press, Shanghai.

Zhou, X.N., 2009. Technical Protocols for Preservation of the Parasitic Germplasm Resources. Shanghai Science and Technology Press, Shanghai (in Chinese).

Zhou, X.N., Lin, J.J., Cao, J.P., Huang, B., Chen, S.H., Zhang, Y., Hu, W., 2006. Theory and practices of the infrastructure establishment on parasite germplasm resources in China. Chin. J. Parasitol. Parasit. Dis. 24 (S1), 1–10 (in Chinese).

Zhou, X.J., Zheng, B., Yi, F.Y., Xiong, Y.H., Zhang, M.Q., 2015. Analysis on research projects supported by the national natural science foundation of China at the national institute of parasitic diseases during 2003-2013. Zhongguo Ji Sheng Chong Xue Yu Ji Sheng Chong Bing Za Zhi 33 (2), 151–153.

CHAPTER THIRTEEN

Contribution of NIPD-CTDR to the parasitic diseases control and elimination in China: Memory of the 70th anniversary for NIPD-CTDR

Ning Xiao*[,†], Shi-Zhu Li[†], Men-Bao Qian[†], Zhi-Gui Xia[†], Qing Yu[†], Qin Liu[†], Shan Lv[†], Xiao-Nong Zhou*

National Institute of Parasitic Diseases, Chinese Center for Disease Control and Prevention; Chinese Center for Tropical Diseases Research; Key Laboratory of Parasite and Vector Biology, Ministry of Health; National Center for International Research on Tropical Diseases, Ministry of Science and Technology; WHO Collaborating Centre for Tropical Diseases, Shanghai, People's Republic of China

*Corresponding authors: e-mail address: xiaoning@nipd.chinacdc.cn; zhouxn1@chinacdc.cn

Contents

1. Introduction 402
2. Achievements of parasite disease control programmes in China 404
3. Challenges for parasitic disease control 409
4. Roadmap for parasitic diseases control 413
5. Role and how to influence the global situation of the new roadmap from China 420
6. Conclusion and perspectives 422
References 423
Further reading 427

Abstract

China has achieved a great success in control and elimination of key parasitic diseases. In 2007, the elimination of lymphatic filariasis was verified by WHO. The schistosomiasis incidence and snail-distributed areas have reduced to the lowest level in the history. The transmission and disease burden of echinococcosis have been contained largely, and the populations infected with soil-transmitted trematode and food-borne parasites have also shown a significantly declining trend. Because of rapid globalization and climate changes, however, many new challenges have arisen. In his paper, the 2020–2030 roadmaps towards the control and elimination of these key parasitic diseases are

[†] These authors equally contribute to this paper.

Advances in Parasitology, Volume 110
ISSN 0065-308X
https://doi.org/10.1016/bs.apar.2020.02.003

described. Moreover, China is actively implementing its global health strategy, and will be more and more engaged into global health affairs, in which a series of China-Africa health cooperation projects have been in planning with a wish of making a greater contribution to the SDGs.

1. Introduction

Parasitic diseases have been one of the most important global public health issues. Because these diseases are closely linked with many soc-economic factors such as poor health knowledge, poor sanitation facilities, and unhealthy life and production ways as well as poverty, the infected populations are mostly living in tropical, sub-tropical and remote areas where are far behind developed. The parasitic diseases severely influence the children's growing, the labour force of the adults, and health of all age groups even life lost, which causes a very high disease burden. Since the establishment of the People's Republic of China (China), under the leadership of Chinese Communist party and governments, the stage-specific strategies and measures have been taken including local conditions based and integrated interventions in China. In 2007 lymphatic filariasis has been eliminated nationwide. Since then, no locally-transmitted cases have been reported yet. The incidence and snail-distributed areas of schistosomiasis have gone down to the lowest point in the history. The transmission and disease burden of echinococcosis have been contained primarily. The populations affected by the soil-transmitted trematode have significantly declined in numbers and density. The infections with food-borne parasites show a down trend as well. Currently three prioritized parasitic diseases, schistosomiasis, malaria and echinococcosis, have been in the list for well control and elimination with a definitive time. According to the China national health action plans and goals, malaria will be eliminated by 2020, and schistosomiasis will be eliminated by 2030, while echinococcosis, soil-transmitted helminths and food-borne parasitic diseases be effectively under control.

Schistosomiasis caused by *Schistosoma japonicum* was one of major infectious diseases in China and endemic over 2100 years (Zhou et al., 2005). This disease caused a plenty of deaths and destroyed many rural villages. The disease was most endemic in the valley of Yangtze River and the south, including 12 provinces, i.e., Hubei, Hunan, Jiangxi, Anhui, Jiangsu, Zhejiang, Yunnan, Sichuan, Fujian, Guangdong, Guangxi and Shanghai. It was estimated that more than 12 million patients (with 5% of advanced cases) in 322 counties and 100 million people were at risk of schistosomiasis in mid-1950s (Ross et al., 1997). In addition, one million cattle were recorded with the infection. The areas inhabited by *Oncomelania hupensis*, the intermediate host of *S. japonicum*, were 14.3 billion m^2. Schistosomiasis strikingly threatened the health of human and livestock and hence hindered the socio-economic development.

Malaria was once seriously epidemic in China, mainly in 24 provinces. After the establishment of the People's Republic of China, the government at all levels have taken consistent efforts in malaria control and made remarkable achievements (Tang, 1999; Zhou, 1981). Particularly in 2010, China released the China Malaria Elimination Action Plan (2010 – 2020), which is a milestone from malaria control to elimination in China. The action plan calls for two-steps elimination process. By 2015, most parts achieve malaria-free, and by 2020, all of the country achieves the goal of eliminating malaria. Through the implementing the action plan, China achieved zero report of local malaria cases in 2017, and continues maintaining no locally-transmitted malaria case. China's strategies, measures and roadmap for malaria control and elimination are of great significance to other low and middle-income epidemic countries for reference (Feng et al., 2018; MOH, 2010).

Echinococcosis (hydatid disease) is a vital chronic zoonotic parasitic disease, which is widely endemic in the areas with animal husbandry in the world. China has a high prevalence of human cystic echinococcosis (CE) and one of the highest prevalence of human alveolar echinococcosis (AE) worldwide. A Chinese national echinococcosis survey of 2012–2016 indicated that the prevalence of population was 0.28% and 50 million people at risk, and more than 98% of the cases were distributed in the western agricultural and pastoral areas, Inner Mongolia, Sichuan, Tibet, Gansu, Qinghai, Ningxia and Xinjiang. The prevalence appears a westward spreading trend. Therefore, according to the phase goal for echinococcosis control to be achieved in 2020, the Chinese government has formulated a prevention and control plan of 2016–2020 (NHC, 2016; Wang, 2016; Wu et al., 2018).

Soil-transmitted helminthiases and food-borne helminthiases were historically highly prevalent in China (Qian et al., 2019). Soil-transmitted helminthiases once distributed around the country, but nowadays most cases distribute in less developed southwestern areas (Zhou, 2018). Clonorchiasis is caused by ingestion of raw freshwater fish containing the larvae of *Clonorchis sinensis*. It is still ranking the most important food-borne parasitic diseases in eastern China (Qian et al., 2012, 2013, 2016).

Massive control activities promoted the decrease of *Taenia solium* taeniasis and cysticercosis in China especially in northern areas. However, local endemic focuses are still demonstrated in western China where traditional pig husbandry, poor sanitation and the ingestion of raw pork coexist (Qian et al., 2019).

This paper summarizes the achievements made and the lessons learned during the process of parasitic diseases control in China till now, and afterwards to describe a roadmap and strategy for next decade programmes. It aims to provide a technical support in reaching the goals of the Health China 2030 and the sustainable development goals (SDGs) inclusive of tropical diseases control approved by the United Nations.

2. Achievements of parasite disease control programmes in China

Chinese government have paid high attention to the health issues caused by the key parasitic diseases and consistently taken actions to control. After seven decades of efforts with particular in this century, great achievements have been made in control of the key parasitic diseases such as schistosomiasis, malaria, lymphatic filariasis and echinococcosis. Regarding to lymphatic filariasis with an over 50-years period and three phases of control activities including preparation, control and surveillance, the incidence was made a sharp reduction from over 30 million of cases at the beginning of P.R. China establishment down to zero infection, and got the WHO certification on the 9th May 2007, which marks a successfully complete block of filariasis transmission and realizes the lymphatic filariasis-free nationwide. Afterwards, through sustainable epidemiological surveillance and care for the chronic patients, the elimination status has been effectively maintained.

A great success in schistosomiasis control has also been achieved in China. First, the epidemiology of schistosomiasis declines to the lowest level. By the end of 2018, five provinces, Shanghai, Zhejiang, Fujian, Guangdong and Guangxi, maintain the elimination over 20 years. Sichuan province has reached the criteria of the transmission interruption. The rest of six provinces reached the criteria of the transmission control. Up to now, the elimination or transmission interruption was achieved in 387 (86.0%) out of 450 endemic counties. The remaining 63 counties achieved the criteria of the transmission control. The number of human cases reduced to 37,601 (Zhang et al., 2020). No acute human cases, infected snails or livestock have been reported. Second, the control strategy of infection sources implemented since 2006 has been effective. Only 14 parasitologically-diagnosed patient were found from 8.4 million of populations in 2017, which reduced by 97.67% compared to 2016 (Zhang et al., 2017, 2018a,b). A total of 737,016 cattle were registered in endemic areas in 2017. Only one out of 454,830 examined cattle was found infected. No infected snail has been found since 2013 across the country. Third, the integrated measures were conducted smoothly. A total of 184 thousand of cattle were fenced in endemic areas (Zhang et al., 2017, 2018a,b). There 1.8 million acres of paddy field were transformed into dry land. The area of newly established fishing ponds was 1.4 million acres. A total of 2679 km of irrigation channels and 1004 km of rivers were consolidated to destroy the snail habitats. A supply of safe water covered 3.9 million of people in the endemic areas. The forest for prohibiting infestation of snails reached 1.4 billion m^2. Fourth, the knowledge rate was significantly elevated. The knowledge rate of schistosomiasis control among students and residents were over 90%.

As one of the most important tropical diseases in the history, the main achievements in malaria prevention and control are as follows: first, the number of malaria cases decreased significantly from more than 30 million a year in the 1940s to zero in 2007. In the early days of the founding of the People's Republic of China, the number of the cases ranked first among all kinds of infectious diseases. Through nearly 70 years of continuous efforts from control to elimination, the epidemic situation has been effectively controlled (Yin et al., 2014). Second, the malaria types and the mortality gradually decreased. In 2017, China reported no locally-transmitted cases, while in 2018 China reported 2678 malaria cases, including 5 clinical, 397 *P. vivax*, 1765 *P. falciparum*, 83 *P. malariae*, 376 *P. ovale* and 52 mixed infections, but all of them were the imported cases either. It has been the second consecutive year of no reported local infections since 2017, indicating a remarkable achievements made (Zhang et al., 2018a,b, 2019).

To date, although it is comparatively slow, echinococcosis control has made an obvious progress in China. First, since the national echinococcosis control programme started in 2006, the project implementation counties have extremely expanded from 10 counties in 2006 to 370 in 2019, with a total of 278.5×10^4 person-time accumulated screening on populations by type-B ultrasonic examination, 2.23×10^4 person-time accumulated treatment with albendazole drugs, 2.66×10^4 accumulated patients operated by surgery, 190.1×10^6 times dogs population to be given deworming, and 4.73×10^4 registered patients (data till to 2018) during this periods. Furthermore, the a great effect has been achieved that the prevalence rate of the residents living in the endemic areas showed 74.07% decreasing from 1.08% in 2004 to 0.28% in 2016 by combined efforts of multi-sectors including agriculture, water conservancy and education at all the levels (NHC, 2006; Wu et al., 2018; Yu et al., in press). To ensure the national echinococcosis control programme to be implemented as expected, according to incomplete statistics during 2005 to 2019, the central government only has invested a total of 1.5 billion Yuan (RMB) on comprehensive measures consisting of the patients' screening, treatment and management, dogs' deworming and containment, surveillance of endemic situation and factors, and the capacity building inclusive of laboratory construction accordingly. Second, the nationwide surveillance network has been set up to monitor the prevalence and related risky factors in all the endemic counties. In 2015, 20 sentinel sites covering Inner Mongolia, Sichuan, Yunnan, Tibet, Gansu, Qinghai, Ningxia, Xinjiang and Xinjiang Production and Construction Corps (XPCC) were selected for pilot, while in 2016, Chinese Center for Disease Control and Prevention (China CDC) released the national surveillance guideline (as 2016 edition) and the surveillance counties expanded to 140. According to the surveillance data in 2017, it indicated that a higher prevalence of echinococcosis appeared in the areas of the Qinghai-Tibet plateau giving a prevalence rate of 0.56% in human population, a positive rate of 3.21% in the dogs' faecal antigen test, a positive infection rate of 2.08% in domestic animals (yaks or sheep) and a positive infection rate of 2.00% for rodents including pikas. Based on the results collected by this surveillance network, consequently, the National Health Committee of China decides to set up sentinel sites throughout all the endemic counties in 2019 (NHC, 2010). Third, the Provincial Reference Laboratory for diagnosis and test of the echinococcosis pathogens have been established. To strengthen the capacity for the pathogen detection, an establishment of the network of the reference laboratory at provincial level has been in processing in the endemic provinces since 2015. Main tasks are to improve equipment in laboratory and professional training. By 2019, five the reference laboratories have been functioning in Qinghai, Gansu, Sichuan, Xinjiang Ningxia, respectively. Four, based on the effects and key role of the joint control mechanism with focus on control at adjacent zone of the severe endemic between different provinces/autonomous regions as done for schistosomiasis, malaria and other parasitical diseases, the same cooperation mechanism was first carried out in the five provinces of Tibet, Qinghai, Sichuan, Gansu and Yunnan, in which the synchronous activities are done for dog deworming, patient screening, health education as well as information sharing and communications. Moreover, in 2017 the national expert committees for echinococcosis control and clinical treatment were set under the leadership of the National Health Committee, which strengthens technical supports for the intervention and management of canines infectious source, patients' treatment by drugs, surgery and others. Five, along with the demonstration areas setting up, many innovative techniques for interventions and diagnostic tools have been developed. A good research cooperation mechanism has been established among varies of institutions and universities in China, which most are funded by the National nature and science funds and the national key project, respectively. In particular, to guarantee the target of 2020 to be achieved, the National Health Committee attempts to establish several comprehensive demonstration areas of echinococcosis control and prevention, with a pilot in Shiqu County of Sichuan Province in 2015. And the areas have subsequently been expanded to 28 areas in 2019. In these areas, a series of techniques or products have been developed for field use. The electronic collar and ID identification system for dogs management, an instant diagnosis tool with high sensitivity, specificity and convenience, a iris recognition techniques in combination with fingerprint identification for population screening, molecular detection for the infections of livestock and rodents, praziquantel-laced baits for deworming of stray dogs and wild canines delivered by the unmanned aerial vehicle (UAVs), as well as varieties of health education products have been developed and tested at these zones. All in all, a group of innovations have played a key role not only in the demonstration areas but also in other endemic areas when have been transformed and applied (Jiang et al., 2017).

Soil-transmitted helminthiases were historically highly endemic in China (Qian et al., 2019). Three national surveys have been implemented in China, which promote the mapping as well as subsequent interventions. During 1988–1992, the first national survey on important parasitic diseases was carried out, in which the overall prevalence of soil-transmitted helminthiases reached 53.6%, namely, 646 million of people under infection (Xu et al., 1995). High infection was demonstrated in whole country. The prevalence of *Ascaris lumbricoides*, *Trichuris trichiura* and hookworm infections were 47.0%, 18.8% and 17.2%, respectively. Then, the second national survey was conducted in 2001 and 2004, according to which the prevalence of soil-transmitted helminthiases declined to 19.6% overall and about 129.0 million of persons were estimated to be infected (Coordinating Office of the National Survey on the Important Human Parasitic Diseases, 2005). Correspondingly, the prevalence of *A. lumbricoides*, *T. trichiura* and hookworm infections was 12.7%, 4.6% and 6.1%, respectively. Most of the infections distributed in southern areas. The third national survey was carried out during 2014–2015, in which the prevalence of soil-transmitted helminthiases further dropped to 4.5% and there was an estimation of 29.1 million of people under infection (Zhou, 2018). The prevalence of three species was 1.4%, 1.0% and 2.6%, respectively. Nowadays, soil-transmitted helminthiases are majorly endemic in southwestern China, especially in Sichuan, Yunnan and Guizhou provinces. Clonorchiasis is currently the most important food-borne trematodiasis in China (Qian et al., 2016). Compared to the significant control of many other parasitic diseases in China, the epidemiological status of food-borne clonorchiasis fluctuates (Qian et al., 2019). The prevalence of clonorchiasis was 0.31% in the first national survey between 1988 and 1992 (Chen et al., 2012). High endemicity was demonstrated in the southeast and northeast areas, as well as some central areas. Because of the more accessibility to freshwater fish owing to aquatic development, the prevalence of clonorchiasis increased to 0.58% in the second national survey in 2001–2004 (Chen et al., 2012). During 2001–2004, another special survey was conducted for clonorchiasis in 27 endemic provinces, in which an estimated prevalence reached 2.4% (Chen et al., 2012). Therefore, 12.5 million of people were estimated under infection with *C. sinensis*. Most cases came from four provinces in southeastern and northeastern areas, namely, Guangdong, Guangxi, Heilongjiang and Jilin. Then, according to the updated survey conducted in 2014–2015, the prevalence decreased to 0.47% and about 6.0 million people were under infection (Zhou, 2018). Most cases distributed still in the above four provinces. The significant decrease is attributed to recent implementation of control activities in some endemic areas. However, the number under infection might be underestimated.

Historically, *T. solium* Taeniasis and cysticercosis were also high endemic in China, especially in northern parts (Qian et al., 2019). Owing to the awareness on the importance of *T. solium* cysticercosis, large-scale intervention was adopted between 1970s and 1990s. Core measures included the deworming for persons with taeniasis, inspection of pork, management of human faeces and pigs, and treatment and management of infected pigs (Ma et al., 1992). Owing to these measures, *T. solium* cysticercosis had decreased significantly in China, especially in northern areas. As to the three national surveys on important parasitic diseases in China, the estimated population infected with *Taenia* spp. decreased from 1.3 million in 1988–1992 to 0.55 million in 2001–2004 and further to 366,200 in 2014–2015 (Hotez et al., 1997; Yu et al., 1994; Zhou, 2018). However, all three species of *Taenia* spp. (*T. solium*, *T. saginata* and *T. asiatica*) are endemic in China, and most of the cases belong to *T. saginata* and *T. asiatica* (Li et al., 2013a,b). Although the accurate map for *T. solium* cysticercosis is not available, it is nowadays majorly endemic in western China where there exist traditional pig husbandry, poor sanitation and ingestion of raw pork (Zang et al., 2019a,b).

3. Challenges for parasitic disease control

It is obvious that great achievements on control of parasitic diseases have been made in China currently. However, because a lot of changes in public health system and heath policy followed by the economic reform and opening to the outside world, many new challenges have arisen, particular in control of schistosomiasis, malaria and echinococcosis.

The challenges to schistosomiasis control include: (1) The transmission of schistosomiasis is complex with many mammals involved as the reservoirs of *S. japonicum* (Chen, 2014). Although schistosomiasis has well controlled in human and livestock in China, many risk factors still remain. (2) It is very difficult in shrinking the distribution of snail due to a remarkably large range and complex environment. The distribution of snail was around 0.36 billion m^2 in last decades (Wang et al., 2015). In addition, re-emergency of snails are frequently observed in some endemic areas (Liu et al., 2008). (3) The infections among high-risk population are still a big concern in some region. A survey conducted in 2013 indicated that the prevalence of schistosomiasis among fishing and boating people was 5.62%, while the acute infections in human accidentally occurred, such as 13 and 9 acute human cases were reported in 2012 and 2013, respectively (Li et al., 2013a,b, 2014). (4) It is difficult in maintaining the achievement in newly transmission-interrupted areas. Some new infections and re-occurrence of snails were observed in abovementioned areas. As the environments suitable for snail infestation are not completely changed, the transmission of schistosomiasis is thus possible. (5) The capacity of professional agencies and the surveillance system is weakened due to declining investment as well.

The main challenges to malaria prevention and elimination are as follows. First, the number of imported malaria cases is increasing, especially for increasing of the number of the imported cases from Africa year by year, which brings severe challenges to the elimination of malaria in China. At present, there are nearly 3000 imported malaria cases in China every year, including *P. falciparum*, *P. vivax*, *P. ovale*, *P. malariae* and *P. knowlesi,* posing a potential risk of malaria transmission in areas with vectors' distribution (Yin et al., 2013). Second, after the elimination of malaria, the prevention and control capacity might reduce but the risk of re-transmission remains. Therefore, we still need emphasizing the importance of malaria surveillance and response, and maintaining the essential institutions and capacity, strengthening government leadership, multisectoral cooperation, mobilizing the whole society's participation, and enhancing the joint efforts along the border. Otherwise, we may fail to consolidate the malaria-free achievement (Li et al., 2016).

The challenges for echinococcosis control are as follows. First, it is hard to control definitive hosts' infections. Regarding to a huge number and wide distribution of domestic and stray dogs as well as wild animals (wolves and foxes), it is a big challenge to deal with about 3 million dogs in the endemic areas, which cause environment contaminated seriously (Liu et al., 2017; Xiao et al., 2004, 2006). In addition, under the plateau climatic conditions, it extends the survival of echinococcial eggs, which benefits to maintain the transmission. It is also make the situation worse with a longer breeding cycle in large number of yaks, sheep and other livestock and habit to use animal organs feeding on canines animals (Xiao et al., 2003). Second, a large number of patients have been detected with a huge disease burden and need to be treated effectively. According to national surveys and population screening in the endemic areas, more than 40,000 patients have recorded in the whole country, and since 2011, there has nearly been 3000 new patients diagnosed every year, all of whom wait for medical assistance. Third, the residents have poor knowledge of echinococcosis control and prevention. Based on the relevant surveys, the results indicated that the local residents' awareness of the prevention, hygienic conditions and living habits and customs are very poor, particular in the Qinghai-Tibet plateau region. Fourth, innovation is very crucial for speed-up the disease control. Echinococcosis is one of chronic diseases with a long-term incubation period (up to 30 years). It is very difficult in definitive diagnosis at an early-developed stage due to lack of good tools. Surgery is a radical treatment approach but reports with 15–20% of recrudescence (Eckert and Deplazes, 2004; McManus et al., 2003; Schantz et al., 2003; Torgerson et al., 2010; Xiao et al., 2012; Zhang et al., 2015).

Although soil-transmitted helminthiases continue decreasing in China, there exist some challenges for further control. First, a high imbalance is demonstrated in geographic distribution of soil-transmitted helminthiases in China (Qian et al., 2019). Significant control has been achieved in developed eastern China and the prevalence even approaches to zero in some provinces, but a high endemicity still occurs in southwestern China, which impacts the national control progress. Second, the composition of dominant species has changed. In first national survey, ascariasis and trichuriasis were the most important (Xu et al., 1995). Currently hookworm infections, however, have become dominant, and this challenges the traditional tools and strategies which are more effective against ascariasis and trichuriasis. However, social and economic development and subsequent improvement of water and sanitation are crucial contributors to further control and even transmission interruption of soil-transmitted helminthiases in China (Qian et al., 2015).

Unlike the significant control of many other helminthiases owing to massive control activities and economic development, the determinants of clonorchiasis persist in the endemic areas in China. First, aquatic development provides the high affordability of raw freshwater fish, that's why the endemicity of clonorchiasis even worsens in some areas (Qian et al., 2014). New techniques promote the cultivation of freshwater fish in many big natural water bodies, which increases the challenges in management of human and animal faeces as transmission sources. Second, the ingestion of raw freshwater fish is rooted in habitants in endemic areas, which is very difficult to change (Qian et al., 2016). Thus, although chemotherapy is high effective, it is usually unsustainable due to high re-infection. Third, compared to other parasitic diseases, the research and development on tools and strategies for clonorchiasis lag behind, which could not meet the need for control (Qian and Zhou, 2019). Thus, on the one hand, more researches are expected to develop new tools and strategies. On the other hand, high priority and more resources should be advocated to control clonorchiasis.

Many determinants drive the persistent endemicity of *T. solium* cysticercosis in local areas in western China (Zang et al., 2019a,b). There still exists traditional pig husbandry, and pigs could roam free in villages. Less economic development in remote areas leads to the inadequate accessibility to sanitation, and thus there exists open defecation. Thus, roaming pig could access to human faeces. The pork inspection is still inadequate in remote markets. Furthermore, the ingestion of raw pork is still popular in some areas. These factors drive local completion of the cycle of human taeniasis, *T. solium* cysticercosis in human beings and pigs. However, national poverty alleviation programme is improving the living conditions including husbandry and sanitation in these less developed areas, which contributes to the control of *T. solium* cysticercosis there.

Additionally, China is also facing an increasing threat of imported parasitic diseases. Some rarely reported or even non-epidemic diseases have been detected, which brings great challenges to the prevention and control of these parasitic diseases in China. Human African trypanosomiasis (HAT) is one of the representatives. Human African trypanosomiasis, also known as sleeping sickness, is a human zoonosis which is transmitted by tsetse fly. The disease is mainly prevalent in 36 sub-Saharan African countries (WHO, 2013) and caused by two sub-species of *Trypanosoma brucei*-*T. brucei gambiense* causes chronic disease in western and central Africa and *T. b. rhodesiense* is associated with acute disease in eastern and southern Africa (Büscher et al., 2017). Since 2014, three cases of imported HAT have been diagnosed and treated in China (Chen et al., 2019c; Liu et al., 2018; Wang et al., 2018). The first case was a 45 years old man who acted as a sailor working on a freighter in Gabon from July 8, 2010 to July 15, 2014. During this period, he often shuttled in tropical jungle and valley with the bites of mosquitoes and flies. In October 2012, he got fever and skin itchy in Gabon and was treated locally pertain to his symptoms. On July 15, 2014, he got fever, drowsiness and mental symptoms and generalized lymphadenopathy. On September 9, 2014, Trypanosoma was detected in the blood smear of the patient and the white blood cell in the cerebrospinal fluid was 90. So, according to the clinical manifestations, epidemiology and laboratory test results it was confirmed that was the first *T. brucei gambiense* infection case in China, and it was in the second stage. The second case was a 41 years old women, who was visited Kenya and Tanzania from July 22 to August 6, 2017. During this process, she was bitten by tsetse fly. After returning to China on August 8, she had a high fever (40.1 °C) along with symptoms of dizziness, fatigue and rigours. On August 11, 2017, Trypanosoma was detected in the blood smear of the patient and the white blood cell in the cerebrospinal fluid was 2. According to the clinical manifestations, epidemiology and laboratory test results, this case was the first imported *T. brucei rhodesiense* infection in China. And it was in the first stage. The third case was 59 year old men, who was also a sailor working in Gabon from 2009 until 2017. One year ago (in 2016), he was bitten by insect like tsetse fly in Gabon and developed fever (up to 40 °C) and fatigue, headache. Since then the patient's body temperature has been repeated with a high fever, mainly in the afternoon, accompanied by night sweats. Two weeks ago, the patient slept in a coma. On August 30, 2017, Trypanosoma was detected in the patient's blood smear and the white blood cell in the cerebrospinal fluid was 119. So, according to the clinical manifestations, epidemiology and laboratory test results it was confirmed that was the second *T. brucei gambiense* infection case in China, and it was in the second stage.

All these cases were treated by the related drugs which delivered from WHO. Two cases were cured after 2 years follow up, but it was a pity that the first case died due to treatment failure. In the process of treatment of the above cases, there was a thorny problem that the effective drugs such as eflornithine, pentamidine, suramin, nifurtimox and others. Those mentioned above are not reserved in China and it was very complicated for applying them for entry to China, which will waste time and brings huge risks to save the life of the critical patients.

4. Roadmap for parasitic diseases control

Every 5 years, the Chinese central government formulates the control plans regarding to different groups of key parasitic diseases. These plans provide the governments at all levels with strategic direction and guidance to take action towards the goals.

Schistosomiasis: the national conference on schistosomiasis control held by the State Council in November 2014 set forth the goal of elimination, which symbolizes a new stage of schistosomiasis control. Three years later the planning framework of Health China 2030 was issued by the Central Committee of the Communist Party and the State Council of China definitively declares to eliminate schistosomiasis by 2030. The 13th Five-Year Plan of Schistosomiasis Control and the Special Three-Year Tough Plan of Endemic Diseases Control (2018–2020) also clarified a roadmap of schistosomiasis control in China (Fig. 1). According to the plan, the elimination or transmission interruption will be achieved in 437, accounting for 96.5% of counties by 2020; namely, no new infection in human, livestock or snail in those counties was found over five consecutive years. The schistosomiasis elimination strategy is mainly based on following

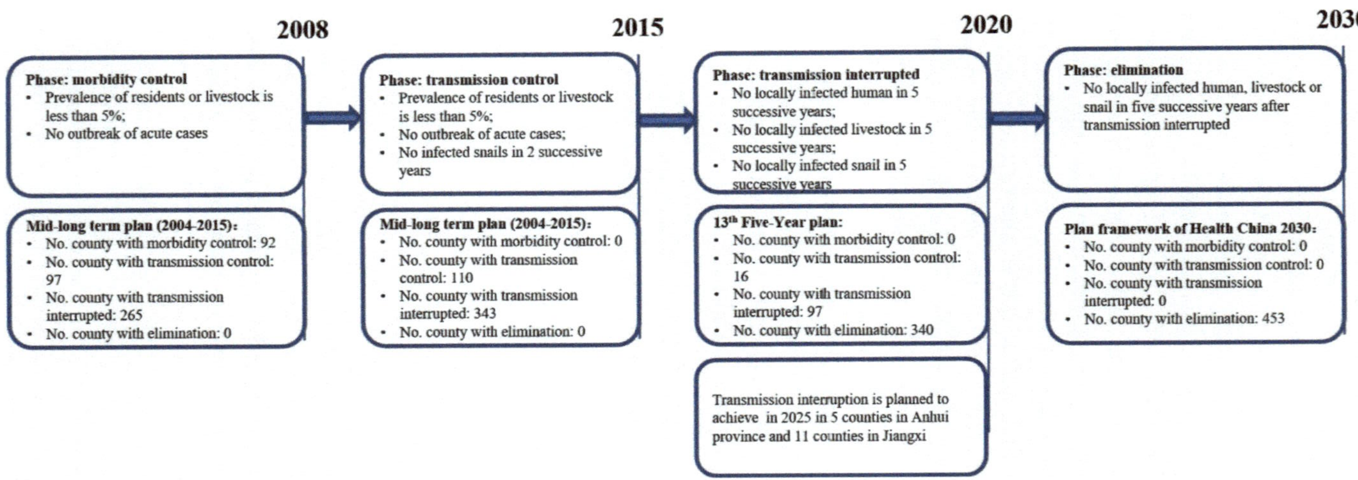

Fig. 1 Roadmap of schistosomiasis control and elimination.

principles: (1) the government leads schistosomiasis control and multiple departments participate, (2) the prevention is first and the treatment is supplementary, (3) the strategy is differentiated for endemic types and the management is differentiated in provinces, (4) multiple measures and goal-oriented management are implemented. Six approaches are recommended to the goal of elimination as following:

First, to strengthen case management and precision strategy. Treatment of symptomatic patients is taken into consideration in the National Program of Poverty Alleviation. Schistosomiasis is listed among diseases supported by the national basic health insurance and assistance plan. The patients including asymptomatic infections will be registered and followed up by community health facilities.

Second, to strengthen implementation of integrated measures focusing on the management of infection source. In endemic area the local government promotes the stable breeding of livestock and forbid the grazing in grassland infested by *O. hupensis*. Replace of farming cattle with machine will be financially supported by government. The utilization of faeces container on boat will be encouraged. The snail control will rely on the joint projects of water conservancy, agriculture, land management and forestry.

Third, to strengthen monitoring and evaluation (M&E). M&E will cover all the endemic communities and run in an informationized system. The M&E information in schistosomiasis control will be shared through the management system of Health and Medical Assistance Program for Poverty Alleviation.

Fourth, to strengthen the mobilization and knowledge delivery. Diversified materials of health education will be developed and forward through multiple media, particularly using popular platforms, e.g., wechat and blog.

Fifth, to strengthen capacity building. The capacity will be maintained or enforced in the counties by the national surveillance and annual national match on pathogen detecting techniques. The national network of labs has also been established to conduct annual training.

Sixth, to enhance research and translation. The following field will be given priorities: elimination roadmap and the verification tools, crucial technology in elimination of schistosomiasis in livestock, spatio-temporal evolution of control and elimination roadmaps, the national survey and collection of specimens of *S. japonicum* and *O. hupensis*, testing of compatibility between *S. japonicum* and *O. hupensis*, sensitive diagnostics for mild infection and asymptomatic infections, development and evaluation of security and effective molluscicide. Establishment of the bio-bank of

S. japonicum, *O. hupensis* as well as positive serum and faeces from the patients is in planning to provide the materials for further research.

Malaria: according to China Malaria Elimination Action Plan (2010–2020) and the report on malaria epidemic situation from 2006 to 2008, the country is divided into the following four categories by county: Type 1 county with local infection in 3 years and the incidence rate is greater than or equal to 1/10,000. Type 2 county, at least 1of 3 years with local infection and the incidence less than 1/10,000. Type 3 counties, no local cases for 3 years. Type 4 counties, non-malaria-endemic areas. Different malaria elimination roadmaps and strategies have been developed according to the type of endemic counties. In Type 1 counties, to strengthen the source of infection control and vector control measures to reduce the incidence of malaria. In Type 2 counties, to remove the source of malaria infection and block the local transmission of malaria. In Type 3 counties, to strengthen surveillance and imported cases disposal to prevent secondary transmission. In Type 4 counties, to take a quick response to any imported cases for detection and treatment. Based on the elimination roadmap, the Type 3 counties achieve the goal of eliminating malaria by 2015, while all the Type 2 counties and the Type 1 counties to achieve no local cases of malaria by 2015 and the goal of eliminating malaria by 2018, with an exception of some border areas in Yunnan. Type 1 counties in the border areas in Yunnan achieve the incidence of malaria less than 1/10,000 by 2015, no local malaria cases by 2017, and the goal of eliminating malaria by 2020 (Fig. 2) (Tang, 2016).

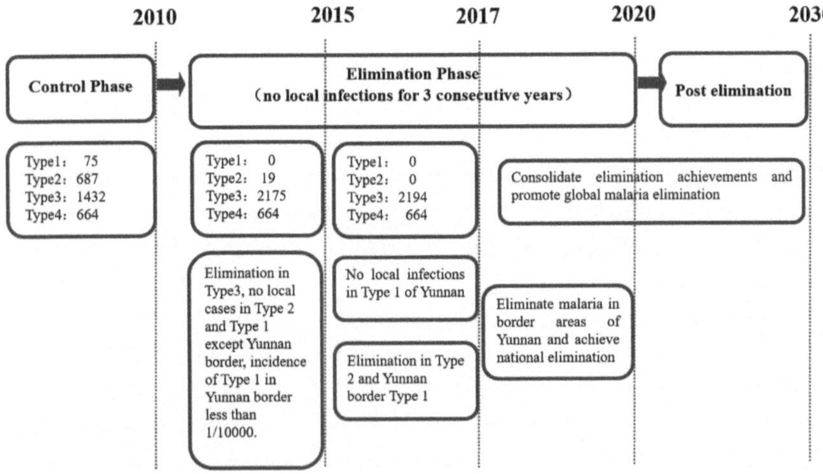

Fig. 2 Roadmap of malaria elimination in China.

From 2020 to 2030, it will continue to strengthen capacity of the public health system and malaria surveillance system, timely take action to the threat of imported malaria, and respond to the construction of human destiny community and the "healthy silk and road" initiative, actively participate in global health cooperation, provide technical support to Africa and southeast Asia in malaria control and elimination, share the Chinese experiences and lessons, and promote the process of global malaria elimination (Lai et al., 2017).

Echinococcosis: A national long-term control programme through comprehensive interventions has been implemented to target the morbidity reduction by the year of 2020 in endemic areas, in which human echinococcosis prevalence rate and dog infection rate at least in 70% endemic counties will achieve <1% and <5%, respectively. Furthermore, the echinococcosis infection rate in livestock should achieve less than 8% under age of 2 years old, and 90% coverage of the patients' management, surveillance on disease endemics and factors, and other control index. Meanwhile, safe water supply will ensured with a coverage of 95% in residents' settlements in endemic areas.

By 2030, it will achieve: (1) echinococcosis prevalence rate in primary school children is <1‰; (2) dog infection rate of *Echinococcus* spp. <1%; (3) echinococcosis infection rate in livestock <1%; and (4) echinococcal infection rate in rodents <1‰. To reach the objectives above, it will depend on implementing the comprehensive strategies and measures, consisting of strengthening control of the infection source of canines and intermediate hosts, slaughtering management and livestock immunization, decreasing of rodent density, clean water supply, health education and development of innovation tools (Fig. 3) (NHC, 2016; PR China, 2016).

Soil-transmitted helminthiases and food-borne helminthiases: Based on the national control plan, the prevalence of soil-transmitted helminthiases by 2020 targets to be decreased by 20% compared to that in 2015 in major endemic provinces and that of clonorchiasis by 30% meanwhile (NHC, 2016). The target of soil-transmitted helminthiases is promising owing to continuing economic development and further improvement in provision of clean water and sanitation (Qian et al., 2019). To achieve the target for clonorchiasis is challenging because of high popularity in eating raw freshwater fish and limited control activities. Nowadays, the plans for soil-transmitted helminthiases and food-borne helminthiases are not yet available for 2025 and beyond. However, it is expected to further control soil-transmitted helminthiases nationally. It is argued to implement hierarchical targets (Fig. 4): to achieve infection control in southwestern China

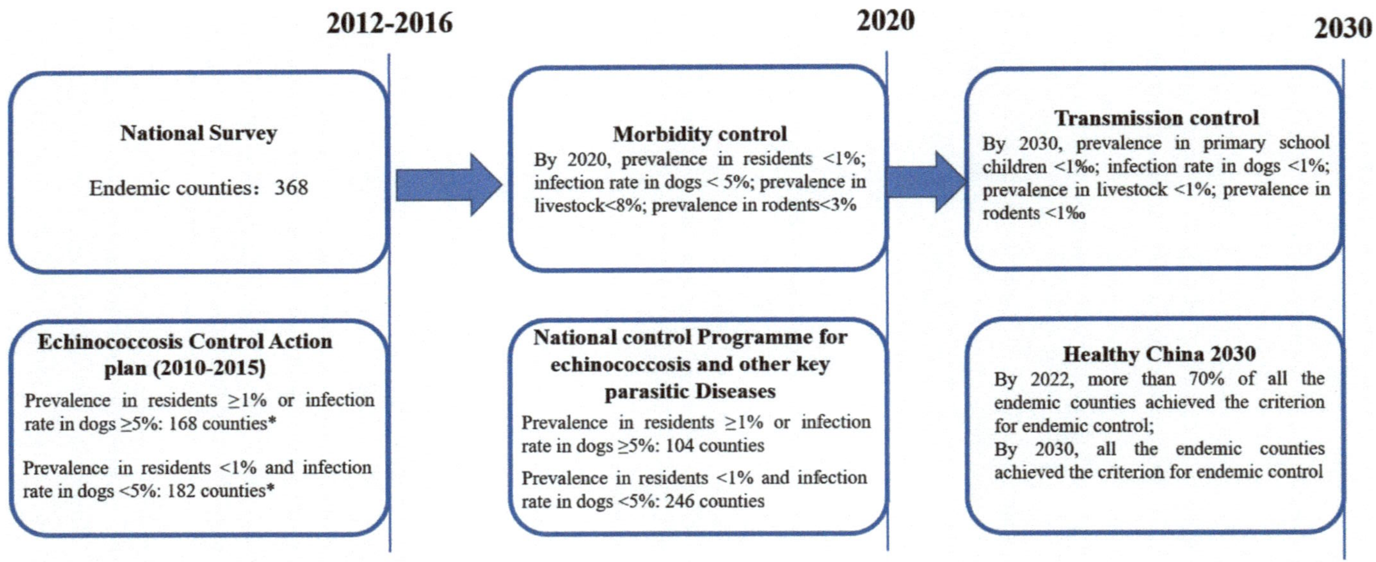

Note: *National survey in 2012

Fig. 3 Roadmap for echinococcosis control and prevention in China.

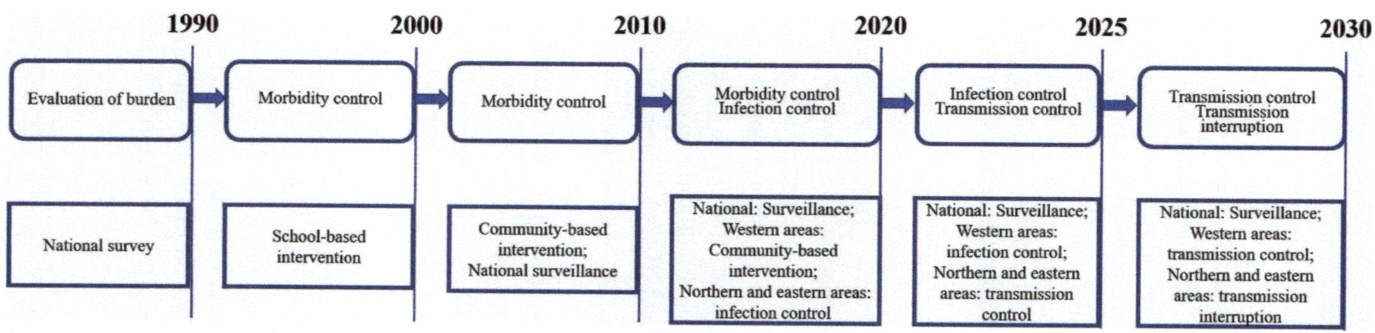

Fig. 4 Roadmap for control of soil-transmitted helminthiases in China.

and transmission control in northern and eastern China by 2025. Then, by 2030 more ambitious targets should be set, namely, transmission control in western China and transmission interruption in northern and eastern China. As to clonorchiasis, the overall target is argued to further control the morbidity in major endemic areas (Qian et al., 2019). Some pilots for *T. solium* cysticercosis are being implemented to explore the control strategy in western areas and elimination strategy in other low-endemic areas (Qian et al., 2019).

5. Role and how to influence the global situation of the new roadmap from China

Great achievements on control of parasitic diseases made in China greatly attribute to government's leadership, multisectoral cooperation, scientific and characteristic control strategy at different periods, phases and regions, community mobilization and experts' technical support. Because Chinese experiences also are highly regarded as the part of global knowledge, it is worth sharing it with other endemic countries. China is actively implementing its global health strategy, by which all-round international cooperation on health has highlighted with emphasizing the countries along one belt and one road (B&R). Through strengthening of the south-south cooperation and of the capacity building on the systems related to key diseases control, a series of China-Africa health cooperation projects have been designed and carried out. Fully utilizing high-level dialogue mechanism between nations, China is actively engaged into global health governance and make its impacts on formulation of the international criteria, regulations and guidelines, all of which will make great contribution to the SDGs (Fig. 5).

For schistosomiasis, great achievements in schistosomiasis control have been made in China. The experience from China is being recognized by international researchers and organizations. The China Government emphasizes the health cooperation with African countries. The first trial of schistosomiasis control is ongoing in Zanzibar with mass drug administration (MDA) and snail control. The primary results imply that the experience is working in the trial. We believe that the China's experiences could contribute more to achieving the goal of the global control of schistosomiasis.

For malaria, in order to further promote malaria control process globally, the world health organization (WHO) put forward the global

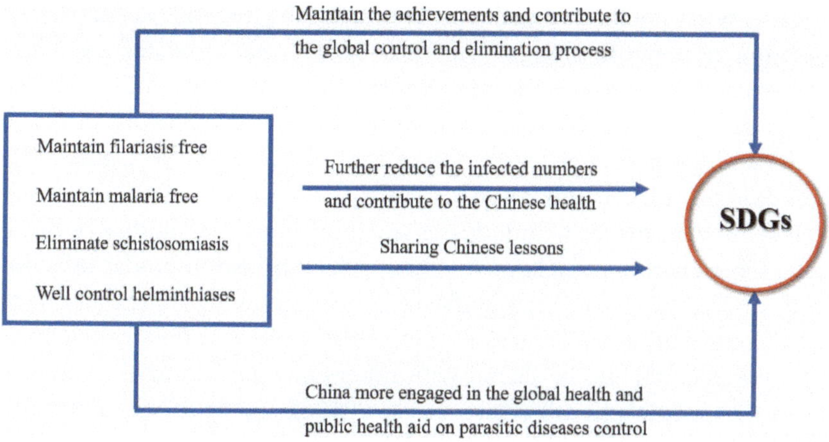

Fig. 5 China's contribution towards the global tropical diseases.

malaria technical strategy for 2015–2030, with the goals set: by 2030, the global malaria morbidity and mortality will decrease by more than 90%, and at least 35 countries achieve malaria elimination and prevent malaria re-transmission. China's malaria elimination path planning provides evidence support to the national malaria elimination goals, strategies, technical measures and timeframe, which ensures the realization of the goal. With China's success in controlling malaria, other endemic countries are looking forward to learning from China's experiences and lessons, while China will assume more international obligations and responsibilities. China's action to set up anti-malaria centers in 30 African countries has aroused the expectation of African countries. China will push forward the China-Africa cooperation at a higher level and provide greater help to African countries in combination with the investment on production export and technological research and development besides its experience on malaria control. It is anticipated that there will be more international cooperation projects in the future, which will also require China to send more CDC professionals oversea to provide technical support (WHO, 2015; Xia et al., 2014, 2017).

For echinococcosis, a national survey in 2012 and 2016 indicated that China has the highest prevalence of echinococcosis and the disease burden, accounting for 40% estimation of the world. It is known that the target against echinococcosis has been set by 2020 and 2030, and great progress made by means of implementing the comprehensive control strategy and measures. Based on the experiences accumulated, it would be very helpful to share with neighbour endemic countries and regions under the co-operation framework along "The Belts and Roads" initiated by China. Moreover, it will also benefit other endemic countries by sharing the technology and products, by which the process of the disease control will be accelerated (Angela et al., 2016; Chen et al., 2019a,b,c; PR China, 2016; Xiao et al., 2013; Yu and Zhou, 2019).

For soil-transmitted helminthiases and food-borne helminthiases, historically, China harboured a large number of the populations under infection. Thus, on the one hand, the significant control in China contributes to the global control progress (Qian et al., 2015, 2019). On the other hand, China's experiences could also be shared with other endemic countries to contribute to the people's health (Chen et al., 2019a,b,c). As to soil-transmitted helminthiasis, the experiences could be learned and applied in Africa, Southeast Asia and Latin America. Although the prevalence of clonorchiasis is still high in China, many tools and techniques have been developed (Qian and Zhou, 2019). Because clonorchiasis and another human liver fluke-*Opisthorchis viverrini* infection are highly endemic in Southeast Asia, thus these useful tools and techniques could be tried and adopted in other endemic countries. The successful control of *T. solium* cysticercosis in northern China could also provide experiences for other endemic countries.

6. Conclusion and perspectives

Parasitic diseases and infections are accompanied with poverty, unhealthy living styles and production patterns as well as environmental contamination. Therefore, with greatly developing the economy and promoting civilized society, China has formulated the Planning Outline of the Health China 2030, proposed the concept of guaranteeing Chinese residents all life-cycle health, and implemented the Three-year Action Tackling Key Endemic Diseases and Health China Action (2019–2030), which aim to carry out a series of health-specific poverty alleviation projects with focus on the populations living at remote, ethnic minority, border and endemic areas. From 2020 to 2030, under the strong governmental leadership, multisectoral cooperation and community participation, China will further strengthen the capacity building of its public health system and diseases surveillance. Through these efforts, China strives for maintaining the achievements of lymphatic filariasis and malaria elimination, effectively dealing with the potential risk caused by the imported pathogens and vectors. Simultaneously China continues to promote eliminating schistosomiasis threat, controlling echinococcosis transmission, getting the infections further down at scale and degree, and reducing health damage from food-borne parasitic diseases.

As only national institute for parasitic diseases control and research and national center for tropical diseases research, the National Institute of Parasitic Diseases (NIPD) of Chinese Center for Disease Control and Prevention will consistently play a leading role in developing control and elimination strategies, scientific and technological innovation, professional training and degree education. NIPD will effectively serve as a key advisory institution for the National Health Commission of China and make greater contribution to achieving eliminating parasitic disease threat to health. Based on previous accumulated cooperation experiences on disease control and public health aid, China will be more and more engaged into global health affairs, deepen and widen the cooperation network and partnership, share the Chinese experiences and lessons on tropical diseases control and elimination. All of these efforts will greatly promote global progress of parasitic diseases control and elimination, and contribute to building a community with a shared future for mankind.

References

Angela, M., Cadavid, R., Yang, Y.R., et al., 2016. The landscape epidemiology of echinococcoses. Infect. Dis. Poverty 5, 13.

Büscher, P., Cecchi, G., Jamonneau, V., et al., 2017. Human African trypanosomiasis. Lancet 390, 2397–2409.

Chen, M., 2014. Assessment of morbidity due to *Schistosoma japonicum* infection in China. Infect. Dis. Poverty 3, 6.

Chen, Y.D., Zhou, C.H., Xu, L.Q., 2012. Analysis of the results of two nationwide surveys on *Clonorchis sinensis* infection in China. Biomed. Environ. Sci. 25, 163–166.

Chen, J., Bergquist, R., Zhou, X.N., Xue, J.B., Qian, M.B., 2019a. Combating infectious disease epidemics through China's belt and road initiative. PLoS Negl. Trop. Dis. 13, e0007107.

Chen, J., Ding, W., Li, Z., et al., 2019b. From parasitic disease control to global health: new orientation of the National Institute of Parasitic Diseases, China CDC. Acta Trop. 201, 105219. https://doi.org/10.1016/j.actatropica.2019.105219.

Chen, N., Jin, K., Xu, J., Zhang, J., et al., 2019c. Human African trypanosomiasis caused by *Trypanosoma brucei gambiense*: the first case report in China. Int. J. Infect. Dis. 79, 34–36.

Coordinating Office of the National Survey on the Important Human Parasitic Diseases, 2005. A national survey on current status of the important parasitic diseases in human population. Zhongguo Ji Sheng Chong Xue Yu Ji Sheng Chong Bing Za Zhi 23, 332–340, (in Chinese).

Eckert, J., Deplazes, P., 2004. Biological, epidemiological and clinical aspects of echinococcosis: a zoonosis of increasing concern. Clin. Microbiol. Rev. 17, 107–135.

Feng, J., Zhang, L., Huang, F., Yin, J.H., Tu, H., Xia, Z.G., Zhou, S.S., Xiao, N., Zhou, X.N., 2018. Ready for malaria elimination: zero indigenous case reported in the People's Republic of China. Malar. J. 17, 315.

Hotez, P.J., Zheng, F., Long-qi, X., Ming-gang, C., Shu-hua, X., Shu-xian, L., Blair, D., McManus, D.P., Davis, G.M., 1997. Emerging and reemerging helminthiases and the public health of China. Emerg. Infect. Dis. 3, 303–310.

Jiang, B., Zhou, X.N., Zhang, H.B., Tao, Y., Huo, L.L., Liu, N., 2017. Slow-release praziquantel for dogs: presentation of a new formulation for echinococcosis control. Infect. Dis. Poverty 6, 140.

Lai, S.J., Li, Z.J., Nicola, A.W., Sun, J.L., Michael, G.H., Huang, Z.J., et al., 2017. Malaria in China, 2011–2015: an observational study. Bull. World Health Organ. 95, 564–573.

Lei, Z.L., Zheng, H., Zhang, L.J., Zhu, R., Xu, Z.M., Xu, J., Fu, Q., Wang, Q., Li, S.Z., Zhou, X.N., 2014. Endemic status of schistosomiasis in People's Republic of China in 2013. Chin. J. Schisto. Control 26, 591–597.

Li, S.Z., Zheng, H., Gao, J., Zhang, L.J., Zhu, R., Xu, J., Guo, J.G., Xiao, N., Zhou, X.N., 2013a. Endemic status of schistosomiasis in People's Republic of China in 2012. Chin. J. Schisto. Control 25, 557–563.

Li, T., Chen, X., Yanagida, T., Wang, H., Long, C., Sako, Y., Okamoto, M., Wu, Y., Giraudoux, P., Raoul, F., Nkouawa, A., Nakao, M., Craig, P.S., Ito, A., 2013b. Detection of human taeniases in Tibetan endemic areas, China. Parasitology 140, 1602–1607.

Li, Z., Zhang, Q., Zheng, C., Zhou, S., Sun, J., Zhang, Z., Geng, Q., Zhang, H., Wang, L., Lai, S., Hu, W., Clements, A.C., Zhou, X.N., Yang, W., 2016. Epidemiologic features of overseas imported malaria in the People's Republic of China. Malar. J. 15, 141.

Liu, H.X., Zhang, Y., Hu, L., Zhu, D., Wu, Y., Lv, S., Guo, J.G., Wu, X.H., Zhou, X.N., 2008. Analysis of re-emergence of *Oncomelania hupensis* in area where schistosomiasis transmission-interrupted in China. J. Pathogen. Bio. 3, 440–442.

Liu, H., Xiao, N., Yang, S.J., Wang, D., Peng, J., 2017. Epidemiological characteristics of canine *Echinococcus* infection in Qinghai-Tibet Plateau of China. Chin. J. Schisto. Control 29, 129–138.

Liu, Q., Chen, X.L., Chen, M.X., et al., 2018. Trypanosoma brucei rhodesiense infection in a Chinese traveler returning from the Serengeti National Park in Tanzania. Infect. Dis. Poverty 7, 50.

Ma, Y.X., Xu, C.B., Shao, L., Zhu, Y.G., Xu, J.T., Kang, Q.D., 1992. Achievements, status and countermeasures on control of taeniasis and cysticercosis during past 40 years in China. Zhongguo Gong Gong Wei Sheng 8, 244–246, (In Chinese).

McManus, D.P., Zhang, W., Li, J., Bartley, P.B., 2003. Echinococcosis. Lancet 362, 1295–1305.

MOH, 2010. National Malaria Elimination Action Plan (2010–2020). Beijing, Ministry of Health.

National Health and Family Planning Commission of the People's Republic of China, 2016. National Control Plan on Echinococcosis and Other Key Important Parasitic Diseases in China (2016–2020). http://www.nhc.gov.cn/jkj/s5873/201702/dda5ffe3f50941a29fb0aba6233bb497.shtml, (in Chinese).

National Ministry of Health of the People's Republic of China, 2006. National Control Plan on the Prevention and Control of Key Parasitic Diseases (2006–2015). http://www.moh.gov.cn/mohbgt/pw10604/200804/27592.shtml.

National Ministry of Health of the People's Republic of China, 2010. National Action Plan for the Prevention and Control of Echioncoccosis (2010–2015). http://www.nhfpc.gov.cn/zwgkzt/wsbysj/201012/50041.shtml.

Qian, M.B., Zhou, X.N., 2019. Human liver flukes in China and ASEAN: time to fight together. PLoS Negl. Trop. Dis. 13, e0007214.

Qian, M.B., Chen, Y.D., Liang, S., Yang, G.J., Zhou, X.N., 2012. The global epidemiology of clonorchiasis and its relation with cholangiocarcinoma. Infect. Dis. Poverty 1, 4.

Qian, M.B., Chen, Y.D., Yan, F., 2013. Time to tackle clonorchiasis in China. Infect. Dis. Poverty 2, 4.

Qian, M.B., Chen, Y.D., Yang, Y.C., Lu, M.F., Jiang, Z.H., Wei, K., Wei, S.L., Zhou, C.H., Xu, L.Q., Zhou, X.N., 2014. Increasing prevalence and intensity of foodborne clonorchiasis, Hengxian County, China, 1989-2011. Emerg. Infect. Dis. 20, 1872–1875.

Qian, M.B., Xia, S., Zhou, X.N., 2015. Soil-transmitted helminths in China. Lancet Infect. Dis. 15, 1262–1263.

Qian, M.B., Utzinger, J., Keiser, J., Zhou, X.N., 2016. Clonorchiasis. Lancet 387, 800–810.

Qian, M.B., Chen, J., Bergquist, R., Li, Z.J., Li, S.Z., Xiao, N., Utzinger, J., Zhou, X.N., 2019. Neglected tropical diseases in the People's Republic of China: progress towards elimination. Infect. Dis. Poverty 8, 86.

Ross, A.G.P., Li, Y.S., Sleigh, A.C., McManus, D.P., 1997. Schistosomiasis control in the People's Republic of China. Parasitol. Today 13, 152–155.

Schantz, P.M., Wang, H., Qiu, J., et al., 2003. Echinococcosis on the Tibetan plateau: incidence and risk factors for cystic and alveolar echinococcosis in Tibetan populations in Qinghai Province, China. Parasitology 127, S109–S120.

Tang, L.H., 1999. Chinese achievements in malaria control and research. Chin. J. Parasitol. Parasit. Dis. 17, 257–259, (in Chinese).

Tang, L.H., 2016. Goal, strategies and roadmap of malaria elimination in China. China Trop. Med. 16, 301–304, (in Chinese).

The People' republic of China, 2016. Healthy China 2030. [EB/OL]. [2016-10-25]. http://www.nhfpc.gov.cn/xcs/wzbd/201610/21d120c917284007ad9c7aa8e9634bb4.shtml.

Torgerson, P.R., Keller, K., Magnotta, M., Ragland, N., 2010. The global burden of alveolar echinococcosis. PLoS Negl. Trop. Dis. 4, e722.

Wang, G.Q., 2016. Epidemiological Survey on Echinococcosis in China. Shanghai Science and Technology Press.

Wang, Q., Xu, J., Zhang, L.J., Zheng, H., Ruan, Y., Hao, Y.W., Li, S.Z., Zhou, X.N., 2015. Analysis of endemic changes of schistosomiasis in China from 2002 to 2010. Chin. J. Schisto. Control 27, 229–234.

Wang, X., Ruan, Q., Xu, B., et al., 2018. Human African trypanosomiasis in emigrant returning to China from Gabon, 2017. Emerg. Infect. Dis. 24, 400–404.

WHO, 2013. Control and surveillance of African trypanosomiasis. Report of a WHO Expert Committee. World Health Organ. Tech. Rep. Ser. 881, 1–114. I-VI.

WHO, 2015. Global Technical Strategy for Malaria 2016–2030. World Health Organization, Geneva.

Wu, W.P., Wang, H., Wang, X., Zhou, X.N., Wang, L.Y., Zheng, C.J., Cao, J.P., Xiao, N., Wang, Y., Zhu, Y.Y., 2018. A nationwide survey on echincoccosis in China during 2012–2016. Chin. J. Parasitol. Parasit. Dis. 36, 1–14.

Xia, Z.G., Wang, R.B., Wang, D.Q., Feng, J., Zheng, Q., Deng, C.S., et al., 2014. China–Africa cooperation initiatives in malaria control and elimination. Adv. Parasitol. 86, 319–337.

Xia, Z.G., Xu, M., Zhao, J.K., Lai, S.J., Zhou, X.N., 2017. An overview on cooperation strength between China and the Global Fund to Fight AIDS, Tuberculosis and Malaria in developing multilateral malaria control projects. Chin. J. Parasitol. Parasit. Dis. 35, 520–526, (in Chinese).

Xiao, N., Qiu, J.M., Nakao, M., Nakaya, K., Yamasaki, H., Sako, Y., Mamuti, W., Schantz, P., Craig, P.S., Ito, A., 2003. Identification of *Echinococcus* species from a yak in the Qinghai-Tibet Plateau region of China. Am. J. Trop. Med. Hyg. 69, 445–446.

Xiao, N., Li, T.Y., Qiu, J.M., Nakao, M., Chen, X.W., Nakaya, K., Yamasaki, H., Schantz, P.M., Craig, P.S., Ito, A., 2004. The Tibetan hare *Lepus oiostolus*: a novel intermediate host for *Echinococcus multilocularis*. Parasitol. Res. 92, 352–353.

Xiao, N., Nakao, M., Qiu, J.M., Budke, C.M., Giraudoux, P., Craig, P.S., Ito, A., 2006. Dual infection of animal hosts with different *Echinococcus* species in the eastern Qinghai-Tibet plateau region of China. Am. J. Trop. Med. Hyg. 75, 292–294.

Xiao, N., Zhou, Z.J., Chen, X.W., Zeng, Y.Z., Wang, Z.Y., Lei, Y., Tian, Y., Huang, Y., Yi, D.Y., Li, Y.Q., Wu, W.P., 2012. Assessment of health education against echinococcosis in Tibetan region of Sichuan Province, China. Chin. J. Parasitol. Parasit. Dis. 30, 6–11.

Xiao, N., Yao, J.W., Ding, W., Giraudoux, P., Craig, P.S., Ito, A., 2013. Priorities for research and control of cestode zoonoses in Asia. Infect. Dis. Poverty 2, 16.

Xu, L.Q., Yu, S.H., Jiang, Z.X., Yang, J.L., Lai, L.Q., Zhang, X.J., Zheng, C.Q., 1995. Soil-transmitted helminthiases: nationwide survey in China. Bull. World Health Organ. 73, 507–513.

Yin, J.H., Yang, M.N., Zhou, S.S., Wang, Y., Feng, J., Xia, Z.G., 2013. Changing malaria transmission and implications in China towards National Malaria Elimination Programme between 2010 and 2012. PLoS One 8, e74228.

Yin, J.H., Zhou, S.S., Xia, Z.G., Wang, R.B., Qian, Y.J., Yang, W.Z., et al., 2014. Historical patterns of malaria transmission in China. Adv. Parasitol. 86, 1–19.

Yu, Q., Zhou, X.N., 2019. Discussion on strategies of health literacy promotion for chinococcosis control in China. Chin. J. Schisto. Control 31, 94–97.

Yu, S., Xu, L., Jiang, Z., Xu, S., Han, J., Zhu, Y., Chang, J., Lin, J., Xu, F., 1994. Report on the first nationwide survey of the distribution of human parasites in China. Regional distribution of parasite species. Zhongguo Ji Sheng Chong Xue Yu Ji Sheng Chong Bing Za Zhi 12, 241–247, (in Chinese).

Yu, Q., Xiao, N., Wu, W.P., Tian, T., Wang, L.W., Wang, Y., Han, S., Yang, S.J., Zhou, X.N., 2020. Effectiveness over 10 Years (2004–2014) of the Echinococcosis Control Progamme in China. Acta Trop. in press.

Zang, X.Z., Li, H.Z., Liu, H.K., Liu, Y.H., Zhou, C.H., Chen, Y.D., Qian, M.B., Li, S.Z., 2019a. Epidemiological analysis of new cases of cysticercosis in Dali Prefecture, Yunnan Province from 2014 to 2017. Zhongguo Xue Xi Chong Bing Fang Zhi Za Zhi 31, 143–147, (in Chinese).

Zang, X.Z., Li, H.Z., Qian, M.B., Chen, Y.D., Zhou, C.H., Liu, H.K., Liu, Y.H., Li, S.Z., 2019b. Extensive disseminated cysticercosis: a case report in Yunnan province, China. BMC Infect. Dis. 19, 535.

Zhang, W., Zhang, Z., Wu, W., Shi, B., Li, J., Zhou, X., et al., 2015. Epidemiology and control of echinococcosis in Central Asia, with particular reference to the People's Republic of China. Acta Trop. 141, 235–243.

Zhang, L.J., Xu, Z.M., Qian, Q.J., Dang, H., Lv, S., Xu, J., Li, S.Z., Zhou, X.N., 2017. Endemic status of schistosomiasis in People's Republic of China in 2016. Chin. J. Schisto. Control 29, 669–677.

Zhang, L., Feng, J., Zhang, S.S., Xia, Z.G., Zhou, S.S., 2018a. The progress of national malaria elimination and epidemiological characteristics of malaria in China in 2017. Chin. J. Parasitol. Parasit. Dis. 36, 201–209, (in Chinese).

Zhang, L.J., Xu, Z.M., Dai, S.M., Dang, H., Lv, S., Xu, J., Li, S.Z., Zhou, X.N., 2018b. Endemic status of schistosomiasis in People's Republic of China in 2017. Chin. J. Schisto. Control 30, 481–488.

Zhang, L., Feng, J., Zhang, S.S., Xia, Z.G., Zhou, S.S., 2019. Epidemiological characteristics of malaria and the progress towards its elimination in China in 2017. Chin. J. Parasitol. Parasit. Dis. 37, 241–247, (in Chinese).

Zhang, L.J., Xu, Z.M., Dai, S.M., Dang, H., Lv, S., Xu, J., Li, S.Z., Zhou, X.N., 2020. Endemic status of schistosomiasis in People's Republic of China in 2018. Chin. J. Schisto. Control 31, 576–582, accepted.

Zhou, Z.J., 1981. The malaria situation in the People's Republic of China. Bull. World Health Organ. 59, 931–936.
Zhou, X.N., 2018. Report on the National Survey of Important Human Parasitic Diseases in China (2015). People's Medical Publishing House, Beijing, (in Chinese).
Zhou, X.N., Wang, L.Y., Chen, M.G., Wu, X.H., Jiang, Q.W., Chen, X.Y., Zheng, J., Utzinger, J., 2005. The public health significance and control of schistosomiasis in China—then and now. Acta Trop. 96, 97–105.

Further reading

Wang, Q., Huang, Y., Huang, L., Vuitton, D.A., Giraudoux, P., Craig, P.S., Wu, W.P., 2014. Review of risk factors for human echinococcosis incidence on the Qinghai-Tibet Plateau, China: a prospective for control options. Infect. Dis. Poverty 3, 3.
Yu, Q., Xiao, N., Yang, S.Y., Han, S., 2017. Deworming of stray dogs and wild canines with praziquantel-laced baits delivered by an unmanned aerial vehicle in areas highly endemic for echinococcosis in China. Infect. Dis. Poverty 6, 117.
Yu, Q., Liu, H., Xiao, N., 2018. Unmanned aerial vehicles: potential tools for use in zoonosis control. Infect. Dis. Poverty 7, 49.

FSC
www.fsc.org
MIX
Papier | Fördert gute Waldnutzung
FSC® C083411

Kolibri 360 GmbH
Ferdinand-Jühlke-Straße 7
99095 Erfurt, Deutschland
produktsicherheit@kolibri360.de